Geophysics and Geosequestration

Geosequestration involves the deep geological storage of carbon dioxide from major industrial sources, providing a potential solution for reducing the rate of increase of atmospheric concentrations of carbon dioxide and mitigating climate change. This volume provides an overview of the major geophysical techniques and analysis methods for monitoring the movement and predictability of carbon dioxide plumes underground. Comprising chapters from eminent researchers, the book is illustrated with practical examples and case studies of active projects and government initiatives, and discusses their successes and remaining challenges. A key case study from Norway demonstrates how governments and other stakeholders could estimate storage capacity and design storage projects that meet the requirements of regulatory authorities. Presenting reasons for embracing geosequestration, technical best practice for carbon management, and outlooks for the future, this volume provides a key reference for academic researchers, industry practitioners, and graduate students looking to gain insight into subsurface carbon management.

Thomas L. Davis is Professor Emeritus at the Colorado School of Mines and founder of the Reservoir Characterisation Project (RCP), which received the Distinguished Achievement Award from the Society of Exploration Geophysics (SEG) in 2014. He is an active member of the SEG as an organizer for conferences, workshops, and education programs, and has served as SEG's Second Vice President, Technical Program Chairman, and Distinguished Lecturer. He has received the C. J. Mackenzie Award from the Engineering College of the University of Saskatchewan, the Milton B. Dobrin Award from the University of Houston, and the Deans Excellence and Melvin F. Coolbaugh Memorial Awards from the Colorado School of Mines.

Martin Landrø is Professor at the Norwegian University of Science and Technology. Past positions include working as a research geophysicist and manager at IKU Petroleum Research and as a specialist at Statoil's research centre in Trondheim. He has won numerous awards including the Norman Falcon Award, Louis Cagniard Award, and Conrad Schlumberger Award from the European Association of Geoscientists and Engineers; the Norwegian Geophysical Award; Statoil's Research Prize; the SINTEF Award for outstanding pedagogical activity; the ENI Award (New Frontiers in Hydrocarbons); and the IOR Award from the Norwegian Petroleum

Directorate. His research interests include seismic inversion, marine seismic acquisition, and 4D and 4C seismic surveys.

Malcolm Wilson has worked within government, academia, and the petroleum research industry, and currently serves on the advisory council for the Reservoir Characterization Project (Colorado School of Mines) and on the International Advisory Board for the New Energy Coalition in the Netherlands. He was instrumental in the creation of the IEA Greenhouse Gas Weyburn CO_2 Monitoring and Storage Project – an extensive research program for monitoring the storage of CO_2 in an oilfield. He was also part of the Intergovernmental Panel on Climate Change (IPCC) special report on capture, transport, and storage of CO_2, and of the IPCC team that received the Nobel Peace Prize in 2007. Dr. Wilson is an adjunct professor at the Faculty of Engineering and Applied Science at the University of Regina, and has received the Alumni of Distinction Award from the University of Saskatchewan.

Geophysics and Geosequestration

Edited by

Thomas L. Davis
Colorado School of Mines

Martin Landrø
Norwegian University of Science and Technology, Trondheim

Malcolm Wilson
New World Orange BioFuels

CAMBRIDGE
UNIVERSITY PRESS

CAMBRIDGE
UNIVERSITY PRESS

University Printing House, Cambridge CB2 8BS, United Kingdom

One Liberty Plaza, 20th Floor, New York, NY 10006, USA

477 Williamstown Road, Port Melbourne, VIC 3207, Australia

314–321, 3rd Floor, Plot 3, Splendor Forum, Jasola District Centre, New Delhi – 110025, India

79 Anson Road, #06–04/06, Singapore 079906

Cambridge University Press is part of the University of Cambridge.

It furthers the University's mission by disseminating knowledge in the pursuit of education, learning, and research at the highest international levels of excellence.

www.cambridge.org
Information on this title: www.cambridge.org/9781107137493
DOI: 10.1017/9781316480724

© Cambridge University Press 2019

First published 2019

Printed in the United Kingdom by TJ International Ltd. Padstow Cornwall

A catalogue record for this publication is available from the British Library.

ISBN 978-1-107-13749-3 Hardback

Contents

Contents

Contributors

Zaki Bassiouni, Louisiana State University

Robert A. Bauer, Illinois State Geological Survey

Peter Bergmann, GFZ German Research Centre for Geosciences

Michael Commer, Lawrence Berkeley National Laboratory

Thomas M. Daley, Lawrence Berkeley National Laboratory

Thomas L. Davis, Colorado School of Mines

Jessica Dongas, University of Calgary

Ola Eiken, Quad Geophysical Company

Alessandro Ferretti, TRE ALTAMIRA

John Gale, IEA Green House Gas R&D Programme

Erika Gasperikova, Lawrence Berkeley National Laboratory

Sallie E. Greenberg, Illinois State Geological Survey

Sissel Grude, Norwegian University of Science and Technology

Eva K. Halland, Norwegian Petroleum Directorate

William Harbert, University of Pittsburgh

Monika Ivandic, Uppsala University, Sweden

Christopher Juhlin, Uppsala University, Sweden

Martin Landrø, Norwegian University of Science and Technology

Donald C. Lawton, University of Calgary

David Lumley, University of Texas at Dallas

Stefan Lüth, GFZ German Research Centre for Geosciences

Marie Macquet, University of Calgary

Shawn Maxwell, Itasca-Image Company

Kirk Osadetz, CMC Research Institutes Inc.

Trevor Richards, Denbury Resources

Alessio Rucci, TRE ALTAMIRA

Amin Saeedfar, CMC Research Institutes Inc.

Sergey V. Samsonov, Canada Centre for Mapping and Earth Observation, Natural Resources Canada

Donald W. Vasco, Lawrence Berkeley National Laboratory

Scott Wehner, Denbury Resources

Steven G. Whittaker, Illinois State Geological Survey

Don White, Geological Survey of Canada

Robert Will, Schlumberger Carbon Services

Malcolm Wilson, New World Orange BioFuels

Mark Zumberge, Scripps Institute of Oceanography

Preface

Two famous scientists in the late nineteenth century laid the groundwork for this book. Svante Arrhenius developed the concept of the greenhouse effect based on the chemistry of the atmosphere and initiated our understanding of the results of changing atmospheric chemistry. Emil Johann Weichert, the first geophysics professor, was examining the use of "artificial earthquakes" to look into the subsurface. Over the last dozen decades, the science of geophysics has moved forward substantially, primarily as a result of demand from the resource industry, and the current state of the art is examined in this book. The issue of the greenhouse effect, now termed global warming, has taken a different path. Concern about global warming came to public attention with the publication in 1987 of *Our Common Future* by the Brundtland Commission and the Toronto Conference of 1988 that tried to develop a pathway for emissions reductions. In subsequent years, the importance of global warming and the need to reduce its impact on economies and ecosystems, including human ecosystems, was recognized by the award of part of the 2007 Nobel Peace Prize to the scientists working on the United Nations reports on climate change, the Intergovernmental Panel on Climate Change [IPCC].

Since the late 1980s much effort has gone into the development of ways to reduce the emissions of greenhouse gases, particularly carbon dioxide (CO_2), into the atmosphere. One of the ways identified is to inject CO_2 either into the deep subsurface as a solvent for enhanced oil recovery (Sacroc Field in Texas has been receiving CO_2 for more than 40 years) or into deep saline aquifers (the first of these projects is offshore Norway, the Sleipner project, started in 1996). This book contains case studies of both types of geosequestration of CO_2. Indeed, the International Energy Agency (IEA) estimates that by 2050, some 13% of reductions of greenhouse gas emissions will need to be realized through geosequestration or the cost of emissions reductions will increase significantly.

The enthusiasm for sequestration in the subsurface is tempered by fear of the unknown. What happens if the CO_2 leaks into my basement, underground parking, or other such confined spaces? This is the same question that plagued the nuclear industry and the storage of spent nuclear fuel in manmade caverns. As an example, Shell was unable to move ahead with a proposed storage project in Barendrecht, Netherlands, some years ago. Fear of the CO_2 (which is known to have asphyxiated people in old coal mines, for example) and concerns about housing prices prevented this venture. In areas where the oil and gas industry is important and people understand enhanced oil recovery (EOR), the issues are far less pronounced. Hence Shell was able to go ahead with the Quest project in Alberta and there were only insignificant problems with EOR in Weyburn, Saskatchewan.

The key to public acceptance is the ability to demonstrate the safe sequestration of the CO_2 in the subsurface. In addition, the regulator must be able to confirm the volumes of CO_2 sequestered to be able to calculate national emissions. Voluntary standards are in the process of being developed in recognition of this need for confirmation. We can measure the CO_2 being injected into the subsurface with a great deal of accuracy, but we need to confirm that the CO_2 stays where it is intended. This important fact underlies the significance of this volume: using geophysics, particularly seismic surveying, as the only way to "watch" the CO_2 in the subsurface and ensure its stable sequestration in its intended injection location. From the perspective of the resource sector, geophysics has proven irreplaceable as a technique for improving the efficiency with which CO_2 is used to extract oil resources. Regardless of environmental issues, CO_2 has a cost and is, therefore, an expensive solvent to put in the ground. Understanding where the CO_2 is going is important to minimizing CO_2 utilization and effectively recovering the oil. This is vitally important for sustaining high-quality jobs in what are often under-developed parts of the world.

The focus of this volume is, therefore, to evaluate the state of the art in seismic work to determine the location and movement of CO_2 in the subsurface. This is a truly exciting venture, not just from the perspective that Weichert would have recognized – the use of seismic surveys to ensure effective and safe exploitation of fossil fuel resources – but also to meet a growing and important need to ensure to the regulator and the public that this is being done safely. In this latter regard, the technology is meeting the goal of the Brundtland Commission to reduce emissions to the atmosphere. Chapter 3 by Daley and Harbert examines in more detail why we need to monitor what is happening in the subsurface.

This volume discusses the technology as it exists today (for example, the excellent review by Lumley in Chapter 2) and then puts the basics of seismic and other geophysical techniques work into the real world with a series of important case studies including Sleipner and Snohvit, among others, as pure sequestration projects (Landrø, one of the editors and contributors, has played a significant role in this work) and also a number of EOR projects where the CO_2 can be monitored in oil fields (Davis, as editor and contributor, has a long history with pushing the limits of geophysics in EOR projects). The authors of the various chapters have all developed distinguished reputations in the field of geophysics, and are worthy successors to the likes of Weichert, with a particular focus on the use of CO_2 in the fossil fuel industry or sequestration in the deep subsurface. Chapter 16 by Lawton et al. also examines our ability to track small volumes of CO_2 at shallow subsurface levels should the CO_2 escape from its storage location (this is also mentioned in Chapter 13 by Eiken on the Sleipner storage project), thereby demonstrating our ability to take action should the remote probability of leakage actually occur. Chapter 19 by Bauer et al. shows the sequestration of CO_2 under an area with significant surface activity, demonstrating not only an ability to survey under less than optimal conditions, but also the acceptance by the public in a situation where care is taken to understand the processes deep beneath the public's feet.

This volume will be a major contribution to students and practitioners of geophysics and to those concerned with the issue of global warming and taking concrete action to reduce emissions to the atmosphere in a safe and sustainable manner. We thank you for reading the book.

Acknowledgments

Publishing a book gives one a new appreciation for not just the science but also the human factor that goes into its preparation. From the beginning, the people at Cambridge University Press have been very supportive. We would like to acknowledge Susan Francis and Zoe Pruce for working with us and providing the encouragement and guidance to see the project through.

To our authors from around the world we say thanks. Your willingness to present your work and then to follow through as you have is quite remarkable and very much appreciated. There is little reward to offer other than "job well done."

To those who helped us review the manuscripts and to provide timely reviews and encouragement to our authors to make their contribution better, we say thanks. Specifically, we would like to recognize Malcolm Wilson, Sean O'Brien, and Jared Atkinson. Malcolm volunteered to do the final editing before the production phase. Thanks, Malcolm, for stepping forward and taking one for the team.

Abbreviations

B	magnetic field
CSEM	controlled source electromagnetics
CO_2	carbon dioxide
DC	direct current
E	electric field
EM	electromagnetic

ERT	electrical resistance tomography
H	magnetic field ($B = \mu H$)
HED	horizontal electric dipole
TDS	total dissolved solids
VMD	vertical magnetic dipole

Climate Change and the Role of Carbon Capture and Storage in Mitigation

John Gale and Malcolm Wilson

1.1 Climate Change and the Need for Adaptation and Mitigation

The atmosphere of the Earth contains, among others, gases that are referred to as the greenhouse gases (GHGs). These gases are so called because, in the atmosphere, they both absorb and emit radiation. It is this process of absorption and emission of radiation in the atmosphere that is the fundamental cause of what is termed the greenhouse effect (Houghton *et al.*, 1990). The main GHGs in the Earth's atmosphere are carbon dioxide (CO_2), water vapor, methane (CH_4), nitrous oxide (N_2O), and ozone. Collectively the greenhouse gases significantly affect the Earth's temperature; scientists predict that without them, the temperature at the Earth's surface would average about 33°C colder than the present average of 14°C (Le Treut *et al.*, 2007).

Since the late eighteenth century, with the beginning of the Industrial Revolution in 1750, the atmospheric concentrations of these GHGs, CO_2 in particular, have risen substantially (Le Treut *et al.*, 2007). Since that time, the atmospheric concentration of CO_2 has increased from 280 to more than 400 parts per million (ppm). Since 1957, the atmospheric concentration of CO_2 has been measured at the Mauna Loa Observatory in Hawaii and is presented in what is known as the Keeling Curve. In May 2016, the National Oceanographic and Atmospheric Administration of the USA reported the highest ever monthly level of CO_2 in the air: 407.7 ppm. The increase in the concentration of CO_2 in the atmosphere is attributed to the increased use of fossil fuels, combined with the extensive deforestation observed since the Industrial Revolution. Atmospheric concentrations of CO_2 have not been observed at such levels for millennia. Scientific analyses suggest that atmospheric CO_2 levels reached as much as 415

ppm during the Pliocene Epoch, between 5 and 3 million years ago. In that period, global average temperatures have been estimated to be 3–4°C and as much as 10°C warmer at the poles than current levels. Sea levels have been estimated to have ranged between 5 and 40 m higher than today.

While CO_2 has the highest atmospheric concentration of the GHGs and its contribution to total radiative forcing is the largest of all the gases, that of the other GHGs cannot be ignored. The atmospheric concentrations of other GHGs have also risen significantly since the late 1980s. The atmospheric concentrations of CH_4 and N_2O in 2011 were 1803 ppb and 324 ppb, respectively, exceeding preindustrial levels by about 150% and 20% (IPCC, 2013).

The Intergovernmental Panel on Climate Change (IPCC) was established by the United Nations Environment Programme (UNEP) and the World Meteorological Organization (WMO) in 1988 to provide the world with a clear scientific view on the current state of knowledge of climate change and its potential environmental and socioeconomic impacts (IPCC, 1988). The IPCC reviews and assesses the most recent scientific, technical, and socioeconomic information produced worldwide relevant to the understanding of climate change. It does not conduct any research nor does it monitor climate-related data or parameters.

The IPCC publishes its results in a series of Assessment Reports, the latest of which is Fifth Assessment Report (AR5), which was published in early 2014 (IPCC, 2014).

The AR5 indicates that global climate change has already had observable effects on the environment. Points highlighted by the report include:

- Warming of the climate system is unequivocal.
- Each of the past three decades has been successively warmer than any preceding decade since 1850.

- Over the last two decades, the Greenland and Antarctic ice sheets have been losing mass, glaciers have continued to shrink almost worldwide, and Arctic sea ice and Northern Hemisphere spring snow cover have continued to decrease in extent.
- The rate of sea level rise since the mid-nineteenth century has been larger than the mean rate during the preceding two millennia.

Over the coming decades, it is expected that

- Global surface temperature changes for the end of the twenty-first century will likely exceed 1.5°C relative to 1850 to 1900 and may even exceed 2°C.
- Warming over the twenty-first century will cause nonuniform responses in the global water cycle, increasing the contrast in precipitation between wet and dry regions and between wet and dry seasons.
- Continued warming of the global ocean will affect ocean circulation as heat penetrates from the surface to the deep ocean.
- The Arctic sea ice cover and the Northern Hemisphere spring snow cover extents, and the global glacier mass will all decrease further as global mean surface temperature rises.
- Increased ocean warming and loss of mass from glaciers and ice sheets will cause global mean sea level to rise at a rate that will very likely exceed that observed from 1971 to 2010.
- Climate change will affect carbon cycle processes in a way that will exacerbate the increase of CO_2 in the atmosphere. Further uptake of carbon by the oceans will increase ocean acidification.

A recent report coauthored by the International Geosphere–Biosphere Programme (IGBP), the Intergovernmental Oceanographic Commission (IOC-UNESCO), and the Scientific Committee on Oceanic Research (SCOR) highlights the impacts of increased ocean acidification, the economic impact of which could be substantial.

Ocean acidification causes ecosystems and marine biodiversity to change. It has the potential to affect food security and limits the capacity of the ocean to absorb CO_2 from human emissions.

The Stern Review (2006) stated that climate change is the greatest and widest-ranging market failure ever seen, presenting a unique challenge for economics. The Review's main conclusion was that the costs and benefits of strong and early action on climate change far outweigh the costs of not acting. According to the Review, without action the overall costs of climate change will be equivalent to losing at least 5% of global gross domestic product (GDP) each year, now and in all the years to come. Including a wider range of risks and impacts could increase this to 20% of GDP or more, also indefinitely. Stern believed at the time that 5–6°C of temperature increase was "a real possibility."

We then have two options to follow: mitigation and adaptation. Mitigation addresses the root causes by reducing greenhouse gas emissions while adaptation seeks to lower the risks posed by the consequences of climatic changes (IPCC, 2007b). In reality, we will follow a twin process because we expect that some 1 to 1.5°C of warming is already "locked in" (World Bank, 2014). Humans have been adapting to climatic changes throughout their evolution, but we probably now face a greater challenge to adapt than ever before. This chapter, however, concentrates on the issue of mitigation.

1.2 What Mitigation Options Are Needed?

The largest sources of global emissions are the power and industry sectors, which represent 56% of global GHG emissions, including fugitive emissions from fossil fuel mining, refining, and transportation. Of the global greenhouse gases, CO_2 accounts for 65% of these emissions, primarily from the use of fossil fuels. The discussion that follows therefore concerns the mitigation of emissions of CO_2 from the use of fossil fuel in the power and industrial sectors. It is important at this junction to stress that there are multiple mitigation courses of action in both of these sectors, and the aim of this discussion is not to select individual ones but to recognize that all the low-carbon technology options will be needed in combination to meet strict emission targets. It also follows that the portfolio and balance of low-carbon technology options will vary in different regions to account for national considerations. To be clear, there is no single low-carbon technology option that will reduce GHGs sufficiently on its own and there is no "one size fits all" low-carbon technology portfolio option either.

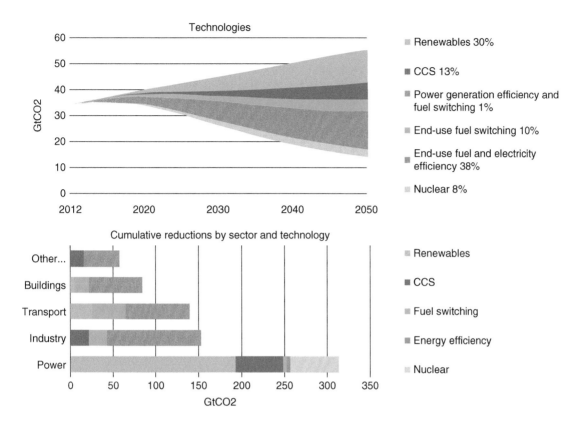

Figure 1.1 Key technology options to reduce global greenhouse gas emissions from the power sector for the IEA's 2-degree scenario.

To assess the mitigation needs, the international community has set itself temperature targets to limit the impact of global warming. These targets are framed and agreed on through the United Nations Framework Convention on Climate Change (UNFCCC), whereby countries meet annually at the Conference of the Parties (COP) to agree on international targets for climate change mitigation. Prior to COP21 in Paris, France in November 2015, the internationally agreed on target for limiting the increase in global average surface temperature rise target was 2 degrees centigrade (2°C Scenario [2DS]). The 2°C goal is achieved by limiting the concentration of GHGs in the atmosphere to around 450 ppm of CO_2.

Considerable work has been done by many organizations that have modeled the energy sector and looked at the technology options that could meet such a temperature target. One such organization that has been very active is the International Energy Agency (IEA). For example, 2DS is the main focus of IEA's Energy Technology Perspectives (ETP) book series. ETP's annual analyses aim to set out sustainable energy transition pathways that incorporate detailed and transparent quantitative analysis and thus provide targeted reading for experts in the energy field, policy makers, and heads of governments.

The 2DS lays out an energy system deployment pathway and an emissions trajectory that are consistent with an at least 50% chance of limiting the average global temperature increase to 2°C. The 2DS limits the total remaining cumulative energy-related CO_2 emissions between 2015 and 2100 to 1000 $GtCO_2$. The 2DS reduces CO_2 emissions (including emissions from fuel combustion and process and feedstock emissions in industry) by almost 60% as compared to 2013 levels by 2050, with carbon emissions being projected to decline after 2050 until carbon neutrality is reached.

An example of the modeling outputs as provided in ETP 2016 is given in Figure 1.1. The figure shows the key technologies that can be used to mitigate GHG emissions from the power sector and the relative contribution between 2013 and 2050 to meet the 2DS compared to a business-as-usual scenario (6DS). As indicated earlier, to achieve the required GHG emission savings a portfolio of technologies will be deployed that include nuclear power, fuel switching and energy efficiency, carbon capture and storage (CCS), and various renewable options.

For industry, there is a similar picture that CO_2 emissions in the 2DS will be achieved through a combination of lower carbon fuels, energy efficiency improvements, and CCS.

The IEA's analysis is not a unique one. Another example is the recent Global Energy Assessment (GEA) 2012, which also works from setting a "business–as–usual" base case but uses a different assessment approach than the IEA. The GEA report shows that there are many combinations of energy resources, end-use, and supply technologies that can simultaneously address the multiple sustainability challenges. One of the report's key findings is that energy systems can be transformed to support a sustainable future through (a) radical improvements in energy efficiency, especially in end use; and (b) greater shares of renewable energies and advanced energy systems with CCS for both fossil fuels and biomass (Johansson *et al.*, 2012).

At a recent COP, COP21 in Paris, the delegates agreed to a new temperature rise target of less than 2°C, with the intent of limiting global surface temperature rise to 1.5°C. This sets an even stronger need to reduce emissions, and the technologies discussed previously will all be required. Discussions in the post-Paris environment indicate that negative emission technologies (NETs) will also be required. These are technologies that remove more emissions from the atmosphere than they put in. A variety of NET options are being considered that include afforestation, enhanced weathering, ocean fertilization, BioChar, direct air capture, and BioCCS, among others. However, unlike those discussed under the 2DS model, most of the NET technologies have not been tested at any realistic scale and thus require considerable development in the coming centuries if they are to be successful in significantly mitigating GHG emissions.

1.3 CCS as a Global Mitigation Option before and after the Paris Agreement

CCS is the process of capturing produced carbon dioxide CO_2 from large point sources, such as fossil fuel power plants, transporting it to a storage site by pipeline or ship, and injecting it where it will be prevented from entering the atmosphere, normally in an underground geological formation. In this way release of large quantities of CO_2 into the atmosphere can be prevented (IPCC, 2015).

The analyses by IEA and GEA discussed earlier have shown that CCS is a key low-carbon technology and has a strong role to play as part of the global low-carbon technology portfolio in reducing global GHG emissions.

The IPCC Fifth Assessment report made two strong points regarding the importance of CCS (IPCC, 2014). First, it showed that most of the global assessment models could not reach the 2°C temperature rise target without the inclusion of CCS in the global technology portfolio. Second, the IPCC analysis indicated that the cost of meeting the 2°C temperature target would be 138% higher without the incorporation of CCS.

With the discussion beginning to unfold regarding a 1.5°C target and the need for negative emissions, both BioCCS and Direct Air Capture (DAC) become relevant to future demand for CCS. Recent work to assess the global potential for BioCCS has suggested the technical potential is large and, if deployed, could result in negative emissions up to 10 Gtonnes of CO_2 equivalents annually. The key obstacle to the implementation of the technology is the absence of a price for stored biomass-based CO_2. There is, therefore, a need for policy developments in this area to assist global adoption of the technology (IEAGHG, 2011a). The implementation of DAC will require the utilization of geological storage space to store CO_2 that has been directly removed from the atmosphere. Consequently, the development of the transportation and storage components of CCS will aid in advancing this technology.

1.4 Geological Storage of CO_2

Geological storage of CO_2 is generally accomplished by injecting it in dense form into rock formations below the Earth's surface (it may well be possible to inject the CO_2 into shallower formations as a gas as well; certainly publications such as Chapter 17 by Juhlin *et al.* in this book demonstrate the safety of shallower traps). Porous rock formations that hold or, as in the case of depleted oil and gas reservoirs, have previously held fluids, such as natural gas, oil, or

Table 1.1 Storage capacity for several geological storage options

Reservoir type	Lower estimate of storage capacity (GtCO$_2$)	Upper estimate of storage capacity (GtCO$_2$)
Oil and gas fields	675[a]	900[a]
Unminable coal seams (enhanced coal-bed methane)	3–15	200
Deep saline formations	1000	Uncertain but possibly 10^4

From Johansson et al. (2012).

[a] These numbers would increase by 25% if "undiscovered" oil and gas fields were included in this assessment. Source: IPCC SRCCS 2005.

brines, are potential candidates for CO_2 storage. Suitable storage formations can occur in both onshore and offshore sedimentary basins (natural large-scale depressions in the Earth's crust that are filled with sediments). Coal beds also may be used for storage of CO_2 where it is unlikely that the coal will later be mined and provided that permeability is sufficient (IPCC, 2005).

The IPCC Special Report on CO_2 Capture and Storage (IPCC SRCCS, 2005) undertook the first review of the global potential for CO_2 storage in geological formations (IEAGHG, 2011a). The IPCC SRCCS considered in detail three types of geological formations that had at that time received extensive consideration for the geological storage of CO_2. The three options were storage in oil and gas reservoirs, deep saline formations, and unminable coal beds.

At the time of the IPCC SRCCS several other possible geological formations or structures were considered, such as basalts, oil or gas shales, salt caverns, and abandoned mines. However, it was believed at the time that these represented only niche opportunities or had not been sufficiently studied at that time to assess their potential. This conclusion is still largely valid today, although interest in shale formations for CO_2 storage is growing.

The estimates of the technical potential for different geological storage options from the IPCC SRCCS are summarized in Table 1.1. While there have been numerous studies on the individual storage options

since the IPCC SRCCS, the fact that the largest CO_2 storage potential globally lies in deep saline formations still remains a core conclusion to this day.

Since the IPCC SRCCS, our knowledge on how these storage resources can be developed has advanced. Gas fields hold a greater storage potential than oil fields. Compared to deep saline formations, both gas and oil fields are much better explored and have a background data set of both geological and operational/production data. On this basis, they both should be more suitable for early application of CO_2 storage than deep saline formations. Storage in oil fields is typically carried out as part of enhanced oil recovery operations (EORs). In such systems, the injection of CO_2 is used to maximize oil production, not for storage of the injected CO_2 (see, e.g., Chapter 14 in this book by Davis et al.). However, incidental storage does occur within the reservoir that can amount to 90% of the CO_2 injected (Whittaker and Perkins, 2013).

From the time of the release of the IPCC SRCCS there has been little research in the potential for geological storage in coal seams. In the past few years, interest in using shale has also surfaced as a potential storage option. Again, it is too early to decide whether this is a promising option for the future or not (IEAGHG, 2013a).

While deep saline formations represent a tantalizing resource for global storage of CO_2, they remain relatively unexplored in most regions of the world. Deep saline formations require much more extensive characterization because, in general, they are "virgin" formations not previously investigated. Because of this, they require much longer lead times, potentially up to 15 years of preexploration, to be considered as viable for geological storage of CO_2 (IEAGHG, 2011b). Chapter 12 by Halland in this book demonstrates the importance of having a catalog of storage opportunities in anticipation of the need for geosequestration in the future.

Storage efficiency, for example, depends on the characteristics of the storage aquifer and confining caprock, operational characteristics of CO_2 storage, and regulatory constraints. Based on these combined factors, storage efficiency can vary widely, with values ranging from less than 1% to greater than 10%. This wide variation in storage efficiency correspondingly impacts storage capacity. In the IPCC SRCCS storage capacity estimates were based on a conservative 1% estimate of pore

Table 1.2 IEA GHG CO_2 monitoring selection tool

				deep	shallow	Plume location / migration	Fine scale processes	Leakage	Quantification
Seismic			3D/4D surface seismic	■		■	■		
			Time lapse 2D surface seismic	■		■			
			Multi-component surface seismic	■			■	■	
	Acoustic imaging		Boomer/sparker profiling		■			■	
			High resolution acoustic imaging		■			■	
			Micro-seismic monitoring			■			
	Well based		4D cross-hole seismic	■			■		
			4D vertical seismic profiling	■					
Sonar bathymetry			Sidescan sonar		■				
			Multi beam echo sounding		■				
Gravimetry			Time lapse surface gravimetry						
			Time lapse well gravimetry						
Electric/electromagnetic			Surface EM	■	■				
			Seabottom EM	■	■				
			Crosshole EM	■					
			Permanent borehole EM	■					
			Crosshole ERT	■					
			Electric spontaneous potential			■	■		
Geochemical	Fluids	Downhole /Springs	Downhole fluid chemistry	■					
			pH measurements	■					
			Tracers	■		■			
		Marine	Sea water chemistry		■				
	Gases	Atmosphere	Bubble stream chemistry		■			■	
			Short closed path (NDIRs & IR)		■			■	
			Short open path (IR diode lasers)		■			■	
			Long open path (IR diode lasers)		■			■	
			Eddy covariance		■			■	
		Soil gas	Gas flux		■			■	
			Gas concentrations		■			■	
Ecosystems			Ecosystems studies						
Remote sensing			Airborne hyperspectral imaging		■				
			Satellite interferometry	■		■			
			Airborne EM		■				
Others			Geophysical logs	■			■		
			Downhole Pressure / temperature	■				■	
			Tiltmeters	■					

Dark pink = method suitable; pink = less suitable; white = not applicable.
From IEA GHG (2010).

volume utilization. Research related to pressure buildup and brine displacement has demonstrated that injection-induced pressure changes can propagate in the injection aquifer much farther than the CO_2 plume itself and can therefore limit storage capacity. Geomechanical effects of pressure buildup can include microseismicity and ground deformation, but pressure management strategies exist. The knowledge base on capillary, or residual, trapping of CO_2 has increased substantially. Laboratory observations confirm that CO_2 will occupy at least 10%, and more typically 30%, of the pore volume. Dissolution of CO_2 at the interface between CO_2 and aquifer water can be significantly accelerated. Taken together, these factors have the effect of reducing the amount of free-phase mobile CO_2 and increasing storage security because dissolved CO_2 is no longer buoyant and prone to leakage.

In summary, the reviews in the Special Issue updating the IPCC SRCSS indicate that CO_2 storage is by and large a safe operation if storage sites are properly selected, characterized, and managed. This should go a long way to alleviating concerns by policy makers, the public, and other stakeholders' concerns that geological storage is a safe and permanent option from the removal of CO_2 from the atmosphere.

1.5 Ensuring Storage Integrity and the Need for Monitoring of Injected CO_2

Storage integrity has been and will continue to be a critical consideration for the storage of CO_2 in the subsurface. In part this is a public issue, as people have concerns about the safety of storage, and in part this relates to the ability to demonstrate predictable conditions in the subsurface prior to any transfer of the project to the public sector in the future. To do this, good monitoring of the project is needed to determine the fate of the injected CO_2 (IPCC SRCSS, 2005).

Table 1.2, taken from the IEA Greenhouse Gas R&D Programme, is a general list of the techniques available for monitoring the state of the CO_2 plume and, in particular, evaluating possible leakage from the storage horizon. The table identifies a number of points salient to this book. The first is the usefulness of geophysical techniques both onshore and offshore. The second is the lack of applicability of all the techniques when it comes to quantifying the amount of CO_2 that exists in the subsurface. In other words, the amount of CO_2 entering the subsurface can be measured with

a high degree of accuracy, but quantification of the amount in place in the subsurface is difficult, if not impossible, to measure with any accuracy. These techniques do, however, allow for the identification of the CO_2 plume and will provide a high degree of certainty on the movement of the plume and hence the predictability of the plume migration in the subsurface for eventual handover to public authorities.

The first requirement of any monitoring program is to create a baseline such as was undertaken at Sleipner (Chadwick et al., 2009) and Weyburn (Wilson and Monea, 2004) against which future surveys can be compared. This is quite consistent with standard oil field practices using a variety of techniques to understand the processes operating in the subsurface and applying history matching to predict future production. Indeed, with enhanced oil recovery, the use of time-lapse seismic surveys has become relatively routine (Weyburn, Wilson and Monea, 2004) and this type of process has been applied in storage areas such as Sleipner (Chadwick et al., 2009). This process is clear in the chapters, particularly the case studies, in this book.

Recognizing the need to ensure storage integrity requires that appropriate monitoring takes place and, as noted, that all changes are referenced back to a baseline set of data. Table 1.2 identifies the monitoring technologies that can be used. These technologies include surface techniques to measure possible CO_2 leakage at the surface (ocean bottom or land surface), shallow CO_2 presence (particularly in potable water zones), and remote sensing of the deeper subsurface to identify leaks and leakage pathways early. These pathways are identified in the IPCC SRCCS 2005 report. While direct measurement techniques can be applied at the surface and near-surface, this is not effective in the deeper subsurface. It is here in the subsurface that geophysical techniques, and particularly seismic surveys, demonstrate their value to ensure that we have storage integrity and that we can predict movement of the CO_2 in the subsurface. Seismic technology also allows for an evaluation of potential leaks and leakage pathways, particularly understanding such features as sealed versus open faults, thinning of caprocks, and other potential natural hazards that may result in CO_2 crossing formational boundaries and migrating toward the surface, economic zones, or potable water zones. Active monitoring is a way of understanding the flow path of the CO_2 and the potential for it reaching man-made or natural routes out of the storage formation.

1.6 Practical Experience in Monitoring CO_2 Storage Formations

At the time of the IPCC SRCCS there were three operational commercial scale CO_2 injection projects globally: Sleipner in the North Sea, the Weyburn CO_2 EOR Project (with an associated research project, the IEAGHG Weyburn CO_2 Capture and Storage Project) in Canada, and the In Salah Project in Algeria. Two of these, Sleipner and InSalah, were injecting CO_2 into deep saline formations, while Weyburn was a CO_2 flood in a depleting oil field (IEAGHG, 2011a). All three of these projects had substantive monitoring projects running alongside their commercial operations.

Since that time the number of CO_2 injection projects has grown considerably. A recent analysis undertaken by the IEA Greenhouse Gas R&D Programme has shown that there were, as of mid-2012, 45 small-scale injection projects and 43 large-scale projects (IEAGHG, 2013b). Small-scale projects were considered to be those injecting less than 100 000 tonnes, though many projects inject considerably less (less than 15 000 tonnes). Large-scale projects were injecting more than 100 000 tonnes/year.

Figure 1.2 provides an overview of the global distribution of the small or pilot CO_2 injection projects.

While there are several individual projects in Australia, China, Europe, and Japan, the majority of these projects are in North America, principally in the USA. The reason for the large number of projects in the USA is that in 2003, the U.S. Department of Energy (DOE) awarded cooperative agreements to seven Regional Carbon Sequestration Partnerships (RCSPs). The seven RCSPs were tasked to determine the best geological and terrestrial storage approaches and apply technologies to safely and permanently store CO_2 for their specific regions (Rodosta, 2016).

The RCSP Initiative has been implemented in three phases:

1. Characterization Phase (2003–2005): Initial characterization of their region's potential to store CO_2 in different geological formations
2. Validation Phase (2005–2011): Evaluation of promising CO_2 storage opportunities through a series of small-scale (less than 500 000 metric tons CO_2) field tests to develop understanding of injectivity, capacity, and storability of CO_2 in the

various geological formations within a wide range of depositional environments
3. Development Phase (2008–2018+): Implementation of large-scale field testing involving at least 1 million metric tons of CO_2 per project

The RCSP program, as of April 2010, has six operational 1 Mtonne CO_2 injection projects, with two more in preparation. This is the largest geological storage demonstration program in the world (IEAGHG, 2013b).

Together the pilot projects around the world have played a fundamental role in enhancing our substantive knowledge of monitoring the storage integrity of the CO_2 storage component of the CCS system. Once again, the experience gained in monitoring over the last 10 or so years has been summarized in the Special Issue of the *International Journal of Greenhouse Gas Control* (IJGGC, 2015). The present book seeks to enhance our understanding of the seismic aspects of monitoring still further.

The Special Issue of IJGGC has shown that

1. Monitoring and verification have developed many shallow monitoring methods in parallel with the assessment of environmental impacts, reflecting societal concerns about leakage to the near-surface.
2. Very significant progress has been made in the deep-focused monitoring techniques, particular examples being marine seismic monitoring at Sleipner and the combination of pressure and seismic imaging at Snøhvit.
3. In the case of Sleipner, both conformance and containment have been convincingly demonstrated by innovative analysis of an impressive data set. Snøhvit is a textbook case of pressure monitoring detecting nonconformance and a mitigation strategy being successfully adopted.
4. Another success for monitoring of reservoir-level processes was at In Salah, where ground surface displacements were detected by the new (in terms of application to CCS) method of interferometric synthetic-aperture radar (InSAR), the measurement of ground surface displacement from satellite platforms. The interpretation of those results in terms of geomechanical processes was subsequently shown to be consistent with microseismic observations and time-lapse seismic imaging,

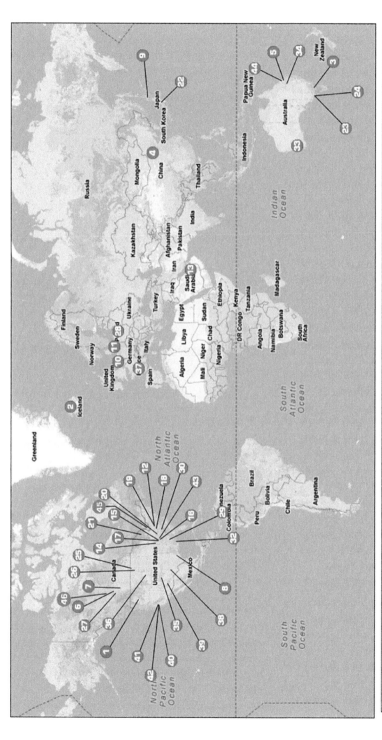

Figure 1.2 Global distribution of pilot-scale CO$_2$ injection projects as of December 2013.

1 BSCSP Basalt	16 MGSC Sugar Creek EOR Phase II
2 Carbfix	17 MGSC Tanquary ECBM Phase II
3 CarbonNet	18 Mountaineer PVF
4 CIDA China	19 MRCSP Appalachian Basin (Burger) Phase II
5 CS Energy Callide Oxyfuel Project	20 MRCSP Cincinnati Arch (East Bend) Phase II
6 CSEMP	21 MRCSP Michigan Basin Phase II
7 Fenn/Big Valley	22 Nagaoka Pilot CO2 Storage Project
8 Frio, Texas	23 Otway I (Stage I)
9 JCOP Yubari/Ishikari ECBM Project	24 Otway II Project (Stage 2A,B)
10 K12B	25 PCOR Lignite
11 Ketzin	26 PCOR Williston Basin -Phase II (NW McGregor Field)
12 Marshall County	27 PennWest Energy EOR Project
13 Masdar/ADCO Pilot project	28 Recopol
14 MGSC loudon Field EOR Phase II	29 SECARB - Black Warrior Basin Coal Seam Project
15 MGSC Mumford Hills EOR Phase II	30 SECARB - Central Appalachian Coal Seam Project

32 SECARB - Mississippi Saline Reservoir Test Phase II	
33 South West Hub (Collie South West Hub)	
34 Surat Basin CCS Project (Previously Wandoan)	
35 SWP San Juan Basin Phase II	
36 Teapot Dome, Wyoming	
37 Lacq-Rousse	
38 West Pearl Queen	
39 WESTCARB Arizona Pilot (Cholla)	
40 WESTCARB Northern California CO$_2$ Reduction Project	
41 WESTCARB Rosetta-Calpine test 1	
42 WESTCARB Rosetta-Calpine test 2	
43 Western Kentucky	
44 Zerogen Project	
45 East Canton Oil Field	
46 PCOR Zama Field Validation Project	

all suggesting initiation and reactivation of fractures.

The common theme to these examples, which has now emerged from many projects at all scales, is the ability of the available techniques for monitoring and interpretation to test containment and conformance. Shallow-focused monitoring methods have also been exploited extensively and have played an important role in countering leakage allegations at Weyburn and providing assurance that environmental impacts of hypothetical leakages are undetectable above natural variability in key parameters.

The pilot projects have led to the production of several best practice documents and guidelines, which vary in scope and technical detail. A number of non-site-specific best practice guides have also been produced, such as the National Energy Technology Laboratory (NETL)'s risk assessment and site selection manuals and the World Resources Institute (WRI)'s CCS guidelines that outline the entire process. There are also best practice guidelines that consider learnings taken from particular projects, such as the Saline Aquifer CO_2 Storage Project (SACS) best practices for the storage of CO_2 in saline aquifers, which uses learnings from the Sleipner storage site in the North Sea. Other examples of best practice guides are the QUALSTORE best practice guide and the EU Guidance documents. There are several documents outlining issues regarding public communication including guidelines from NETL and WRI. The Global CCS Institute commissioned CO2CRC to produce a summary of best practice guides, including a summary of the varying areas of coverage and technical detail (CO2CRC, 2011).

Pilot projects have played a key role in helping build public confidence in geological storage (Romanak et al., 2013). It is safety/integrity of storage sites that principally gets raised in public debates on CCS. These pilot projects have assisted through

- Establishing visitor centers at sites so the public can get first-hand experience of storage operations
- Enabling direct local dialogue with farmers and other key stakeholders
- Enabling the public to meet the scientists involved so people can learn and speak openly about their concerns
- Providing the opportunity to disseminate information at a local level

Through these actions the pilot projects have helped build public confidence in CCS at a local/regional level, which is important for the success of projects and CCS globally.

References

Chadwick, R. A., Arts, R., Bentham, M., et al. (2009). Review of monitoring issues and technologies associated with the long-term underground storage of carbon dioxide. In Special Publications, Vol. **313**: 257–275. London: Geological Society.

CO2CRC. (2011). A review of existing best practice manuals for carbon dioxide storage and regulation. A desktop study prepared for the Global CCS Institute by CO2CRC.

ENERGY.GOV. (2012). Regional partnership U.S. Department of Energy Office of Fossil Energy.

European Council (2014). Conclusions on 2030 climate and energy policy framework. Copenhagen: European Environment Agency.

Houghton, J. T., Jenkins, G. J., and Ephraums, J. J. (Eds.). (1990). *Climate Change:* The IPCC Scientific Assessment Report prepared for Intergovernmental Panel on Climate Change by Working Group I. Cambridge University Press, Cambridge and New York.

IEAGHG. (2011a). Potential for biomass and carbon dioxide capture and storage. Report 2011/06. Cheltenham: IEAGHG.

IEAGHG. (2011b). Global storage resource gap analysis for policy makers. Report 2011/10. Cheltenham: IEAGHG.

IEAGHG. (2012). Quantification techniques for CO_2 leakage. Report 2012/02. Cheltenham: IEAGHG.

IEAGHG. (2013 a). Potential implications on gas production from shales and coals for geological storage of CO_2. Report 2013/10. Cheltenham: IEAGHG.

IEAGHG. (2013b). The process of developing a CO_2 test injection: Experience and best practice. Report 2013/13. Cheltenham: IEAGHG.

IEAGHG. (2015). Carbon capture and storage cluster projects: Review and future opportunities. Report 2015/03. Cheltenham: IEAGHG.

IGBP, IOC, SCOR (2013). Ocean acidification summary for policymakers. In Third Symposium on the Ocean in a High-CO_2 World: International Geosphere-Biosphere Programme. Stockholm: Ocean Acidification International Coordination Centre.

IJGGC (2015). Special Issue commemorating the 10th year anniversary of the publication of the Intergovernmental Panel on Climate Change Special Report on CO_2 Capture and Storage. Edited by J. Gale, J.C. Abanades,

S. Bachu, and C. Jenkins. *International Journal of Greenhouse Gas Control*, **40**: 1–458.

IPCC. (1988). Organization. www.ipcc.ch/organization/organization.shtml#.UvyrqmJ_v_E

IPCC. (2005). IPCC Special Report on Carbon Dioxide Capture and Storage. Prepared by Working Group III of the Intergovernmental Panel on Climate Change: Cambridge University Press, Cambridge, United Kingdom and New York, NY, USA.

IPCC. (2007a). Summary for Policymakers. In *Climate Change 2007: Impacts, Adaptation and Vulnerability. Contribution of Working Group II to the Fourth Assessment Report of the Intergovernmental Panel on Climate Change* [M. L. Parry, O. F. Canziani, J. P. Palutikof, P. J. van der Linden, and C. E. Hanson (eds.)]. Cambridge University Press, Cambridge, United Kingdom and New York, NY, USA.

IPCC. (2007b). Summary for Policymakers. In *Climate Change 2007: Mitigation. Contribution of Working Group III to the Fourth Assessment Report of the Intergovernmental Panel on Climate Change* [M. L. Parry, O. F. Canziani, J. P. Palutikof, P. J. van der Linden, and C. E. Hanson (eds.)]. Cambridge University Press, Cambridge, United Kingdom and New York, NY, USA.

IPCC. (2013). Summary for Policymakers. In *Climate Change 2013: The Physical Science Basis. Contribution of Working Group I to the Fifth Assessment Report of the Intergovernmental Panel on Climate Change* [T. F. Stocker, D. Qin, G.-K. Plattner, M. Tignor, S. K. Allen, J. Boschung, A. Nauels, Y. Xia, V. Bex, P. M. Midgley (eds.)]. Cambridge University Press, Cambridge, United Kingdom and New York, NY, USA.

IPCC. (2014). Summary for Policymakers. Fifth Assessment Report (AR5). Cambridge University Press, Cambridge, United Kingdom and New York, NY, USA.

Johansson, T. B., Nakicenovic, N., Patwardhan, A., Gomez-Echeverri, L., eds. (2012). *Global Energy Assessment (GEA): Toward a Sustainable Future*. Cambridge: Cambridge University Press.

Le Treut, H., Somerville R., Cubasch U., *et al.* (2007). Historical Overview of Climate Change. In *Climate Change 2007: The Physical Science Basis. Contribution of Working Group I to the Fourth Assessment Report of the Intergovernmental Panel on Climate Change* [Solomon, S., D. Qin, M. Manning, Z. Chen, M. Marquis, K. B. Averyt, M. Tignor, and H. L. Miller (eds.)]. Cambridge University Press, Cambridge, United Kingdom and New York, NY, USA.

NOAA. (2016). Comment on recent record-breaking CO_2 concentrations. https://scripps.ucsd.edu/programs/keelingcurve/2016/04/20/comment-on-recent-record-breaking-co2-concentrations/#more-1406

OECD, IEA (2012). *Energy technology perspectives 2012: Pathways to a clean energy system*. Energy Technology Perspectives. Paris: OECD Publishing.

OECD/IEA. (2016). *Energy technology perspectives 2016: Towards sustainable urban energy systems*. Energy Technology Perspectives. Paris: OECD Publishing.

Romanak, K., Sherk, G.W., Hovorka, S., and Yang, C. (2013). Assessment of alleged CO_2 leakage at the Kerr Farm using a simple process-based soil gas technique: Implications for carbon capture, utilization, and storage (CCUS) monitoring. *Energy Procedia*, GHGT-11 **37**: 4242–4248.

Stern, N. (2006). Stern review on the economics of climate change (pre-publication edition). Executive Summary). HM Treasury, London. Archived from the original on January 31, 2010. http://mudancasclimaticas.cptec.inpe.br/~rmclima/pdfs/destaques/sternreview_report_complete.pdf

Whittaker, S., and Perkins, E. (2013). Technical aspects of CO_2 enhanced oil recovery and associated carbon storage. Docklands, VIC: Global CCS Institute.

Wilson, M., and Monea, M., eds. (2004). IEA GHG Weyburn CO_2 Monitoring & Storage project summary report 2000–2004: 7th International Conference on Greenhouse Gas Control Technologies, Vancouver, Canada. Vol. 3.

World Bank Group. (2014). *Turn down the heat: Confronting the new climate normal*. Washington, DC: World Bank. https://openknowledge.worldbank.org/handle/10986/20595

The Role of Geophysics in Carbon Capture and Storage

David Lumley

Introduction

Geosequestration

Geosequestration is the capture, transport, injection, and long-term storage of industrial carbon dioxide (CO_2) in deep geological reservoirs, thus accelerating the Earth's natural carbon cycle, while simultaneously reducing manmade CO_2 emissions to the atmosphere. Geosequestration represents one of the famous "climate stabilization wedges" proposed by Pacala and Socolow (2004), and may be able to capture and store approximately 15% or more of the global manmade CO_2 emissions required to keep atmospheric CO_2 levels below 500 ppm. More details on the driving forces and methods associated with geosequestration in specific, and carbon capture and storage (CCS) in general, can be found in other chapters of this book.

The basic approach for geosequestration is to capture CO_2 at industrial point sources (e.g., coal-fired power plants, mineral and agricultural processing plants, cement plants, liquefied natural gas facilities, etc.) and inject it into nearby deep geological formations. Subsurface storage reservoirs include depleted hydrocarbon reservoirs and saline aquifers. An international process is underway to map and rank subsurface storage potential for CO_2 sequestration among the world's major sedimentary basins, and future work will require detailed geological and geophysical site characterization of these basins, reservoirs, and saline aquifers in terms of defining storage capacity (volume, porosity, etc.), injectivity (permeability, pressure regime, etc.), and sealing efficiency (structural and stratigraphic traps, impermeable geological layers and faults, pore-space CO_2 trapping mechanisms and capillary pressures, geochemistry, etc.). For example, Australia has recently become the first country in the world to open up offshore exploration leases specifically for the purpose of finding and testing subsurface CO_2 storage reservoirs in preparation for a potential future global market in CO_2 sequestration (Figure 2.1).

The Role of Geophysics

In addition to the initial storage site and reservoir characterization requirements, there will be a major demand for ongoing geophysical monitoring and verification of CO_2 injection during the life of CO_2 storage projects. As operators inject CO_2 into the subsurface, there will be a requirement to image and monitor the evolving CO_2 plume to ensure that it is being injected at the correct depths and locations, that it fills the storage reservoir efficiently as predicted, that it does not interfere with other subsurface resources (e.g., groundwater, hydrocarbons, geothermal, etc.), that it does not flow toward high-risk areas such as major faults with uncertain seismic and flow properties, and that it remains sealed in the storage reservoir over time. Furthermore, in order for an operator to effectively transfer any potential long-term liability associated with a CO_2 sequestration project to a government agency, operators will likely be required to develop a 3D subsurface flow model that accurately predicts and matches the CO_2 injection and monitoring data for a significant period of time (e.g., 25–50 years), and thus "close the loop" in terms of reconciling the subsurface geology, flow simulations, and geophysical images of the subsurface CO_2 distribution. This will require advances in CO_2 sequestration geophysics, including core measurements of the fluid-flow and geochemical effects of CO_2 on rock properties, computational simulations and imaging algorithms to predict and image the effects of CO_2 injection in the subsurface, and inversion of geophysical data to make accurate quantitative estimates of the amount of CO_2 in the ground. Owing to the complexity of CO_2 geophysics, achieving these goals will likely require combining the powerful

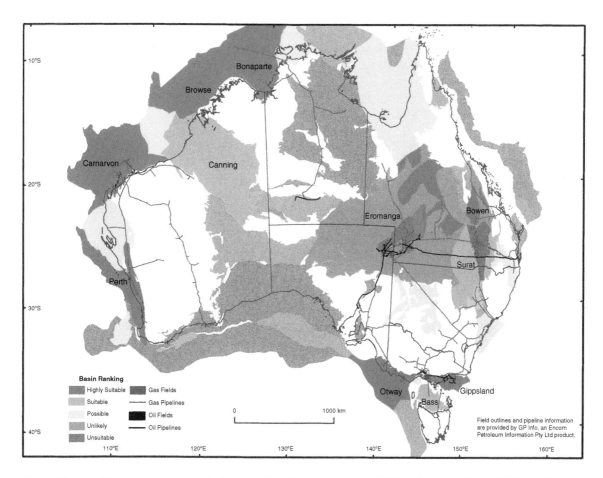

Figure 2.1 Map of Australian sedimentary basins ranked for CO_2 storage potential (Carbon Storage Taskforce, DRET, Australia, 2009). http://hub.globalccsinstitute.com/publications/potential-impacts-groundwater-resources-co2-geological-storage/21-assessments-potential

subsurface monitoring technique of time-lapse seismology with complementary geophysical methods such as passive seismology, electromagnetics (EM), gravity, and interferometric synthetic aperture radar (InSAR) satellite imaging.

Detailed CO_2 injection case studies are presented in other chapters of this book, which show specific examples of various topics presented in this overview chapter on seismic monitoring.

Time-Lapse Seismology

This chapter is intended to provide a comprehensive overview of the subsurface monitoring method known as time-lapse (or 4D) seismology; the reader is referred to the following publications for additional detailed review material (Lumley, 1995, 2001; Calvert,

2005; Johnson, 2013; other chapters in this book). Time-lapse seismology involves repeating active-source seismic surveys in time-lapse mode, or continuously recording passive seismic (micro) earthquake and ambient noise data, in order to image and estimate physical property changes in the subsurface over time. To first order, seismic body waves measure the compressibility of the subsurface porous rock–fluid system. Thus, in order to create a time-lapse seismic signal, the fluid-saturated rock compressibility will likely change over time. In general, the more compressible the rock matrix is (e.g., unconsolidated sand), and the larger the compressibility contrast between the fluids of interest (e.g., water vs. supercritical CO_2), or the larger the change in injection pore-pressure/stress, the larger the resulting time-lapse seismic signal will be. Conversely, stiff rocks,

13

fluids with similar compressibility, and small injection pressure/stress changes tend to produce weak or nondetectable time-lapse signals. The dependency of the time-lapse seismic signal response to changes in rock and fluid properties caused by CO_2 injection will be reviewed in this chapter, and additional details can be found in other chapters of this book.

As with the measurement of any physical phenomena, the ability to detect a time-lapse signal in field data depends both on the magnitude of the signal and on the noise level in the data. The variable physical properties of the injected CO_2 with reservoir pressure and temperature, and the variable properties of the reservoir rocks saturated with CO_2–fluid mixtures after injection, determine the strength of the time-lapse seismic signal. The quality of the time-lapse seismic data and images, especially the level of nonrepeatable noise and complexity of wavefields, determines whether the CO_2 can be accurately detected, imaged, and quantified. The largest source of noise in time-lapse seismic data tends to be nonrepeatable noise, which results from the fact that we cannot perfectly repeat the seismic imaging experiment and its environmental conditions from one survey to the next. Issues related to nonrepeatability in time-lapse seismic data acquisition, image processing, and inversion will be comprehensively reviewed in this chapter.

Outline for This Chapter

This chapter is intended to provide a comprehensive overview and discussion of time-lapse seismology as it specifically relates to monitoring CO_2 injection and storage, and summarizes my experiences gained from working on 200+ time-lapse seismic projects (of all types) during the past 25+ years. As much of this book focuses on CO_2 injection projects in Europe and North America, I have included a significant amount of Australian content in this chapter to add variety, having spent the previous decade as Professor and Chair in Geophysics at the University of Western Australia.

The outline for the remainder of the chapter proceeds as follows:

1. Review the fundamentals of **time-lapse seismic theory** that form the mathematical and physical basis for time-lapse seismology.
2. Review the key **time-lapse rock and fluid physics** aspects of CO_2 injection, and the relationships to key time-lapse seismic physical properties.
3. Discuss how the foregoing can be used to perform **time-lapse simulation/modeling** of seismic data for a wide variety of CO_2 injection scenarios.
4. Review key aspects of **time-lapse seismic acquisition and imaging** in terms of suppressing sources of nonrepeatable noise, and enhancing the detection of time-lapse seismic signals generated by CO_2 injection.
5. Review multiple strategies for **time-lapse seismic inversion**, and discuss their relative strengths and weaknesses in terms of making accurate quantitative estimates of physical properties associated with CO_2 injection (fluid saturation, pressure, etc.).
6. Discuss recent advances in **time-lapse passive seismology** that may allow us to monitor CO_2 injection and storage reservoirs without the need for manmade active source efforts, by instead using seismic data from (semi) permanent sensor arrays that continuously record passive (micro) earthquake energy and (ocean) ambient noise.

Theory

It is helpful to review the general mathematical and physical theory that forms the basis of time-lapse seismic monitoring. We begin with a simple equation that describes how we can model synthetic seismic data:

$$d = F\{m\} \tag{2.1}$$

where $d(\underline{x}, t)$ are the seismic data traces for the source and receiver positions located at $\underline{x} = (x, y, z)$, and recording time t. The forward modeling operator F generates seismic data at the receivers given a seismic source and an earth model m. F can be as simple as a 1D linear convolutional operator that convolves a seismic wavelet with a reflection coefficient series, or F can be a fully nonlinear 3D computational operator that solves a complex wave equation including acoustic, elastic, viscoelastic, anisotropic, etc. seismic properties of the earth (e.g., Aki and Richards, 1980).

The earth model $m(\underline{x})$ contains the spatial distribution of rock and fluid properties in the earth required to generate seismic data. These rock and fluid properties typically include the elastic bulk modulus K (stiffness, incompressibility), elastic shear modulus G (shear strength), and the density ρ. These three fundamental seismic properties can be recombined to calculate P- and S-wave propagation velocities V_p and V_s, P- and S-wave impedances I_p and I_s, and other seismic physical properties (Mavko *et al.*, 2009):

$$V_p = \sqrt{\frac{K+4G/3}{\rho}} \;\; ; \;\; V_s = \sqrt{\frac{G}{\rho}} \;\; ; \;\; I_p = \rho V_p \;\; ; \;\; I_s = \rho V_s.$$

For time-lapse seismic monitoring purposes, we consider that the earth model m is not static, but instead may vary with calendar time τ (to distinguish from recording time t). We can thus write the time-varying earth model as $m(x, \tau)$. There are many reasons that the rock and fluid properties of the earth can vary over time, including but not limited to

- Changes in pore fluid type, saturations, or mixtures with fluid flow (e.g., CO_2 injection)
- Changes in pore fluid pressure, changes in tectonic or confining stress
- Changes in temperature due to advective heat/fluid flow, diffusion, geothermal activity
- Geomechanical deformation, compaction, dilation, shearing, or fracturing
- Geochemical reactions due to changes in the rock–fluid chemical equilibrium
- And other physical phenomena . . .

The foregoing sources of time-varying rock and fluid properties can be either natural, or manmade, and both may occur together. When we inject CO_2 and other fluids into deep geological rock formations, we can change the rock and fluid properties in a time-varying manner, especially the pore fluid saturations, pore pressure, and fluid temperature. CO_2 injection may also cause subsurface changes due to geochemical and geomechanical effects.

As an aside, it may be useful to mention that Eq. (2.1) can be rewritten more generally as

$$(F + e_F)\{m + e_m\} = d + e_d, \qquad (2.2)$$

where e_* are the respective errors in the earth model m, the forward modeling operator F, and the resulting data d. If we have the correct earth model, and generate synthetic seismic data to compare with real data, we know that the data sets may not match because of noise present in the real data. However, Eq. (2.2) also shows that even with noise-free data and the correct model, the synthetic and real data sets may still not match because of errors and approximations (e_F) we make in the physics and

numerical representation of the forward modeling operator (e.g., acoustic versus elastic, convolution vs. finite difference, etc.), and also errors and approximations (e_m) we make in representing and parameterizing the earth model (e.g., choice of model mesh size, petro-physical relationships, etc.). This leads to the important topic of uncertainty analysis, in which errors and nonuniqueness in the modeling domain often make it challenging to compare synthetic data to real data (signal + noise) in the data domain (e.g., Lumley, 2006). A full discussion of error analysis, uncertainty analysis, and nonuniqueness is beyond the scope of this chapter. However, it is helpful to keep in mind that these concepts arise in many important ways throughout time-lapse seismic data analysis, especially when trying to determine the validity of a reservoir flow model by trying to match the observed time-lapse monitoring data at a specific CO_2 injection and storage site.

Since the earth model is time-varying, we can therefore record different seismic data sets at different calendar times:

$$\begin{aligned} d_1 &= F\{m(x, \tau_1)\} \\ d_2 &= F\{m(x, \tau_2)\} \\ &\cdots \\ d_n &= F\{m(x, \tau_n)\} \end{aligned} \qquad (2.3)$$

and hence form a variety of time-lapse *difference* data sets:

$$\begin{aligned} \Delta d_{2,1} &= d_2 - d_1 \\ &\cdots \\ \Delta d_{k,j} &= d_k - d_j. \end{aligned} \qquad (2.4)$$

Note the time-lapse polarity convention used in Eq. (2.4); *I always recommend forming time-lapse data differences by subtracting an earlier survey from a later survey* . . . this convention proves to be very useful for subsequent time-lapse interpretations and analysis. As Figure 2.2 shows, *I also recommend that positive differences be colored blue, and negative differences be colored red.* Then, with these time-lapse polarity and color conventions, a positive/blue difference typically implies that the reservoir rock has hardened, for example, via increased water saturation (hence blue), and a negative/red difference typically implies that the reservoir rock has softened, for example, via increased pore pressure, or increased gas/CO_2 saturation (hence red).

In addition to *modeling* data, we are also interested in *imaging* and *inversion* of seismic data. Seismic

15

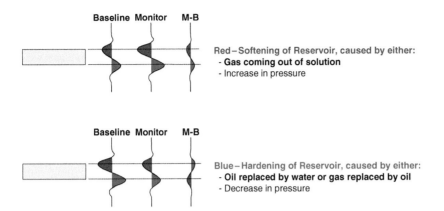

Baseline Monitor M-B

Red – Softening of Reservoir, caused by either:
- **Gas coming out of solution**
- Increase in pressure

Baseline Monitor M-B

Blue – Hardening of Reservoir, caused by either:
- **Oil replaced by water or gas replaced by oil**
- Decrease in pressure

Figure 2.2 Recommended time-lapse seismic polarity convention (Lumley, 1995, 1996). Graphic prepared in collaboration with Guy Duncan, BHP Billiton.

imaging is the process of using seismic data to image geological and fluid-flow features within the earth, for example the storage containment reservoir and the injected CO_2 plume. Seismic *inversion* is the process of using seismic data to make quantitative estimates of the physical properties of rocks and fluids within the earth, for example, the amount of residual CO_2 trapped within the storage reservoir, and its spatial and temporal distribution.

We can mathematically invert Eq. (2.1) to estimate an earth model from the seismic data as follows:

$$m = F^{-1}\{d\} \qquad (2.5)$$

where F^{-1} is known as the inverse operator. For simple forward modeling problems, for example when modeling a set of data points y_i using the equation of a line $y_i = mx_i + b$ with model parameters slope m and intercept b, we can easily find an inverse operator F^{-1} that accurately estimates the model parameters (m, b) from the data. Unfortunately, for most realistic seismic data problems it is impossible to explicitly estimate F^{-1} because the seismic problem is too large, ill-posed, underconstrained, non-unique, and thus the inverse operator may not in fact exist. For these reasons, we instead implicitly determine the inverse operator by finding a model m that generates synthetic seismic data d^{mod} which matches, in some sense, the observed real data d^{obs}. A common approach to solving this inverse problem uses one of many variations of the least-squares (L_2 norm) data fitting method:

$$E(m) = [d^{obs} - d^{mod}(m)]^2 + constraints \qquad (2.6)$$

in which the data misfit error $E(m)$ is iteratively minimized until a model m is found such that the synthetic data matches the real data, within some error tolerance. Other optimization methods can be used instead of least-squares; for example, nonlinear techniques based on L_p norms (e.g., L_1), random model realizations (e.g., Monte Carlo), statistical methods (e.g., Bayesian), genetic or evolutionary algorithms, particle swarm optimization, etc. These and related optimization approaches form the basis for many seismic inversion methods, ranging from simple 1D poststack acoustic impedance inversion to complex 3D and 4D wave-equation full waveform inversion (FWI).

A classical solution to the least-squares problem defined in Eq. (2.6) can be written as

$$m = (F^*F)^{-1}F^*d = H^{-1}g \qquad (2.7)$$

where F^* is known as the adjoint operator to the forward modeling operator F. Any adjoint operator F^* is formally defined by the representation theorem $< F^*d, \ m \ > \ = \ < Fm, \ d \ >$, where $< a, \ b >$ is the mathematical inner (dot) product between two quantities a and b. For seismic wave propagation, the adjoint operator F^* propagates seismic data in reverse time from the surface receivers back into the earth. In optimization theory, the term F^*d is known as the gradient g of the misfit energy $E(m)$, and the term F^*F is known as the Hessian H (curvature) of $E(m)$. If the adjoint operator F^* is a reasonable approximation in some sense to the inverse operator F^{-1}, then H^{-1} is approximately unitary, and Eq. (2.7) reduces to

$$m \approx F^* d. \qquad (2.8)$$

The equation above shows that the adjoint F^* is some sort of seismic imaging operator which uses the data d to form an estimated "image" of the earth model m. It can be shown that F^* makes an estimate of only the high-wavenumber components (reflectivity) of the full earth model, and thus only an image of the reflective boundaries caused by geological layer boundaries (e.g., a CO_2 storage reservoir) and fluid contacts (e.g., an injected CO_2 plume) in the earth. The imaging operator F^* cannot directly recover the low-wavenumber components (smooth velocity) of the full earth model, but the imaging process can be repeated in an iterative manner until a smooth velocity model is found which generates a well-focused seismic image. This iterative process forms the basis for modern integrated seismic image processing and velocity analysis.

Given multiple time-lapse seismic data sets, it becomes evident that there are multiple strategies available for time-lapse imaging and inversion, including the following methods (Lumley, 1995, 1996; Lumley *et al.*, 2003a; Kamei and Lumley, 2017):

- "Parallel" imaging/inversion
- "Double Difference" imaging/inversion
- "Bootstrap" or "Sequential" imaging/inversion
- "Simultaneous" imaging/inversion

For the Parallel method, the time-lapse data sets are imaged/inverted independently in parallel using an initial reference velocity model m_0, after which the desired image/model differences are created by subtracting the results:

$$\begin{aligned} m_1 &= F^*\{d_1, m_0\} \\ m_2 &= F^*\{d_2, m_0\} \\ \Delta m_{2,1} &= m_2 - m_1 \end{aligned} \qquad (2.9)$$

where it is henceforth understood that F^* is the imaging operator for seismic imaging, or the inverse operator F^{-1} for seismic inversion. The Parallel method can accommodate nonrepeatable source and receiver acquisition geometries for d_1 and d_2, but is highly sensitive to nonrepeatable seismic noise conditions. We can see this by decomposing

the seismic data into signal and noise, $d = s + n$; if the noise sources n_1 and n_2 are not repeatable (equal), then the errors they cause in the image/model results will not cancel, and in fact generally become statistically additive. This means that with the Parallel method, the error in the difference image/model Δm will be larger than the individual image/model errors caused by nonrepeatable seismic noise.

For the Double Difference (DD) method, the time-lapse data sets are not imaged/inverted separately; instead, the difference of the data sets is imaged/inverted directly:

$$\Delta m_{2,1} = F^*\{\Delta d_{2,1}, m_0\}. \qquad (2.10)$$

The DD method is more robust to nonrepeatable noise than the Parallel method, especially if the noise is random and uncorrelated. In this case the imaging/inversion operator is able to somewhat "see through" the nonrepeatable noise and extract the correct image/model difference by taking advantage of the constraint that most of the image/model remains unchanged between the time-lapse surveys. However, the DD method is highly sensitive to changes in time-lapse acquisition, especially nonrepeatable source-receiver positioning and environmental conditions. The DD method assumes that the time-lapse data sets share identical geometries, and identical wave paths, in order to form accurate difference data $\Delta d_{2,1}$. In practice for many real-life scenarios these assumptions are not valid; for example, marine streamer positioning is notoriously nonrepeatable, and land data surveys can be subject to changing near-surface conditions (dry, wet), and permit variations regarding where sources and receivers can be deployed from survey to survey.

For the Bootstrap method, the time-lapse data sets are imaged/inverted in a special sequential manner, by estimating the incremental changes between each seismic survey, while simultaneously constraining the overall image/model to be similar:

$$\begin{aligned} m_1 &= F^*\{d_1, m_0\} \\ m_2 &= F^*\{d_2, m_1\} \\ \Delta m_{2,1} &= m_2 - m_1. \end{aligned} \qquad (2.11)$$

The Bootstrap method has proven to be fairly robust to both nonrepeatable seismic data noise, and nonrepeatable acquisition geometry conditions, as will be shown later.

Finally, the Simultaneous method solves for the time-lapse image/model estimates using the time-lapse data sets in a simultaneous manner:

$$m_1, m_2 = F^*\{d_1, d_2, m_0\}$$
$$\Delta m_{2,1} = m_2 - m_1. \tag{2.12}$$

The Simultaneous method is the most robust of all the methods described earlier because it performs a highly constrained global imaging/inversion procedure using all of the time-lapse data sets together, rather than a localized step-by-step imaging/inversion procedure one data set at a time. The Simultaneous method is not widespread but has been used successfully in time-lapse image processing for at least a decade (Lumley *et al.*, 2003a), and is an active area of time-lapse inversion research today (e.g., Kamei and Lumley, 2017, among others).

Rock and Fluid Physics

The time-lapse seismic responses to changes in the earth caused by CO_2 injection are controlled primarily by the physical properties of the dry and fluid-saturated rock, and of the injected CO_2 at reservoir pressure and temperature conditions. The three primary *elastic* rock properties to be considered are the bulk modulus (stiffness, incompressibility) K, the shear modulus (shear strength) G, and the density ρ. These three elastic parameters result from an approximation that rocks behave like a linear isotropic elastic material, which by definition returns to its original state after weak deformation by an applied force (e.g., a seismic wave). Additional seismic rock and fluid physics properties may also change over time, in both a linear elastic and nonlinear nonelastic manner (e.g., anisotropy parameters, the attenuation factor Q, etc.), but for CO_2 injection these changes are often second order in comparison to (K, G, ρ) and thus will not be discussed in much detail here (but are of great interest in the time-lapse seismic research community).

For CO_2 injection projects, the most likely sources for changes in the primary rock and fluid physical properties will be caused by *changes in*

- Rock pore fluid saturations and mixtures due to CO_2 injection
- Rock pore fluid pressure and rock stress conditions due to CO_2 injection
- Rock and fluid temperature due to (cold) CO_2 injection into a (hot) reservoir

- Rock geomechanical properties caused by CO_2 injection
- Rock and fluid geochemical reactions caused by CO_2 injection

To first order, seismic waves primarily measure the elastic stiffness (or compressibility) of the subsurface porous rock–fluid system. Therefore, in order to create a time-lapse seismic signal, the fluid-saturated rock compressibility is likely to change over time. In general, the more compressible the rock matrix (e.g., unconsolidated sand), the larger the pore pressure change, and the larger the compressibility contrast between the fluids of interest (e.g., water and injected CO_2), the larger the resulting time-lapse seismic signal. Conversely, stiff rocks, small pore pressure changes, and/or fluids with similar compressibility properties, produce a much weaker time-lapse signal. Injected CO_2 tends to be much more compressible than formation water, but may be of a similar compressibility to hydrocarbon gas (methane) and high-GOR (live) oil. In situations where storage reservoir rocks are sufficiently soft, CO_2 injection into saline aquifers is likely to produce a strong time-lapse signal, whereas CO_2 injection into depleted hydrocarbon reservoirs is more complex and may produce a much weaker time-lapse signal.

The dry and saturated rock properties of a CO_2 storage reservoir can be determined by core and fluid lab measurements, geophysical borehole logging measurements, and theoretical rock and fluid physics relationships.

Seismic Properties of Dry Rocks

The dry rock bulk and shear moduli (K, G) are a strong function of the rock's porosity φ and grain cementation c, among other factors. For narrow porosity ranges K and G are approximately linear with porosity $(K \sim K_{max} - k\varphi)$, and can be accurately estimated from core measurements and well logs, per rock facies type. As shown in Figure 2.3, for large porosity ranges K and G are approximately inversely proportional to porosity $(K \sim k/\varphi)$ and fall between the Reuss lower bound [an iso-stress harmonic average $(\Sigma 1/K)$ of the constituent K values] and the upper Voigt bound [an iso-strain arithmetic average (ΣK) of the constituent K values]. For rocks with somewhat spherical grain shapes, the Hashin–Shtrikman

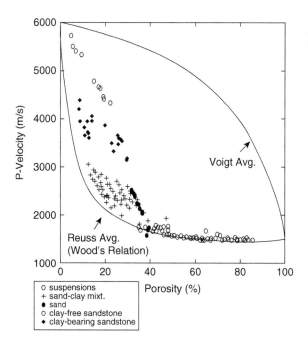

Figure 2.3 Plot of P-wave velocity versus Porosity, showing various core sample measurements falling within the with Voigt and Reuss bounds (Mavko *et al.*, 2009).

material physics bounds for K and G can provide a more narrow and accurate set of bounds than the Voigt–Reuss approach. The Hashin–Shtrikman method can also accommodate the inclusion of grain cement, as can the improved cementation model by Dvorkin and Nur (1996). Further details regarding various rock physics K, G bounds can be found in Mavko *et al.* (2009).

The elastic dry rock bulk and shear moduli (K, G) can also vary with effective (differential) stress σ_{ij}, or more simply, effective pressure $P_{\mathrm{eff}} = (P_{\mathrm{conf}} - P_{\mathrm{pore}})$, where P_{conf} is the confining (overburden) pressure, and P_{pore} is the pore pressure of the fluid in pore space. A rock tends to weaken/soften with increasing pore pressure such that K (and G) follow nonlinear (approximately exponential) relationships of the form

$$K(P_{\mathrm{eff}}) \sim K_{min} + \alpha K_{max}[1 - exp(P_{\mathrm{eff}}/P_{max})],$$

where K_{min} is the K value at the highest pore pressure when the rock begins to hydraulically fracture (the "frack point"), and K_{max} is the K value at the lowest pore pressure when all of the micro-cracks in the rock have been closed under confining pressure such that the rock cannot become elastically stiffer. Saul and Lumley (2013) give a good discussion of the pressure sensitivity of reservoir rocks, both elastic and nonelastic, and provide a new formulation that fits a wide range of core lab data much more accurately than

previous models such as Hertz–Mindlin and others. It is worth noting that injection of highly compressible supercritical CO_2 is unlikely to cause increases in pore pressure much greater than 0–5 MPa, and thus the pressure effects will typically be much smaller than water injection, which can easily create pore pressure increases of 5–35 MPa or more.

The dry rock density ρ is approximately linear with porosity, and a function of the volume-weighted arithmetic average of the individual density values of the mineral grain constituents. The dry rock density is easily measured in a core lab, or via borehole density and saturation logs, and can be accurately fit with a linear trend versus porosity. To first order, the dry rock density does not vary much with pressure; however, there are often exceptions when the rock is extremely unconsolidated and compressible (e.g., Meadows *et al.*, 2005; Saul and Lumley, 2013).

Seismic Properties of Pure and Impure CO$_2$

For most CO_2 storage reservoirs, the pressure and temperature regime will be, by design, such that the injected CO_2 will exist in the supercritical state of its phase diagram, thus having the complex supercritical physical properties of both a gas and a fluid. Supercritical CO_2 is approx. 500 times denser than gaseous CO_2 and thus much more efficient for CO_2

Table 2.1 PCC CO_2 mixture composition

Fluids	Mole fractions
CO_2	0.938
SO_2	0.005
Nitrogen	0.038
Argon	0.007
Oxygen	0.01
Methane	0.002

Composition of CO_2 + impurities resulting from an end-member example of a low-efficiency Post-Combustion Capture (PCC) process.

storage. The critical-point pressure and temperature of pure CO_2 occurs at (P_c, T_c) = (7.39 MPa, 31.10°C), and can vary somewhat for CO_2 mixtures that contain small amounts of impurities. For typical geothermal gradients (approx. 25°C/km) and hydrostatic pressure gradients (approx. 10 MPa/km), CO_2 storage reservoirs must therefore be deeper than approx. 1 km depth in order to ensure that the injected CO_2 remains in a dense supercritical fluid state.

A liquid natural gas (LNG) plant, using a liquid amine capture process, results in a CO_2 stream that is typically >95% pure CO_2. The bulk modulus and density (and other physical properties) of pure CO_2, can be calculated using Equation of State (EoS) methods with algorithms publicly available from National Institute of Standards and Technology (e.g., SUPERTRAPP, Span-Wagner), GERG-2008, and the more recent EoS–CG method developed specifically for CCS applications. A coal-fired power plant, using a low efficiency post-combustion capture (PCC) process, may generate a CO_2 stream that contains residual impurities such as SOx, NOx, nitrogen, argon, and methane (e.g., Table 2.1). The physical properties of CO_2 mixtures that contain such impurities can be calculated most accurately using the new EoS–CG model (Gernert and Span, 2016), which is a recently published improvement to the GERG-2008 EoS model.

Figure 2.4 shows plots of the CO_2 bulk modulus K_{CO2} and CO_2 density ρ_{CO2} versus fluid pressure and temperature calculated using the EoS–CG method, for two realistic end-member CO_2 capture cases: pure CO_2 representing an ideal liquid-amine capture process for an LNG plant, and post-combustion capture (PCC) representing a low-efficiency coal-fired power plant capture process. Note that the compressibility and density of pure CO_2 can vary as much as one order of magnitude across the storage reservoir pressure–temperature range. CO_2 is much more compressible than water ($K_{water} \sim 2$ GPa) but can have a similar large range of variability as hydrocarbon oil and gas, and the density of CO_2 varies from gas-like to oil or water-like conditions depending on the exact pressure and temperature. Adding impurities such as SOx, NOx, and methane to pure CO_2 shifts and blurs the critical point of the CO_2 mixture properties in the detailed phase diagram, from (P_c, T_c) = (7.39 MPa, 31.10°C) for pure CO_2 to (8.32 MPa, 28.39°C) for CO_2 with the PCC impurities listed in Table 2.1, but the overall range of compressibility and density variation remains similar.

Seismic Properties of Injected CO_2 + Fluid Mixtures

When CO_2 is injected into a storage reservoir, it mixes with the preexisting pore fluids (brine, oil, gas, etc.) The bulk modulus K_{fluid} of the resulting CO_2 + fluid mixture can be calculated using Wood's Law, which is a harmonic average of the saturation-weighted constituent fluid values ($1/K_{fluid} = \sum S_i/K_i$), and the density of the CO_2 + fluid mixture ρ_{fluid} can be calculated by the arithmetic average of the constituent fluid densities ($\rho_{fluid} = \sum S_i \rho_i$).

Typical maximum values of residual/trapped CO_2 saturation are on the order S_{CO2} = 0.2–0.6.

CO_2 is likely to be injected and stored in saline aquifers containing brine (very salty water), or in depleted hydrocarbon (HC) reservoirs containing a mixture of residual oil, water, and/or HC gas. In the case of saline aquifers, the injection of CO_2 into brine will make the resulting fluid mixture much more compressible and somewhat less dense, depending on the exact pressure and temperature. Because time-lapse seismic is sensitive to the fluid compressibility *contrast* of the *in situ* fluid versus the injectant, CO_2 injection into saline aquifers is highly favorable for seismic monitoring given sufficiently soft reservoir rocks (e.g., Sleipner).

In depleted HC reservoirs, the situation is more complex because oil has widely variable physical properties, and CO_2 reacts with oil over short time scales to change the physical properties of the residual oil, which is why CO_2 injection is a well-established method for enhanced oil recovery (EOR). Because

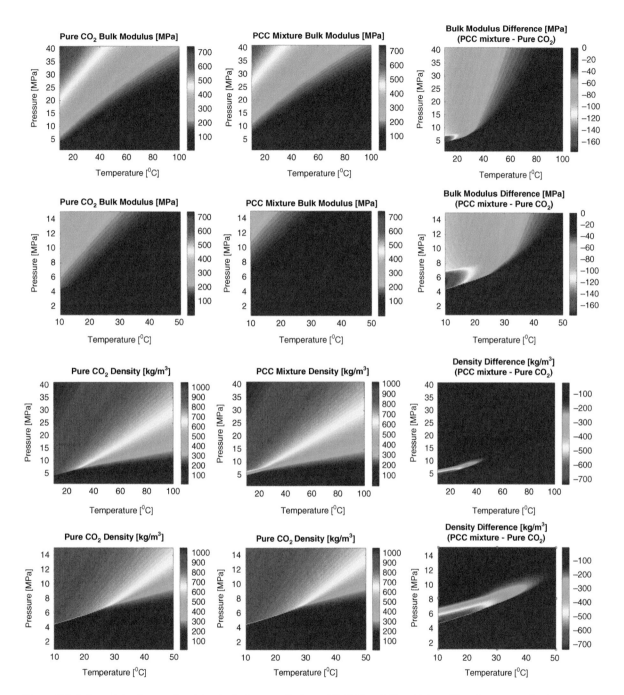

Figure 2.4 Bulk modulus (top rows) and density (bottom rows) of pure CO_2 (left), and CO_2 with impurities after post-combustion capture (PCC) (center), and the difference (right), as a function of pressure and temperature. The second and fourth rows are zooms of the first and third rows respectively. (Figures prepared in collaboration with Paul Connolly and Eric May, University of Western Australia.)

time-lapse seismic is sensitive to fluid compressibility contrasts, the most favorable conditions for seismic monitoring of CO_2 in depleted HC reservoirs are when the residual oil has a low GOR value, for example, at the Weyburn project (GOR = solution gas/oil ratio; i.e., the amount of dissolved HC gas in the oil).

In contrast, conditions which are less favorable for seismic monitoring of CO_2 in a depleted HC reservoir occur when the residual oil is high-GOR and thus may be of a similar compressibility to the injected CO_2, or when there is highly compressible residual HC gas saturation in pore space (e.g., the CO_2CRC Otway Stage 1 project).

Seismic Properties of Rocks Saturated with Mixtures of CO_2 and *In Situ* Fluids/Gas

Once we have calculated the bulk modulus K_{fluid} of the CO_2 + fluid mixture, the saturated rock bulk modulus K_{sat} can be calculated by combining K_{fluid} with the dry rock properties using the well-known Gassmann equation (e.g., Mavko *et al.*, 2009). The saturated density is simply $\rho_{sat} = \rho_{dry} + \varphi\,\rho_{fluid}$. Gassmann theory works well when the reservoir rock fabric and fluid distributions are fairly homogeneous, for example, a clean well-sorted sandstone with a uniform distribution of injected CO_2. In addition, or

alternately, saturated (versus dry) core samples can be measured directly in the lab while being core-flooded with CO_2 to simulate the CO_2 injection process, but this requires an advanced lab setup and expertise that is commonly found only in research environments. Figure 2.5 shows P- and S-wave velocity measurements versus effective pressure for an Otway core sample under dry, brine-saturated, and CO_2-saturated lab conditions; note that V_p decreases approx. 5% from 3550 m/s at the original brine-saturated conditions, to 3390 m/s after core-flooding to a maximum CO_2 saturation of $S_{CO2} = 0.5$ (Lebedev, 2012).

As we have discussed, time-lapse seismic is primarily sensitive to the elastic compressibility of the dry rock frame, and the compressibility contrast of the pore fluid properties. A strong time-lapse seismic signal requires rocks with high values of porosity and dry-frame compressibility (i.e., low K_{dry}). High porosity increases rock compressibility and the volume of injected CO_2 in the rock to better affect

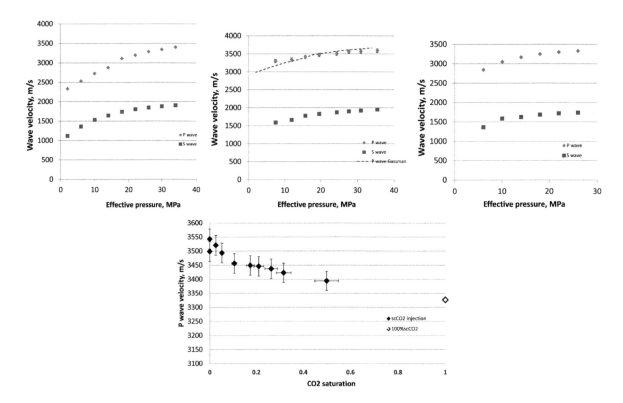

Figure 2.5 Otway core lab measurement data showing P-wave velocity of dry rock versus effective pressure (top left), brine-saturated rock versus effective pressure (top center; dashed line is Gassmann theory prediction from dry measurements), CO_2-saturated rock versus effective pressure (top right), and CO_2-saturated rock versus CO_2 saturation values (bottom); (Lebedev 2012).

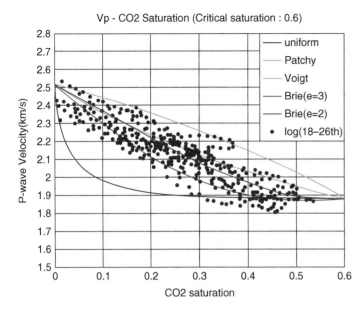

Vp - CO2 Saturation (Critical saturation : 0.6)

Legend:
— uniform
— Patchy
— Voigt
— Brie(e=3)
— Brie(e=2)
• log(18–26th)

y-axis: P-wave Velocity(km/s)
x-axis: CO2 saturation

Figure 2.6 P-wave velocity versus CO_2 saturation. Time-lapse geophysical well log data (blue dots) measured during the Nagaoka CO_2 sequestration project injection, exhibiting patchy saturation behavior (Lumley, 2010; Caspari *et al.*, 2011).

the seismic response, and large dry-frame compressibility allows the seismic waves to better sense (via fluid compressibility contrast) the type of fluid in pore space. Rocks that are favorable for time-lapse monitoring of CO_2 injection include unconsolidated sands, turbidites, and heavily fractured material, for example. At the Sleipner CO_2 project, the injection of supercritical CO_2 into brine-saturated unconsolidated sand creates a huge P-wave velocity (V_p) decrease of up to 60%. Rocks that are less favorable for time-lapse monitoring of CO_2 injection include well-cemented sandstones, low-porosity tight sands, and hard carbonates, for example. At the Weyburn CO_2 project, the injection of CO_2 into low-GOR, residual-oil-saturated carbonate rocks creates a small V_p decrease of only a few percent, which is at or near the time-lapse seismic noise level for detection. At the CO_2CRC Otway Stage 1 project, the depleted HC reservoir rocks were moderately favourable, but the residual HC gas in pore space made it difficult to detect injected CO_2 with time-lapse seismic due to the weak fluid compressibility contrast between the injected CO_2 and the residual HC methane gas.

Additional Complications . . .

It is worth mentioning a few important complications to be aware of, with specific regard to time-lapse seismic monitoring of CO_2 injection.

Patchy Saturation

A first major complication is the issue of patchy saturation: the fact that the compressibility of a rock as a function of CO_2 saturation depends on the spatial size of the CO_2 fluid "patches" in the rock pore space compared to the wavelength of the seismic measurement. When the seismic wavelength is much larger than the average CO_2 patch size (i.e., low-frequency limit), the saturated rock compressibility tends to follow the highly nonlinear Reuss lower bound. When the seismic wavelength is much smaller than the average CO_2 patch size (i.e., high-frequency limit), the saturated rock compressibility follows a quasi-linear upper bound; and when the average CO_2 patch size is similar to the seismic wavelength, the saturated rock compressibility plots along a "patchy saturation" curve in between the two bounds, as Figure 2.6 shows for time-lapse log data measured during the Nagaoka CO_2 injection project (Lumley, 2010; Caspari *et al.*, 2011). The patchy-saturation values for K_{sat} can be calculated using Hill's equation (Mavko *et al.*, 2009), which tends to produce a fairly linear relationship between the initial and final K_{sat} values as a function of injected CO_2 saturation. It is still an open research question as to whether patchy saturation effects measured in the lab at MHz frequencies, and in log data at 10–100 kHz, are a significant effect in surface seismic data at 10–100 Hz.

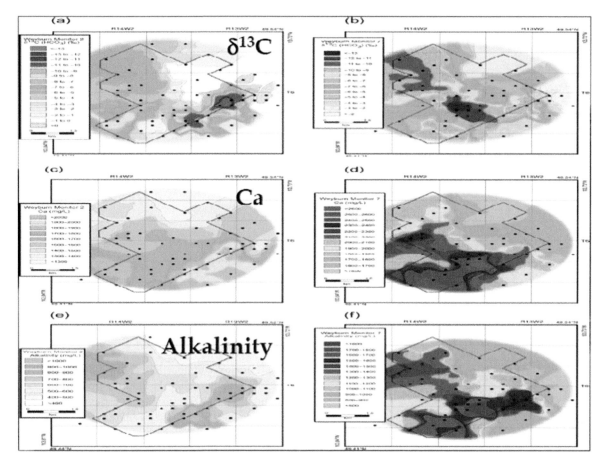

Figure 2.7 Geochemical fluid sample monitoring at 45 boreholes pre- (a, c, e) and post- (b, d, f) CO_2 injection at the Weyburn project: (a, b) Carbon-13 isotope ratio (‰); (c, d) dissolved calcium (mg/L); (e, f) alkalinity (mg/L). These plots show clear evidence that major geochemical changes are occurring due to CO_2 injection into the Weyburn carbonate reservoir, causing significant dissolution of calcium and possibly other minerals (Emberley et al., 2004).

However, quantitative calibration of the velocity-saturation relationship is very important for designing high-quality monitoring experiments, interpreting field data results, and quantifying the amount of CO_2 present in the subsurface from time-lapse seismic images and inversions.

Geochemical Effects

A second major complication arises because CO_2 is not an inert fluid. My own personal experiences monitoring CO_2 injection for EOR and CO_2 storage projects over the years have shown that CO_2 can react in the pore space to alter or dissolve the rock matrix or pore-filling diagenetic material on very short timescales (weeks or months), thus creating secondary

porosity and weakening the cementation and/or dry frame modulus. This is especially true for carbonate reservoir rocks (e.g., Figure 2.7; Emberley et al., 2004), but may also be important for clastic reservoir rocks that contain carbonate grains and/or cement (e.g., Figure 2.8; Vanorio, 2015). The injected CO_2 can react with calcium carbonate ($CaCO_3$) or other minerals that may be present in the reservoir rock to produce soluble bicarbonate [$Ca(HCO_3)_2$] which is a much weaker material than the original hard carbonate. In addition, the CO_2 can react to produce carbonic acid (H_2CO_3), which can further dissolve minerals and grain contact cement to weaken the reservoir rock. Recent lab measurements on CO_2-flooded cores (e.g., Vialle and Vanorio, 2011; Vanorio, 2015) seem to support our field observations

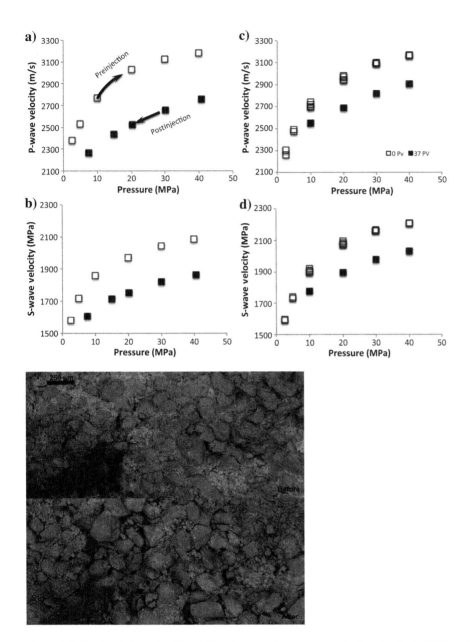

Figure 2.8 Geochemical reaction effects in Tuscaloosa sandstone core samples from the Cranfield CO_2 project: P- and S-wave velocity versus effective pressure for (a, b) shaly sandstone and (c, d) cleaner sandstone; (e) SEM image of sandstone before (top) and after (bottom) CO_2 injection (Vanorio, 2015). The open and closed symbols refer to the velocities of the dry frame measured, respectively, before and after the injection of 37 PV of aqueous CO_2. Note that the injection of CO_2 causes dissolution and removal of iron-rich chlorite from the pore space, thus weakening the sandstone rock matrix.

that CO_2 can be geochemically reactive with the reservoir rock over seismic monitoring timescales, and thus cause significant decreases in dry frame stiffness. Figure 2.8 shows that CO_2 injection can cause the dissolution and removal of grain contact cement and diagenetic pore-filling material, thus weakening the rock matrix of Tuscaloosa sandstone cores associated with the Cranfield CO_2 storage project. These combined geochemical effects can make it difficult to isolate the effects of CO_2 fluid saturation alone, which in

25

turn makes it difficult to accurately quantify the amount of supercritical CO_2 present in pore space using only seismic images. Many time-lapse seismic data sets at CO_2 injection projects seem to exhibit geochemical reaction effects in the reservoir, which may be why observed time-lapse CO_2 anomalies are sometimes quite different (typically larger) than predicted in advance using non-geochemically reactive modeling methods.

Geomechanical Effects

A third major complication can arise due to geomechanical effects; reservoir rocks may undergo irreversible nonlinear *nonelastic* plastic, ductile, or brittle mechanical deformation during CO_2 injection due to changes in pore pressure, *in situ* stress conditions, and/or geochemical reactive effects. Geomechanical effects include porosity loss or increase, grain rotation, displacement or cracking, cement bond damage, compaction, dilation, stress arching, shearing, fracturing, and other nonelastic phenomena. Perhaps the best examples of geomechanical effects that have been observed in time-lapse seismic data are in the North Sea, where both geomechanical and geochemical effects of seawater injection into carbonate (especially chalk) reservoirs can cause significant reservoir compaction, often accompanied by ductile stretching of the overburden rocks, and subsidence at the seafloor (e.g., Guilbot and Smith, 2002; Hatchell and Bourne, 2005).

Nonlinearly Coupled and Combined Time-Lapse Seismic Responses to CO_2 Injection

Often the individual rock and fluid physics effects discussed in the foregoing become combined in a nonlinearly coupled manner to produce the resulting full time-lapse seismic response. To provide a more intuitive understanding for the potential complexities of the various time-lapse seismic responses, let us consider the following hypothetical scenario of supercritical CO_2 injection into a saline aquifer storage reservoir.

- **Repeatability**
  - Let's assume that the time-lapse seismic surveys are sufficiently repeatable in terms of acquisition geometry, noise and near-surface/environmental conditions such that the

seismic difference data sets can achieve NRMS values of <0.4. In this case, from my experience, reservoir P-impedance changes >5% are often clearly visible in the time-lapse seismic data sets.

- **Reservoir properties**
  - Let's assume that the saline aquifer is at a depth of approx. 2 km, with a normal hydrostatic pore pressure of approx. 20 MPa and a reservoir temperature of approx. 60°C. Figure 2.4 shows that the injected CO_2 will clearly be in its supercritical phase.
  - Let's further assume that the reservoir rocks are clastic (e.g., sandstone) and sufficiently compressible (e.g., K_{dry}, G_{dry} <10 GPa) and porous (e.g., φ>0.2) such that supercritical CO_2 saturations S_{CO2} >0.3 will create P-impedance changes >5%.

Given the foregoing assumptions, we may expect the following time-lapse seismic responses.

- **Saturation effects**
  - Per Figure 2.4, the injected supercritical CO_2 will be much more compressible (approx. 600%) than the saline water, and somewhat less dense (approx. 30%). In regions of maximum CO_2 saturation, especially near the injection well, the saturation effect will cause a significant "softening" (P-impedance and V_p decrease >5%) of the reservoir rock, a moderate density decrease, and a small/negligible shear strengthening (S-impedance and V_s increase) due to the density effect.
  - If a significant volume of the injected CO_2 should flow upward past the top seal to a depth shallower than 1 km, it would no longer remain in its supercritical phase, and would "flash" from a supercritical liquid to a highly compressible low-density gas; this would cause an extremely bright time-lapse seismic response in the overburden due to a major decrease in I_p, V_p, and density (>10–20%).

- **Pressure/stress effects**
  - CO_2 injections are highly likely to be pressure-controlled so that important stress thresholds are not exceeded (e.g., at the Snøhvit and Aquistore projects, the injection rate is constrained such that the pore pressure does

not exceed 90% of the reservoir and cap rock fracture pressure values). In addition, because supercritical CO_2 is highly compressible compared to water, only small pore pressure changes are anticipated due to CO_2 injection. However, unexpected pressure changes sometimes can, and do, occur due to unanticipated geological complexity, decreased injectivity, reservoir compartmentalization, etc.

. In general, there will likely be a modest (0–5 MPa) pressure increase near the injection well, which may moderately (<2%) or negligibly soften the reservoir rock. If there are water production (withdrawal) wells to help manage reservoir pressure, and/or to steer the injected CO_2 plume, then modest (0–5 MPa) pressure decreases can be expected near the water production wells, but these will tend to have a negligible effect because rocks are asymmetrically much more sensitive to pressure increases than pressure decreases.

. If the reservoir rock was initially fractured, or if the maximum horizontal confining stress is much greater than the minimum horizontal stress, then the reservoir may exhibit initial azimuthal anisotropy (HTI or orthorhombic). The change in pore pressure due to CO_2 injection can open or close fracture sets, and/or modify the effective confining stresses, which can thus cause the seismic anisotropy to change over time. These time-lapse anisotropy effects are most likely to be observed in S-wave data, especially shear-wave splitting, or in azimuthal amplitude versus offset (AVO)/A responses at large offsets/angles which are sensitive to shear impedance changes. This topic is an area of active research, and few field data examples have been observed to date (e.g., Davis *et al.*, 2003).

- **Temperature effects**
  - . Industrial CO_2 is likely to be pressurized and heated (if necessary) to become a supercritical fluid at its capture source point, and then transported in special pipelines to an injection and storage site. Thus for energy efficiency purposes, the temperature of the supercritical CO_2 at the injection point in the storage reservoir is likely to be slightly above the critical point, i.e., about 32–35°C.

. Since the saline aquifer at 2 km depth will be at a normal geothermal gradient temperature of approx. 60°C (see earlier), the injected CO_2 will be approx. 25–30°C colder than the reservoir at the injection point, and will progressively heat up as it flows away into the distant parts of the reservoir.

. Injection of "cold" CO_2 into a "hot" reservoir may cause thermal fracturing of the rock near the injection point, which may be visible as a time-lapse "bulls-eye" anomaly centered on the injection well, especially in S-wave or AVO/A data. This may explain the anomalous S-wave results in the time-lapse cross-well data acquired by Daley *et al.* (2008) for the Frio CO_2 injection project. Such time-lapse bulls-eyes are often seen in 4D seismic data at water injection wells that inject very cold seawater into very hot hydrocarbon reservoirs.

. Injection of cold CO_2 into hot saline aquifer water will advectively cool the brine, but this fluid temperature effect on its own is likely to be weak or negligible (<1–2%).

. Injection of cold CO_2 into a hot reservoir will diffusively cool the rock near the injection well (beyond the thermal fracture radius), but thermal diffusion is a very slow and gradual process and so the time-lapse effect is also expected to be weak or negligible (<1–2%).

- **Geochemical effects**
  - . The injection of CO_2 into a saline aquifer may cause geochemical reactions if the reservoir sandstone contains carbonate minerals, and/or carbonate (versus quartz) grain cement. In this case the dry rock frame may weaken due to dissolution of grain contact cement, carbonate mineral grains and/or carbonate pore-filling material (e.g., Vanorio, 2015). These geochemical reaction effects, if they occur, could easily cause a significant time-lapse decrease in the P-wave and S-wave velocities of 5–10% or more, which would be readily visible in time-lapse seismic data.

- **Geomechanical effects**
  - . The injection of CO_2 into a saline aquifer is highly unlikely to cause significant geomechanical stress effects since the injection rate will likely be controlled to not increase the

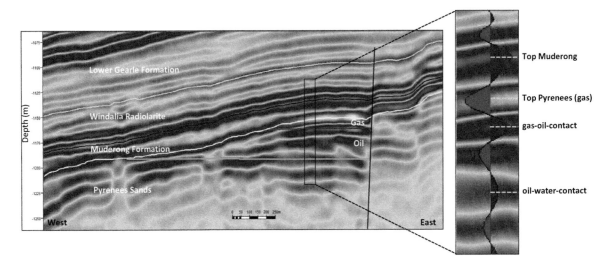

Figure 2.9 Idealized example of a seismic well tie. (Left) Marine seismic image of a hydrocarbon reservoir offshore Australia. (Right) Extracted seismic trace overlain on background seismic image. (Figure courtesy of Guy Duncan and colleagues, BHP Billiton.)

original reservoir pressure by more than 0–5 MPa, and certainly to not exceed 90% of the rock fracture pressure value.

. However ... sometimes unanticipated geological reservoir complexity and fault compartmentalization can lead to strong pressure transients, which may be detected in time-lapse seismic data as decreases in P- and S-wave velocity, and wide-aperture TL AVO/A responses. Such unanticipated pressure transients may also induce seismicity (micro-earthquakes) on any faults within the pressurized region of the injected plume, if the faults are originally close to the "critical stress point," i.e., the critical stress value at which a particular fault is ready to slip and release seismic energy.

. If injected CO_2 geochemical effects have significantly weakened the reservoir rock matrix per the preceding discussion, the altered reservoir rock may be much more susceptible to geomechanical stress effects such as reservoir compaction and overburden stretching in the area of the CO_2 plume, reservoir dilation near the CO_2 injectors, and perhaps even ground/seafloor subsidence or uplift. For example, extreme reservoir compaction and seafloor subsidence (>10 m over 40+ years) has been observed in

time-lapse seismic data at the Ekofisk Field in the Norwegian North Sea due to injection of seawater into a chalk reservoir and production of hydrocarbons (Guilbot and Smith, 2002); and measurable surface ground uplift of approx. 10 mm/year or more near CO_2 injection wells, and ground subsidence of approx. 5 mm/year near gas production wells, has been observed in time-lapse InSAR satellite data for the CO_2 injection project at In Salah, Algeria (Mathieson et al., 2009).

• **Combined effects**

. Any of the foregoing individual effects can be coupled and combined nonlinearly in any number (few or all), and can vary in both space and time, during the life of a CO_2 injection and storage project. Hopefully this simple hypothetical saline aquifer example provides some intuition for the potential complexity and nonuniqueness of the full time-lapse seismic response to CO_2 injection.

Time-Lapse Seismic Modeling

Given the preceding discussion regarding the physical properties of the storage reservoir rocks, the *in situ* pore fluids, and the injected CO_2, it is evident that seismic modeling to predict the expected time-lapse response can be performed for virtually any CO_2 reservoir injection scenario. Time-lapse seismic modeling can be done at a single point in the reservoir ("zero-D" modeling), at a well location (1D modeling), along a reservoir cross section (2D modeling), and in a full 3D reservoir model coupled with the results from a flow simulation (e.g., Lumley, 1996), and optionally may include flow-coupled geochemical and geomechanical simulations. Seismic modeling can range from simple 1D convolution of a seismic source wavelet with a reflection coefficient series, to full-blown Finite Difference (FD) or Finite Element (FEM) modeling, and can represent the physics of first-order acoustic wave propagation (P-waves only), higher order elastic waves (P- and S-waves), viscoelasticity (including Q attenuation), and various forms of anisotropy (polar, azimuthal, orthorhombic, etc.).

A vital starting point to the time-lapse seismic modeling process is to ensure that a high-quality "seismic well tie" is achievable. A seismic well tie combines the geological overburden and reservoir model, the rock and fluid property relationships from core and log data, and a seismic forward modeling method, to generate synthetic seismic data that matches or "ties" with the actual seismic data at a given physical location (usually a borehole). If a good quality well tie is obtained, this builds confidence that the earth model and physical property relationships are reasonably correct, and that the seismic data have been imaged correctly. If the synthetic seismic does not match the real data, this highlights that there are major discrepancies to be resolved in the geological model, physical properties, core and log data, or seismic image processing. Figure 2.9 shows an example of an extracted seismic volume trace representing an idealized seismic well tie from a reservoir offshore Western Australia.

Figure 2.10 shows an example of 2D FD viscoelastic seismic wavefield data generated from an earth model based on Sleipner reservoir rock and CO_2 properties, respectively containing no CO_2, one layer of CO_2, and two layers of CO_2, near the top of the Utsira unconsolidated sand reservoir. The panels show the preinjection seismic shot gather, the difference between the preinjection and one-layer CO_2 gather, and the difference between the preinjection and two-layer CO_2 gather. Note the rich and complex wavefield scattering differences generated by just one

layer of CO_2, and two layers of CO_2, respectively. An accurate time-lapse imaging algorithm would need the ability to correctly image these complex time-lapse coda waveforms into a single (or double) layer of CO_2 ... quite a challenge!

Figure 2.11 shows the result of applying a state-of-the-art prestack depth migration (PSDM) image-processing flow to the synthetic data for the single CO_2 layer (blue arrow), including elastic wavefield separation, multiple suppression, depth migration velocity analysis, etc. The synthetic seismic difference image is compared to the real marine data (PSTM) seismic difference image at Sleipner, the latter of which is interpreted to contain many layers of CO_2. The blue dot is the location of the Sleipner CO_2 injection point. The imaging artifacts in the synthetic data, which look similar to the seismic differences observed in the real data, are caused by strong multiple wave-scattering and mode-conversion effects (time-lapse coda) that are not properly handled by conventional wave-equation imaging operators, and thus can be easily misinterpreted as false layers of CO_2, and therefore may contaminate the image of the true CO_2 distribution (Lumley *et al.*, 2008; Lumley, 2010).

Figure 2.12 shows an example of 3D FD modeling using publicly available Sofi3D software to predict time-lapse synthetic seismic data with realistic additive noise for Stage 2C of the CO_2CRC Otway CO_2 injection project. The modeling process includes building a detailed 3D geological model, and generating a flow simulation of the injected CO_2 plume. The 3D synthetic shot gathers replicate the identical source and receiver geometries of the real seismic surveys, and realistic noise is estimated from the real data to be added to the synthetic data. The synthetic data sets are then passed through the same image processing flow as the real data in order to make realistic predictions and comparisons regarding the ability to detect, image and interpret the injected CO_2 plume at various volumes of injection, and under various noise conditions (Glubokoskikh *et al.*, 2016).

Figure 2.13 shows an example of 3D FD anisotropic viscoelastic modeling using custom developed UWA software to generate full 9-component (9C) 3D seismic elastic wavefield data sets. This example uses a full seismic source moment tensor, both for manmade surface sources, and for natural or induced (micro) earthquakes at depth, in a 3D geological

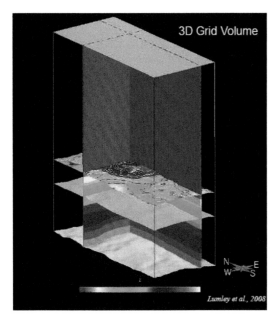

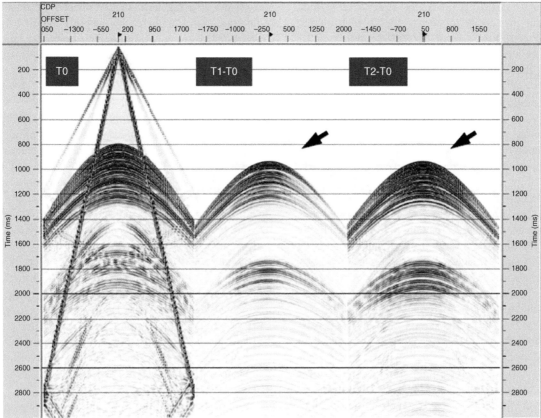

Figure 2.10 Reservoir model (top), and finite-difference elastic wavefield shot gathers (bottom) at same scale generated from the reservoir model containing (left) no CO_2, (center) one layer of CO_2, and (right) two layers of CO_2, near the top of an unconsolidated sand storage reservoir. The left panel shows the preinjection seismic shot gather; the center panel shows the difference between the preinjection and one-layer CO_2 gather; and the right panel shows the difference between the preinjection and two-layer CO_2 gather. Note the rich and complex wavefield scattering differences generated by only one layer of CO_2 (center) and two layers of CO_2 (right), respectively. An accurate time-lapse imaging algorithm would require the challenging ability to correctly image these complex time-lapse wave coda into a single (or double) layer of CO_2 (Lumley *et al.*, 2008; Lumley, 2010).

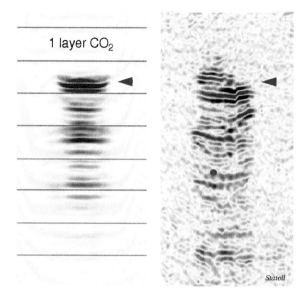

1 layer CO$_2$

Statoil

Figure 2.11 (Left) Synthetic PSDM time-lapse seismic difference image from FD data using a reservoir model with only a single CO$_2$ layer (blue arrow). (Right) Marine data PSTM time-lapse seismic difference image at Sleipner interpreted to contain multiple layers of CO$_2$. The blue dot is the location of the true CO$_2$ injection point. The imaging artifacts in the left panel are caused by strong multiple wave-scattering and mode-conversion effects (time-lapse coda) that are not properly handled by conventional wave-equation imaging operators, and thus may be misinterpreted as false layers of CO$_2$, or may contaminate the image of the real CO$_2$ distribution (Lumley et al., 2008; Lumley, 2010).

model consisting of more than 5 billion grid points, for the South West Hub CO$_2$ storage project, Western Australia (Lumley et al., 2016). This type of realistic complex 3D modeling requires a massive computational effort, and thus the resources of a Top 100 world-wide high-performance computation (HPC) facility, like the Pawsey Centre in Perth, Australia, which was used to create the Otway Stage 2C and South West Hub project seismic simulations.

Time-Lapse Seismic Acquisition

The ability to image injected CO$_2$ with time-lapse seismic data depends on the magnitude of the seismic response to CO$_2$ (i.e., the time-lapse signal), in comparison to the time-lapse seismic noise level. The largest sources of time-lapse seismic noise are typically caused by nonrepeatability in the acquisition of the time-lapse seismic data surveys. Seismic acquisition nonrepeatability can be caused by source–

receiver positioning differences between surveys; natural variations in the near-surface conditions on land (wet/dry, water table, ground coupling, etc.); natural variations within the water column at sea (tides, wave heights, water temperature, salinity, etc.); and variations in source waveforms, receiver functions, and equipment hardware specifications.

Many of the manmade nonrepeatability factors can be reduced or eliminated by careful time-lapse seismic survey planning and data preprocessing (e.g., source and receiver positions). Many of the natural nonrepeatability factors cannot be similarly controlled, but can be quantitatively measured and monitored during each survey, and then used to deterministically compensate or correct the data in processing (e.g., tidal variations). However, it is important to be aware that there will always be some nonrepeatablity factors that we cannot easily control or quantify (e.g., ambient noise, weather, overburden variations, etc.) and therefore there will always be some level of nonrepeatable noise that we will have to include and manage as part of our time-lapse seismic imaging, inversion, and interpretation analysis.

Nonrepeatable (NR) noise can be quantified in many ways, but the most common measure is known as "normalized RMS" or NRMS:

$$NRMS = \frac{2 * RMS(Data2 - Data1)}{RMS(Data2) + RMS(Data1)}$$

where RMS is the standard definition of "root-mean-squares." Kragh and Christie (2002) provide a very good discussion of NRMS with respect to measuring nonrepeatable noise in time-lapse seismic surveys. Nonrepeatability can be measured at any stage of time-lapse seismic acquisition and image processing, starting from raw shot gather traces, to final processed image volumes. From my monitoring experience, a reasonable rule of thumb is that the final processed image volume data can be expected to have an NRMS value that is approximately half of the initial shot gather NRMS values (with very basic filtering applied).

The normalization inherent in the NRMS calculation is intended to provide an unbiased estimate between data sets, but in fact the NRMS values can be significantly biased depending on the specific parameters chosen by the user: time window, location, length, width, and data contents (reflection signal and noise strengths). From my experience, I recommend choosing a time window approx. 250–500 ms

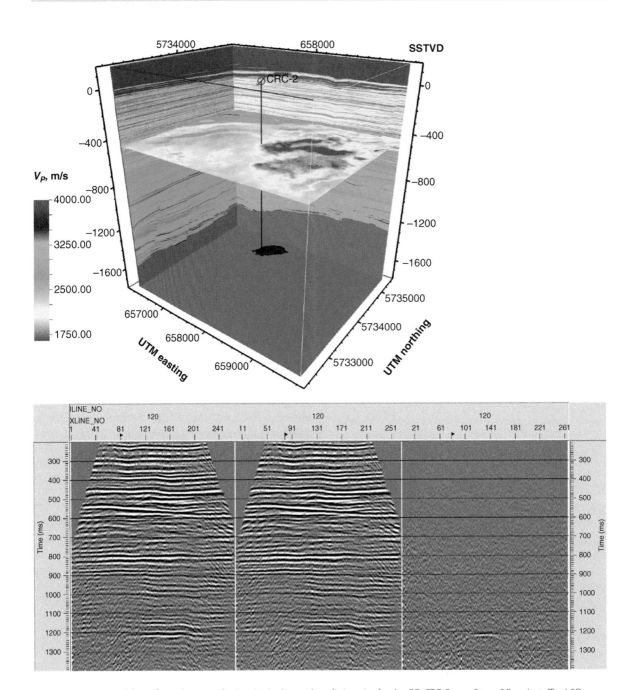

Figure 2.12 3D modeling of time-lapse synthetic seismic data with realistic noise for the CO₂CRC Otway Stage 2C project. (Top) 3D geological model and flow simulation of injected CO_2 plume. (Bottom) Synthetic seismic images from 3D earth model. (Left) Preinjection seismic image. (Center) Postinjection seismic image. (Right) Difference seismic image showing injected CO_2 plume and background nonrepeatable noise level (Glubokoskikh et al., 2016).

long for typical 10–100 Hz seismic data, in a windowed zone just above the reservoir that is unlikely to have changed between surveys (beware of unexpected real fluid/pressure leakage effects or geomechanical compaction/dilation effects in the overburden), and that contains similar reflection event characteristics and strengths as in the reservoir zone. In this case, NRMS usually provides a quantitative value that is a

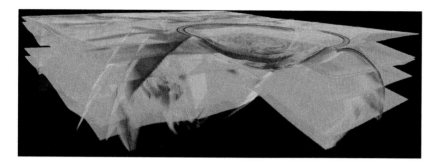

Figure 2.13 Finite difference viscoelastic modeling of the 3D seismic wavefield generated by a seismic source in the 3D earth model, South West Hub CO_2 storage project, Western Australia (Lumley *et al.*, 2016). (Figure prepared in collaboration with Taka Miyoshi.)

Table 2.2 NRMS values versus 4D data repeatability

NRMS	Comment
<0.1	Outstanding repeatability (permanent arrays)
<0.2	Excellent repeatability (very low NR noise)
0.2–0.4	Very good repeatability (common for marine surveys)
0.4–0.6	Reasonable repeatability (common for land surveys)
0.6–0.8	Poor repeatability (but may be ok for qualitative work)
0.8–1.2	Highly nonrepeatable (time-lapse data may not be useable)
1.40	Equivalent to two random data sets!
2.00	Data sets are identical, but polarity reversed

very good indication of the (non)repeatability of the time-lapse seismic data sets and their amenability for reservoir analysis. Table 2.2 summarizes my rules of thumb for time-lapse data quality versus NRMS value (Lumley, 1996).

For marine time-lapse seismic surveys, source–receiver positioning differences are a dominant source of nonrepeatable noise. With differential GPS (DGPS), we can often locate and repeat seismic source locations to within a few meters; however, the nonrepeatability of receiver (hydrophone group) locations can often be on the order of 10s to 100s of meters, generally increasing proportional to source–receiver distance. For this reason, modern time-lapse marine streamer surveys often employ over-sampling methods (more cables, wider arrays) for each successive survey, and take advantage of steerable streamers to help reduce variable cable feather. Two-boat surveys which separate the source boat independently from the streamer array boat, also make it easier to repeat the surveys in the presence of variable and often conflicting current, tides, and wave directions (Ridsdill-Smith *et al.*, 2008). Ocean bottom cables (OBC), nodes (OBN), and broadband seismographs (OBS), deployed by subsea remotely operated vehicles (ROVs), can greatly increase the repeatability on the receiver side of the equation to within a few meters. (Semi) permanent arrays, which can be deployed long term by burying cables/sensors in the seafloor mud, provide the maximum receiver repeatability available today. At the Valhall project in the Norwegian North Sea, BP and partners used a permanently entrenched OBC seafloor array to achieve repeatability levels of 0.05–0.10 for raw (unprocessed) seismic repeat shot gather data (van Gestel *et al.*, 2008).

For land time-lapse seismic surveys, near-surface variations are a dominant source of nonrepeatable noise. As demonstrated in Figure 2.14 by Pevzner *et al.* (2011), variations between surveys in the near-surface moisture conditions (wet, dry, frozen, mud ...) can greatly affect the coupling of the sources and receivers to the ground, and therefore the source strength and waveform, the frequency content of the data, and the strength of Rayleigh surface wave "groundroll" noise, among other factors. At the land-based Weyburn and Aquistore CO_2 projects in Canada, White *et al.* (2015, and this book) were able to reduce the raw shot gather data nonrepeatability from average NRMS values of

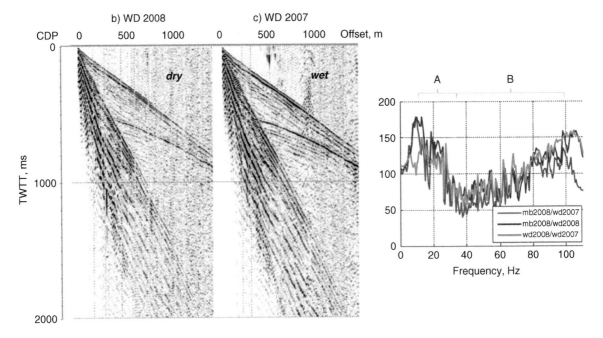

Figure 2.14 Acquisition nonrepeatability example: repeated weight-drop source (WD) shot gathers at the Otway CO_2 injection test site. (Left) Repeated shot gather during dry surface conditions in 2008. (Center) Original shot gather during wet surface conditions in 2007. (Right) Nonrepeatability NRMS (%) plotted versus seismic data frequency (green curve corresponds to the two shot gathers displayed here). (Extracted from Pevzner *et al.*, 2011.)

1.20 to 0.50, by switching from surface-based geophone arrays to geophones buried at 20 m depth, in order to suppress the variable near-surface layer effects. At the land-based CO_2CRC Otway project in Australia, Pevzner *et al.* were able to reduce the prestack shot gather nonrepeatability NRMS values within the main 20- to 80-Hz seismic passband from approx. 0.75 in Stage 1 (Figure 2.14), to 0.20 in Stage 2C, by burying a 900+ geophone array (Figure 2.15) at just 4 m below the surface within the permanent groundwater-saturated zone (Shulakova *et al.*, 2015; Pevzner *et al.*, 2017).

For borehole seismic surveys, the major sources of nonrepeatability are source and receiver positioning, source and receiver instrumentation, and the coupling between the sources and/or receivers with the borehole fluid, casing, or formation. In many instances, the best results for time-lapse vertical seismic profiling (VSP) and cross-well surveys have been achieved by cementing in the receiver arrays, although interestingly some observations have been reported of significant changes over time due to suspected long-term curing effects of cement in the borehole.

The development of highly repeatable seismic acquisition systems for both land and marine environments (sources, receivers, permanent arrays, high-precision navigation, quantitative measurements of environmental meta-data, etc.), is providing a continuously improving ability to detect and image very weak time-lapse signals generated in the earth. In the near future, we may not only be able to image and monitor the injected CO_2 plume as it evolves over time, but we may also be able to image more subtle phenomena such as the weak injection pressure/stress front as it propagates away from the CO_2 injection well and ultimately defines the regulatory Region of Influence (ROI) for the overall CO_2 storage and containment project. The increasing use of (semi)permanent arrays is also allowing us to develop and use innovative passive seismic and ambient noise techniques to image the storage reservoir, as will be shown later in this chapter, thus potentially minimizing the need for active manmade source efforts, and thereby reducing the "environmental footprint" of the seismic surveying operations that are required to monitor a CO_2 injection project over its long life.

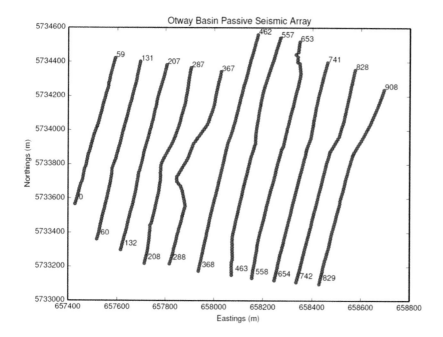

Figure 2.15 Buried geophone array at the CO₂CRC Otway Stage 2C project site (coordinates courtesy of R. Pevzner and CO₂CRC). Prestack shot gather repeatability was improved from approx. NRMS = 0.75 to NRMS = 0.2 within the main 20- to 80-Hz seismic passband by burying the 900+ geophone array at just 4 m below the surface within the permanent groundwater-saturated zone (Shulakova *et al.*, 2015; Pevzner *et al.*, 2017).

Time-Lapse Seismic Imaging

Time-lapse image noise, errors, and artifacts can be created during the process of seismic imaging due to (sometimes unavoidable) variations in the image processing algorithms and parameters, the approximate physics of certain imaging operators that do not account for full wave propagation effects, and the sensitivity of certain nonlinear statistical and data-dependent signal processing and imaging operators to the presence of nonrepeatable seismic noise.

The theory for time-lapse seismic imaging is discussed in the Theory section of this chapter. In general, I recommend that time-lapse seismic data sets be imaged using prestack depth migration (PSDM) that is amplitude-preserved such that the image gathers have the correct angle-dependent reflectivity (AVO/A) in depth. Often time-lapse seismic data sets are imaged in vertical travel time (PSTM), as these algorithms are faster and cheaper; however, when we get to the detailed analysis of fluid flow features within the reservoir, we inevitably find ourselves asking whether the images we are interpreting have the correct structure, closure, stratigraphy, flowpaths, etc. in

depth, and whether the time-to-depth conversion process is accurate. For these reasons, I recommend that time-lapse data sets be imaged directly in depth with amplitude-preserved algorithms.

Figure 2.16 shows a comparison of the resulting time-lapse difference images for various image-processing and data combinations. The use of an explicit time-lapse image processing flow (e.g., the Parallel method), applied to two time-lapse data sets, produces a good quality difference image of the real fluid anomaly, with modest levels of nonrepeatable noise in the background. Using a single 3D seismic data set (i.e., no time-lapse data!), with two similar, but nonidentical, 3D image processing flows (e.g., somewhat different algorithm and parameter choices by different data processors), produces a nonzero difference image with artifacts that could be falsely interpreted as real fluid anomalies. Using two time-lapse seismic data sets, but with a nonidentical image processing flow and parameters, produces a poor-quality difference image in which background noise and image artifacts obscure the true fluid anomaly.

4D processing flow,
4D datasets (Mon-Base)

Similar 3D processing flows,
same 3D dataset (no 4D!)

Similar 3D processing flows,
4D datasets (Mon-Base)

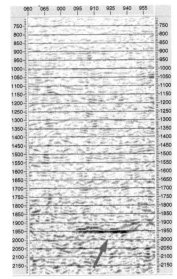 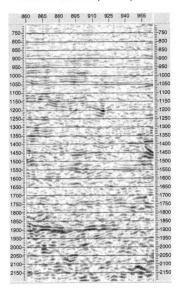 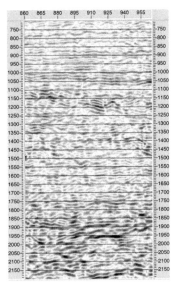

Real 4D fluid flow anomaly

Figure 2.16 Comparison of time-lapse (4D) difference images for various time-lapse image-processing flows. (Left) Using a time-lapse consistent image processing flow with two time-lapse data sets produces a good quality difference image of the real fluid anomaly, with modest nonrepeatable noise in the background. (Center) Using a single 3D seismic data set (i.e., no time-lapse data!), with two similar, but not identical, 3D image processing flows (e.g., different parameter choices by different data processors), produces a nonzero difference image with artifacts that could be falsely interpreted as real fluid anomalies. (Right) Using the two time-lapse seismic data sets in the left panel, but the nonidentical image processing flow parameters of the center panel, produces a poor-quality difference image in which background noise and artifacts obscure the true reservoir fluid anomaly. (Figure adapted from an image prepared by Jim Keggin of BP, while on assignment at Statoil.)

This set of results shows that, if time-lapse image processing is not done specifically and carefully to optimize difference images, real time-lapse anomalies can be obscured, and false anomalies can be created, by the image artifacts created by nonrepeatable image processing methods.

In general, a time-lapse specific image processing flow should avoid statistical data-dependent operators, and focus on deterministic and physical model-based operators, in this general order of priority:

1. First, apply all imaging corrections using deterministic methods and auxiliary (meta) data measurements (e.g., navigation data, tidal corrections, etc.).
2. Second, calculate and apply further imaging corrections using physical model-based operators (e.g., surface-consistent source and receiver corrections, migration velocity analysis, Radon multiple suppression, etc.).

3. Lastly, calculate and apply statistical data-dependent operators only if absolutely necessary, and only for residual "clean-up" (e.g., FXY or prediction error filter [PEF] noise suppression). These data-dependent operators are highly risky for time-lapse imaging, especially "Match filters," as they can remove real anomalies, and also create false anomalies (Rickett and Lumley, 2001). In my experience, a high-quality time-lapse image processing flow will never (rarely) need to rely on image processing operators that fall within this category (#3).

The quality of the time-lapse processing flow and results should be monitored step-by-step with various quality control measures (NRMS, 4D SNR, Coherency etc.) to ensure that (1) the individual seismic images are well focused and (2) the time-lapse difference images contain minimal NR noise (e.g., Lumley *et al.*, 2003a, among others). In the Acquisition section

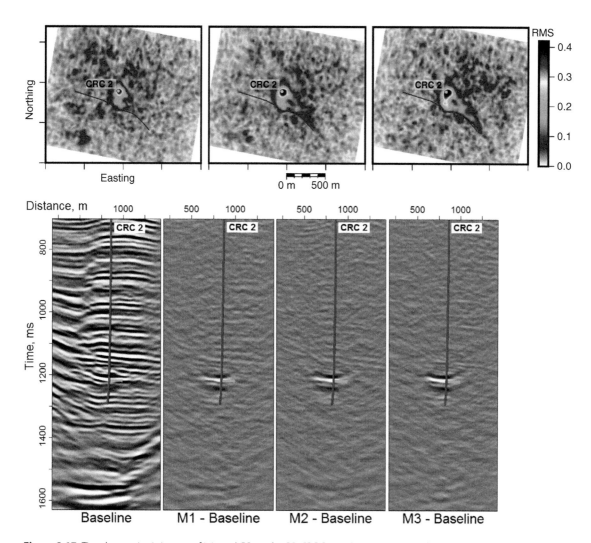

Figure 2.17 Time-lapse seismic images of injected CO_2 at the CO_2CRC Otway Stage 2C project. The top panels are 4D seismic map views at reservoir level showing the injected CO_2 plume growing over time, and being contained by a sealing fault. The bottom panels are 4D seismic cross-section images for a vertical plane passing through the CRC-2 injection well, showing preinjection baseline data and three difference volumes corresponding to 5000, 10 000, and 15 000 tons of injected CO_2/CH_4 (Pevzner et al., 2017a,b). These 4D seismic field data images can be compared to the modeled data images in Figure 2.12 predicted as part of the original feasibility study.

of this chapter, I provide advice on how to calculate NRMS to obtain reasonable unbiased estimates of NR noise, and to be wary of real time-lapse changes above the reservoir due to fluid/pressure leakage or geomechanical compaction/dilation effects.

There are many excellent examples of time-lapse seismic images of CO_2 injection from around the world, especially in Europe and North America, presented as case studies in other chapters of this book. In keeping with the theme of highlighting Australian content, Figure 2.17 shows a spectacular recent

example of time-lapse seismic images for CO_2 injection at the CO_2CRC Otway Stage 2C project. The top panels are 4D seismic map views at reservoir level showing the injected CO_2 plume growing over time, and being contained by a sealing fault. The bottom panels are 4D seismic cross-section images for a vertical plane passing through the CRC-2 injection well, showing preinjection baseline data and three difference volumes corresponding to 5000, 10 000, and 15 000 tons of injected CO_2/CH_4 (Pevzner et al., 2017a, b). These 4D seismic field data images can be

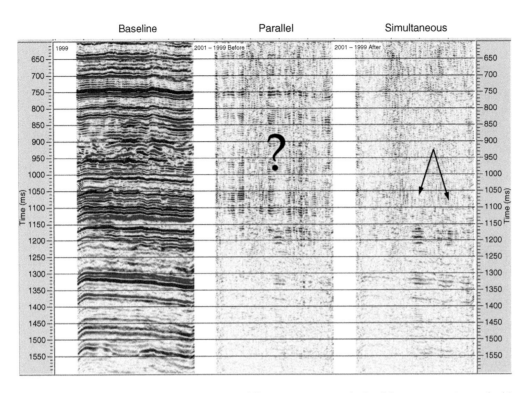

Figure 2.18 Cross-section view of time-lapse seismic difference images using the Parallel image-processing method (center) compared to the Simultaneous image-processing method (right). Left panel is the 3D baseline (preinjection) seismic image. Note the improvement in the imaging of the weak CO_2 injection anomalies in the right panel compared to the more conventional Parallel image processing result in the center panel (Lumley *et al.*, 2003a; Lumley, 2010).

compared to the modeled data images in Figure 2.12 predicted as part of the original feasibility study.

As discussed previously, the strength of the time-lapse signal is related to the compressibility of the reservoir rocks, and the compressibility contrast between pore fluid types. Weak time-lapse signals resulting from CO_2 injection can be due to hard rocks (e.g., Weyburn) and/or weak fluid compressibility contrast (e.g., Otway Stage 1), and in these cases extra effort must be invested to optimize repeatability in order to suppress time-lapse noise and thus enhance the detection of a weak CO_2 signal. Figures 2.18 and 2.19 show examples for a weak time-lapse signal (Weyburn), and compare the Parallel and Simultaneous time-lapse imaging methods. Note the significant improvement in the imaging of the CO_2 injection anomalies using the Simultaneous method, compared to the Parallel method result (Lumley *et al.*, 2003a; Lumley, 2010). In the opposite case of very strong time-lapse responses to CO_2 (e.g., Sleipner), the presence of injected CO_2 is easily detected above the

nonrepeatable noise level. However, the wavefields generated may be complicated by strong internal scattering and wave-mode conversions that are not properly handled by current image-processing operators, and thus can lead to artifacts in the resulting time-lapse seismic images (Figure 2.11).

Most time-lapse seismic imaging is performed using only the P-wave components of the data. However, most seismic sources generate both P- and S-waves, and even pure acoustic or compressive sources generate converted P–S waves along the propagation path. Because S-waves are first-order sensitive to pressure/stress changes rather than fluid saturation changes, using both P- and S-wave data information can allow for more accurate determination of both fluid saturation and pressure/stress change due to CO_2 injection. Figure 2.20 shows a multicomponent time-lapse seismic example, with map-view time-lapse seismic images of the Weyburn CO_2 project at reservoir depth. Horizontal CO_2 injection wells are labelled as red forked lines, horizontal producing wells as white lines, and vertical wells as

Parallel   Simultaneous

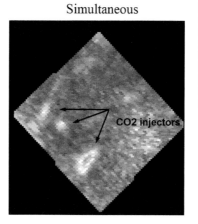

Figure 2.19 Map view of time-lapse seismic difference images (Figure 2.18) at reservoir depth using the Parallel image-processing method (left) compared to the Simultaneous image-processing method (right). Note the clear improvement in the imaging of the weak CO_2 injection anomalies in the right panel compared to the conventional image processing result in the left panel (Lumley *et al.*, 2003a).

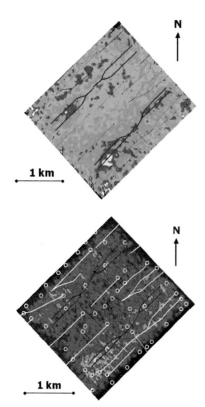

Figure 2.20 Multicomponent time-lapse seismic data example. Map view time-lapse seismic images for the Weyburn CO_2 project. Horizontal CO_2 injection wells (red forked lines), horizontal producing wells (white lines), and vertical wells (white circles). P-wave (top) and S-wave (bottom) time-lapse seismic amplitude difference maps at reservoir depth (Davis *et al.*, 2003).

white circles. The P-wave (top) time-lapse amplitude difference image is sensitive to both pore pressure and CO_2 saturation changes, whereas the S-wave (bottom)

amplitude difference maps are primarily sensitive to pressure and stress changes only, caused by CO_2 injection (Davis *et al.*, 2003).

In this chapter we are primarily considering surface-seismic surveying methods, for both marine and land environments. However, as we have seen, it can be very advantageous to place the seismic sources and/or receivers in the subsurface closer to the reservoir for improved signal-noise conditions and resolution. The ultimate example of this is using borehole seismic surveys, especially VSP and cross-well (well to well) geometries. Figure 2.21 shows a crosswell tomographic image of the CO_2 plume during injection at the Frio CO_2 project. The CO_2 plume is imaged by the time-lapse change in the P-wave velocity, which matches well with the RST (saturation) logs displayed for each borehole. The RST logs were repeated over time during CO_2 injection and the changes are shown in yellow (Daley *et al.*, 2008).

There is still an ongoing need to develop improved seismic imaging and inversion algorithms specifically for time-lapse objectives. A concept that has shown very encouraging results is to treat the image processing of a series of N time-lapse seismic data surveys as a global optimization problem by addressing all N of the seismic data sets simultaneously – as with the Simultaneous approach – constrained by the fact that some areas of the subsurface are changing whereas most other areas are not (e.g., Lumley *et al.*, 2003a; Kamei and Lumley, 2017). This contrasts with the current common approach of processing N time-lapse data sets as a series of paired data sets, two-by-two, using the same or similar processing flows and parameters for each (analogous to parallel processing by local optimization).

39

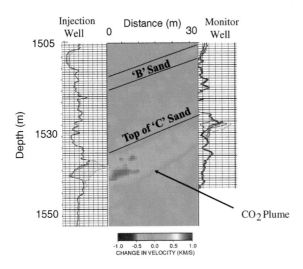

Figure 2.21 Cross-well tomographic image of the injected CO_2 plume at the Frio CO_2 project. The CO_2 plume is imaged by time-lapse change of the P-wave velocity, displayed along with the RST (saturation) logs for each well. The RST logs had multiple runs and the changed zones are highlighted in yellow (Daley *et al.*, 2008).

Another major opportunity and challenge is to extract more of the information contained in the time-lapse wavefields, especially using both the P- and S-wave arrivals (e.g., Davis *et al.*, 2003), as well as the full complex scattered wavefield and the associated time-lapse "4D coda" (e.g., Lumley and Shragge, 2013; Shragge and Lumley, 2013; Kamei and Lumley, 2017).

Time-Lapse Seismic Inversion

Time-lapse Seismic Inversion to Quantify CO_2 Saturation

Whereas seismic *imaging* is the process of using seismic data to image geological and fluid-flow features within the earth, for example the storage containment reservoir and the injected CO_2 plume; seismic *inversion* is the process of using seismic data to make quantitative estimates of the physical properties of rocks and fluids within the earth, for example the amount of residual CO_2 trapped within the storage reservoir, and its spatial and temporal distribution.

The ability to quantify the amount of injected CO_2 present in the reservoir using time-lapse seismic data would be extremely useful. However, this quantification goal faces several challenges, including

- The effects of saturation changes and injection pressure in the reservoir, and the possible presence of multiple phases of CO_2, are combined in the total seismic response and thus not easy to separate uniquely.
- Pressure–saturation effects may be complicated by CO_2 miscibility and reactive effects on the rock matrix (e.g., dissolution of minerals and grain cement) or in pore fluids (e.g., miscible oil fractionation).
- Rock and fluid physics analysis shows that the seismic response is not very sensitive to CO_2 saturation levels beyond S_{CO2} approx. 30% when present as a supercritical "fluid," and beyond S_{CO2} approx. 10% when present as a supercritical "gas."
- The density effect of injected CO_2 is difficult to isolate and estimate from seismic data alone unless there are very large amounts of CO_2 present, and the seismic data is exceptionally clean (low NR noise environment) and contains ultrafar offsets and wide-aperture reflection angles.
- The time-lapse seismic response may be significantly complicated by additional geochemical and geomechanical effects.
- The full time-lapse seismic response to CO_2 can be highly nonlinear and nonunique, thus making it difficult to extract specific information accurately from prestack seismic data or images, which may be required as input for many CO_2 inversion algorithms.

Despite these challenges, there are several different approaches to performing time-lapse seismic inversion in order to make quantitative estimates of the injected CO_2 distribution.

AVO/AVA Inversion

A first broad category of time-lapse inversions is based on the time-lapse changes in elastic reflection amplitude versus source-receiver offset or reflection angle (AVO/A). In general, the reflection coefficients for a seismic wave are nonlinear and depend on gradients of the earth's rock and fluid properties. In practice, a more simple linearized (Bortfeld) form of the full nonlinear Zoeppritz equations is often used (e.g., Aki and Richards, 1980):

$$R(\underline{x}, \varphi) \approx c_1(\varphi) \, m_1(\underline{x}) + c_2(\varphi) \, m_2(\underline{x}) + c_3(\varphi) \, m_3(\underline{x})$$

where $R(\underline{x}, \varphi)$ is the angle-dependent P-wave reflectivity at a given point $\underline{x}$ in the earth, $(c_1 \; c_2 \; c_2)$ are

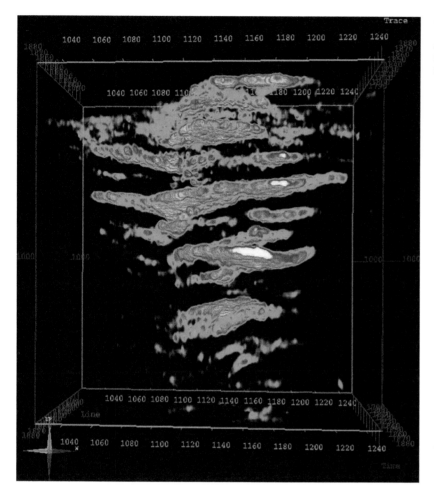

Figure 2.22 Time-lapse seismic inversion estimate of the 3D CO_2 saturation distribution in the Utsira sand storage reservoir for the Sleipner project (4th Wave Imaging).

trigonometric functions that include the geometry of the P-wave reflection angle φ at the reflection point $\underline{x}$, and $m = (m_1, m_2, m_3)$ is a set of three elastic properties at the reflection point, for example relative contrasts in P- and S-wave impedances I_p, I_s, and density ρ. For AVO/A methods, it is assumed that the $R(\underline{x}, \varphi)$ values can be accurately estimated from angle-dependent amplitude-preserved seismic image gathers that are correctly imaged in depth.

In the rock physics section of this chapter we discussed how CO_2 injection can cause time-lapse changes in fluid saturation and pore pressure, among other physical parameters, which can in turn create time-lapse changes in rock and fluid properties such as bulk modulus K, shear modulus G, density ρ, seismic velocities V_p and V_s, and seismic impedances I_p and I_s. If the seismic surveys are reasonably repeatable in terms of source, receiver and wavepath

geometry, the foregoing equation can be rewritten as a time-lapse change in reflectivity ΔR:

$$\Delta R(\underline{x}, \varphi) \approx c_1(\varphi)\Delta m_1(\underline{x}) + c_2(\varphi)\Delta m_2(\underline{x}) + c_3(\varphi)\Delta m_3(\underline{x})$$

where the time-lapse changes in rock physics properties Δm are explicit functions of the time-lapse changes in CO_2 injection fluid-flow properties ΔS_{CO2}, ΔP_p, etc. Given the equations above, and time-lapse seismic data measurements of ΔR, an inverse problem can be defined in several ways to estimate the Δm values, and hence the time-lapse ΔS_{CO2}, ΔP_p, etc. fluid-flow property changes, that fit the observed time-lapse seismic data responses.

Time-lapse AVO/A inversion methods were initially developed in the mid–late 1990s by Lumley, Landrø, and others (e.g., Lumley, 1995b; Tura and Lumley, 1998, 1999; Landrø, 1999, 2001; Cole *et al.*, 2002; Lumley *et al.*, 2003b,c). For a single seismic

41

survey, the AVO response has long been used to make estimates of the coupled rock type and pore fluid content at a given reservoir reflection location (e.g., Ostrander, 1984). For time-lapse seismic data sets, the time-lapse (TL) changes in the AVO response (TL AVO, or 4D AVO) can be used to estimate the time-lapse changes in fluid saturation and pore pressure, as summarized in Figure 2.2. For example, by injecting CO_2 into a saline aquifer storage reservoir, the increased CO_2 saturation with respect to the original water will cause the rock to soften, and the small injection pore pressure increase will likely have a negligible seismic effect; the combined TL AVO response would therefore be a negative amplitude change (P-impedance decrease) at zero offset/angle that brightens with offset/angle. Such TL AVO responses can thus be used as a basis to develop inversion methods that separately estimate the changes in CO_2 saturation and pore pressure at every reflection point along, or within, the reservoir. Figure 2.22 shows an example of a time-lapse seismic AVO/A inversion estimate of the 3D CO_2 saturation distribution in the Utsira sand storage reservoir for the Sleipner project.

Impedance Inversion

A second broad category of time-lapse inversions is based on the 1D convolutional P-wave reflection seismic model approximated as

$$D(t) \approx S(t) \, ¤ \, R(t) \approx S(t) \, ¤ \, \partial_t \, I_p(t)$$

where D is a seismic reflection image data trace, S is the source wavelet, R is the reflectivity (reflection coefficient series), I_p is the P-impedance log converted from depth to time, ∂_t is the time derivative operator, and "¤" is the convolutional operator. This equation can be rewritten in the frequency (w) domain as

$$D(w) \approx S(w) \, R(w) \approx (iw) \, S(w) \, I_p(w)$$

And solved or "inverted" for I_p as

$$I_p(w) \approx D(w) \, / \, [(iw) \, S(w)]$$

and then inverse Fourier-transformed back to the time domain:

$$I_p(t) \approx \int_0^t \check{D}(\tau) \, d\tau.$$

The approximate equations above form the basis for all "impedance inversion" methods, in which the P-wave (or "acoustic") impedance $I_p(t)$ is estimated

from the seismic image data $D(t)$, via a local 1D solution to a convolutional model (not a wave equation), and is essentially a causal integration of the source-deconvolved seismic reflection image data $\hat{D}$. A more accurate form involves the logarithm of the impedance derivative and an exponentiated integral. These equations can be generalized to include angle-dependent reflection data $D(t, \varphi)$, both in time or in depth, in which case three independent seismic properties can be estimated, for example (I_p, I_s, ρ), which is known as "prestack" or "elastic" impedance inversion.

Because the seismic source is band limited, and the earth is naturally attenuative, the seismic data and hence the foregoing impedance estimates are missing both low- and high-frequency information. The source deconvolution in the impedance inversion equation attempts to recover some of the high-frequency impedance information, within seismic noise limits. Several methods can be used to recover the low-frequency background impedance trend, for example, by using constraints from well-log information and seismic imaging velocities. For impedance inversion methods, it is implicitly assumed that the reflectivity image data D can be accurately estimated from angle-dependent amplitude-preserved seismic imaging methods correctly in depth.

The foregoing impedance inversion method can be applied to time-lapse seismic data sets using any of the approaches discussed in the Theory section (Parallel, Double Difference, Bootstrap, Simultaneous, etc.). Figure 2.23 shows an example of time-lapse impedance inversion to estimate the water saturation change (ΔS_w) in a hydrocarbon reservoir that is being produced with horizontal wells. Each panel shows a depth slice map of the time-lapse change in P-impedance, with blue indicating increased water saturation, starting at the gas–oil contact (GOC) at 1295.5 m depth, and stepping downwards in approx. 3 m depth increments to the original oil–water contact (OOWC) at 1307.7 m depth. There has been significant vertical water movement of approx. 6 m from the OOWC depth of 1307.7 m up to 1301.5 m depth, and additional water movement in some areas of approx. 12 m up to the level of the GOC at 1295.5 m depth. Although this is a hydrocarbon production example, it is not difficult to imagine that similar maps could be made for a CO_2 injection project, to monitor supercritical CO_2 that has buoyantly migrated upwards from a deep injection point through a saline aquifer or depleted hydrocarbon storage reservoir, toward the top seal.

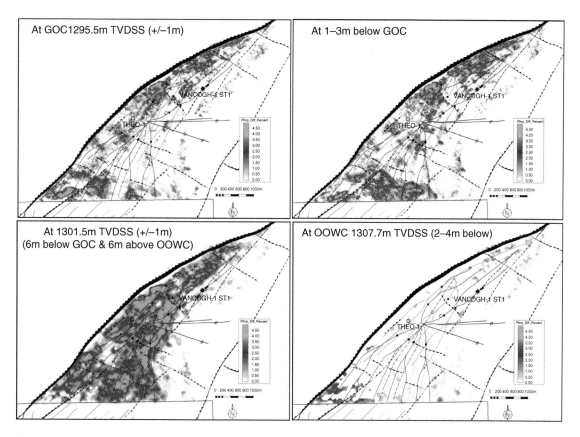

Figure 2.23 Time-lapse seismic impedance inversion example. Each panel shows a depth slice map of the time-lapse change in P-impedance, with blue indicating increased water saturation, starting at the gas–oil contact (GOC) (top left) at 1295.5 m depth, and stepping downwards in approx. 3 m depth increments to the original oil–water contact (OOWC) at 1307.7 m depth (lower right). There has been significant vertical water movement of approx. 6 m from the OOWC depth of 1307.7 m up to 1301.5 m depth, including additional water movement in some areas right up to the level of the GOC at 1295.5 m depth. (In collaboration with and courtesy of Paul Bouloudas and colleagues, Quadrant Energy.)

Full Waveform Inversion

A third broad category of time-lapse inversion is based on solving a wave equation such that an estimated model of the time-varying earth properties $m(x, \tau)$ creates synthetic seismic shot gathers that match the real data shot gathers to within some specified error tolerance. The theory that forms the basis for these time-lapse full waveform inversions (TL FWI, or 4D FWI) is discussed in the Theory section of this chapter.

TL FWI methods are currently the subject of much active research in my group and others; they are not yet mature. FWI methods are also approx. 10–100 times more computationally expensive than the AVO/A and impedance inversion methods (including their prerequisite PSDM/PSTM costs). However, TL

FWI methods are very promising because they can include far more realistic nonelastic wave propagation and scattering effects, do not require a local 1D earth assumption, and have the potential to achieve much higher resolution than the AVO/A and impedance-inversion methods, all of which are extremely important for detailed reservoir monitoring objectives.

Figure 2.24 shows an example of time-lapse full waveform inversion (TL FWI) for a near-surface microbubble injection monitoring experiment (Kamei *et al.*, 2017). The microbubble injection method is a new technique being developed and tested in Japan, in which water infused with tiny bubbles is injected into shallow sediments to help "cushion" them in order to suppress earthquake surface-wave damage to buildings and infrastructure, and help

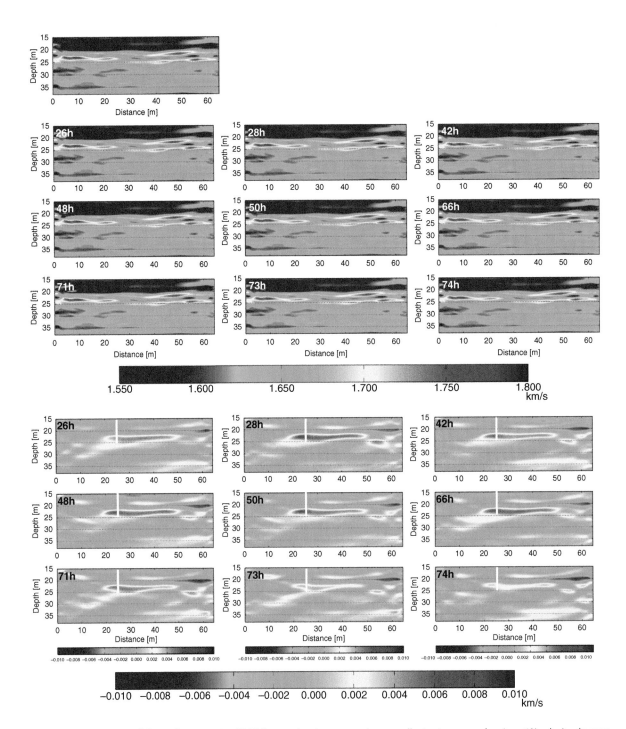

Figure 2.24 Time-lapse full waveform inversion (TL FWI) example using repeated cross-well seismic surveys, showing <1% velocity changes as injected plume evolves during 74 hours of microbubble injection (injector location marked in white). Top panels are individual FWI velocity inversion results; bottom panels are time-lapse FWI *velocity-change* inversion results (Kamei *et al.*, 2017, and unpublished work).

avoid loss of life. Under certain conditions, the injection of compressible microbubbles into the original formation water may be similar to injecting compressible CO_2 into a saline aquifer storage reservoir. For this research project, 15 time-lapse cross-well seismic surveys were acquired over a period of 74 hours (3 days) before, during, and after microbubble injection (the injection well is shown in white). Figure 2.24 shows the results of our TL FWI for the baseline preinjection survey, and 9 of the 14 repeated time-lapse surveys. The top panels are the individual FWI velocity estimates for each cross-well seismic survey, the bottom panels are the time-lapse *velocity-change* estimates obtained using the TL FWI Bootstrap method. These spectacular results show that we are clearly able to see the effects of the microbubble injection plume evolving over time to within less than 1% P-wave velocity variation (Kamei *et al.*, 2017, unpublished work).

Integrated Inversion with Nonseismic Data

To help overcome the limitations of using time-lapse seismic alone to quantify the CO_2 saturation distribution in the storage reservoir, it is likely that seismic inversions would benefit from being combined with, and constrained by, complementary geophysical data sets such as electromagnetic (EM), gravity, and InSAR satellite radar. EM data are highly sensitive to CO_2 saturation when the CO_2 is electrically resistive compared to the brine formation water. Gravity data are approximately linearly dependent on CO_2 saturation because the density of the injected CO_2 is less than that of water. InSAR satellite data measures the ground deformation at the Earth's surface caused by geomechanical deformations within and above the reservoir, where small surface ground displacements on the order of a few millimeters or centimeters can be related to the injection pressure and stress changes in the reservoir via coupled flow-simulation and geomechanical modeling.

One strategy forward for enhanced seismic CO_2 inversion would be to combine the higher resolution but more qualitative estimates of CO_2 saturation from seismic, with the lower resolution but more quantitative CO_2 saturation estimates from EM and gravity, and the high-resolution ground surface deformation data measured by InSAR data. These in turn can be coupled with, and constrained by, fluid-flow and geomechanical modeling, in order to obtain high-

resolution quantitative estimates of the volume and mass of CO_2 present in the storage reservoir. These are topics of much active research in the geophysical community today, as partly discussed in other chapters of this book, and thus deserve a more detailed discussion that is beyond the scope of an overview chapter on seismic monitoring.

Passive Seismic Monitoring

In this final section, I will present recent "hot off the press" research advances in time-lapse passive seismology that may allow us, in the future, to monitor CO_2 injection and storage reservoirs without the need for active manmade source efforts, by instead using seismic data from (semi)permanent sensor arrays that continuously record (micro) earthquake energy and (ocean) ambient noise.

Up until this point in the chapter, we have been considering time-lapse seismic surveying techniques that use *active* manmade sources, for example dynamite shots or Vibroseis trucks on land, or compressed airgun arrays at sea. For each time-lapse seismic survey, thousands of source locations are typically required to form high-quality 2D and 3D images of the Earth. This manmade source effort requires a major mobilization of crew personnel, equipment, and resources, and can have a significant financial cost and environmental impact on the survey area (e.g., by cutting hundreds of source-line access swaths through fields or farm land). For many CO_2 storage projects, time-lapse seismic surveys will need to be acquired every 1–10 years depending on the stage in the project's life. In many cases, the stakeholders who control the surface access rights to the CO_2 storage site may not welcome this magnitude or frequency of seismic survey impact, and the project operators may not be able to afford the significant financial costs of repeated active-source seismic surveying.

One possible strategy to minimize the environmental impact, and the expense, of repeating time-lapse seismic surveys at a CO_2 storage site, may be to deploy a (semi)permanent receiver array, and also to avoid the use of manmade sources. In this scenario, a (semi)permanent sensor array could be deployed at the site as a one-time event, buried in the shallow subsurface/seafloor zone (e.g., the top 1–10 m); this not only keeps the array secure from interference with, and damage by, manmade and natural activities, but also typically enhances the time-lapse seismic data

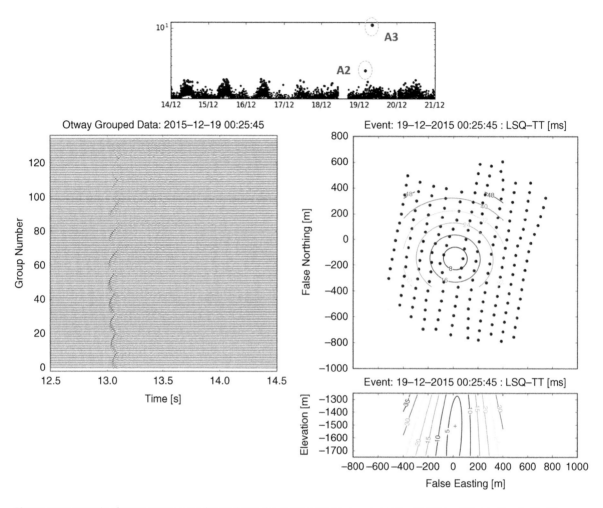

Figure 2.25 Example of microseismic event detected with the passive seismic array during the 15 000-ton CO_2 injection for the Otway Stage 2C project. The top panel shows two detected microseismic events (A2 and A3) compared to the background noise level (A3 event S/N>>10) during 1 week of recording (December 14–21, 2016). The lower left panel shows the high-quality seismic trace data for the A3 microseismic event. The lower right panel shows the A3 estimated microseismic source location, which is very close to both the CO_2 injection well located at the center of the array (star in Figure 2.26), and to the CO_2 injection point at approx. 1500 m depth (Drew, Saygin, and Lumley, unpublished work, 2017).

quality by 20 dB or more (van Gestel *et al.*, 2008; Pevzner *et al.*, 2017). And, instead of imaging with seismic waves generated by active manmade sources, seismic images of the subsurface can be made by using (micro) earthquake seismic energy, and/or interferometric images created from (ocean) ambient noise, continuously recorded in *passive* "listen" mode by the permanent sensor array (Claerbout, 1968; Schuster, 2009; Wapenaar *et al.*, 2010; Issa and Lumley, 2015; Issa *et al.*, 2017).

Advantages of using a passive seismic array for monitoring of CO_2 injection projects may include

- Low environmental impact on the survey area
- Low cost of repeating many time-lapse seismic surveys
- Ability to provide continuous, real-time monitoring and imaging of the reservoir
- Ability to detect weak (unfelt) seismicity and adjust the CO_2 injection program to avoid the

possibility of inducing felt earthquake events, and thus maintain public confidence and support (social license) for the project

- Ability to use *microseismic events* to monitor the shape, location, and progress of the injected CO_2 plume and pressure front over time
- Ability to use *ambient seismic noise* to image the shape, location, and progress of the injected CO_2 plume and pressure front over time

Disadvantages of using a passive seismic array for monitoring of CO_2 injection projects may include the following:

- Initial cost and environmental impact of deploying the sensor array may be high.
- Ongoing power, access, maintenance, and repair requirements for a (buried) permanent array may become increasingly challenging over long periods of time.
- No ability to control the source frequency spectrum of the passive seismic energy, and thus the image resolution at reservoir depth, as it depends on the microseismic events and ambient noise naturally present in the area.
- May require very long recording times (days, weeks, months, or even years) to accumulate enough passive seismic body-wave energy to form high-quality images at the reservoir depth.

Harnessing passive seismic energy to monitor subsurface reservoirs is an active area of research in my group, among others. Over the past five years, my group has been deploying various (semi) permanent sensor arrays at basin scales (100+ km spacing), regional scales (10–100 km spacing), and reservoir monitoring project scales (0.1–10 km spacing), using high-sensitivity exploration geophones with an approximate 1–500 Hz flat spectral response, and also broadband seismographs with a 0.01–200 Hz flat spectral response. With these arrays, we have been continuously recording large distant and moderate regional earthquakes of magnitude M_w >2, microseismic earthquakes of magnitude M_w >−2, and ambient seismic noise of all types, especially strong ocean ambient noise with spectral peaks at about 0.05–0.2 Hz.

Figure 2.15 shows a buried geophone array at the Otway Stage 2C project site, deployed by Roman Pevzner *et al.* at Curtin University and the CO_2CRC. Pevzner *et al.* were able to reduce the average *prestack* shot-gather NRMS values from approximately 0.75

for a previous surface geophone array, to NRMS less than 0.20 within the main 20–80 Hz seismic passband by burying the 900+ geophone array at just 4 m below the surface within the permanent groundwater-saturated zone (Shulakova *et al.*, 2015; Pevzner *et al.*, 2017). As part of the Otway Stage 2C research project funded by the CO_2CRC, my research group has been using passive seismic data recorded with this array for long continuous periods (weeks, months) to try and monitor the Otway Stage 2C injection of 15 000 tons of CO_2/CH_4, using both micro-seismic energy, and ocean ambient noise energy.

Figures 2.25 and 2.26 show examples of our ongoing research results for microseismic events detected during CO_2 injection using the Otway Stage 2C passive seismic array data (Drew, Saygin, and Lumley, unpublished work, 2017). Figure 2.25 shows two detected microseismic events (A2 and A3), which are stronger than the background noise levels (A3 event S/N $\gg$ 10), during a week of continuous recording (December 14–21, 2016). The lower left panel shows the high quality of the passive seismic data for the A3 microseismic event. The lower right panel shows the estimated microseismic source location, which is very close to the CO_2 injection well located at the center of the array (star in Figure 2.26), and is estimated to be at approx. 1500 m depth, which is the true CO_2 injection depth. Figure 2.26 shows the source mechanism estimated for the A3 microseismic event. The estimated event magnitude is $M_w = 0.0$, and the detection noise threshold of the array is estimated to be approx. M_w >−1.5. Given the low CO_2 injection pressures, this relatively large $M_w = 0$ value suggests this is a fault reactivation event for a fault that intercepts the reservoir, and that is at or near critical stress. The event waveform energy and polarity is plotted in an overlay on the passive seismic array; note the event polarity reversal and nodal plane across the array. This microseismic source mechanism corresponds to a SW–NE striking normal fault dip–slip motion, which is also consistent with weak fault reactivation induced by CO_2 injection.

Figures 2.27–2.29 show our ongoing research results related to ambient seismic noise imaging and time-lapse reservoir monitoring of CO_2 injection using the Otway Stage 2C passive seismic array. Figure 2.27 shows an interferometric ambient seismic noise image created from 1 hour of continuous recording with an initial linear 2D test array, plotted against two vintages of conventional active-source 3D

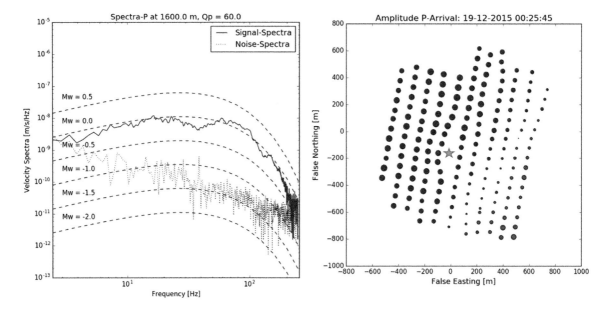

Figure 2.26 Source mechanism estimated for the A3 microseismic event, detected with the passive seismic array during the Otway Stage 2C CO_2 injection. (Left) estimated magnitude of $M_w = 0.0$ for the A3 microseismic event, versus a detection noise threshold of $M_w > -1.5$ in the main frequency band of 10–100 Hz. Given the small injection pressures, this relatively large M_w value suggests a fault reactivation event for a fault at or near critical stress. (Right) Waveform energy and polarity of the microseismic event arrivals overlain on the passive seismic array (note the polarity reversal and SW–NE nodal plane across the array); this source mechanism corresponds to a SW–NE normal fault dip-slip motion, also consistent with fault reactivation (Drew, Saygin, and Lumley, unpublished work, 2017).

seismic data, at the Otway CO_2 project site (Issa and Lumley, 2015; Issa *et al.*, 2017). Note that the 2D ambient noise image has good signal to about 2–3 km depth (deeper than the CO_2 injection reservoir at approx. 1500 m depth), but is limited to a maximum frequency content of approx. 25 Hz due to the nature of the ocean storm-wave noise sources. Figure 2.28 shows beam-steering velocity-azimuth "radar" plots of ambient seismic noise recorded with the full 3D passive seismic array (Issa and Lumley, unpublished work, 2016). The top panel shows 2-Hz surface waves arriving from the SW direction of the Southern Ocean and propagating horizontally along the surface at approx. 700 m/s. The bottom panel shows 14-Hz body waves arriving from the same SW ocean direction but scattering off of the nearby coastline, propagating at approx. 3000 m/s and approx. 30 degrees from vertical incidence, as determined from the array apparent velocities, and local 1D sonic log velocities.

Figure 2.29 shows preliminary results of time-lapse reservoir monitoring with ambient seismic noise interferometry at the Otway Stage 2C CO_2 injection project (Saygin and Lumley, unpublished work,

2017). The top panel shows that interferometric Green's function seismic traces are highly repeatable before CO_2 injection, but start to change significantly just before (well pad activity) and during CO_2 injection (especially in the later parts of the waveform coda). The bottom panel shows that the Green's function coherency (repeatability) is close to 100% before CO_2 injection, and then significantly decreases during the activity immediately prior to, and during, CO_2 injection.

These passive seismic results are "hot off the press" at the time of writing this book chapter, and are the subject of our ongoing active research. The results are certainly highly encouraging and interesting; we look forward to reporting further passive seismic monitoring results in journal publications and future editions of this book.

Acknowledgments

The author of this Chapter (David Lumley) kindly acknowledges the support of his current employer as Professor at the University of Texas at Dallas (UTD), and the generous permanent endowment made by Cecil and Ida Green (founders of Texas

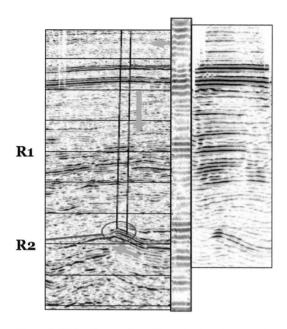

R1

R2

Figure 2.27 Interferometric ambient seismic noise image (center) created from 1 hour of continuous recording, compared to conventional active-source 3D seismic data images, at the Otway CO_2 project site. Note that the ambient noise image has good signal to about 2–3 km depth (deeper than both reservoirs), but is limited to a maximum frequency content of approx. 25 Hz due to the nature of the ocean storm wave noise sources (Issa and Lumley, 2015; Issa et al., 2017).

Instruments), which supports the author's current position as the Green Chair in Geophysics at UTD. The author also kindly acknowledges the generous support of Woodside Energy, and Chevron, for partial support of the author's previous position as Professor and Chair in Geophysics at the University of Western Australia (UWA), where a significant portion of the research and teaching material discussed in this chapter was developed.

The author wishes to acknowledge research financial assistance provided through Australian National Low Emissions Coal Research and Development (ANLEC R&D). ANLEC R&D is supported by Australian Coal Association Low Emissions Technology Limited and the Australian Government through the Clean Energy Initiative.

The author wishes to acknowledge the CO_2CRC for their research financial assistance, and leadership in the conceptualization, design, and field operations associated with the Otway 2C Project (especially Roman Pevzner et al.), which received CO_2CRC funding through its industry members and research partners, the Australian Government under the CCS Flagships Programme, the Victorian State Government and the Global CCS Institute.

Portions of this research material were created using the high-performance computing (HPC) resources of the Pawsey Supercomputing Centre in Perth, Australia.

Finally, the author wishes to kindly acknowledge the many research collaborations and discussions with colleagues, staff, and graduate students over the years that have contributed to some of the material in this chapter, including but not limited to (apologies in advance to anyone I may have inadvertently forgotten to mention): The University of Western Australia (especially Paul Connolly, Julian Drew, Nader Issa, UGeun Jang, Rie Kamei, Eric May, Taka Miyoshi, Matt Saul, Erdinc Saygin, Jeff Shragge et al.), CSIRO (especially Andrew King, Joel Sarout, Linda Stalker et al.), Curtin University (especially Boris Gurevich, Maxim Lebedev, Roman Pevzner et al.), WAERA, WA-DMP (especially Dominique van Gent, Sandeep Sharma et al.), ANLEC (especially Rick Causebrook, Kevin Dodds et al.), CO_2CRC (especially Charles Jenkins, Matthias Raab, Max Watson et al.), Geoscience Australia (especially Clinton Foster et al.), Woodside (especially Neil Kavanagh, André Gerhardt, Andrew Lockwood, Tom Ridsdill-Smith et al.), Chevron (especially Ron Behrens, Steve Doherty, Ray Ergas, Gary Hampson, Fred Herkenhoff, Harry Martin, Jason McKenna, Don Sherlock, Joe Stefani, Mark Trupp, Wendy Young, Zee Wang et al.), BHP Billiton (especially Guy Duncan, Robin Hill et al.), Quadrant Energy (especially Paul Bouloudas, Rob Kneale, Fred Wehr et al.), Shell (especially Rodney Calvert, Peter Chia, Paul Hatchell, Jan Stammeijer et al.), 4th Wave Imaging (especially Don Adams, Steve Cole, Dave Markus, Mark Meadows, Rich Wright et al.), Geological Survey of Canada (especially Don White et al.), Stanford University (especially Jon Claerbout, Jack Dvorkin, Gary Mavko, Amos Nur, Lynn Orr, Tiziana Vanorio et al.), Statoil (especially Ola Eiken, Lars Kristian Strønen, Per Digranes et al.), NTNU (especially Martin Landrø et al.), LBNL (especially Tom Daley et al.), JOGMEC (especially Masashi Nakatsukasa, Takuji Mouri, Ayato Kato, Mamoru Takanashi, Yuki Nakamura et al.), and many other colleagues I have worked with on 200+ time-lapse 4D seismic projects over the past 25+ years.

Figure 2.28 Beam-steering azimuth-velocity "radar" plots of ambient seismic noise arriving at the Otway CO_2 project site. (Top) 2-Hz surface waves arriving from the (SW) Southern Ocean at approx. 700 m/s. (Bottom) 14-Hz body waves arriving from the same SW direction at approx. 3000 m/s and approx. 30 degrees from vertical incidence, with complex azimuthal scattering off the irregular coastline (Issa and Lumley, University of Western Australia).

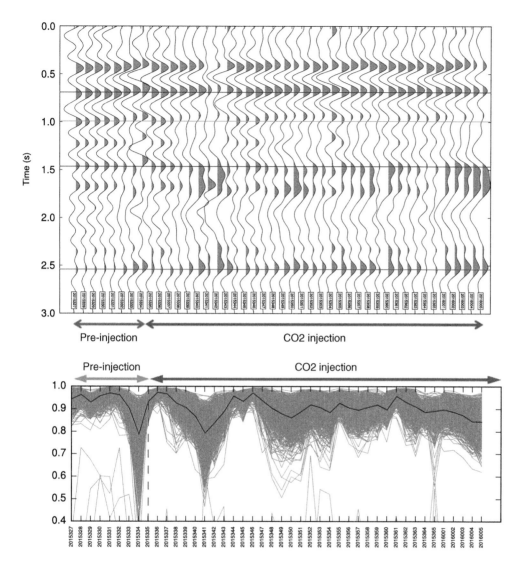

Figure 2.29 Time-lapse reservoir monitoring with ambient seismic noise interferometry at the Otway Stage 2C CO_2 injection project (45 days of passive seismic data displayed; 8 days preinjection and 37 days during injection). (Top) Interferometric Green's function seismic traces are highly repeatable before CO_2 injection, then start changing significantly (especially the later parts of the coda) just prior to and during CO_2 injection starting at red dash line. (Bottom) Green's function coherency (repeatability) is close to 100% before injection, and then significantly changes during the well pad activity just prior to, and during the CO_2 injection (Saygin and Lumley, unpublished work, 2017).

References

Aki, K., and Richards, P. G. (1980). *Quantitative seismology: Theory and methods.* San Francisco: W. H. Freeman. DOI:10.1017/S0016756800034439

Calvert, R. (2005). Insights and methods for 4D reservoir monitoring and characterization. *Society of Exploration Geophysics.* www.amazon.com/Insights-Methods-4D-Reservoir-Charterization/dp/1560801360

Carbon Storage Taskforce. (2009). National carbon mapping and infrastructure plan. Department of Resources, Energy and Tourism (DRET), Australia.

Caspari, E., Müller, T. M., and Gurevich, B. (2011). Time-lapse sonic logsreveal patchy CO_2 saturationin-situ. *Geophysics Research Letters,* **38**: L13301. DOI:10.1029/2011GL046959.

Claerbout, J. (1968). Synthesis of a layered medium from its acoustic transmission response. *Geophysics,* **33**(2): 264–269. https://doi.org/10.1190/1.1439927

Cole, S., Lumley, D., Meadows, M., and Tura, A. (2002). Pressure and saturation inversion of 4D seismic data by rock physics forward modeling. Technical Program Expanded Abstracts. *Society of Exploration Geophysicists*, **2475**–2478. DOI:10.1190/1.1817221.

Daley, T. M., Myer, L. R., Peterson, J. E., Majer, E. L., and Hoversten, G. M. (2008). Time-lapse crosswell seismic and VSP monitoring of injected CO_2 in a brine aquifer. *Environmental Geology*, **54**(8): 1657–1665. DOI:10.1007/s00254-007-0943-z.

Davis, T., Terrell, M. J., Benson, R. D., Cardona, R., Kendall, R. R., and Winarsky, R. (2003). Multicomponent seismic characterization and monitoring of the CO_2 flood at Weyburn Field, Saskatchewan. *Leading Edge*, **22**(7): 696–697. DOI:10.1190/1.1599699.

Dvorkin, J., and Nur, A. (1996). Elasticity of high-porosity sandstones: Theory for two North Sea data sets. *Geophysics*, **61**: 1363–1370. DOI:10.1190/1.1444059.

Emberley, S., Hutcheon, I., Shevalier, M., Durocher, K., Gunter, W. D., and Perkins, E. H. (2004). Geochemical monitoring of fluid-rock interaction and CO_2 storage at the Weyburn CO2-injection enhanced oil recovery site, Saskatchewan, Canada. *Energy*. https://ideas.repec.org/a/eee/energy/v29y2004i9p1393-1401.html

Gernert, J., and Span, R. (2016). EOS–CG: A Helmholtz energy mixture model for humid gases and CCS mixtures. *Journal of Chemical Thermodynamics*, **93**: 274–293. https://doi.org/10.1016/j.jct.2015.05.015

Glubokovskikh, S., Pevzner, R., Dance, T., *et al.* (2016). Seismic monitoring of CO2 geosequestration: CO2CRC Otway case study using full 4D FDTD approach. *International Journal of Greenhouse Gas Control*, **49**: 201–216. https://doi.org/10.1016/j.ijggc.2016.02.02

Guilbot, J., and Smith, B. (2002). 4D constrained depth conversion for reservoir compaction estimation: Application to Ekofisk Field. *Leading Edge*, **21**(3): 302–308. http://dx.doi.org/10.1190/1.1463782

Hatchell, P., and Bourne, S. (2005). Rocks under strain: Strain-induced time-lapse time shifts are observed for depleting reservoirs. *Leading Edge*, **24**(12): 1222–1225. http://library.seg.org/doi/abs/10.1190/1.2149624

Issa, N., and Lumley, D. (2015). Passive seismic imaging at depth using ambient noise fields recorded in a shallow buried sensor array. *Australian Society of Exploration Geophysics*. library.seg.org/doi/pdf/10.1071/ASEG2015ab135

Issa, N., Lumley, D., and Pevzner, R. (2017). Passive seismic imaging at reservoir depths using ambient seismic noise recorded at the Otway CO_2 geological storage research facility. *Geophysical Journal International*, **209** (3): 1622–1628. https://doi.org/10.1093/gji/ggx109

Johnston, D. H. (2013). Practical applications of time-lapse seismic data. *Society of Exploration Geophysicists*. http://dx.doi.org/10.1190/1.9781560803126

Kamei, R., and Lumley, D. (2017). Full waveform inversion of repeating seismic events to estimate time-lapse velocity changes. *Geophysical Journal International*, **209** (2): 1239–1264.

Kamei, R., Jang, U., Lumley, D., *et al.* (2017). Time-lapse full waveform inversion for monitoring near-surface microbubble injection. Expanded Abstracts, *European Association of Engineering Geosciece (EAGE)*, Paris, France. DOI:10.3997/2214-4609.201700956.

Kragh, E., and Christie, P. (2002). Seismic repeatability, normalized RMS, and predictability. *Leading Edge*, **21** (7): 640–647. DOI:10.1190/1.1497316.

Landrø, M. (1999). Discrimination between pressure and fluid saturation changes from time-lapse seismic data. In Expanded Abstracts, 69th Annual International Meeting. *Society of Exploration Geophysicists*, 1651–1654.

Landrø, M. (2001). Discrimination between pressure and fluid saturation changes from time-lapse seismic data. *Geophysics*, **66**(3): 836–844. http://dx.doi.org/10.1190/1.1444973

Lebedev, M. (2012). Geophysics laboratory: Otway rock physics tests. In B. Evans, R. Rezaee, V. Rasouli, *et al.*, Milestone Report for ANLEC Project #3-1110-0122. http://anlecrd.com.au/projects/predicting-co-sub-2-sub-injectivity-properties-for-application-at-ccs-sites/

Lumley, D. (1995a). *Seismic time-lapse monitoring of subsurface fluid flow*. PhD thesis, Stanford University.

Lumley, D. E. (1995b). Seismic monitoring of hydrocarbon fluid flow. *Journal of Mathematical Imaging and Vision*, 5(4): 287–296. DOI:10.1007/BF01250285.

Lumley, D. E. (2001). Time-lapse seismic reservoir monitoring. *Geophysics*, **66**: 50–53.

Lumley, D. E. (2006). Nonlinear uncertainty analysis in reservoir seismic modeling and inverse problems: Technical Program Expanded Abstracts, *Society of Exploration Geophysicists*, 2037–2041. DOI:10.1190/1.2369936.

Lumley, D. (1996–present). *4D seismic reservoir monitoring*: Course book.

Lumley, D. (2010). 4D seismic monitoring of CO_2 sequestration. *Leading Edge*, **29**(2): 150–155. DOI:10.1190/1.3304817.

Lumley, D., and Shragge, J. (2013). Advanced concepts in active and passive seismic monitoring using full wavefield techniques. Extended Abstracts, *Australian Society of Exploration Geophysicists*, 2013: 1–4. https://doi.org/10.1071/ASEG2013ab167

Lumley, D., Adams, D. C., Meadows, M., Cole, S., and Wright, R. (2003a). 4D seismic data processing issues and examples. Technical Program Expanded Abstracts,

Society of Exploration Geophysicists, 1394–1397. DOI:10.1190/1.1817550.

Lumley, D., Adams, D., Meadows, M., Cole, S., and Ergas, R. (2003b). 4D seismic pressure-saturation inversion at Gullfaks Field, Norway. *First Break*, 21, September, *European Association of Geoscientists and Engineers* (EAGE).

Lumley, D., Meadows, M., Cole, C., and Adams, D. (2003c). Estimation of reservoir pressure and saturations by crossplot inversion of 4D seismic attributes. Technical Program Expanded Abstracts, *Society of Exploration Geophysicists*, 1513–1516. DOI:10.1190/1.1817582.

Lumley, D., Adams, D., Wright, R., Markus, D., and Cole, S. (2008). Seismic monitoring of CO_2 geo-sequestration: Realistic capabilities and limitations. Technical Program Expanded Abstracts, *Society of Exploration Geophysicists*, 2841–2845. DOI:10.1190/1.3063935.

Lumley, D., King, A., Pevzner, R., *et al.* (2016). Feasibility and design for passive seismic monitoring at the SW Hub CO_2 Geosequestration Site. *Australian National Low Emissions Council* (ANLEC) R&D Project Number 7-0212-0203. http://anlecrd.com.au/reports_storage/

Mathieson, A., Wright, I., Roberts, D., and Ringrose, P. (2009). Satellite imaging to monitor CO2 movement at Krechba, Algeria. *Energy Procedia*, **1**(1): 2201–2209. DOI:10.1016/j.egypro.2009.01.286.

Mavko, G., Mukerji, T., and Dvorkin, J. (2009). *The rock physics handbook: Tools for seismic analysis of porous media*. Cambridge: Cambridge University Press.

Meadows, M., Adams, D., Wright, R., Tura, A., Cole, S., and Lumley, D. (2005). Rock physics analysis for time-lapse seismic at Schiehallion Field, North Sea. *Geophysical Prospecting*, 53: 205–213.

Ostrander, W. (1984). Plane-wave reflection coefficients for gas sands at non-normal angles of incidence. *Geophysics*, **49**(10): 1637–1648. http://dx.doi.org/10.1190/1.1441571

Pacala, S., and Socolow, R. (2004). Stabilization wedges: Solving the climate problem for the next 50 years with current technologies. *Science*, **305**: 968–972. DOI:10.1126/science.1100103.

Pevzner, R., Shulakova, V., Kepic, A., and Urosevic, M. (2011). Repeatability analysis of land time-lapse seismic data: CO_2CRC Otway pilot project case study. *Geophysical Prospecting*, **59**: 66–77. DOI:10.1111/j.1365-2478.2010.00907.

Pevzner, R., Urosevic, M., Tertyshnikov, K., *et al.* (2017a). Stage 2C of the CO2CRC Otway Project: Seismic monitoring operations and preliminary results. *Energy Procedia*. www.sciencedirect.com/science/article/pii/S1876610217317344

Pevzner, R., Urosevic, M., Popik, D., *et al.* (2017b). 4D surface seismic tracks small supercritical CO2 injection into the subsurface: CO2CRC Otway Project. *International Journal of Greenhouse Gas Control*, **63**: 150–157. https://doi.org/10.1016/j.ijggc.2017.05.008

Rickett, J. E., and Lumley, D. E. (2001). Cross-equalization data processing for time-lapse seismic reservoir monitoring: A case study from the Gulf of Mexico. *Geophysics*, **66**, Special Section, 1015–1025. DOI:10.1190/1.1487049.

Ridsdill-Smith, T., Flynn, D., and Darling, S. (2008). Benefits of two-boat 4D acquisition, an Australian case study. *Leading Edge*, **27**(7): 940–944. DOI:10.1190/1.2954036.

Saul, M., and Lumley, D. (2013). A new velocity-pressure-compaction model for uncemented sediments. *Geophysical Journal International*, **193**(2): 905–913. DOI:10.1093/gji/ggt005.

Schuster, G. T. (2009). *Seismic interferometry*. Cambridge: Cambridge University Press.

Shragge, J., and Lumley, D. (2013). Time-lapse wave-equation migration velocity analysis. *Geophysics*, **78**(2): S69–79.

Shulakova, V., Pevzner, R., Dupuis, J. C., *et al.* (2015). Burying receivers for improved time-lapse seismic repeatability: CO2CRC Otway field experiment. *Geophysical Prospecting*, **63**: 55–69.

Tura, A., and Lumley, D. E. (1998). Subsurface fluid-flow properties from time-lapse elastic-wave reflection data. In 43rd Annual Meeting, *SPIE*, Proceedings, 125–138.

Tura, A., and Lumley, D. E. (1999). Estimating pressure and saturation changes from time-lapse AVO data. Technical Program Expanded Abstracts, *Society of Exploration Geophysicists*, 1655–1658.

van Gestel, J-P., Kommedal, J. H., Barkved, O. I., Mundal, I., Bakke, R., and Best, K. D. (2008). Continuous seismic surveillance of Valhall Field. *Leading Edge*, **27**(12): 1616–1621. DOI:10.1190/1.3036964.

Vanorio, T. (2015). Recent advances in time-lapse, laboratory rock physics for the characterization and monitoring of fluid-rock interactions. *Geophysics*, **80**(2): WA49–WA59. DOI:10.1190/geo2014-0202.1.

Vialle, S., and Vanorio, T. (2011). Laboratory measurements of elastic properties of carbonate rocks during injection of reactive CO_2-saturated water. *Geophysics Research Letters*, **38**: L01302. DOI:10.1029/2010GL045606.

Wapenaar, K., Draganov, D., Snieder, R., Campman, X., and Verdel, A. (2010). Tutorial on seismic interferometry: Part 1 – Basic principles and applications. *Geophysics*, **75**(5): 75A211–227. https://doi.org/10.1190/1.3457445

White, D. J., Roach, L. A. N., and Roberts, B. (2015). Time-lapse seismic performance of a sparse permanent array: Experience from the Aquistore CO_2 storage site. *Geophysics*, **80**(2): WA35–WA48. http://library.seg.org/doi/abs/10.1190/geo2014-0239.1

Goals of CO₂ Monitoring

Why and How to Assess the Subsurface Changes Associated with Carbon Capture and Storage

Thomas M. Daley and William Harbert

Why Monitor?

Nearly all subsurface engineering activities begin with an effort to characterize subsurface properties important to the activity and have a goal to assess (and reduce) risk. Over the last century, many methods needed to characterize the subsurface have been developed, largely in support of extraction of resources from the Earth (e.g., minerals or hydrocarbons). As the extraction methods became more complex and efficient, the characterization efforts merged into monitoring efforts in which measurements were repeated over time. For example, active seismic imaging started in the 1930s as two-dimensional (2D) cross sections of the Earth, expanded to three-dimensional (3D) images in the 1980s and then became 4D in the 1990s when 3D surveys were repeated over time (with time as the fourth dimension). Often these early 4D data sets had only two or three time samples (with each full repeat of a 3D survey being a time sample). Early success in detecting spatially localized changes in subsurface properties (e.g., mapping where oil was being depleted or left in place) led to routine monitoring of the subsurface with 4D seismic. Thus, 4D seismic is now considered a key monitoring technology in oil and gas production (e.g., Johnston, 2013). The initial characterization of a new subsurface project will impact later monitoring, especially monitoring that needs a baseline. In fact, modern subsurface activities should view site characterization as an iterative process linked to and updated by repeated monitoring, and this is true of carbon capture and storage (CCS) projects.

Geological carbon storage (GCS or geosequestration) of carbon dioxide (CO_2) will require subsurface characterization and monitoring. It is important to understand the goals of monitoring to guide decision-making. The specific subsurface properties to be characterized, along with the technology used, are selected based on the goals for characterizing and monitoring; i.e., they are chosen after answering the question: why are we monitoring? Therefore, understanding the goals of monitoring is essential. The site-specific goals for each project will vary, but will follow from the main drivers for monitoring: performance assurance, regulatory requirements, public assurance, and an overarching desire for project risk reduction. As we describe each of these goal drivers, it is important to remember that minimizing monitoring costs will also be an overarching goal – though not always the most important. While this chapter discusses the goals and some of the techniques of monitoring, other chapters look at case studies of projects storing CO_2 in the subsurface and some issues encountered in monitoring as well as successes.

Performance Assurance

A goal for operators of a geological storage project will be to confirm that the injection and growth of a CO_2 plume is progressing as expected and predicted by their modeling of flow and transport within the reservoir – this is a key aspect of performance assurance. An example of performance assurance is monitoring injection pressures and volumes, which are typically monitored on the surface at hourly to weekly intervals. The correct performance of surface equipment (e.g., pumps and compressors) can be directly tied to these measurements, and subsurface changes near the injection well may be reflected quickly. Initial performance monitoring at the start of injection will likely focus on surface measurements. The subsurface is of course heterogeneous and, as the CO_2 plume grows, changes in reservoir performance are possible and likely. The operator will want to understand how the entire reservoir, including the seal, is performing over time. For this purpose, the monitoring will be focused on the predicted extent of the CO$_2$ plume at each monitoring time interval, as well as the distribution of CO$_2$ within the plume. For performance assurance, the monitoring will be expected to provide quantitative values of properties, such as CO$_2$ saturation in a given volume, which can be compared with modeled values.

A component of performance assurance is early and definitive leakage detection and attribution (i.e., is the detected CO$_2$ attributable to the storage project?). Monitoring technologies need to be optimal for leakage detection and, if necessary, they will need to guide mitigation. It is likely that a separate set of monitoring technologies will be deployed in the event of leakage and mitigation efforts. For example, Lawton *et al.* (this volume) look at a specific pilot project to examine the identification of leakage in a shallow subsurface zone.

Regulatory Compliance

Regulatory compliance is a first-order goal for monitoring. The importance of regulation as a driver of monitoring and verification (MV) was recognized by the intergovernmental panel on climate change (IPCC) in their 2005 summary of CCS (IPCC, 2005) by the statement that the MV cost estimate "covers pre-injection, injection and post-injection monitoring, and depends on the regulatory requirements" (IPCC, 2005). The United States Environmental Protection Agency (US EPA) and other regulators require that carbon capture and storage projects implement monitoring, reporting, and verification (MRV) systems that detect leakage of CO$_2$ from either the containment reservoir or away from the storage site (EPA, 2016). Beginning in 2011, the US EPA Underground Injection Control (UIC) program established the Class VI CO$_2$ injection well category for the purposes of protecting underground sources of drinking water (USDW). The US EPA requires Class VI wells to include a monitoring and testing plan during the operational phase of the project and to set minimum technical criteria for injection permits that include geological site characterization, area of review (AOR), well mechanical integrity testing (MIT), monitoring, postinjection site care (PISC), and site closure. Both AOR and PISC are examples of regulatory needs for site characterization and monitoring, possibly including monitoring ground water quality above the confining zone and plume and pressure front tracking. AOR and PISC are defined as follows:

- *Area of review*: The region surrounding the geological carbon storage/sequestration project where USDWs may be endangered by CO$_2$. The AOR is likely delineated by computational modeling, based on geological and geophysical site characterization that accounts for injected and displaced fluids.
- *Postinjection site care*: Appropriate monitoring and other actions needed following cessation of CO$_2$ injection to ensure that USDWs are not endangered.

Further EPA regulation, building upon the UIC program for geological sequestration wells, is subpart RR associated with the Clean Air Act which requires a US EPA-approved MRV plan and reporting of the amount of CO$_2$ sequestered using a mass balance approach (EPA, 2015, 2016).

Within the European Commission framework, Directive 2009/31/EC (EU, 2009), the postinjection or closure/decommissioning activities for large storage sites include the development of a monitoring plan with targets and methods, postclosure monitoring, updated site characterization and risk assessment, and inspections by the Competent Authority (CA) following the site closure. Before transfer of the site to the CA, long-term containment of CO$_2$ must be verified, the site sealed, and monitoring and assessment demonstrated for 20 years.

Storage projects will need to identify the monitoring technologies to meet the goals of regulatory requirements. Examples include defining and monitoring the entire AOR (while possibly reducing the AOR based on characterization or monitoring results), and selecting long-term monitoring for PISC while potentially reducing the PISC time-line.

Public Assurance

The need to assure the general public that a CCS project is operating safely may require different monitoring strategies and activities than those used for technical or regulatory requirements; thus public assurance can be a separate goal. Public outreach programs for GCS projects will interact with local constituencies and may lead to monitoring that specifically addresses local concerns. For example, storage under public (or private) land with water wells may necessitate geochemical well sampling to assure the specific water well owners, separate from the required regulatory monitoring requirements. As another example, atmospheric sampling of CO_2 for some locations may not be effective from a technical leakage-detection cost/benefit analysis but may provide public assurance value greater than the cost.

Risk Reduction

A site-specific risk assessment is likely to guide monitoring. Reduction of the overall project risk will be a goal for the monitoring program. A quantitative risk assessment will also allow a cost/benefit or value of information analysis to be used for selecting monitoring technologies. The benefit from a given technology is the detection ability, both spatially and temporally, i.e., how large a plume of CO_2 is detectable and how quickly you can detect it, while the cost of each monitoring technology includes data acquisition (including permitting) and data processing/analysis. An example is a site with known abandoned boreholes that could be leakage pathways, and thus a technology such as soil gas monitoring at or near those boreholes could be high benefit and also help reduce the risk profile. This type of monitoring may also address goals of performance assurance. Risk assessment for subsurface activities often has high uncertainty because of geological heterogeneity and uncertainty in characterization. Harbert *et al.* (2016) discuss the importance of monitoring strategies for risk reduction and outline an approach to integrate monitoring strategies with risk assessment and reduction. Even research can be driven by risk reduction, as one of their conclusions is that "the need for field experiment testing that is risk-driven" (p. 260). While monitoring for performance assurance is focused on demonstrating a predicted outcome, monitoring for risk reduction will be more focused on the unexpected or unwanted outcomes.

Characterization versus Monitoring

The goals of monitoring overlap with the goals of site characterization. Subsurface characterization is an iterative process based on, and including, specific monitoring and geoengineering objectives related to injection. This process begins with conceptual and numerical models being refined to produce the most accurate possible quantitative model of the subsurface in the area of study. This model should include structure, material properties, pore fluid characteristics, pressure and stress, fracture and fault characteristics, and material anisotropy characteristics within the area of study. These and other properties are the initial site characterization, typically including data from an initial characterization well. After the initial characterization has been completed, baseline data can be acquired and the characterization refined. In addition to determining the suitability of a storage site, characterization is used in the design of monitoring to constrain the expected response of target formations to CO_2 injection activities. Geophysical survey techniques, such as those listed in Table 3.1, have parameters that can be tuned to better image specific spatial and depth zones, or alternatively, if not properly designed, can introduce noise into geophysical data sets through inappropriate acquisition geometries. These acquisition noise components can be difficult to remove, or misinterpreted as subsurface behavior. Site characterization allows acquisition noise to be minimized by careful survey design, yielding higher quality time-lapse results. Only then, and as a distinct step, should the acquisition of additional monitoring data occur. It is important to view the characterization process as a closed loop, beginning and ending with the goal of an accurate and thorough knowledge of the reservoir performance behavior with respect to the project objectives. Initial characterization can serve as baseline monitoring, but ideally baseline surveys would be designed after initial characterization was complete.

For geophysical monitoring, characterization provides bounds to the behavior of key parameters and the expected geophysical response of monitoring methods in the presence of realistic noise conditions. Once these characterizations have been completed they can be incorporated into the design of baseline surveys and assessments to yield the maximum value of information. Monitoring is quite different from either characterization or baseline surveys. In monitoring, detecting subtle subsurface changes from the baseline case is the principal focus of analysis and interpretation (Fgure 3.1 from Calvert [2005]). In the best case, an observed change will have a high value of information because it ensures that the objectives are being successfully obtained (performance assurance), or because in a situation with anomalous behavior the observation can inform rapid operational changes as needed (risk reduction).

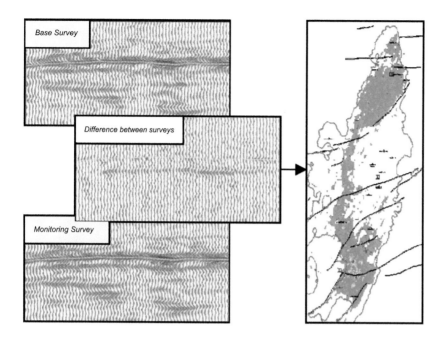

Figure 3.1 The ability of carefully designed monitoring and baseline geophysical surveys to generate a quality difference image of reservoir formation changes using reflection seismic methods is shown in this figure taken from Calvert (2005). The Draugen Field is located approximately 160 km offshore of Norway and was discovered in 1984. These data were interpreted to better understand water flooding of a target reservoir related to enhanced oil recovery (EOR). The region of time lapse change (interpreted with respect to the water flood changes) is shown in map view on the right. The spatial topology of injection related changes, formation compartmentalization, and intensity of change were constrained by this approach.

Monitoring can be considered a continuous sub-surface characterization during the CO$_2$ injection activity, as shown in Figure 3.2. Characterization and monitoring activities include initial baseline surveys, thorough reservoir characterization, risk assessment, developing specific injection engineering protocols, managing the CO$_2$ injection, accurately modeling the expected plume configuration, developing metrics for monitoring evolving risk, monitoring the injection and planning for site closure, and decommissioning (EU, 2009; Wright *et al.*, 2009; Wright, 2011).

What Properties to Monitor?

To address each of the goal drivers discussed in the foregoing, a different set of subsurface properties may be targeted. Deciding what properties to monitor requires understanding how they are measured, and then the process of choosing the properties and monitoring tools can be considered. The properties can be measured in different ways. In general, subsurface measurements can be considered as direct or indirect. Direct measurement monitoring tools include subsurface determinations of pressure or temperature, measured at a specific *in situ* location of the subsurface. Downhole fluid analysis would be another direct measurement approach, in which formation fluids from a specific location and depth are sampled and analyzed. In contrast, indirect measurements typically observe a proxy, such as seismic velocity or electrical resistivity, which is physically related to the subsurface property of interest, but is not that property itself. These observations are then analyzed and interpreted to estimate a subsurface property of interest (such as pressure or gas saturation). For example, electrical or seismic monitoring are indirect monitoring methods; the actual geophysical observations, such as elastic wave velocity or electrical conductivity, are interpreted to derive the parameters of interest (see Chapter 2 by Lumley, this volume, for a more detailed discussion). Direct monitoring is limited to the specific subsurface locations available from boreholes, while indirect monitoring, typically using geophysical methods, can survey large volumes of the subsurface with varying spatial resolution.

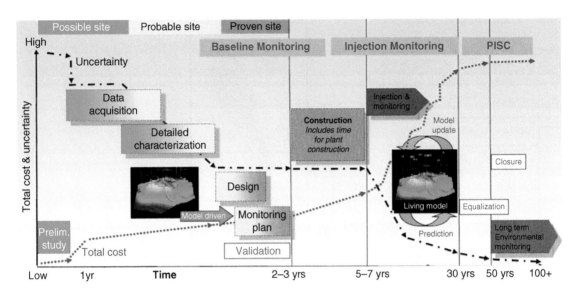

Figure 3.2 Role of monitoring in the lifecycle of a commercial CO_2 storage site. Using relative cost and uncertainty scales, it is suggested that uncertainty decreases with time (after Peters 2007). That is, within a framework of minimum monitoring requirements, measurements such as geophysical monitoring are made to assure decreasing uncertainty in performance of the GCS site and geological carbon storage.

Geophysical Monitoring

Geophysics is a fundamental part of subsurface observation, change detection, and reservoir analysis. Geophysical technologies can be used to observe and monitor various subsurface characteristics relevant to CO_2 operations. Two common geophysical technologies are seismic (Lumley, 2001, 2010; Lumley *et al.*, 2008; Lumley, Chapter 2, this volume) and electrical (Bergmann *et al.*, 2012; Gasperikova and Commer, Chapter 9, this volume) surveys. A number of chapters in this volume discuss the application of these techniques in field cases. Elastic wave (seismic) based geophysical methods are sensitive to subsurface elastic moduli, which can be used to determine porosity, density, and fracture geometries, among other properties. The determination of these reservoir characteristics is referred to as seismic lithology. Electrical or electromagnetic geophysical methods are sensitive to conductivity, which can be used to determine porosity and accurately estimate the geochemical composition of formation fluids. Conductivity can also be impacted by reactive geochemistry, formation permeability, and/or seal integrity, thus allowing the system to be better understood via monitoring. Table 3.1 summarizes geophysical approaches, the physical property measured, and the subsurface properties they may constrain.

It is important to understand the physical parameters that affect each geophysical technology. For example, seismic methods are dependent on effective stress, effective pressure, elastic properties, and density of solid and fluid components of rock materials. The elastic moduli are ratios of applied stresses and the resulting strain in linear elastic materials and vary with rock framework composition, porosity topology, pore-filling phase, and effective pressure conditions (Sayers, 2006, 2010). Because of their sensitivity to these and other properties, seismic imaging methods are considered important tools for subsurface characterization and monitoring.

As mentioned earlier, following the characterization of the potential site and the careful design of a seismic survey to accurately image the target formations, seismic tools are used for improving and adding detail to the structural and stratigraphic model and confirming the expected baseline seismic responses. An initial 3D reflection seismic survey can also serve as a structural framework for the spatial location of other monitoring data and as a baseline for determining temporal (i.e., time-lapse, 4D) changes in seismic properties (Johnson, 2013) and their temporal uncertainty (Kragh and Christie, 2002; Cantillo, 2011). A key application of 3D/4D reflection seismic is the mapping of changes in reflector amplitude (δA),

Table 3.1 Summary of geophysical methods used for subsurface CO_2 monitoring

Method	Principle	Physical property measured	Interpreted property
Electromagnetic			
Airborne electromagnetic frequency domain (FDEM)	Detects frequency domain-based variation in electromagnetic flow in subsurface materials	Electrical conductance and inductance	Geological structure[a] Groundwater geometry[b]
Airborne electromagnetic time domain	Detects time domain-based variation in electromagnetic flow in subsurface materials	Electrical conductance and inductance	Geological structure Groundwater geometry
			Subsurface pressure changes CO₂ geometry
Ground-based electrical and electromagnetic 2D and 3D surveys (resistivity sounding, spontaneous potential, induced polarization, electromagnetic, very low frequency, long offset time domain electromagnetics, and magnetotelluric)	Detects electromagnetic flow in subsurface materials	Electrical conductance and inductance Electrical resistivity and capacitance (induced polarization and magneto-telluric methods) Potential differences (spontaneous potential method)	Geological structure Groundwater geometry CO₂ geometry
Crosswell electromagnetics	Detects electromagnetic flow in subsurface materials	Electrical conductance and inductance	Geological structure Groundwater geometry CO₂ characteristics and geometry
Potential Fields			
Surface microgravity, downhole microgravity	Detects variations in the Earth's gravitational field	Mass density	Geological structure Changes in pore-filling fluid phases
Seismic			
Surface survey-based reflection seismic; 2D, 3D, and 4D, including 3-component	Detects reflected seismic P- or S-waves	Elastic moduli and density	Geological structure Pore fluid identification Geological structure Pore-pressure estimation Fracture geometry
Vertical seismic profiling (VSP)	Detects refracted seismic P- or S-waves	Elastic moduli and density	Geological structure Pore fluid identification Pore-pressure estimation
Crosswell tomography	Detects seismic P and S ray-path changes	Elastic moduli and density	Geological structure Pore fluid identification Pore-pressure estimation

Table 3.1 (cont.)

Method	Principle	Physical property measured	Interpreted property
Microseismic monitoring	Detects small failure events related to injection, production, or hydrofracturing	Elastic moduli, density, stress	Geological structure Pore fluid location Fracture geometry *In situ* stress, hydraulic diffusivity, failure surface orientation
Surface Imaging			
InSAR	Satellite-based system generates and detects changes in reflected radar frequency electromagnetic waves.	Ground displacement Phase shift of reflected wave	High-resolution topography Variation in topography related to subsidence or inflation
Tilt measurement	Change in orientation of ground (surface or borehole)	Ground displacement	Variation in topography related to subsidence or inflation

[a] Geological structure includes geologic strata geometry, heterogeneity, and fault geometry.

[b] Groundwater geometry includes presence or absence of groundwater and variation in groundwater conductivity that can be related to interaction with CO_2 and changes in pH.

reflector arrival time (δt), poststack seismic attributes, pore fluid properties such as water saturation (ΔS_w), and variations in effective pressure. Seismic data are analyzed for these properties both before imaging (prestack) and after imaging (poststack) (Figure 3.3). Changes in effective pressure (ΔP_{eff}), defined as the difference between confining and pore pressure, can influence poroelastic characteristics that are resolvable using seismic methods. Perturbations in pore pressure have a variety of impacts, such as opening compliant cracks or pores that can be observed using seismic methods. Such pore-pressure perturbations have been related to induced seismicity (e.g., Mavko et al., 2009; Shapiro, 2015). There is a large body of knowledge related to estimating formation pore fluids and pressure using calibrated seismic data (Landrø et al., 2001, 2003; Sayers, 2010).

A variety of advanced seismic techniques can be used to calculate seismic attributes from well log calibrated 3D reflection seismic volumes (Simm and Bacon, 2014). These techniques can estimate acoustic or elastic impedance and can determine material properties, such as Lamè parameters or combinations of parameters such as $\lambda\rho/\mu\rho$ that are proxies for other important

properties (Goodway et al., 1997). In addition, prestack amplitude versus offset (AVO) analysis (Castagna and Backus, 1993; Castagna and Swan, 1997; Castagna et al., 1998) uses relationships that contain both rock framework and fluid terms, allowing these properties of the injection reservoir to be investigated in detail. In general, prestack tools can efficiently utilize properties of the near and far offset reflection seismic data prior to the reflection stacking and imaging operations.

An example associated with CO_2 injection at a Permian basin site had prestack AVO analysis identifying a useful proxy for subsurface CO_2 (Purcell et al., 2009, 2010; Harbert et al., 2011). In this analysis the prestack AVO attribute proxy was defined within the injection reservoir and constrained at the top by a caprock unit. When extracted as a geobody from the data set and interpreted, this AVO-anomaly feature was observed to have a flat bottomed AVO attribute expression through the reservoir and was interpreted to represent a pore-phase variation, based on comparison with a rock-calibrated petrophysical model, microcomputer tomography (µCT) analysis, and well logs (Mur et al., 2011).

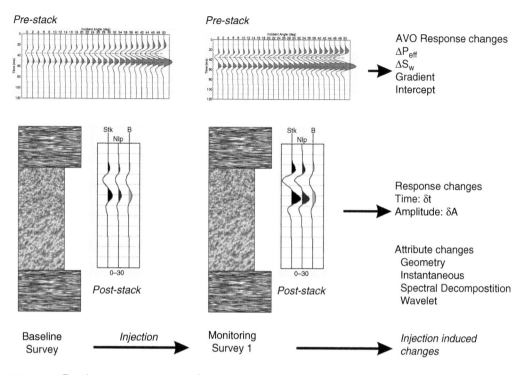

Figure 3.3 Time-lapse seismic monitoring of an injection. Both prestack and poststack reflection amplitude versus offset (AVO) data can be analyzed to detect changes in the subsurface. Time-lapse data should be interpreted within a framework of petrophysical models, differential stress variation, and production effects (Duffaut and Landrø, 2007).

Well Logging

Well-based logging tools (e.g., wire-line well logging) provide highly detailed measurements directly within the formations of interest and are critical for AVO and seismic inversion analyses and to calibrate and better understand all other reservoir associated geophysical data, such as electrical, electromagnetic, crosswell, VSP, and reflection seismic (Chapter 11 by Bassiouni, this volume). Log-based measurements include conductivity, pressure, temperature, acoustic velocity, nuclear magnetic resonance (NMR), electrical and electromagnetic responses, borehole images, and formation fluid composition (Kirksey, 2012). Processing and analysis of various acoustic, electrical, and nuclear measurements lead to estimates of other rock and fluid properties, such as porosity, permeability, salinity, and anisotropic geomechanical properties (Schlumberger, 1989, 1991). Analysis of pulsed neutron logging allows indirect estimation of CO$_2$ saturation and is commonly used in CO$_2$ storage and EOR projects, although results are highly dependent on quality of the cement bond (Nelson and Guillot, 2006). Because the observations of logging tools are localized in their sensitivity to the well region, these estimates are commonly extrapolated to populate formation models in correlation with seismic lithology inversion. Boreholes, and their well log data, can be horizontal in orientation, allowing lateral formation variation and zoning to be estimated. Incorporating structural data from borehole log data has been shown to considerably improve geophysical electrical inversion results (Doetsch et al., 2012, 2013).

Geochemical Monitoring

Geochemical monitoring provides insight into both reservoir containment and active geochemical processes that could impact injection efficiency. Geochemical tracer techniques provide a direct measure of subsurface connectivity between the points of tracer injection and measurement in the subsurface. Geophysical monitoring techniques are more accurately understood when constrained by the geochemistry of the subsurface systems. Geochemical alteration of the rock matrix can cause notable error in geophysical analysis that assumes

61

constant frame properties. Above-storage-zone fluid sampling is a direct measurement for storage security, and knowledge of fluid composition is important to calibrate and aid in the interpretation of electrical and electromagnetic geophysical monitoring.

As an example of field testing the geochemical and electrical geophysical monitoring of a leakage scenario, Dafflon *et al.* (2012) used laboratory and field experiments to explore the sensitivity of time-lapse complex resistivity response, and associated geochemical transformations, to a dissolved CO_2 plume at Plant Daniel, Mississippi. Results showed that electrical resistivity and phase responses correlated well with dissolved CO_2 injection processes. Specifically, resistivity initially decreased in response to an increase of bicarbonate and dissolved species, after which the resistivity rebounded toward the initial conditions due to transition of bicarbonate into non-dissociated carbonic acid. Laboratory studies confirmed that field observations of electrical resistivity, similar to crosswell electrical resistivity tomography, could detect dissolved CO_2. Both laboratory and field experiments demonstrate the potential of field-based geophysical methods for remotely monitoring changes in subsurface groundwater quality (geochemistry) due to CO_2 leakage.

An innovative technology, the U-tube fluid sampler (Freifeld *et al.*, 2005) is a tool for direct fluid sampling that can be deployed at multiple depths as a permanent installation to allow a multiyear sampling program. An additional innovative type of analysis, which can significantly aid geophysical monitoring, is downhole fluid analysis (DFA) (Mullins, 2008). Downhole fluid analysis (or direct sampling via wireline) draws fluids directly from the reservoir of interest for immediate geochemical analysis and allows complexities of reservoir fluid composition, reservoir compartmentalization and delineation, and the potential for reactive geochemical processes that could alter formation permeability, porosity, or seal integrity, to be better understood.

Joint Inversion from Multiple Methods

When data sets from multiple technologies are collected for monitoring, we are often faced with very different measures of subsurface properties, requiring balancing of uncertainty. For example, an integrated analysis of hydrological and geophysical monitoring data requires joint or coupled inversions to estimate the "best" fit to the data and must address spatial sampling variations between data sets and models. Traditional single well hydrological data provide high-resolution measurements for properties immediately surrounding boreholes. In the interwell region, geophysical cross-borehole measurements are sensitive to subsurface properties over larger regions but can have lower spatial or temporal resolution and can be difficult to interpret quantitatively.

Doetsch *et al.* (2012, 2013) combined the advantages of hydrological and geophysical data in a fully coupled hydrogeophysical inversion. Time-lapse seismic and electrical resistivity tomography (ERT) data (Carrigan, *et al.*, 2009) from the Cranfield site were inverted using cross-gradient joint inversion. This joint inversion approach ensures structural similarity while not assuming any specific petrophysical relations between model parameters. Inversion of seismic and ERT data for simultaneous changes in seismic velocity and resistivity from baseline surveys to nine months after the start of the CO_2 injection were successful in detecting changes in formation pore fluids. In this method, the individual time-lapse data sets are initially inverted: seismic inversion for changes in seismic velocity and electrical inversion for changes in conductivity – both related to CO_2 gas saturation. The joint inversion results show a more clearly focused CO_2 plume: in particular, the scatter plots indicate a link between change in velocity and ERT-derived gas saturation. A joint inversion can estimate the gas saturation that best matches both the seismic data and the electrical data (Kowalsky *et al.*, 2016). The results of the ERT time-lapse inversion were then combined with the hydrological data (e.g., gas sampling and tracer tests) to test different conceptual aquifer models. Among the outcomes was the ability to estimate width of fluid flow paths that connect the injection and monitoring wells.

How to Choose Monitoring Tools and Parameters

Cost/Benefit Assessment of Tools

Once the goals of monitoring are chosen, monitoring tools can be assessed for their ability to aid in achieving those goals, their benefit, and for their site specific costs. A large number of monitoring tools are

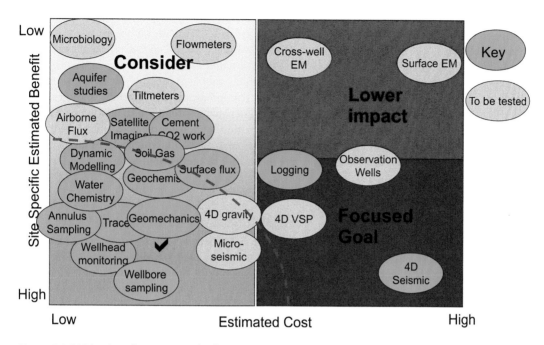

Figure 3.4 Initial review of monitoring technologies presented in a Boston Square for the Krechba Reservoir at In Salah, Algeria (modified after Mathieson *et al.*, 2010). The green box in the lower left were technologies identified in the initial planning as having high estimated benefit and lower costs. On the Boston Square plot these positions were not static during the project and the final techniques used included dynamic modeling, surface flux, geochemistry, cement CO$_2$ work, well logging, water chemistry, geomechanics, aquifer studies, 4D seismic, annulus sampling, geochemical tracers, wellhead CO$_2$ monitoring, wellbore sampling and satellite (InSAR) imaging. Technologies tested included microbiology, tilt meters, and microseismic. The final benefit/cost locations can be reviewed in Mathieson *et al.* (2010).

available, as shown in Table 3.1, and reducing cost will always be a goal, so assessing cost/benefit is important. The initial selection of monitoring tools to consider for cost/benefit analysis can be daunting. The International Energy Agency Greenhouse Gas (IEAGHG) program has developed an on-line selection process that can be a first step (IEAGHG, 2017).

One method to compare a large number of monitoring techniques, using their estimated benefits and costs for a specific site, is plotting them on a Boston Square (or Boston Chart) (Hannas, 2013).

Figure 3.4 shows the initial format of such an analysis by Mathieson *et al.* (2010) where high-benefit, low-cost monitoring options are in a quadrant labeled "just do it." These techniques were reviewed and an updated approach shown in Mathieson *et al.* (2010) was then used in monitoring. In this analysis, high-cost/low-benefit methods were avoided, low-cost/high-benefit methods routinely included, and methods in the high-cost/high-benefit quadrant were given careful

consideration. In this Boston Square, repeat 3D reflection seismic (4D seismic) was located in the high-cost/high-benefit quadrant but was included in the monitoring. Because CO$_2$ reservoirs can be complexly heterogeneous, a high-resolution approach that includes reservoir characterization and modeling is preferred (Mathieson *et al.*, 2010). In addition to understanding the monitoring techniques that reside in the "just do it" quadrant for a site, two important additional concerns are the time line for postinjection site care and determining the impact of volume and rate of CO$_2$ injection on the monitoring design.

As reviewed earlier, monitoring technologies are divided naturally between direct and indirect methods. Geophysical methods in general are indirect methods of monitoring. Direct measurement monitoring tools include subsurface determinations of pressure or temperature, measured at a specific *in situ* location of the subsurface. Indirect measurements, well constrained by physical principals and models, are analyzed and interpreted to estimate the

63

subsurface material property of interest. Electrical or seismic monitoring methods can be interpreted to derive subsurface fluid saturation (between brine and supercritical CO_2) using laboratory calibrated variations in electrical conductivity or bulk modulus along with petrophysical models. A value of information (VOI) study by Trainor-Guitton *et al.* (2013) provides another example in which calibrated electrical resistivity was analyzed to accurately estimate subsurface water characteristics such as total dissolved solids (TDS) and pH.

In addition to direct and indirect methods, monitoring techniques can be characterized as being active, methods that require a source, or passive, methods that simply observe the system. An example of a high-resolution active monitoring approach is borehole seismic methods such as vertical seismic profiles (VSP), 3D-VSP (Cheng *et al.*, 2010), and continuous active-source seismic monitoring (CASSM) (Daley *et al.*, 2008). These methods have higher spatial resolution, better signal to noise characteristics, easier direct calibration, and more rapid time-lapse monitoring near wellbores (Coueslan, 2013) compared with surface seismic surveys. Technology developments such as fiber optic sensing for VSP (e.g., Daley *et al.*, 2015) are improving the cost/benefit ratio of active source methods.

An example of a passive monitoring approach would be the use of microgravity. Microgravimetry is a cost-effective and relatively rapid means of observing changes in density distribution in the subsurface, particularly those changes caused by the migration of fluids with differing densities. This is a passive monitoring technique because the variation within the earth system itself creates a time lapse anomaly observable without active probing. Time-lapse gravity has been used since 1961 (Allis and Hunt, 1986), but substantial improvements in gravimeter technology, and the advent of highly precise Global Positioning Satellite (GPS) systems, have led to rapid growth in microgravity applications. This technology has been successfully applied to carbon sequestration at Sleipner (Alnes *et al.*, 2008), aquifer recharge studies in Utah and elsewhere (Chapman *et al.*, 2008; Davis *et al.*, 2008), and to hydrocarbon EOR surveillance in Alaska (Ferguson *et al.*, 2007). This method gives direct estimates of densities and its implementation cost is low among geophysical methods. Drawbacks are

that sensitivity decreases with depth and the solution is nonunique. Microgravimetry is most useful when combined with other methods such as surface deformation and seismic.

Borehole methods require wellbore access for receiver placement and provide information at generally higher spatial resolution in a more limited subsurface region. When the goals of monitoring need a localized high-resolution method, perhaps a specific depth zone for risk reduction, borehole techniques are warranted.

What Depths/Zones to Monitor?

One aspect of developing monitoring goals is determining what area and depths to monitor. For performance assessment the reservoir depth zone is one target. The reservoir depths are defined by the top of the permeable/porous storage zone – or alternately as the base of the impermeable seal above the reservoir – and by the bottom of the reservoir formation (or the base of expected CO_2 plume). There are likely to be goals requiring deeper monitoring, below the reservoir, such as monitoring fluid pressure change in basement rocks which may cause induced seismicity on unknown faults. Also, there is likely to be above zone monitoring intervals (AZMI, e.g., Hovorka, 2013), including multiple seals above the reservoir defined as part of a storage zone. For performance assessment, the spatial area (the "footprint") of reservoir monitoring will be determined by reservoir/injection modeling, with the expected plume dimensions being a likely minimum size of monitoring. The depths monitored for performance assessment could also include AZMI. For regulatory compliance, the area and depths will be determined by regulatory definitions such as the US EPA's AOR, which is also initially tied to reservoir modeling. The AOR is largely controlled by the pressure footprint, as well as the CO_2 plume, with the pressure area being larger. Groundwater depths, especially those depth zones with usable drinking water, are also likely to need monitoring to meet regulatory requirements. For public assurance, areas of surface development, housing, industry, etc., may extend the area or cause more detailed effort on some subset of the footprint. The risk reduction goals will have site specific areas of focus such as existing wellbores or fault locations.

The size of the monitoring area may also change with time. Initial characterization may be larger than the expected plume footprint to allow for unexpected flow and to assess the area of the pressure plume (which will be larger than the CO$_2$ plume). Repeat monitoring, such as surface seismic, may start with a smaller area and expand to larger areas as the CO$_2$ plume and pressure fronts grow. The goals will determine whether CO$_2$ presence or pressure or some other factor defines the areal extent of monitoring. The choice of monitoring tools used will consider the depth and areal extent of required monitoring as well as the target. For example, groundwater monitoring is more likely to use geochemical sampling and electrical geophysics, as opposed to seismic or deformation monitoring (of course site-specific conditions and cost/benefit are deciding factors between tools).

Quantitative versus Qualitative Monitoring

When addressing the goals of monitoring with specific monitoring tools, one important differentiator is quantitative versus qualitative monitoring. An example of the difference is simple detection of CO$_2$ at any concentration (qualitative) versus estimating CO$_2$ mass in a given volume (quantitative). In general, qualitative monitoring is likely to have lower costs. Qualitative monitoring is also likely to be indirect monitoring, while direct monitoring (e.g., geochemical sampling) is more likely to provide quantitative information. For some goals, qualitative monitoring will be sufficient and will have lower cost/benefit ratio. The same data may be used for both qualitative and quantitative assessment with the difference being analysis effort. For example, 4D surface seismic monitoring can provide qualitative detection with less costly processing. Quantitative assessment of CO$_2$ saturation per unit volume can also be obtained from 4D seismic but more intensive processing and analysis is required, often including other information such as core studies and calibrated rock physics models of the reservoir rock. Ultimately, the most certain quantitative assessment involves drilling wells to have subsurface access for direct measurements such as geochemical sampling and pressure measurement. Indirect or remote monitoring will have larger uncertainty.

An example of using qualitative versus quantitative monitoring is leakage assessment. Definitions of leakage will probably be project specific and have regulatory definitions. Migration of CO$_2$ outside the initial injection zone is not necessarily defined as leakage, especially if the storage zone includes multiple seals. Here leakage means migration of CO$_2$ beyond predetermined project boundaries. Leakage assessment is a likely goal for risk reduction, performance assessment, regulatory compliance, and public assurance. Initial monitoring for leakage assessment can be qualitative – is a leak detected – rather than quantitatively monitoring leakage amounts. However, if a leak is detected then the monitoring will need to be more quantitative in order to guide decisions. Mitigation of leakage (if required) will have designs dependent on measurement of leakage rates, mass, volume, location, etc. Monitoring assessment of leakage mass flow rates will have a value of information that may be considered with the cost of monitoring tools to decide on both monitoring deployment and intervention techniques to stop/reduce the leakage.

How Often to Monitor?

The temporal sampling of monitoring results is an important consideration. Monitoring tools can be separated between those that are essentially continuous and those that have some discrete time period between measurements. For those tools with discrete time intervals, the goals of monitoring will inform the decision of how often to monitor. Because essentially all modern data collection is digital with some discrete time sampling, continuous monitoring can be defined as data collected frequently enough to capture all changes related to processes of interest. For example, a downhole (or surface) pressure gauge is digitally sampled, typically at intervals of seconds to minutes, with all changes in pressure expected to happen at intervals at least an order of magnitude longer than the sampling interval. With modern digitizing and recording systems, there is essentially no cost savings for longer intervals of sampling so we expect pressure to be a continuous recording. Another example would be tilt meters used to measure surface deformation. We expect tilt meters to record data at intervals at least an order of magnitude smaller than any induced surface deformation (which is typically hours to months).

Noncontinuous monitoring will require selection of a time interval. Two examples are surface seismic and geochemical fluid sampling. Both require active use of instrumentation and both have some time required for data processing and analysis. These types of monitoring are likely to have variable periods between data collection. Initially some interval will be selected, perhaps selected because of regulatory requirement or public assurance goals. When monitoring results conform to predictions of reservoir models (or other process models) the sampling interval will be maintained or may be extended to reduce cost. If some anomaly is detected, frequent repeated monitoring may be used to confirm and understand the anomalous results.

In many areas of monitoring, the separation between continuous and discrete measurements is being reduced or eliminated in current research. For example, the use of borehole fluid sampling with a U-tube device (Freifeld *et al.*, 2005) has reduced the time interval and incremental cost of downhole samples by using a permanent or semi-permanent installation which allows sampling on hourly intervals. Monitoring tools in the oil and gas industry are being considered in the context of "permanent reservoir monitoring." In this context, permanent typically means the instrumentation is permanent but the data collection interval may still be variable depending on goals and costs. A permanent installation of instrumentation does facilitate more frequent monitoring and, often, higher quality results (an example for surface seismic monitoring of injected CO_2 is in Urosevic *et al.* [2010] and Pevezner *et al.* [2011, 2017]). Advances in technology, such as distributed acoustic sensing with fiber optics, have the potential to greatly reduce the cost of repeat surveys via low-cost permanent sensing cables (Daley *et al*, 2015).

Some monitoring may have its own protocol for time intervals that can depend on other monitoring data. For example, fluid sampling in an observation well may trigger seismic monitoring if CO_2 is detected when not expected. Because the cost of seismic monitoring is typically higher, this is also an example of cost-driven monitoring intervals.

Because the selection of a monitoring interval is likely cost constrained, the goals of the project will be important factors in considering the cost/benefit of time intervals. In some cases, regulatory requirements will define the initial monitoring interval for various tools. For example, regulatory compliance for postinjection site care can determine time intervals that are required for the project and site. More frequent monitoring may be used because of public assurance, performance assessment, or risk reduction goals. For performance assessment, the monitoring interval will be considered in an iterative loop with the reservoir model – as monitoring results confirm or modify the model assumptions, the monitoring interval and model parameters will change. Thus, the monitoring interval, like the monitoring method used, is subject to change over the life of a storage project.

References

Allis, R. G., and Hunt, T. M. (1986). Analysis of exploitation-induced gravity changes at Wairakei Geothermal Field. *Geophysics*, **51**(8): 1647–1660.

Alnes, H., Eiken, O., and Stenvold, T. (2008). Monitoring gas production and CO_2 injection at the Sleipner field using time-lapse gravimetry. *Geophysics*, **73**(6): WA155–WA161.

Bergmann, P., Schmidt-Hattenberger, C., Kiessling, D., *et al.* (2012). Surface-downhole electrical resistivity tomography applied to monitoring of CO_2 storage at Ketzin, Germany. *Geophysics*, **77**(6): B253–B267.

Calvert, R. (2005). Insights and methods for 4D reservoir monitoring and characterization. *Society of Exploration Geophysicists Distinguished Instructor Series*, No. 8.

Cantillo, J. (2011). A quantitative discussion on time-lapse repeatability and its metrics: Expanded Abstracts, *Society of Exploration Geophysicists*. http://dx.doi.org/10.1190/segam2012-0719.1

Carrigan, C. R., Ramirez, A. L., Newmark, R. L., Aines, R., and Friedmann, S. J. (2009). Application of ERT for tracking CO_2 plume growth and movement at the SECARB Cranfield site. In 8th Annual Conference on Carbon Capture & Sequestration, Pittsburgh, PA (Vol. 4, No. 7).

Castagna, J. P., and Backus, M. M. (1993). Offset-dependent reflectivity – theory and practice of AVO analysis. *Society of Exploration Geophysicists Investigations in Geophysics* No. 8.

Castagna, J. P., and Swan, H. W. (1997). Principles of AVO crossplotting. *Leading Edge*, **16**(4): 337–344.

Castagna, J. P., Swan, H. W., and Foster, D. J. (1998). Framework for AVO gradient and intercept interpretation. *Geophysics*, **63**(3): 948–956.

Chapman, D.S., Sahm, E., and Gettings, P. (2008). Monitoring aquifer recharge using repeated high-precision gravity measurements: A pilot study in South Weber, Utah. *Geophysics*, **73**(6): WA83–WA93.

Cheng, A., Huang, L., and Rutledge, J. (2010). Time-lapse VSP data processing for monitoring CO$_2$ injection. *Leading Edge*, **29**: 196–199.

Coueslan, M. (2013). Monitoring CO$_2$ injection at the Illinois Basin – Decatur Project: Second monitor survey. Presentation at MGSC Annual Meeting September 2013. http://sequestration.org/resources/PAGOct2013Presentations/09-Time-lapse_3DVSP-Mcoueslan_Sept2013.pdf

Dafflon, B., Wu, Y., Hubbard, S. S., *et al.* (2012). Monitoring CO$_2$ intrusion and associated geochemical transformations in a shallow groundwater system using complex electrical methods. *Environmental Science and Technology*, **47**(1): 314–21. http://dx.doi.org/10.1021/es301260e

Daley, T. M., Myer, L. R., Peterson, J. E., Majer, E. L., and Hoversten, G. M. (2008). Time-lapse crosswell seismic and VSP monitoring of injected CO$_2$ in a brine aquifer. *Environmental Geology*, **54**: 1657–1665. DOI:10.1007/s00254-007-0943-z.

Daley, T. M., Miller, D. E., Dodds, K., Cook, P., and Freifeld, B. M. (2015). Field testing of modular borehole monitoring with simultaneous distributed acoustic sensing and geophone vertical seismic profile at Citronelle, Alabama. *Geophysical Prospecting*, **64**(5): 318–1334. DOI:10.1111/1365–2478.12324.

Davis, K., Li, Y., and Batzle, M. (2008). Time-lapse gravity monitoring: A systematic 4D approach with application to aquifer storage and recovery. *Geophysics*, **73**(6): WA61–WA69.

Doetsch, J., Kowalsky, M. B., Doughty, C., *et al.* (2012). Fully coupled hydrogeophysical inversion of CO$_2$ migration data in a deep saline aquifer. In SEG-AGU Hydrogeophysics Workshop, July 8–11, 2012, Boise State University, Boise, Idaho.

Doetsch, J., Kowalsky, M. B., Doughty, C., *et al.* (2013). Constraining CO$_2$ simulations by coupled modeling and inversion of electrical resistance and gas composition data. *International Journal of Greenhouse Gas Control*, **18**: 510–522.

Duffaut, K., and Landrø, M. (2007). Vp/Vs ratio versus differential stress and rock consolidation: A comparison between rock models and time-lapse AVO data. *Geophysics*, **72**(5): C81–C94.

EPA. (2013). Underground Injection Control (UIC) Program Class VI Well Testing and Monitoring Guidance, Office of Water (4606 M), EPA 816-R-13-001. www.epa.gov/safewater

EPA. (2015). Subpart RR: Geologic sequestration of carbon dioxide, Greenhouse Gas Reporting Program (GHGRP). http://www2.epa.gov/ghgreporting/subpart-rr-geologic-sequestration-carbon-dioxide

EPA. (2016). Greenhouse Gas Reporting Program (GHGRP), 2016. www.epa.gov/ghgreporting

EU. (2009a). Implementation of the CCS Directive, Guidance Documents 1, 2, 3 and 4. http://ec.europa.eu/clima/policies/lowcarbon/ccs/implementation/documentation_en.htm

EU. (2009b). Directive 2009/31/EC of the European Parliament and of the Council of 23 April 2009 on the geological storage of carbon dioxide and amending Council Directive 85/337/EEC, Official Journal of the European Union, 5.6.2009. http://eur-lex.europa.eu/LexUriServ/LexUriServ.do?uri=OJ:L:2009:140:0114:0135:EN:PDF

Ferguson, J. F., Chen, T., Brady, J., Aiken, C. L. V., and Seibert, J. (2007). The 4D microgravity method for waterflood surveillance II: Gravity measurements for the Prudhoe Bay reservoir, Alaska. *Geophysics*, **72**(2): I33–I43.

Freifeld, B. M., Trautz, R. C., Kharaks, Y. K., *et al.* (2005). The U-tube: A novel system for acquiring borehole fluid samples from a deep geologic CO$_2$ sequestration experiment. *Journal of Geophysical Research*, **110**: B10203.

Goodway, B., Chen, T., and Downton, J. (1997). Improved AVO fluid detection and lithology discrimination using Lamé petrophysical parameters; $\lambda\rho$, $\mu\rho$ and λ/μ fluid stack", from P and S inversions. Expanded Abstracts, *Society of Exploration Geophysicists*, 183–186.

Hannas, S. D. (2013). Monitoring the geological storage of CO$_2$. Geological storage of carbon dioxide (CO2). *Geoscience, technologies, environmental aspects and legal frameworks*, **54**. Duxford, UK: Woodhead Publishing Series in Energy, 68–96.

Harbert, W., Purcell, C., and Mur, A. (2011). Seismic reflection data processing of 3D surveys over an EOR CO$_2$ injection. *Energy Procedia*, **4**: 3684–3690.

Harbert, W., Daley, T. M., Bromhal, G. Sullivan, C., and Huang, L. (2016). Progress in monitoring strategies for risk reduction in geologic CO$_2$ storage. *International Journal of Greenhouse Gas Control*, **51**: 260–275. DOI:10.1016/j.ijggc.2016.05.007.

Hovorka, S. D. (2013). Three-million-metric-ton-monitored injection at the Secarb Cranfield Project: Project update. *Energy Procedia*, **37**: 6412–6423. http://dx.doi.org/10.1016/j.egypro.2013.06.571

IEAGHG. (2017). Monitoring selection tool. http://ieaghg.org/ccs-resources/monitoring-selection-tool1

IPCC. (2005). IPCC Special Report on Carbon Dioxide Capture and Storage. Prepared by Working Group III of the Intergovernmental Panel on Climate Change [Metz, B., O. Davidson, H. C. de Coninck, M. Loos, and L. A. Meyer (eds.)]. Cambridge University Press, Cambridge, United Kingdom and New York, NY, USA, 442 pp.

Johnson, D. H. (2013). Practical applications of time-lapse seismic data. *Society of Exploration Geophysics, Distinguished Instructor Series*, No. 16.

Kirksey, J. (2012). Deep well monitoring and verification at the Illinois Basin Decatur Project. Presentation at MGSC Annual Meeting September 2012. http://seques tration.org/resources/PAGSept2012Presentations/06-JimKirksey_PAG2012.pdf

Kowalsky, M. B., Doetsch, J., Commer, M., *et al.* (2016). Coupled inversion of hydrological and geophysical data for improved prediction of subsurface CO_2 migration. Lawrence Berkeley National Laboratory Report, NRAP-TRS-III-004–2016, Level III Technical Report Series.

Kragh, E., and Christie, P. (2002). Seismic repeatability, normalized RMS, and predictability. *Leading Edge*, **21**: 640–647.

Landrø, M. (2001). Discrimination between pressure and fluid saturation changes from time-lapse seismic data. *Geophysics*, **66**(3): 836–844.

Landrø, M., Hafslund Veire, H., Duffaut, K., and Najjar, N. (2003). Discrimination between pressure and fluid saturation changes from marine multicomponent time-lapse seismic data. *Geophysics*, **68**(5): 1592–1599.

Lumley, D. (2001). Time-lapse seismic reservoir monitoring. *Geophysics*, **66**: 50–53.

Lumley, D. (2010). 4D seismic monitoring of CO_2 sequestration. *Leading Edge*, **29**(2): 150–155.

Lumley, D., Adams, D., Wright, R., Markus, D., and Cole, S. (2008). Seismic monitoring of CO_2 geo-sequestration: Realistic capabilities and limitations. Expanded Abstracts, *Society of Exploration Geophysicists*, 2841–2845.

Mathieson, A., Midgley, J., Dodds, K., Wright, I., Ringrose, P., and Saoul, N. (2010). CO_2 sequestration monitoring and verification technologies applied at Krechba, Algeria. *Leading Edge*, **29**(2): 216–222.

Mavko, G., Tukerji, T., and Dvorkin, J. (2009). *The rock physics handbook*, 2nd edn. Cambridge: Cambridge University Press.

Mullins, O. G. (2008). *The physics of reservoir fluids: Discovery through downhole fluid analysis*. Houston: Schlumberger.

Mur, A., Purcell, C., Soong, Y., *et al.* (2011). Integration of core sample velocity measurements into a 4D seismic survey and analysis of SEM and CT images to obtain pore scale properties. *Energy Procedia*, **4**: 3676–3683.

Nelson, E. B., and Guillot, D. (2006). *Well cementing*, 2nd edn. Houston: Schlumberger.

Peters, D. (2007). CO2 geological storage: Methodology and risk management process. Presented at NHA Hydrogen Conference, March 20, 2007.

Pevzner, R., Shulakova, V., Kepic, A., and Urosevic, M. (2011). Repeatability analysis of land time-lapse seismic data: CO2CRC Otway pilot project case study. *Geophysical Prospecting*, **59**: 66–77.

Pevzner, R., Urosevic, M., Popik, D., *et al.* (2017). 4D surface seismic tracks small supercritical CO_2 injection into the subsurface: CO2CRC Otway Project. *International Journal of Greenhouse Gas Control*, **63**: 150–157. https://doi.org/10.1016/j.ijggc.2017.05.008

Purcell, C., Harbert, W., Soong, Y., *et al.* (2009). Velocity measurements in reservoir rock samples from the SACROC unit using various pore fluids and integration into a seismic survey taken before and after a CO_2 sequestration flood. *Energy Procedia*, **1**: 2323–2331.

Purcell, C., Mur, A., Soong, Y., McLendon, T. R., Haljasmaa, I. V., and Harbert, W. (2010). Integrating velocity measurements in a reservoir rock sample from the SACROC unit with an AVO proxy for subsurface supercritical CO2. *Leading Edge*, **29**(2): 192–195.

Sayers, C. M. (2006). An introduction to velocity-based stress changes in sandstones. *Leading Edge*, **24**(12): 1262–1266.

Sayers, C. M. (2010). Geophysics under stress: Geomechanical applications of seismic and borehole acoustic waves. *Society of Exploration Geophysicists Distinguished Instructor Series*, No. 13.

Schlumberger. (1989). Cased hole log interpretation: Principles/applications. www.slb.com/resources/publications/books/ch_lipa.aspx

Schlumberger. (1991). Log interpretation: Principles/applications. www.slb.com/resources/publications/books/lipa.aspx.

Shapiro, S. (2015). *Fluid induced seismicity*. Cambridge: Cambridge University Press.

Simm, R., and Bacon, M. (2014). *Seismic amplitude: An interpreter's handbook*. Cambridge: Cambridge University Press.

Trainor-Guitton, W. J., Ramirez, A., Yang, X., Mansoor, K., Sun, Y., and Carroll, S. (2013). Value of information methodology for assessing the ability of electrical resistivity to detect CO_2/brine leakage into a shallow aquifer. *International Journal of Greenhouse Gas Control*, **18**: 101–113. DOI:10.1016/j.ijggc.2013.06.018.

Urosevic, M., Pevzner, R., Kepic, A., and Wisman, P. (2010). Time-lapse seismic monitoring of CO$_2$ injection into a depleted gas reservoir-Naylor Field, Australia. *Leading Edge*, **29**(2): 164–169.

Wright, I. (2011). In Salah CO$_2$ storage JIP lessons learned. In 10th Annual Conference on Carbon Capture and Sequestration, Pittsburgh, PA, May 2–5, 2011.

Wright, I., Ringrose, P., Mathieson, A., and Eiken, O. (2009). An overview of active large-scale CO$_2$ storage projects. In Society of Petroleum Engineers, SPE 127096, SPE International Conference on CO$_2$ Capture, Storage, and Utilization, November 2–4, 2009, San Diego, CA.

4

Rock Physics of CO$_2$ Storage Monitoring in Porous Media

Thomas M. Daley

Introduction

Remote geophysical monitoring of carbon dioxide (CO$_2$) injection and storage typically starts with qualitative analysis, i.e., changes in monitoring data are interpreted for the presence of CO$_2$ in the subsurface – simply showing that we can "see" the CO$_2$. However, such detection is often insufficient for regulatory or scientific use. Once CO$_2$ is detected the next step is quantitative analysis – deriving estimates of volume and/or mass of CO$_2$ in a given volume of the reservoir. Quantitative analysis requires rock physics – theories and models. The physics and chemistry that impact monitoring the injection and long-term storage of CO$_2$ in rocks are a broad topic, with a full summary deserving its own book length treatment. In this chapter we limit the discussion to rock physics that impacts the geophysical monitoring of CO$_2$ injection and storage in porous media, with specific limiting assumptions described in the text that follows. This topic is a subset of the broad use of geophysics for time-lapse monitoring of subsurface processes. Johnston (2013) presents a good summary of seismic monitoring and defines seismic rock physics as "the link between the static and dynamic properties of a reservoir and the elastic properties of the reservoir rock" (p. 23). This link is what allows us to use remote monitoring data to detect and quantitatively estimate the amount of stored CO$_2$ in the subsurface, and holds for whatever fluid state (liquid or gas) and mixture of fluids is contained in the reservoir. However, with increasing number of fluids and fluid states the rock physics relationships become more complex and uncertainty increases. To understand rock physics relationships, as typically used in storage projects, we first describe the limitations and assumptions used.

In this chapter we restrict ourselves to focusing on realistic CO$_2$ storage scenarios and limit the range of depths considered (and thus also pressures and temperatures) to those most likely for storage, taken here as 0.8–4 km. The minimum depth of approximately 0.8 km is that which would allow supercritical phase CO$_2$ (a limit previously used by IPCC, 2005). Storage of supercritical phase CO$_2$ is assumed because the increased fluid density allows more efficient storage. The maximum depth of storage is a practical limit set by the cost of drilling and storage operations and is stated here, somewhat arbitrarily, as 4 km. Notably, we also limit our discussion with the assumption that the preexisting pore fluid is brine and thus consider only brine–CO$_2$ mixtures following injection. Brine aquifers are the target because they provide the largest potential storage capacity for geological sequestration of CO$_2$ (IPCC, 2005). We are avoiding the issues related to injection in reservoirs with hydrocarbons including preexisting natural gas (typically methane [CH$_4$]). While multiple pore fluids add complexity, much work is available from the oil and gas industry in this area. To date, all large-scale storage projects (not enhanced oil recovery [EOR] projects) have targeted CO$_2$ injection in brine formations, such as the long running Sleipner project (e.g., Arts et al., 2004; Chadwick et al., 2016). The Otway project Stage 1, is one small (60 000-tonne) pilot that did target a depleted methane reservoir for storage, and this work is well documented by Cook (2014). To further constrain the scope of this chapter, we limit ourselves to the rock physics controlling remote geophysical monitoring (e.g., reflection seismic imaging) and do not consider issues specific to core measurements, well logging, or direct CO$_2$ detection (geochemical sampling). However, it is important to note that many of the parameters needed for rock physics calculations, and thus for interpretation of remote monitoring, come from core and well logging studies. Other chapters in this book will consider some aspects excluded here.

The most widely used remote geophysical monitoring technique is active-source seismic (both surface and borehole based measurements) with additional work in electrical monitoring techniques (mostly electrical resistance tomography [ERT], again surface and borehole; see Chapter 9 by Gasperikova and Commer, this volume) and deformation (geomechanical) monitoring, typically with surface-based measurements (e.g., interferometric synthetic aperture radar [InSAR], GPS, and tilt monitoring surveys; see Chapter 6 by Vasco *et al.*, this volume). Rutqvist (2012) has reviewed geomechanical monitoring of CO_2, which looks at deformation due to pore pressure changes. Gasperikova and Hoversten (2008) consider both electrical and gravity monitoring. Gravity monitoring uses changes in the Earth's gravity field due to CO_2 displacing pore fluid of different density (Chapter 7 by Eiken, this volume). Here we will focus on the rock physics important to active-source seismic and electrical monitoring.

Seismic Monitoring: Elastic Moduli

A fundamental measurement in active seismic monitoring is the travel time of a seismic wave propagating between a controlled source and a sensor (typically one of an array of sensors). This travel time is controlled by the elastic wave velocity of the subsurface. A second important measurement is the amplitude of waves reflected from velocity interfaces. Reflection amplitudes are controlled by the impedance contrast, where impedance is the product of formation velocity and density. For wave propagation in a linear elastic medium, the wave velocity is controlled by the elastic moduli, originally introduced by Robert Hooke as Hooke's "Law of Nature" for deforming materials (Hooke, 1678), which linearly related stress and strain. Two elastic moduli for isotropic linear elastic rocks, λ and μ, were defined by G. G. Stokes (1845), and used along with density, ρ, to define the velocity of P- and S-waves. Further developments in elasticity, and the various elastic moduli, are summarized in many books (e.g., Mavko *et al.*, 1998; Aki and Richards, 2002). For seismic monitoring of CO_2, the bulk modulus, K, where $K = \sqrt{(\lambda + 2\mu)}$, is perhaps the most important elastic moduli because of its sensitivity to pore fluid. The P-wave seismic velocity (V_p) of an isotropic rock volume is given by:

$$V_p = \sqrt{\left(K_{\text{sat}} + \frac{4\mu}{3}\right)/\rho} \qquad (4.1)$$

where the bulk modulus K_{sat} is modified by the saturating fluids in the rock pore space, including CO_2.

The shear-wave velocity is given by

$$V_s = \sqrt{\mu/\rho} \qquad (4.2)$$

and depends on the shear modulus of the rock, μ, which is assumed to be unaffected by fluids, and the rock density, ρ, which is affected by fluids in the rock. Thus, understanding the effect of CO_2 on seismic velocity requires understanding the effect of CO_2 on elastic moduli K and ρ.

Common use of elastic moduli for monitoring fluids in rocks began with Fritz Gassmann's pioneering study of elasticity in porous media (Gassmann, 1951) showing the impact of fluid-filled pores on the bulk modulus and the resultant seismic velocity. Analysis of pore fluid changes (the "fluid substitution problem") such as CO_2 displacing brine typically is called "Gassmann substitution" and has been restated and summarized often, including by Berryman (1995), Mavko *et al.* (1998), Wang (2001), and Smith *et al.* (2003). These authors take effort to remind readers of underlying assumptions (such as homogeneous isotropic rock with fully connected pore space and zero viscosity fluid) as well as usage pitfalls; these should be heeded as they often impact quantitative analysis of seismic data. Wang (2001) summarizes the assumptions as follows:

1. The rock (both the matrix and the frame) is macroscopically homogeneous.
2. All the pores are interconnected or communicating.
3. The pores are filled with a frictionless fluid (liquid, gas, or mixture).
4. The rock–fluid system under study is closed (undrained).
5. The pore fluid does not interact with the solid in a way that would soften or harden the frame.

These assumptions are known to be problematic in experimental work (field and laboratory scale). However, in general, Gassmann's assumptions are assumed to hold for unconsolidated clastic rocks at low seismic frequencies (<100 Hz). This is the case for typical surface seismic monitoring of saline storage of CO_2, and thus we follow the Gassmann equation approach to accurately determine the relationship between the material properties of CO_2 (such as compressibility and density) and the corresponding seismic wave velocity of rocks containing CO$_2$. The simplifying assumptions used allow estimation of partial CO$_2$ saturation with only seismic P-wave velocity changes. If more information is available, then more complete (and complex) rock physics theories can be used. For example, the effects of fluid pressure can be important for geophysical monitoring. Recently Pride *et al.* (2016) developed "explicit analytical models for how the seismic velocities, porosity, electrical conductivity and fluid-flow permeability vary with effective stress," where effective stress is the difference between confining stress and pore fluid pressure.

The notation of elastic moduli varies somewhat between authors and problems addressed. We will start with notation following Smith *et al.* (2003). The fundamental relationship in Gassmann fluid substitution expresses the bulk modulus of a saturated rock, K_{sat}, in terms of the "dry frame" bulk modulus, K^*, the bulk modulus of the minerals making the rock frame, K_0, the bulk modulus of the pore fluid, K_{fl}, and the rock porosity, ϕ:

$$K_{sat} = K^* + \frac{\left(1 - \frac{K^*}{K_0}\right)^2}{\frac{\phi}{K_{fl}} + \frac{(1-\phi)}{K_0} - \frac{K^*}{K_0^2}} \quad (4.3)$$

The bulk rock density, ρ_B, is found from the frame or grain density, ρ_g, and the density of the saturating fluids, ρ_{fl}, as: $\rho_B = \rho_g(1 - \phi) + \rho_{fl}\phi$. The shear modulus, μ, is generally assumed to be unaffected by fluid properties, i.e., an assumption of no geochemical alteration of cement or grains by the fluids – an assumption questioned for some CO$_2$ injections and discussed later.

While Eq. (4.3) shows the effect of a single pore fluid on the bulk modulus K_{sat} and on V_p via Eq. (4.1), we need an equivalent K_{fl} and ρ_{fl} for two fluids filling the pore space, such as for CO$_2$ and brine. The fluids will have variable partial saturations S_w (brine) and S_g (CO$_2$), where $S_w + S_g = 1$. Assuming the pore space has two homogeneous, uniformly distributed fluids (brine and CO$_2$), with individual bulk moduli K_w (brine) and K_g (CO$_2$), the Reuss (isostress) average as described by Wood's equation (Mavko *et al.*, 1998) gives the combined-fluid bulk modulus as:

$$K_{fl} = \left[\frac{S_w}{K_w} + \frac{S_g}{K_g}\right]^{-1} \quad (4.4)$$

and similarly, the fluid density is a saturation weighted average:

$$\rho_{fl} = S_w\rho_w + S_g\rho_g \quad (4.5)$$

Equations (4.1)–(4.5) show that changes in P-wave velocity from CO$_2$ displacing brine in the pore space are determined by the bulk modulus and density of brine and CO$_2$ and their partial saturations (other rock properties such as K^* and μ are determined from core or well logs and typically assumed to be static). While supercritical CO$_2$ has a range of compressibility ($1/K_g$) and density, for the typical pressures and temperatures of storage depths the compressibility is most often the dominant factor controlling P-wave velocity change from brine displacement (e.g., as described in Figure 4.1). It is fundamental to seismic monitoring of CO$_2$ storage that changes in CO$_2$ saturation due to injection can be determined from changes in seismic (P-wave) velocity using these equations, within the given assumptions. To apply these relationships for a specific site, knowledge of the brine and CO$_2$ properties is needed.

Brine and CO$_2$ Properties

For any given storage project, with site-specific subsurface conditions, the properties of brine and CO$_2$ in the reservoir need to be understood for rock physics analysis. For brine, the most commonly used estimates come from work published by Batzle and Wang (1992), who developed relationships for the density and sonic velocity of brines as a function of pressure, temperature, and salinity (for sodium chloride [NaCl]). These relationships are Eqs. 4.27b and 4.29 in Batzle and Wang (1992) and are also summarized in Mavko *et al.* (1998). The properties of CO$_2$ are most commonly taken from the database of the National Institute of Standards and Technology (NIST). The NIST results for CO$_2$ density and sonic velocity are available in tabular form from the website: http://webbook.nist .gov/chemistry/fluid/, which is periodically updated with recent improvements in data and equations of state for CO$_2$ and other fluids. An example of calculations of brine and CO$_2$ properties for a storage site is shown in Figure 4.1, which has isothermal density and bulk modulus for a range of pressures corresponding to the Cranfield, Mississippi, DAS project site (Hovorka *et al.*, 2013). In this example we see that density contrast is large and variable over a range of pore pressures, but the bulk modulus contrast is an order of magnitude larger, which is typical for supercritical CO$_2$ at depth.

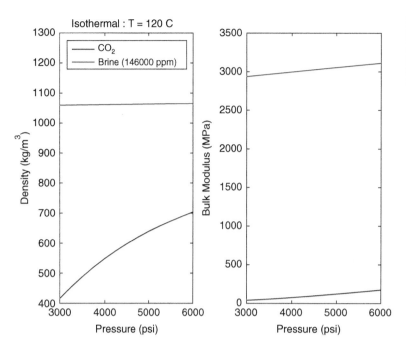

Figure 4.1 Isothermal ($T = 120°C$) density (left) and bulk modulus (right) for brine (200 ppt NaCl) and CO_2. The temperature, brine properties, and pressure range are based on a CO_2 storage site near Cranfield, Mississippi. (Courtesy Jonathan Ajo-Franklin.)

In general, the acoustic properties of brine–CO_2 solutions for conditions of geological storage have limited experimental data. Also, while the presence of pure pore filling phases of brine and CO_2 are common assumptions for Gassmann equation based calculations, it should be noted that CO_2 will dissolve into brines at variable rates, creating a mixed phase. This dissolution is typically considered as part of reservoir modeling of injection and flow, and, importantly, leads to a slightly acidic solution including carbonic acid (H_2CO_3). This can lead to geochemical alteration of the reservoir rock frame (matrix), with implications discussed in the text that follows. Furthermore, some trace amounts of hydrocarbons (gas or liquid) can be found in reservoirs considered "saline," and this can impact the rock physics, and be a "pitfall" in quantitative interpretation. While we are not considering these complications here, they should be assessed for each project site.

Patchy Saturation

As stated earlier, Gassmann fluid substitution has many simplifying assumptions which can be modified when more information is available. One of the most common extensions of the first-order Gassmann analysis described in the foregoing is to consider the impact when brine and CO_2 are not homogeneously distributed in all pore space. This assumption is implicit in Wang's (2001) assumption 3 listed earlier. Rock physics theories that consider inhomogeneous distributions are typically termed "patchy saturation" models because the individual fluids have isolated "patches." As we know that an assumption of homogeneous fluid mixing is not valid for the naturally heterogeneous subsurface, using patchy saturation models is advisable when possible (for a case study, see Chapter 18 by Grude and Landrø, this volume). These models lead to frequency dependence in both seismic velocity and attenuation. Gassmann's assumption is valid only for very long wavelengths (low frequencies) as compared to the scale of fluid patches. Because the patch size is unknown at field scale, more information is needed to constrain the velocity–saturation relationship. Measurement of attenuation along with velocity (and their frequency dependence) is thus a way to obtain more accurate estimates of partial CO_2 saturation from seismic monitoring data. The frequency dependence becomes important when using seismic data with higher frequencies such as crosswell seismic measurements (typically 500–2000 Hz) or sonic logging data (typically 10 000 Hz or higher), rather than surface seismic (typically below 150 Hz).

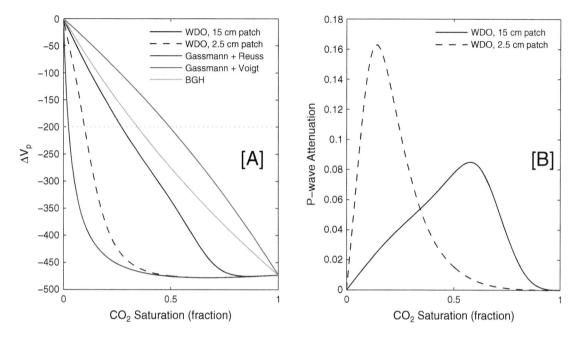

Figure 4.2 (A) Change in P-wave velocity as a function of CO_2 saturation for various rock physics models and (B) P-wave attenuation as a function of CO_2 saturation for two patch sizes in the WDO model. (From Daley *et al.*, 2011.)

An early patchy saturation model was proposed by White (1975), with corrections made by Dutta and Seriff (1979). White's model assumes a homogeneous rock frame with concentric spherical fluid patches, a central sphere of one fluid and an outer spherical shell saturated with a second fluid phase, in our case brine and supercritical CO_2.

White's model also assumes that the seismic wavelength is larger than the characteristic patch (sphere) dimension and the results are dependent on seismic wavelength.

Following the early results of White, the patchy saturation model has been generalized to more realistic fluid distributions. Recent summaries of patchy saturation models include Toms *et al.* (2007) and Müller *et al.* (2010). Pride (2005) and Pride *et al.* (2004) use a characteristic length scale, L, for patches of uniformly spaced but arbitrarily shaped fluid distribution. This model allows for the pressure redistribution between fluids, over a length scale, during compression and dilation of a passing P-wave, without specifying the geometry of the distribution within L. The Pride model also allows for and requires estimates of fluid properties such as viscosity. This length L is then related to the dependence of velocity and

attenuation on wavelength (and thus frequency). A further extension allowing for variable (random) length scales of patch size has been developed by Müller and Gurevich (2004) and Toms *et al.* (2007).

For patchy saturation models the change in V_p due to CO_2 displacing brine will be less than the Gassmann model with Reuss fluid mixing (as described earlier), so Gassmann with Reuss serves as a lower bound on ΔV_p and there is an upper bound defined by "Voigt" fluid mixing (Mavko *et al.*, 1998), where the bulk modulus is given by

$$K_{fl} = S_w K_w + S_g K_g \tag{4.6}$$

which can be compared to Eq. (4.4). A typical comparison of the difference between rock physics models for a CO_2/brine mix is shown in Figure 4.2.

Figure 4.2(A) compares White's model (WDO), with two example patch sizes, to three other poroelastic models used to predict the properties of rocks partially saturated with CO_2 including Gassmann fluid substitution with either Reuss or Voigt effective fluid assumptions (Mavko *et al.*, 1998) and the so-called Biot–Gassmann–Hill model (BGH) (Hill, 1963), which is a quasi-static prediction for

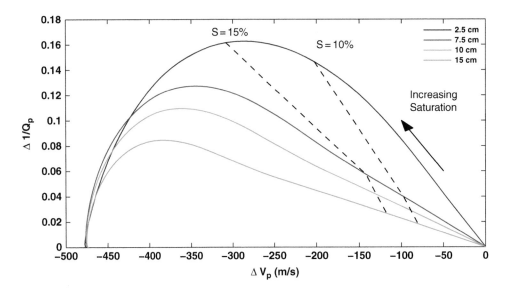

Figure 4.3 Change in P-wave attenuation ($1/Q_p$) as a function of change in P-wave velocity for four patch sizes using White's model. CO_2 saturation has characteristic values as shown by the two dashed lines for saturations of 10% and 15%. Rock properties taken from Frio formation (Daley *et al.*, 2011). (Figure courtesy of Jonathan Ajo-Franklin.)

a partially saturated medium with macroscopic patches. This example comes from monitoring a CO_2 injection test in the Frio formation in southeast Texas (Daley *et al.*, 2007). The elastic properties of the rock frame were selected from log and core information and are stated in Daley *et al.* (2011), with a $V_p =$ 2700 m/s and porosity = 25%. CO_2 and brine properties were calculated for *in situ* reservoir pressures and temperatures ($P = 15$ MPa, $T = 55°C$). The BGH model is the quasi-static prediction for a partially saturated medium with macroscopic patches. All calculations were made for a seismic frequency of 1100 Hz. This frequency is typical for crosswell seismic data, which is higher frequency than surface seismic and lower frequency than sonic well logging.

As seismic monitoring uses higher frequency to obtain better spatial resolution, the need for patchy saturation models is increased. All patchy saturation models invoke some patch size property, such as White's radius or Pride's characteristic length, which then becomes another unknown in the rock physics analysis. In practice, we typically have no prior knowledge of patch size, which leads to considerable uncertainty when attempting to estimate CO_2 saturation from changes in seismic velocity using patchy saturation models. A patch size value could come from calibration experiments on the core or log scale;

however, these scales are limited in their spatial range. Another option is to constrain the patch size by field scale measurement of other complementary geophysical properties such as P-wave attenuation, typically referred to as Q (quality factor; see Mavko *et al.*, 1998). The use of P-wave attenuation is attractive since it has complementary information as shown in Figure 4.2B. Combining the measurement of change in velocity with change in attenuation can lead to a unique value of saturation and patch size from seismic data, as shown in Figure 4.3.

Assuming that the impact of patchy saturation is accounted for, or may not be needed due to use of low-frequency data, there is another assumption from typical fluid-substitution models that can have first-order impacts on seismic monitoring data: rock frame changes.

Rock Frame Changes: Geochemical Alteration

In addition to complications and uncertainty introduced by either the assumption of homogeneous fluid mixing or invoking models for the effects of patchy saturation, the underlying assumption of constant frame properties (Wang's assumption 5 in the foregoing) has been shown to be problematic for CO_2

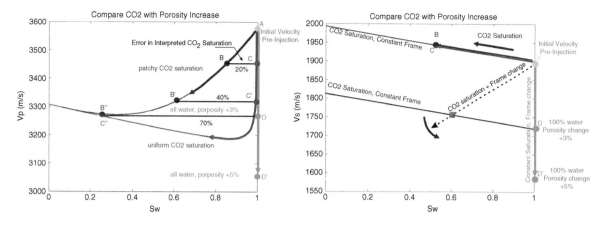

Figure 4.4 A numerical model example of seismic velocity changes from patchy and uniform fluid distribution along with rock frame change for CO_2 displacing water in pore space. (A, left) P-wave velocity (V_p) versus water saturation (S_w). (B, right) S-wave velocity (V_s) versus water saturation (S_w). (From Vanorio et al., 2010.)

injection/storage applications. Recent work has shown that constant frame assumption can lead to significant errors in calculation of elastic moduli and thus in the inverse calculation of saturation from measured moduli (e.g., Vialle and Vanorio, 2011; Lebedev et al., 2014). A summary of recent advances in understanding of CO_2 induced geochemical frame changes is given in Vanorio (2015).

When considering alteration of rock frames for CO_2 storage, we can consider the common rock types of siliciclastic and carbonate. Carbonates are found to be especially reactive to CO_2 mixing with brine as pore filling fluid. Vanorio et al. (2011) consider four geochemical "scenarios" that can alter the carbonate rock frame:

1. Mineral dissolution leading to porosity increase
2. Mineral dissolution leading to matrix cementation decrease
3. Mineral dissolution leading to both porosity increase and cementation decrease
4. Salt precipitation leading to porosity decrease

Each of these has differing, but nonnegligible, impacts on elastic moduli and seismic velocity via the rock frame properties. It is notable that rock frame properties often are derived from preinjection well logs and core samples and therefore do not allow for any of these changes. Laboratory-scale core measurements and/or post-injection well logging are needed to understand geochemical alterations.

Figure 4.4 shows the impact of patchy versus homogeneous fluid saturation along with the impact

of rock frame change on P-wave and S-wave velocities as a function of water saturation (S_w) for a water/CO_2 mixture. Figure 4.4 is from a numerical model of a carbonate reservoir at a depth of 2,580 m. This type of calculation is often used to interpret observed changes in seismic velocity (from time-lapse monitoring data) for CO_2 saturation in each volume of the reservoir, but rarely considers frame alteration.

Starting at point A in Figure 4.4(A) (the initial "preinjection" P-wave velocity) a change of 100 m/s in velocity corresponds to a 20% change in CO_2 saturation for a patchy distribution (A to B, in blue) or <1% CO_2 saturation change for a homogeneous fluid distribution (A to C, in red). Comparing A to B′ versus A to C′, a larger decrease in observed velocity (250 m/s) leads to interpreted CO_2 saturation changing either 40% or 1% for patchy or homogeneous mixing, respectively. Now consider a geochemical rock frame change, an increase in porosity of 3%. The initial, preinjection velocity drops by nearly 300 m/s (from point A to D). This example shows that a 3% change in porosity can lead to a 70% error in estimated saturation, depending on the "patchyness" of the CO_2/water distribution. A 5% change in porosity (A to D′) causes a larger velocity change (550 m/s) than any constant-frame fluid substitution model could explain. As noted by Vanorio et al. (2010), a similar change could also be caused by reduction in frame moduli due to change in grain contact stiffness (e.g., mineral/cement dissolution) without invoking porosity change. Assessing and correcting for changes in frame properties is clearly important for storage monitoring and will require more than just P-wave seismic data.

Use of Multiphysics Monitoring

The size of potential errors in interpreted CO_2 saturation indicated in Figure 4.4 demonstrates the importance of correctly accounting for all the components of rock physics impacting seismic monitoring. Understanding time-varying frame changes requires models of reactive flow and transport to be used with geochemical models and laboratory data to interpret time-lapse geophysical monitoring data. For example, recent work in geochemical modeling of CO_2 mineralization has utilized advanced imaging techniques to understand the reactive surface area of CO_2 reservoir rocks (e.g., Landrot et al., 2012).

Using multiphysics monitoring data can aid in the estimation of frame changes induced by CO_2 storage, as well as provide independent estimates of saturation, and thus improve saturation estimates from monitoring data. Examples of multiphysics monitoring include acquiring seismic P-wave and shear-wave velocities, or electrical resistivity and seismic P-wave velocity. Figure 4.4B shows the impact of having S-wave velocity data. Starting from a preinjection S-velocity at point A, with no frame change, V_S would increase with CO_2 saturation, e.g., approx. 50 m/s from point A to point B/C (the same increase for patchy or homogeneous fluid distribution). This velocity increase is due to the decrease in bulk density due to lower density CO_2 displacing water. However, when we include frame change, for example, a –3% porosity change (point D), the shear velocity would now decrease by about 150 m/s. This drop in V_s would be a very different monitoring observation than a small V_s increase, and thus provides complementary information to V_p by indicating changing frame properties.

An example of field data utilizing V_s has been described by Daley et al. (2008) and Al Hosni et al. (2016a) using time-lapse crosswell seismic velocity tomography at the Frio Brine Pilot Project (Hovorka et al., 2006). Analysis of both crosswell and vertical seismic profile (VSP) P-wave data indicated V_p decreases following CO_2 injection that were larger than could be explained by fluid substitution for the siliciclastic Frio formation (Daley et al., 2008; Al Hosni et al., 2016b). The V_s change measured by crosswell showed a decrease in velocity, not the slight increase predicted by fluid substitution. Al Hosni et al.

developed an interpretation of rock frame change utilizing measured V_s changes (and using supporting geological, well log, and geochemical data) and using a grain cementation model (Dvorkin and Nur, 1996; Avseth et al., 2000). They propose a variable reduction in contact cement (from initial 0.1% to as low as 0.01%) that leads to a reduction in frame moduli (bulk and shear) of 30–40%. From the modified frame moduli, an improved estimate of CO_2 saturation is made using patchy saturation models for a 0.2-m patch size and the 300-Hz seismic data. As stated by Al Hosni et al., "having P- and S-wave velocity time-lapse data is key to improved saturation estimates."

Electrical Conductivity

Electrical conductivity is an important rock property for remote monitoring of CO_2 storage, and it can be measured from a variety of electromagnetic (EM) methods, as summarized by Nabighian (1991) and Gasperikova and Commer (Chapter 9, this volume). EM geophysical methods cover a wide range of frequencies from DC (0 Hz) to gigahertz. Fundamental EM properties of interest include conductivity, σ, and dielectric permittivity, ε, both of which are sensitive to the electrical properties of pore fluids. For earth materials, high-frequency electromagnetic waves propagate following a classic wave equation largely controlled by permittivity, while at low frequencies propagation becomes a diffusion process largely controlled by conductivity. The transition frequency for typical crustal rocks is at about 100 kHz (Mavko et al., 1998). CO_2 storage is at depths typically accessible to EM methods below this frequency. For example, a common high-frequency EM monitoring tool is ground penetrating radar (GPR), typically at 0.1–2 GHz, which is sensitive to permittivity. However, the typical depth of penetration of GPR is centimeters to meters and thus has little application to CO_2 storage monitoring. Lesmes and Friedman (2005) review theoretical and empirical models for electrical methods, with high-frequency permittivity measurements having a focus on hydrogeophysics and application to shallow soils.

For the depths of interest to CO_2 storage (below 800 m), lower frequency methods focus on measuring the change in conductivity due to electrically resistive CO_2 displacing electrically conductive brine in the reservoir pore space. Lower frequency electrical measurements, such as electrical resistivity [ER], and its use for electrical resistance tomography [ERT], along with electromagnetic induction methods, can have a depth of penetration of 10^2 to >10^3 m and have been used for remote monitoring of injected CO$_2$. For CO$_2$ storage applications, EM methods use rock physics developed to assess and monitor reservoir fluids, typically assumed to be initially brine, and then mixed brine and CO$_2$.

The monitoring of reservoir fluids with electrical methods began with well logging (c. 1927). Widespread quantitative application followed the publication by G. E. Archie of "Archie's Law" (Archie, 1942), which is not a law, but an empirical fit of core and well log data for saturated porous rocks. Archie defines a "formation resistivity factor," F, and relates F to the bulk resistivity of a brine saturated rock, R, the resistivity of the brine, R_w, and the rock's porosity, ϕ, as

$$F = \frac{R}{R_w} = \phi^{-m} \qquad (4.7),$$

where m is an empirical exponent fitted at $m = 1.3$ for unconsolidated laboratory sand packs and at $m = 1.9$–2.0 for consolidated sandstone cores. Archie's work has been given more theoretical underpinning, for example, by Sen et al. (1981) and then Berryman (1995), who gives bounds on F and shows, using a differential effective media (DEM) approach, that for nonconducting spherical beads in a conducting fluid $m = 1.5$. Berryman also shows model dependence on porosity with the spherical DEM model is best at porosities of 25–30%, while a self-consistent effective medium theory with needle-shaped pores is best for porosities of approx. 15%.

Importantly, Archie and others following him assume that formation brine has such high conductivity that other formation fluids and gases can be assumed to have no contribution to the conductivity. Archie's data and analysis show that for water saturation, S_w, and a fully saturated rock resistivity, R_0, the *in situ* measured resistivity of a partially saturated rock, R, is given by

$$R = R_0 S_w^{-n} = \phi^{-m} R_w S_w^{-n} \qquad (4.8),$$

where n is termed the saturation exponent and $n =$ approx. 2.0 for typical sandstone reservoir rocks.

For monitoring CO$_2$ in brine-saturated formations, Archie's Eqs. (4.7) and (4.8) form the basis of quantitative interpretation of resistivity, where CO$_2$

saturation $S_{CO_2} = 1 - S_w$, and porosity and brine conductivity are taken from separate measurement, thus

$$R = \phi^{-m} R_w (1 - S_{co_2})^{-n} \qquad (4.9)$$

and

$$S_{co_2} = 1 - \left(\frac{\phi^{-m} R_w}{R} \right)^{\frac{1}{n}} \qquad (4.10)$$

For time-lapse monitoring using electrical conductivity measurements, the calculation is typically simplified further (e.g., Nakatsuka et al., 2010) by considering the ratio of fully brine saturated resistivity and partially saturated resistivity, termed the resistivity index (Guéguen and Palciauskas, 1994),

$$RI = \frac{R}{R_0} = S_w^{-n} \qquad (4.11)$$

Thus

$$S_{co_2} = 1 - \left(\frac{1}{RI} \right)^{\frac{1}{n}} \qquad (4.12)$$

Often n is simply assumed to be 2 based on general use, e.g., Carrigan et al. (2013), who cite Nakatsuka et al. (2010). Bergmann et al. (2012) also follow Nakatsuka et al. in using a ratio of time-lapse resistivity measurement to obtain CO$_2$ saturation with the saturation index n obtained from laboratory experiments at *in situ* conditions ($n = 1.62$).

The use of Archie's equations becomes problematic as the clay content increases. For example, Nakatsuka et al. (2010: 209) state that "Resistivity values from the induction log increased during the CO$_2$ injection period but, when calculated by Archie's equation it shows CO$_2$ saturation of about 10% which is lower than that estimated from neutron logging data. This low saturation is assumed as the effect of clay inclusion in the reservoir."

For reservoirs with measurable clay or shale content, surface conduction mechanisms on the rock frame become important. Various modifications to Archie's equations have been proposed, although there is not yet a fundamental understanding of the mechanism of surface conductance for arbitrary complex frame minerals (S. Pride, personal communication, 2017). To allow for clay/shale content, Archie's Eq. (4.7) is often written as $F = a\phi^{-m}$, where a is near 1 for clean sands but is found empirically for more clay-rich sands (Mavko et al., 1998). Waxman and Smits (1968) proposed a model using the cation-exchange capacity and charge per unit pore volume, Q_v, and an ion mobility term, B, such that $\frac{BQ_v}{F} = \sigma_s$, where σ_s is the surface conductivity. Sen *et al.* (1988), among others, have developed further refinements to B and the Waxman–Smits model, as summarized by Lesmes and Friedman (2005), Worthington (1985) and Mavko *et al.* (1998).

Further extensions include the recent work of Pride *et al.* (2016), who consider the stress and fluid pressure impacts on conductivity in the context of an integrated flow–mechanics–electrical theory. The stress-dependent Pride model is important for crystalline rocks (where fractures have strong sensitivity to effective stress) but is less likely to be important for CO_2 storage applications in porous media. However, geochemical changes in rock frame properties due to CO_2 displacing brine could have an impact on electrical properties, especially surface conductance. Therefore, the use of electrical methods for quantitative estimation of CO_2 saturation should be considered as having uncertainty that is difficult to characterize with existing rock physics models.

In practice, electrical resistivity monitoring, especially crosswell ERT, has proven very useful in monitoring changes in CO_2 saturation within a reservoir (Bergmann *et al.*, 2012; Carrigan *et al.*, 2013). Also, Dafflon *et al.* (2012) demonstrated the use of electrical methods, including ERT, to monitor the migration of groundwater with dissolved CO_2. As the electrical response of rocks is largely independent of the mechanical response, seismic and electrical measurements provide complementary estimates of CO_2 saturation and should be able to reduce uncertainty in a joint inversion methodology.

References

Aki, K. and Richards, P. G. (2002). *Quantitative seismology*. Sausalito, CA: University Science Books.

Al Hosni, M., Caspari, E., Pevzner, R., Daley, T. M., and Gurevich, B. (2016a). Case history: Using time-lapse vertical seismic profiling data to constrain velocity–saturation. *Geophysical Prospecting*, 64(4): 987–1000.

Al Hosni, M., Vialle, S., Gurevich, B., and Daley, T. M. (2016b). Estimation of rock frame weakening using time-lapse crosswell: The Frio Brine Pilot Project. *Geophysics*, 81: B235–B245. DOI:10.1190/GEO2015-0684.1.

Archie, G. E. (1942). The electrical resistivity log as an aid in determining some reservoir characteristics. *Transactions of the American Institute of Mining, Metallurgical, and Petroleum Engineers*, 146: 54–62.

Arts, R., Eiken, O., Chadwick, A., Zweigel, P., van der Meer, L., and Zinssner, B. (2004). Monitoring of CO_2 injected at Sleipner using time-lapse seismic data. *Energy*, 29: 1383–1392.

Avseth, P., Dvorkin, J., Mavko, G., and Rykkje, J. (2000). Rock physics diagnostic of North Sea sands: Link between microstructure and seismic properties. *Geophysical Research Letters*, 27: 2761–2764. DOI:10.1029/ 1999GL008468.

Batzle, M., and Wang, Z. (1992). Seismic properties of pore fluids. *Geophysics*, 57: 1396–1408.

Benson, S. (2008). Multi-phase flow of CO_2 and brine in saline aquifers. Expanded Abstracts, *Society of Exploration Geophysicists*, 27: 2839. DOI:10.1190/ 1.3063934.

Benson, S., Tomutsa, L., Silin, D., Kneafsey, T., and Miljkovic, L. (2005). Core scale and pore scale studies of carbon dioxide migration in saline formations. Lawrence Berkeley National Laboratory Report, LBNL-59082. http://repositories.cdlib.org/lbnl/LBNL-59082

Bergmann, P., Schmidt-Hattenberger, C., Kiessling, D., *et al.* (2012). Surface-downhole electrical resistivity tomography applied to monitoring of CO2 storage at Ketzin, Germany. *Geophysics*, 77(6): B253–B267.

Berryman, J. G. (1995). Mixture theories for rock properties. In T. J. Ahrens (ed.), *Rock physics & phase relations: A handbook of physical constants*. Washington, DC: American Geophysical Union. DOI:10.1029/RF003, 205–228.

Biot, M. A. (1956). Theory of propagation of elastic waves in a fluid-saturated porous solid. 1. Low-frequency range. *Journal of the Acoustical Society of America*, 28(2): 168–178.

Carrigan, C. R., Yang, X., LaBrecque, D. J., *et al.* (2013). Electrical resistance tomographic monitoring of CO_2 movement in deep geologic reservoirs. *International Journal of Greenhouse Gas Control*, 18: 401–408. http:// dx.doi.org/10.1016/j.ijggc.2013.04.016.

Chadwick, R. A., Williams, G. A., and White, J. C. (2016). High-resolution imaging and characterization of a CO_2 layer at the Sleipner CO_2 storage operation, North Sea using time-lapse seismics. *First Break*, 34(2): 77–85.

Cook, P. J., ed. (2014). *Geologically storing carbon: Learning from the Otway Project experience*. Melbourne: CSIRO Publishing.

Dafflon, B., Wu, Y., Hubbard, S. S., *et al.* (2012). Monitoring CO_2 intrusion and associated geochemical transformations in a shallow groundwater system using complex electrical method. *Environmental Science and Technology* 47(1): 314–321.

Daley, T. M., Solbau, R. D., Ajo-Franklin, J. B., and Benson, S. M. (2007). Continuous active-source monitoring of CO$_2$ injection in a brine aquifer. *Geophysics*, **72**(5): A57–A61. DOI:10.1190/1.2754716.

Daley, T. M., Myer, L. R., Peterson, J. E., Majer, E. L., and Hoversten, G. M. (2008). Time-lapse crosswell seismic and VSP monitoring of injected CO$_2$ in a brine aquifer. *Environmental Geology*, **54**: 1657–1665. DOI:10.1007/s00254-007-0943-z.

Daley, Thomas M., Ajo-Franklin, J. B., and Doughty, C. (2011). Constraining the reservoir model of an injected CO$_2$ plume with crosswell CASSM at the Frio-II brine pilot. *International Journal of Greenhouse Gas Control*, **5**: 1022–1030. DOI:10.1016/j.ijggc.2011.03.002.

Dutta, N. C., and Seriff, A. J. (1979). On White's model of attenuation in rocks with partial gas saturation. *Geophysics*, **44**: 1806–1812.

Dvorkin, J., and Nur, A. (1996). Elasticity of high-porosity sandstones: Theory for two North Sea data sets. *Geophysics*, **61**: 1363–1370. DOI:10.1190/1 .1444059.

Gasperikova, E., and Hoversten, G. M. (2008). Gravity monitoring of CO$_2$ movement during sequestration: Model studies. *Geophysics*, **73**(6): WA105–WA112. DOI:10.1190/1.2985823.

Gassmann, F. (1951). On elasticity of porous media. Reprinted in M. A. Pelissier, H. Hoeber, N. van de Coevering, and I. F. Jones (eds.), *Classics of elastic wave theory*. Geophysics Reprint Series No. 24, *Society of Exploration Geophysicists*, 2007.

Guéguen, Y., and Palciauskas, G. (1994). *Introduction to the physics of rocks*. Princeton, NJ: Princeton University Press.

Hill, R. (1963). Elastic properties of reinforced solids: Some theoretical principles. *Journal of the Mechanics and Physics of Solids*, **11**: 357–372.

Hooke, R. (1678). Potentia Restitutiva, or Spring. Reprinted in M. A. Pelissier, H. Hoeber, N. van de Coevering, and I. F. Jones (eds.), *Classics of elastic wave theory*. Geophysics Reprint Series No. 24, *Society of Exploration Geophysicists*, 2007, 55–67.

Hovorka, S. D., Doughty, C., Benson, S. M., *et al.* (2006). Measuring permanence of CO$_2$ storage in saline formations: The Frio experiment. *Environmental Geoscience*, **13**(2): 105–121.

Hovorka, S. D., Meckel, T., and Treviño, R. H. (2013). Monitoring a large-volume injection at Cranfield, Mississippi–Project design and recommendations. *International Journal of Greenhouse Gas Control*, **18**: 345–360.

IPCC. (2005). IPCC Special Report on Carbon Dioxide Capture and Storage. Prepared by Working Group III of the Intergovernmental Panel on Climate Change [Metz, B., O. Davidson, H. C. de Coninck, M. Loos, and L. A. Meyer (eds.)]. Cambridge University Press, Cambridge, United Kingdom and New York, NY, USA, 442 pp.

Johnston, D. H. (2013). Practical applications of time-lapse seismic data. *Society of Exploration Geophysicists Distinguished Instructor Series* No. 16. http://dx.doi.org/10.1190/1.9781560803126

Landrot, G., Ajo-Franklin, J., Yang, L., Cabrini, S., and Steefel, C. I. (2012). Measurement of accessible reactive surface area in a sandstone, with application to CO$_2$ mineralization. *Chemical Geology*, **318**–319: 113–125.

Lebedev, M., Toms-Stewart, J., Clennell, B., *et al.* (2009). Direct laboratory observation of patchy saturation and its effects on ultrasonic velocities. *Leading Edge*, **28**: 24–27.

Lebedev, M., Wilson, M. E. J., and Mikhaltsevitch, V. (2014). An experimental study of solid matrix weakening in water-saturated Savonnieres limestone. *Geophysical Prospecting*, **62**: 1253–1265. DOI:10.1111/1365-2478.12168.

Lemmon, E. W., McLinden, M. O., and Friend, D. G. (2005). Thermophysical properties of fluid systems. In P. J. Linstrom and W. G. Mallard (eds.), *Chemistry web book*. NIST Standard Reference Database Number 69. National Institute of Standards and Technology.

Lesmes, D. P., and Friedman, S. P. (2005). Relationships between the electrical and hydrological properties of rocks and soils. In Y. Rubin and S. Hubbard (eds.), *Hydrogeophysics*. Dordrecht, The Netherlands: Springer.

Mavko, G., Mukerji, T., and Dvorkin, J. (1998). *The rock physics handbook: Tools for seismic analysis in porous media*. Cambridge: Cambridge University Press.

Müller, T. M., and Gurevich, B. (2004). One-dimensional random patchy saturation model for velocity and attenuation in porous rocks. *Geophysics*, **69**(5): 1166–1172. https://doi.org/10.1190/1.1801934

Müller, T. M., Gurevich, B., and Lebedev, M. (2010). Seismic wave attenuation and dispersion resulting from wave-induced flow in porous rocks: A review. *Geophysics*, **75**(5): 75A147–75A164. DOI:10.1190/1.3463417.

Nabighian, M., ed. (1991). Electromagnetic methods in applied geophysics, Vol. 2: Applications. In *Society of Exploration Geophysicists Investigations in Geophysics*. DOI:10.1190/1.9781560802686.

Nakatsuka, Y., Xue, Z., Garcia, H., and Matsuoka, T. (2010). Experimental study on monitoring and quantification of stored CO$_2$ in saline formation using resistivity measurements. *International Journal of Greenhouse*

Gas Control, **4**: 209–216. http://dx.doi.org/10.1016/j .ijggc.2010.01.001

Pride S. R. (2005). Relationships between seismic and hydrological properties. In Y. Rubin and S. S. Hubbard (eds.), *Hydrogeophysics*. Water Science and Technology Library, Vol. 50. Dordrecht: Springer, 253–290. DOI:10.1007/1-4020-3102-5_9.

Pride, S. R., Berryman, J. G., and Harris, J. M. (2004). Seismic attenuation due to wave-induced flow. *Journal of Geophysical Research*, **109**: B01201. DOI : 10.1029/ 2003JB002639.

Pride, S. R., Berryman, J. G., Commer, M., Nakagawa, S., Newman, G. A., and Vasco, D. W. (2016). Changes in geophysical properties caused by fluid injection into porous rocks: Analytical models. *Geophysical Prospecting*, **65**(3). DOI:10.1111/1365-2478.12435.

Rubin, Y., and Hubbard, S., eds. (2005). Hydrogeophysics, Water Science and Technology Library, Vol. **50**. Dordrecht, The Netherlands: Springer.

Rutqvist, J. (2012). The geomechanics of CO_2 storage in deep sedimentary formations. *Geotechnical and Geological Engineering*, **30**(3): 525–551. DOI:10.1007/ s10706-011-9491-0.

Saito, H., Nobuoka, D., Azuma, H., Xue, Z., and Tanase, D. (2006). Time-lapse crosswell seismic tomography for monitoring injected CO_2 in an onshore aquifer, Nagaoka, Japan. *Exploration Geophysics*, **37**: 30–36.

Sen, P. N., Scala, C., and Cohen, M. H. (1981). A self-similar model for sedimentary rocks with application to the dielectric constant of fused glass beads. *Geophysics*, **46**: 781–795.

Sen, P. N., Goode, P. A., and Sibbit, A. (1988). Electrical conduction in clay bearing sandstones at low and high salinities. *Journal of Applied Physics*, **63**: 4832–4840.

Sethian, J. A., and Popovici, A. M. (1999). 3-D traveltime computation using the fast marching method. *Geophysics*, **64**(2): 516–523.

Smith, T. M., Sondergeld, C. H., and Rai, C. S. (2003). Gassmann fluid substitutions: A tutorial. *Geophysics*, **68**: 430–440.

Stokes, G. G. (1845). On the theories of the internal friction of fluids in motion, and of the equilibrium and motion

of elastic solids. Reprinted in M. A. Pelissier, H. Hoeber, N. van de Coevering, and I. F. Jones (eds.), *Classics of elastic wave theory*. Geophysics Reprint Series No. 24. *Society of Exploration Geophysicists*, 2007. 125–161.

Toms, J., Müller, T. M., and Gurevich, B. (2007). Seismic attenuation in porous rocks with random patchy saturation. *Geophysical Prospecting*, **55**(5): 671–678. DOI: 10.1111/j.1365-2478.2007.00644.x.

Vanorio, T. (2015). Recent advances in time-lapse, laboratory rock physics for the characterization and monitoring of fluid-rock interactions. *Geophysics*, **80** (2): WA49–WA59. DOI:10.1190/geo2014-0202.1.

Vanorio, T., Mavko, G., Vialle, S., and Spratt, K. (2010). The rock physics basis for 4D seismic monitoring of CO_2 fate: Are we there yet? *Leading Edge*, **29**: 156–162.

Vanorio, T., Nur, A., and Ebert, Y. (2011). Rock physics analysis and time-lapse rock imaging of geochemical effects due to the injection of CO2 into reservoir rocks. *Geophysics*, **76**(5): 23–33. DOI:10.1190/ geo2010-0390.1.

Vialle, S., and Vanorio, T. (2011). Laboratory measurements of elastic properties of carbonate rocks during injection of reactive CO_2-saturated water. *Geophysical Research Letters*, **38**: L01302. DOI:10.1029/2010GL045606.

Wang, Z. (2001). Fundamentals of seismic rock physics. *Geophysics*, **66**: 398–412.

Wang, Z., Cates, M. E., and Langan, R. T. (1998). Seismic monitoring of a CO_2 flood in carbonate reservoir: A rock physics study. *Geophysics*, **63**: 1604–1617.

Waxman, M. H., and Smits, L. J. M. (1968). Electrical conductivities in oil-bearing shaly sands. *Society of Petroleum Engineers Journal*, **8**: 107–122.

White, J. E. (1975). Computed seismic speeds and attenuation in rocks with partial gas saturation. *Geophysics*, **40**: 224–232.

Worthington, P. F. (1985). The evolution of shaly-sand concepts in reservoir evaluation. *Log Analyst*, **26**: 23–40, SPWLA-1985-vXXVIn1a2.

Xue, Z., Tanase, D., and Watanabe, J. (2006). Estimation of CO_2 saturation from time-lapse CO_2 well logging in an onshore aquifer, Nagaoka, Japan. *Exploration Geophysics*, **37**: 19–29.

Multicomponent Seismic Monitoring

Thomas L. Davis and Martin Landrø

Introduction and Background

The large-scale deployment of carbon capture and sequestration will benefit greatly from new methods of seismic monitoring based on multicomponent seismic data and rock physics modeling. Simply put, our reservoirs and carbon dioxide (CO_2) storage sites are often too complex to monitor without multicomponent seismic data. The rock frame can and does change in many of our reservoirs with injection and production. The interaction between fluid and the rock frame is critical to understanding dynamic changes in the reservoir. As a result, new approaches based on robust measures of velocity and attenuation anisotropy are critical to include in dynamic reservoir characterization. These approaches require a solid foundation based on rock physics and geomechanical modeling.

Multicomponent seismic data involves 3-component geophones that enable the recording of shear or S-waves as well as compressional or P-waves. In marine multicomponent seismic acquisition we measure the omnidirectional pressure component in addition to the three geophone components. This is referred to as 4-component seismic or simply 4C. Recording both waves enables us to quantify elastic properties more fully in the subsurface. These elastic properties are anisotropic in nature by virtue of the media through which seismic waves propagate. Shear waves are especially sensitive to azimuthally anisotropic media as they polarize or undergo birefringence which causes shear wave splitting. This phenomenon is especially useful for seismic monitoring involving CO_2 enhanced oil recovery (EOR) and geosequestration (Figure 5.1).

Properties and Characteristics of Shear Waves

When a shear wave enters an anisotropic medium it splits into two waves, fast (S1) and slow (S2).

The velocity difference between the two waves is a measure of fracture density in a fractured medium. A majority of our reservoirs are fractured and fractures cause the reservoir to be inherently anisotropic. Rock fabric within a reservoir can also cause a reservoir to be anisotropic. Stresses in the subsurface can also induce anisotropy. Thus anisotropy of a reservoir is linked to the rock frame and stresses in the subsurface. Both can change over time due to injection, depletion, or stimulation of a reservoir. Monitoring changes in S-wave azimuthal anisotropy is a very robust means of reservoir monitoring especially in fractured reservoirs.

Shear waves come from shear wave sources, such as horizontal vibrators. They can also be generated by mode conversion from compressional waves impinging on reflection boundaries at nonnormal incidence. This is the case for marine multicomponent acquisition, as it is considered impractical and costly to place the source at the seabed. Both pure shear and converted shear waves exhibit shear wave splitting in anisotropic media. Time and amplitude measures of S-wave splitting can be used as a measure of S-wave azimuthal anisotropy. Time measurements are used if the medium is above seismic resolution (thickness greater than 1/2 wavelength) and amplitude measurements are used for thin beds (below 1/2 wavelength). Warping algorithms are very useful for detecting and visualizing S-wave splitting in thin beds.

What can be done with multicomponent seismic data? One can best link this question back to principles of rock physics. Elastic moduli such as Poisson's ratio can be measured (Figure 5.2). By propagating P- and S-waves through a medium we can measure V_p and V_s and determine Poisson's ratio. The equation written in Figure 5.2 is for an isotropic medium. For an anisotropic medium the symmetry of the medium must be defined and then the shear wave splitting phenomenon can be used to define the symmetry

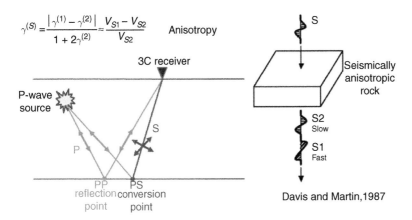

Figure 5.1 Seismic recording with 3-component (3C) receivers or 4C in the case of adding a hydrophone allow for the recording of multicomponent seismic data including P and S body waves.

Vp/Vs What Does it Mean?

Vp^2 = Compressibility + 4/3 Rigidity/Density

Vs^2 = Rigidity/Density

Vp/Vs = Compressibility/Rigidity

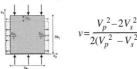

$$v = \frac{V_p^2 - 2V_s^2}{2(V_p^2 - V_s^2)}$$

Figure 5.2 V_p/V_s links to Poisson's ratio, which is a fundamental geomechanical property. Greater emphasis needs to be placed on determining anisotropic elastic parameters, for example, Poisson's ratio in the subsurface.

planes and obtain V_p and V_s in those symmetry planes. Thus two Poisson's ratios are determined and these can be useful for characterizing the medium, a fractured reservoir, for example. They are very useful for linkage to geomechanical properties of the medium. Time-lapse changes in the medium relate to stress change in the subsurface. The dynamic changes are important to monitor especially in the case of CO_2 flood monitoring.

Multicomponent Seismic for Characterization of Fractured Reservoirs

Employing multicomponent seismology improves reservoir characterization because the technology provides the only volume-based method to be able to directly observe the fractures and their changes over time. Multicomponent seismology enables the recording of shear waves that are necessary for the characterization of naturally fractured reservoirs. Shear waves are more sensitive than P-waves to detecting velocity and attenuation changes associated with fractured reservoirs. Multicomponent seismology is suited to the monitoring of fractured reservoirs owing to the influence of pressure changes that cause changes in velocity and attenuation. Both compressibility and rigidity influence the velocity of compressional waves, whereas shear wave velocity is influenced only by rigidity. Natural fractures affect the rigidity of the rock causing velocity anisotropy owing to the oriented nature of fracture sets. The advantage of shear waves over P-waves is the ability to detect azimuthal anisotropy along a singular raypath. With P-waves multiple ray paths are necessary to detect velocity anisotropy. Heterogeneous, anisotropic media above the target level can interfere with P-wave detection and characterization of anisotropy.

Multicomponent seismic data provide for better quantitative interpretation of the subsurface. The principal use of these data is in fracture detection and better imaging and characterization of the subsurface. Multicomponent seismology provides for better porosity determination in carbonates, better imaging especially under gas clouds, improved fault imaging in wrench fault terrains, and finally improved imaging of low-impedance contrast reservoirs. It requires little or no incremental cost to record 3C versus 1C seismic data. The cost of recording 9C data (3-component sources and 3-component receivers) on land is almost always twice the cost of a 1C survey due to the two orthogonal source polarization directions that one needs to impart at each source point.

Shear waves are generated as horizontally polarized waves and because they travel in near normal incidence with horizontal polarizations they are especially sensitive to open natural fractures. The advantage of recording shear waves in three dimensions (3D) and four dimensions (4D) is to see fractures and the connectivity of the fracture networks. Displaying connectivity is important for well location in reservoir development. One wants to minimize the number of wells necessary to develop the resource and not drill into compartments that are already connected. Highly deviated wells enable us to drill across the main fracture trends thereby landing wells with the potential for greater recovery. In some instances, however, fractures can also be conduits for water delivery to these wells so every reservoir is different and one needs to monitor reservoir behavior over time. In wrench fault terrains it is common for sealing faults to occur as they can set up reservoir compartmentalization. An advantage of shear waves is their use in detecting pressure cells associated with these compartments. Pressure cells are zones of weakness or low rigidity in the subsurface. Fluid pressures hold fractures apart and because shear waves are stress sensitive, we can observe pressure compartments.

Developments within Acquisition and Processing of Multicomponent Data

New acquisition and processing technologies are evolving. The increase in channel count and distributed recording systems allows for increased fold and redundancy necessary to carry out successful multicomponent seismic acquisition. New vibrator technology allows for greater bandwidth recording especially on the low end of the frequency spectrum. Increased bandwidth on the low end is especially important for shear wave recording and vector fidelity. New interpretive processing involving better velocity modeling, resolution enhancement through full waveform inversion (FWI), elastic inversion, and better imaging through reverse time migration (RTM) is especially encouraging in creating value through multicomponent seismology. The detection of geopressure and geomechanical changes in the subsurface will be especially important in the future. Quantitative interpretation of rock and fluid properties is necessary and multicomponent seismology can reduce risk and uncertainty.

Completion and production processes induce dynamic changes in reservoir properties including pressure, saturation, and permeability. The most accurate geophysical tool for monitoring these changes is time-lapse multicomponent seismology, specifically in delineating the spatial distribution of the changes. Studies by the Reservoir Characterization Project (RCP) document the power of multicomponent seismology to observe and quantify these dynamic changes. RCP's studies illustrate that more accurate quantitative measurements of reservoir properties can occur through the time-lapse multicomponent seismic monitoring and that these measurements when introduced into dynamic reservoir characterization provide for more accurate prediction of reservoir performance.

Both pressure effects and fluid composition changes associated with a CO_2 injection program can be detected using multicomponent seismic data. This is a key result with major implications for using dynamic reservoir characterization to manage and improve the economics of an EOR project.

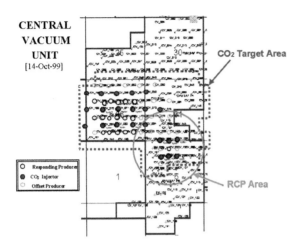

Figure 5.3 Map of Central Vacuum Unit including area of RCP study.

Figure 5.3 shows a map of the Central Vacuum Unit Field in Lea County, New Mexico. The study area is an approximate circle with a radius of 3300 ft. (approx. 1000 m). Four multicomponent 3D surveys were acquired. Both reservoir pressure and fluid composition were altered between the first and second surveys over a small reservoir volume near a single CO_2 injection well. The third and fourth surveys were conducted while reservoir pressure was held approximately constant while a greater volume of the

reservoir was affected by larger quantities of CO_2 injected in multiple wells. The primary measurement tool used was measurements of S-wave anisotropy.

S-wave anisotropy is important. Quantifying S-wave anisotropy relates to the relative amount of open fractures, microfractures and low aspect ratio pore structure within the rock volume being interrogated by the S-waves. S-wave anisotropy can be related to permeability, both magnitude and preferred direction. Providing a quantitative measure of permeability is critical for reservoir characterization. Typically, permeability is not a quantity provided to the reservoir engineering community by geophysicists working with seismic data. S-wave anisotropy measurements within the reservoir can be related to permeability. Coupled with a porosity estimate, we may be able to determine the reservoir storage volume and its preferred permeability pathways for optimum recovery of hydrocarbons. A static estimate is determined from a single multicomponent 3D survey integrated with geological and petroleum engineering data. With repeated surveys, a dynamic picture of the reservoir is possible (dynamic reservoir characterization). Figure 5.4 shows on the left panel the S-wave anisotropy difference between the first and second surveys, before and after the injection of 50 million standard cubic feet in a single well, and on the right panel S-wave anisotropy difference between the third and fourth surveys before and after a billion standard cubic feet of injection into six wells.

Figure 5.4 Anisotropy difference associated with CO_2 flood at Vacuum Field, New Mexico. Phase I is a single-well pilot. Phase II is a six-well pilot.

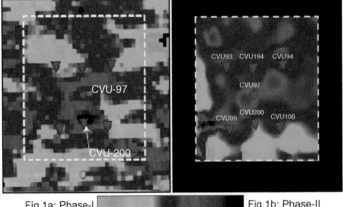

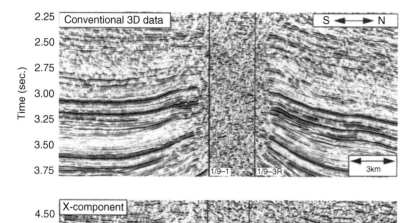

Figure 5.5 Comparison of a PP-section (P-wave down and P-wave up) and a PS-section (P-wave down and converted to S-wave at each interface; or the X-component section). Notice the improved data quality on the X-component data between the two vertical solid lines (the area affected by presence of gas in the overburden). (From Granli *et al.*, 1999.)

Marine Multicomponent Seismic Data

Inspired by results obtained on land shear-wave data acquisition, Statoil embarked on developing 4-component marine seismic data acquisition at the beginning of the 1990s. After some field tests offshore Norway, the first successful example was imaging below the gas chimney at the Tommeliten Field (Granli *et al.*, 1999). Figure 5.5 shows that the X-component data images the area that is distorted by presence of gas better. If we assume that the converted S-waves are close to vertical when they hit the seabed receiver, they will dominate the response at the X-component. This is why we for simplicity assume that the X-component (or radial component) is practically equal to PS-converted data.

In the early days of 4C seabed acquisition nodes were planted into the seabed using remote operating vehicles (ROVs). The node method has been further developed, and today several node-based systems are offered. Nodes are especially attractive for deep water acquisition and for fields with a lot of other seabed installation. Later, four-component receiver cables that were either placed directly at the seabed or trenched into the seabed (for permanent monitoring [PRM] systems were developed). For most marine 4C acquisitions the number of seismic traces is significantly higher than conventional marine streamer acquisitions. As marine 4C acquisition means that there is a regular receiver grid at the seabed and a regular shooting grid at the water surface, it is obvious that the number of measurements will increase significantly compared to towed streamer acquisition. Furthermore, it means that a continuous azimuthal coverage is achieved, which is important for anisotropy analysis. Rognø *et al.* (1999) demonstrated that this increase in azimuthal coverage is crucial for imaging of complex reservoirs, as shown in Figure 5.6.

Multicomponent Marine PRM Systems

Presently, there are several permanent marine multicomponent systems installed. Offshore Norway, PRM systems have been installed at the Valhall, Ekofisk, Snorre, and Grane Fields. Typical 4D sampling time interval for these fields is six months. Offshore Brazil, the Jubarte Field has been instrumented by a PRM system operating at a water depth of 800 m.

It has been decided that the Johan Sverdrup Field (estimated reserves of 3 billion barrels of oil) also will be instrumented with a seabed multicomponent receiver array. There are multiple examples of how these arrays are used in the daily reservoir management of

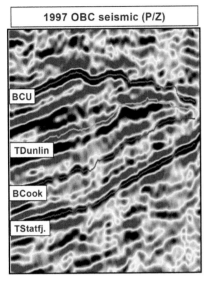

1997 OBC seismic (P/Z)

BCU

TDunlin

BCook

TStatfj.

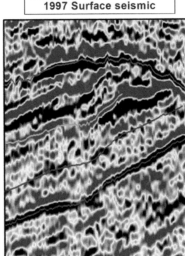

1997 Surface seismic

Figure 5.6 (Left) P/Z seismic data (exploiting both the pressure component and the vertical geophone component and the multiazimuthal coverage). (Right) Conventional 3D streamer data using only the pressure component. Notice the improved definition of the fault blocks at the basal Cretaceous unconformity level (like a staircase effect) on the P/Z data compared to the conventional data to the right. (From Rognø *et al.*, 1999.)

these fields (Gestel *et al.*, 2008; Bertrand *et al.*, 2014; Thompson *et al.*, 2016; Elde *et al.*, 2016).

Landrø *et al.* (2017) suggested that such large permanent arrays can be exploited to detect leakage of gas or CO_2 from the subsurface into the water layer. Conventional air gun arrays emit significant amounts of high-frequency signals (Landrø *et al.*, 2011, 2016) and these signals may be used to image relatively thin chimneys of gas leakage at the seabed. This possibility is also of interest for monitoring CO_2-storage sites, in order to detect unwanted leakage of CO_2 into the water layer as early as possible. As the public acceptance for offshore CO_2-storage is higher than for land-based storage systems, it is crucial to develop cost-effective and reliable monitoring systems for early leakage detection, both close to the storage reservoir and within the overburden, including the water layer.

Repeatability Issues

For all 4D seismic studies, it is crucial to repeat the experiments performed at two different calendar times as accurately as possible. This is challenging and often involves several steps during acquisition and processing to achieve a high degree of repeatability. In a 4C seismic survey, the X-component is crucial because shear waves are more apparent on this component. It is therefore of interest to study the repeatability of both the Z- and the X-component of seismic data. Landrø (1999) used a 3D VSP dataset acquired over the Oseberg Field for this purpose. A five-level VSP-tool recording X-, Y-, and

Z-component geophone data was located at approx. 2 km depth in a well, while an air gun source was used to cover a nearly circular shooting pattern at the surface, as shown in Figure 5.7. Ten thousand shots were fired, and normalized root mean square (NRMS) errors were computed for 70,000 shot pairs. All shot pairs were aligned prior to differencing and NRMS computation. Figure 5.8 shows the average NRMS error as a function of source separation distance for Z- and X-component VSP data. We clearly observe that the NRMS error increase is somewhat larger for the X-component compared to the Z-component. Misaghi *et al.* (2007) showed that the nonrepeatability is caused by lenses and inhomogeneous overburden geology. If we assume that these shallow geobodies have larger contrast in S-wave velocity than the corresponding P-wave velocity, it is reasonable to expect less repeatable data, especially for larger source separation distances, like we observe in Figure 5.8. If this hypothesis is correct, we must expect that the difference in repeatability between S-wave and P-wave data will vary with geological complexity and lithology contrasts in the overburden. Large S-wave velocity contrasts in the overburden reduce the repeatability of data.

Figure 5.9 from Oseberg Field, offshore Norway, shows a comparison between standard 3D streamer data and ocean bottom cable data (4C). We notice that the low frequency content is much higher on the OBC data. We interpret this to be mainly attributed to the fact that when the receivers are lowered to the seabed, the influence of ocean noise related to waves and swell noise is significantly less. This is also one of the motivations

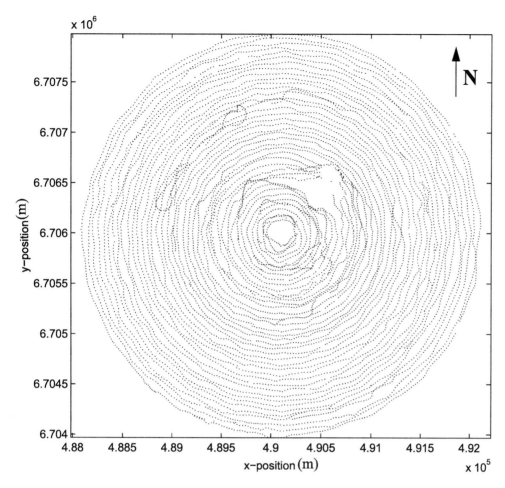

Figure 5.7 Shot positions of the 10,000 shots acquired in the 3D VSP experiment. The hole in the middle is the well position, and the noncircular hole to the northwest of the well is a platform. Typical distance between two adjacent shots is 25 m (from Landrø, *Geophysics*, 1999). The size of the square is 4 by 4 km. Notice that several of the shot positions are very close, enabling repeatability quantification also for source separation distances much less than 25 m.

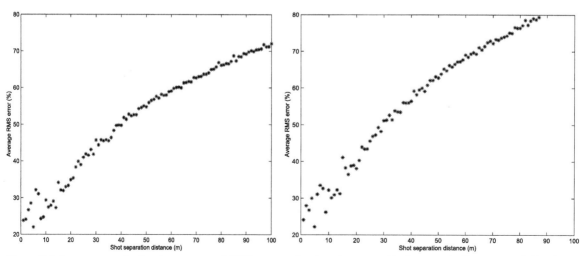

Figure 5.8 Average normalized root mean square (NRMS) error as a function of source positioning errors for Z-component (left) and X-component VSP-data acquired at 2 km depth at Oseberg (North Sea). (Figure from Landrø [1999], *Geophysics*, 64: 1673–1679.) Notice that the repeatability is very similar for source separation distances less than 10 m, and thereafter the repeatability is worse for the X-component data.

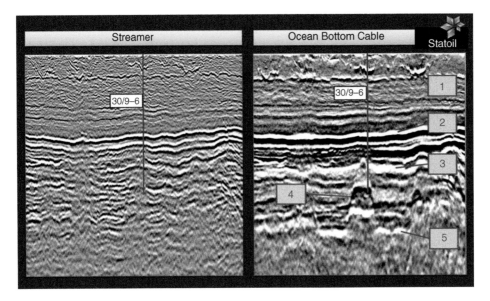

Figure 5.9 Comparison between conventional streamer data and 4-component ocean bottom cable data from the Oseberg Field, North Sea. Notice that the Oligocene sands (1) is better imaged on the 4C seismic data, as well as the strong chalk layer (2) situated above the reservoir (3). Furthermore, the horst-structure (4) is clearer, as well as some potential deeper sand layers (5). Both data sets are P-wave only.

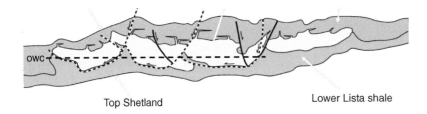

Figure 5.10 Geological cross section of the Grane reservoir (yellow) and shale layers above and below. This example was presented by J. P. Fjellanger *et al.* at the Society of Exploration Geophysicists meeting in New Orleans in 2006, and is presented here with permission from the authors.

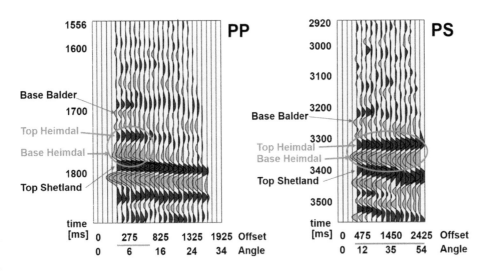

Figure 5.11 PP (left) and PS (right) image gathers from Grane. Notice that the top and base Heimdal events are continuous for a wider range of offsets on the PS data.

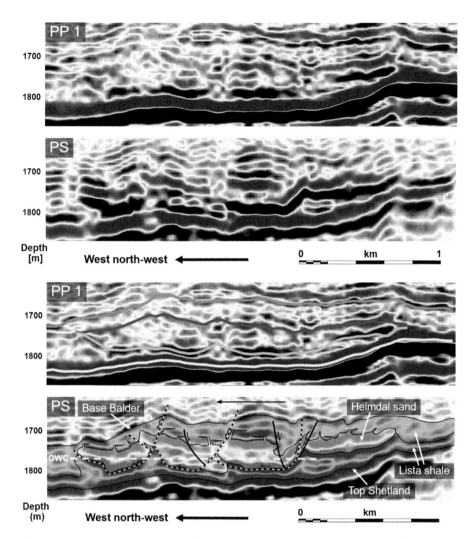

Figure 5.12 PP and PS stack images without interpretation overlaid (top) and the same including interpretation. Notice that the reservoir body (Heimdal sand) is better defined on the PS image.

for towing conventional streamers deeper today compared to what was the standard 10–15 years ago. In Figure 5.9 we clearly see that this richness in lower frequencies results in a far better image: The horst structure which is penetrated by the well is clearly imaged as well as the Oligocene sands in the overburden.

The Grane Field is one of the Norwegian fields that are monitored by a PRM field. The reservoir sand (Heimdal) is complex and it is a challenge to map the reservoir by conventional P-wave seismic. Fjellanger *et al.* (2006) presented an interesting case study using converted wave data acquired over the Grane Field, and we will present some of the key results from this study here. Figure 5.10 shows

a geological depth section, including the reservoir and the surrounding shale layers. Figure 5.11 shows seismic image gathers of the PP and PS data.

The major cause for the improved quality of the PS-seismic data in this case is shown in Figure 5.11, where we observe that the PP-reflectivity dims versus offset for both top and base reservoir. For the PS-gather we notice that this dimming is not observed, yielding a more stable and noise-free stacked signal.

Figure 5.12 shows a comparison of the stacked PP and PS data. The two sections at the top are displayed without the interpretation overlaid, and the two sections at the bottom are displayed with the interpretation overlaid.

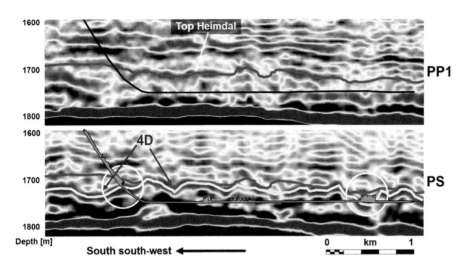

Figure 5.13 Demonstrating the improved definition of the reservoir by using PS seismic data. The solid blue line is the interpretation of top reservoir based on PP data only, while the solid yellow line is the same horizon interpreted based on the PS data. On the lower figure, the Gamma log reading is overlaid with the well path trajectory, clearly showing that the yellow line interpretation is in better accordance with the Gamma log.

We clearly see that the PS section is clearer and gives a better interpretation of the reservoir. This is also confirmed when the Gamma log is added to the well penetrating the reservoir, as shown in Figure 5.13. We notice that the interpretation of top reservoir based on the PP data predicts reservoir before the wells actually enters the reservoir. We also notice that the Gamma log has a strong response in the region between the blue and the yellow lines in the lower section in Figure 5.13. We observe a similar effect closer to the toe of the horizontal section of the well (to the right), where the blue line again is above the yellow line.

References

Bertrand, A., Folstad, P. G., Lyngnes, B., *et al.* (2014). Ekofisk life-of-field seismic: Operations and 4D processing. *Leading Edge*, **33**: 142–148.

Davis, T. L., and Martin, M. A. (1987). Shear wave birefringence: A new tool for evaluating fractured reservoirs. *Leading Edge*, **6**: 22–28.

Elde, R., Roy, S. S., Andersen, C. F., and Andersen, T. (2016). Grane permanent reservoir monitoring – meeting expectations! In 78th *EAGE Conference*, Expanded Abstract, Tu LHR2 02.

Fjellanger, J. P., Bøen, F., and Rønning, K. J. (2006). Successful use of converted wave data for interpretation and well optimization on Grane. Expanded Abstracts, *Society of Exploration Geophysicists*, 1138–1148.

Gestel, J-P, Kommedal, J. H., Barkved, O. I., Mundal, I., Bakke, R., and Best, K. D. (2008). Continuous seismic surveillance of Valhall Field. *Leading Edge*, **27**: 1616–1621.

Granli, J. R., Arntsen, B., Sollid, A., and Hilde, E. (1999). Imaging through gas-filled sediments using marine shear-wave data. *Geophysics*, **64**: 668–677.

Landrø, M. (1999). Repeatability issues of 3-D VSP data. *Geophysics*, **64**: 1673–1679.

Landrø, M., Amundsen L., and Barker, D. (2011). High-frequency signals from air-gun arrays. *Geophysics*, **76**: Q19–Q27.

Landrø, M., Ni, Y., and Amundsen, L. (2016). Reducing high-frequency ghost-cavitation from marine air-gun arrays. *Geophysics*, **81**: P47–P60.

Landrø, M., Hansteen F., and Amundsen, L. (2017). Detecting gas leakage using high-frequency signals generated by air-gun arrays. *Geophysics*, **82**: A7–A15.

Misaghi, A., Landrø, M., and Petersen, S. (2007). Overburden complexity and repeatability of seismic data: Impacts of positioning errors at the Oseberg Field, North Sea. *Geophysical Prospecting*, **55**: 365–379.

Rognø, H., Kristensen Å., and Amundsen, L. (1999). The Statfjord 3-D, 4_C OBC survey. *Leading Edge*, **18**: 1301–1305.

Thompson, M., Andersen, M., Skogland, S. M., Courtial, C., and Biran, V. B. (2016). Time-lapse observations from PRM at Snorre. In 78th EAGE Conference, Expanded Abstract Tu LHR2 01.

Monitoring the Deformation Associated with the Geological Storage of CO$_2$

Donald W. Vasco, Alessandro Ferretti, Alessio Rucci,
Sergey V. Samsonov, and Don White

Introduction

The geological storage of greenhouse gases from a power plant source requires the injection of large quantities of supercritical carbon dioxide (CO$_2$), typically into a small number of wells. A key factor in the Earth's response to such an injection is how fast the injected fluid can redistribute within the volume surrounding the injection wells. For a medium with a well-connected pore space the redistribution is controlled by the hydraulic diffusivity. In particular, the velocity of the fluid $\mathbf{v}$, under the influence of a pressure gradient ∇P, depends on the porosity and permeability of the medium and the viscosity of the fluid. The variation in fluid pressure with time will influence the effective stress buildup and consequently the deformation around the injection well. Heterogeneity can have a major influence on the fluid redistribution and associated pressure variations. For example, high-permeability pathways, such as conductive faults, can concentrate pressure changes into a confined region leading to large pressure changes extending a considerable distance from the injection site. Barriers to flow can lead to the compartmentalization of the fluid buildup and sharp spatial variations in fluid pressure. Heterogeneities in flow properties are typically accompanied by variations in mechanical properties, leading to coupled deformation and flow. Such heterogeneities and fluid pressure variations can lead to issues associated with long-term storage, such as seal damage or potential leakage.

Anomalous pressure buildup around an injection well can also result in deformation that is observable in the overburden and even at the surface. Geodetic methods, techniques for accurately measuring distances and angles between objects on the Earth, are well suited to detect such deformation. These methods have a long history, and geodesy is one of the oldest branches of geophysics. In this chapter we will discuss geodetic observations that are useful for monitoring the deformation associated with the geological storage of CO$_2$. After an overview of methods we will quickly focus on inteferometric synthetic aperture radar (InSAR), as it is the most practical approach for long-term, frequent monitoring, with dense spatial coverage and millimeter-level accuracy.

Overview of Geodetic Techniques

Geodetic techniques have a long history, with methods such as chaining, where an object of known length is used to estimate the distance between points, dating back to ancient times (Smith, 1997; Torge and Muller, 2012). The classic techniques of triangulation and trilateration were used to establish geodetic networks in the seventeenth century. In triangulation, two reference points are used to establish a base distance and the angles of sight-lines to a third point are used to estimate additional distances. In trilateration, distances between three points are used to locate benchmarks with respect to a base point. The distances are usually measured electronically, with an electronic/optical distance meter (EDM). Estimates of deformation follow from repeated surveys over a network of stations. A helpful overview of geodetic methods, examined in relation to monitoring volcanic deformation, is provided in Dzurisin (2007).

Here we shall focus on approaches that are useful for monitoring the fate of CO$_2$ stored within the Earth. For example, to serve as a useful monitoring technology, a geodetic technique should cover a sufficient area, the sampling time between observations should be a few months at most, and the cost of data from a single survey should be of the order of a few hundred to a few thousand dollars to allow for cost-effective, long-term monitoring. Field methods in geodesy can be roughly categorized into individual point measurements and scanning systems. Point measurements are obtained from stand-alone instruments, such as tilt meters (Wright, 1998) and Global Navigation Satellite System (GNSS) receivers (Misra and Enge, 2001, Petrovski and Tsujii, 2012). One characteristic of these devices is fine temporal sampling, often producing observations every few minutes. Scanning systems actively reflect electromagnetic energy from deforming surfaces, differencing the returning signal at various times in order to estimate the displacement of points on the surface. Examples include InSAR (Ferretti, 2014) and light detection and ranging, known as LiDAR (McManamon, 2015). We shall have much more to say about the former technique in the section that follows. Scanning systems typically produce a higher spatial density of sample points, while sacrificing some degree of temporal resolution. Still, large-scale scanning systems can sample the deformation field every few weeks or months, a sufficient rate for monitoring the ground motion associated with the geological storage of CO_2. In addition to techniques that are suitable for dry land, there are methods such as repeat bathymetry and sensitive pressure transducers that function in aquatic environments. Owing to the limited number of large-scale carbon storage sites, few geodetic monitoring techniques have been applied to date. InSAR has been the most commonly used approach and has been applied at a handful of sites, such as the three described in the text that follows. A network of tilt meters was also tested at the In Salah storage site, but it proved too expensive and did not provide a significantly better monitoring capability than InSAR (Mathieson et al., 2011).

Overview of InSAR

Introduction to InSAR

A synthetic aperture radar (SAR) is a coherent, active system capable of recording the electromagnetic energy backscattered to the radar antenna by an illuminated scene and generating a two-dimensional complex-valued image, containing both amplitude and phase information. While the amplitude data depend on the amount of backscattered energy, the phase values of a SAR image are related to the distance from the radar antenna to the objects on ground and they can be used as the basis for geodetic measurements. SAR sensors operate in the microwave

domain, with wavelengths ranging from 3 to 24 cm (for the sensors currently in orbit), 100,000 times longer than those of the visible spectrum. Electromagnetic waves in this frequency band are capable of penetrating clouds, fog, and rain. Being an active system, a radar sensor can operate day and night, throughout the year, independent of solar illumination.

InSAR techniques for geodetic applications rely on the *comparison* of the phase values of two or more SAR images acquired at different times from a similar acquisition geometry to detect and monitor surface deformation. InSAR data can be gathered using both airborne and space-borne sensors. Here we will focus on the latter case and will provide a short introduction to this technology. For a more detailed treatment, the interested reader can refer to the tutorial papers available in the literature (e.g., Bamler and Hartl, 1998; Massonnet and Feigl, 1998; Rosen et al., 2000; Ferretti, 2014).

The basic idea behind satellite InSAR measurements is rather simple. Satellite SAR sensors orbit the Earth along near polar paths, at an altitude ranging from 500 to 800 km. However, the combination of the rotation of the Earth and the motion of satellite makes it possible for SAR sensors to acquire data over the same area, from the same acquisition geometry, but at different times. By definition, radar measures the distance between the antenna source and a target on the ground. InSAR is just a way to monitor range variations (changes in distance) using multiple SAR images from an area gathered at different times. To detect very *small* surface displacements, of the order of a centimeter, compared to the meter-scale resolution of the radar image, we must obtain accurate estimates of the *phase shifts* between two successive SAR acquisitions from the same nominal orbit (Figure 6.1).

To gain some insight into the complex-valued nature of SAR images, we should think of a system illuminating each resolvable cell in the area of interest with a short, monochromatic, sinusoidal, radar pulse, with a frequency equal to the central frequency of the SAR sensor. For the sake of illustration, we suppose that there is just one dominant scatterer in each cell. When the backscattered signal is sampled by the acquisition system, the radar return can be fully characterized by two variables: the amplitude and the phase of the recorded pulse. While the amplitude will depend on the so-called *radar cross section* of the target (Ferretti, 2014), its phase value is related

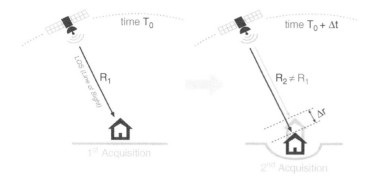
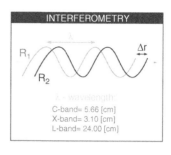

Figure 6.1 A schematic showing the relationship between ground displacement and signal phase shift. The numeric value of the wavelength λ depends on the radar sensor and ranges from 3 cm (X-band) to 24 cm (L-band) for the sensors currently in orbit.

to the sensor-to-target distance, as it is proportional to the time delay between pulse transmission and reception. Of course, the complete process for SAR imaging is much more complicated because it involves concepts such as antenna synthesis, sensor velocity, range compression, signal bandwidth, and sampling frequency (Ferretti, 2014), but our simplistic model can be useful for non-radar specialists and will suffice for this discussion.

A detailed analysis of the phase φ associated with a pixel (picture element) P from a radar image, can help to identify when InSAR will be successful as well as the current limitations of this technology. For example, the phase φ can be modeled as a mixture of four distinct contributions (Rosen *et al.*, 2000; Ferretti *et al.*, 2007; Ferretti, 2014):

$$\varphi(P) = \vartheta + \frac{4\pi}{\lambda} r + \alpha + n \qquad (6.1)$$

where ϑ is a term related to the nature and the location of all elementary scatterers within the resolution cell associated with pixel P. In our simplistic model we have considered a single, dominant scatterer per cell, but typically there are several such scatterers. The term $\frac{4\pi}{\lambda} r$ is an important factor in geodetic applications, because it is associated with the sensor-to-target distance r, the range. The term α, referred to as the atmospheric phase screen (APS), is related to the delay introduced by the Earth's atmosphere, where the propagation speed might change. Amplitude data are nearly immune to atmospheric disturbances, but phase values can be strongly affected by the atmosphere. The last term, n, is a contribution related to system noise, such as thermal vibration, quantization errors, and so on.

The phase values of a single SAR image are of no practical use, as it is impossible to separate the different contributions in Eq. (6.1) without additional information. Moreover, owing to the sinusoidal nature of the representation, the phase values wrap around and are insensitive to factors of 2π. Therefore, the SAR phase matrix looks very much like a set of random numbers ranging from $-\pi$ to $+\pi$. The basic idea of SAR interferometry is to measure the phase *change*, or interference between two radar images, generating an *interferogram* I:

$$I = \Delta\varphi(P) = \Delta\vartheta + \frac{4\pi}{\lambda}\Delta r + \Delta\alpha + \Delta n \qquad (6.2)$$

for the time interval between the two satellite passes. In fact, if we consider an idealized scenario where both target characteristics, such as the nature and location of all elementary scatterers within each cell, and atmospheric conditions are the same in the two acquisitions, then – if noise can be considered negligible – Eq. (6.2) reduces to

$$I = \Delta\varphi(P) = \frac{4\pi}{\lambda}\Delta r \qquad (6.3)$$

that is, interferometric *phase values are proportional to range changes*. In other words, if a point on ground moves during the time interval between the two radar images gathered with the same acquisition geometry, the distance between the sensor and the target may change (unless the motion is orthogonal to the satellite line of sight), creating a phase shift proportional to the displacement. Rather than computing the difference of the phase values, radar specialists would say that an interferogram is computed by multiplying the complex values of the first SAR image, called the master image, by the complex conjugate of the second

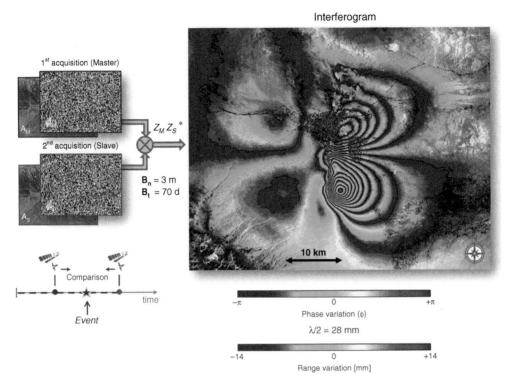

Figure 6.2 Example of SAR interferogram. The two SAR images were acquired by the Envisat ASAR sensor operating at C-band ($\lambda/2 = 28$ mm). The area of interest in southeastern Iran, where on December 26, 2003 an M_w 6.6 earthquake devastated the town of Bam. The images were acquired one before and one after the event. The computation of phase variations highlights the coseismic deformation. Interferometric fringes resemble contour lines of the displacement field. (Adapted from Ferretti, 2014.)

acquisition, called the slave image (Figure 6.2), but this is terminology not necessary for the purpose of our introduction.

Equation (6.3) can explain the extremely high sensitivity of InSAR measurements: the unit of length is the centimeter-scale wavelength, rather than the meter-scale range resolution of the radar image. A displacement of the radar target along the satellite line of sight of $\lambda/2$ creates a phase shift of 2π radians due to the two-way travel path of the radar beam. Thus, a range variation of just 1 mm creates an easily detectable phase shift of more than 20 degrees between two SAR images acquired at X-band ($\lambda = 3.10$ cm). Figure 6.2 is an example of a SAR interferogram (C-band data from Envisat, $\lambda = 5.66$ cm). The color fringes created by the interferometric processing are due to the coseismic displacement field originating from an earthquake in Iran (Funning *et al.*, 2005). It is informative to compare the phase matrix of the master and slave images, exhibiting completely random sets of values, with the coherent fringe pattern of their interferogram.

To retrieve the range change for each pixel of the image, the phase values need to be unwrapped, starting from a reference pixel, assumed to be stationary. In Figure 6.2, each fringe corresponds to 28 mm of range displacement; it is not difficult to estimate about 30 cm of total displacement in the lower right part of the "butterfly" by simply counting the number of cycles with respect to the reference point P_0. Some areas are affected by phase noise, but in general the quality of the interferogram is rather high.

It is important to point out that InSAR data are *differential* measurements, as displacement values are estimated with respect to a reference point that is relatively stable, or a benchmark whose displacement is known. This is the reason why uncertainty in the satellite position can be mitigated: systematic errors affecting the whole area can be removed or strongly reduced thanks to the Double-Difference approach,

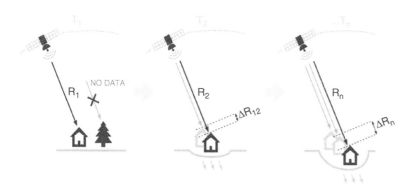

Figure 6.3 A schematic showing the basic idea for measuring ground displacement using successive radar signals and providing a time series of displacement of ground points identified within the illuminated scene. The analysis of range variations is carried out on a subset of image pixels only slightly affected by decorrelation phenomena.

similar to the method used in differential GPS measurements (Ferretti, 2014). Note that InSAR measures only *a single component* of the three-dimensional displacement field: the component along the satellite line of sight. However, a combination of two or more different orbital geometries, such as ascending and descending paths, can be used to obtain two components of displacement, as discussed in a later section.

From Eq. (6.2) one can see that the limitations of InSAR analyses based on just two SAR images are due to (apart from noise) atmospheric effects and possible changes of the nature and the locations of elementary scatterers within each resolvable cell or pixel, creating what are known as "decorrelation" effects. Indeed, over water, snow, vegetation, and agricultural areas, it is difficult, if not impossible, to retrieve good fringe patterns because the radar signature of the target, the first term in Eq. (6.1), changes with time and range variations cannot be estimated accurately. While phase decorrelation, due to changes of radar signature, produces spatially uncorrelated phase changes, atmospheric disturbances are uncorrelated in time but well correlated in space. This feature is exploited by more sophisticated multi-interferogram techniques. In general, conventional InSAR results based on just two SAR images acquired at different times should be treated with care. Without prior information, even considering areas not affected by possible phase unwrapping errors and phase decorrelation, the precision of range displacement data does not exceed 1–2 cm if the area of interest is larger than a few square kilometers, as the impact of atmospheric disturbances increases with the distance from the reference point (Ferretti, 2014). Improved results can be obtained working on a long time series of SAR data, using a multi-interferogram approach.

Multi-interferogram Techniques

As mentioned previously, conventional InSAR, involving the computation of phase variations on a pixel-by-pixel basis between two radar images, can be used successfully only if the radar signature of the reflective target does not change significantly with time, the signal-to-noise ratio (SNR) is high enough, and the atmospheric phase components are negligible in the two SAR images. Multi-interferogram techniques aim to overcome the limitations of conventional InSAR analyses, based on just two SAR acquisitions, by estimating a *time series* of displacement data for those pixels characterized by high SNR values. The PSInSAR technique (Ferretti *et al.*, 2000, 2001), developed in the late 1990s, was the one of the first such InSAR algorithms. The basic idea is to compare many SAR images and to focus the analysis on the high-quality radar targets (Figure 6.3), usually referred to as permanent scatterers (PSs). That is, we focus on those pixels that are only slightly affected by phase decorrelation, such as buildings and bare outcrops. Such targets, identified by signal processing algorithms, exhibit very stable radar signatures and a high SNR, and allow for the implementation of sophisticated filtering procedures for estimating and removing atmospheric disturbances. Thus, for the selected set of scatterers, very accurate displacement data can be obtained. As in conventional InSAR, range variations are computed with respect to a reference point that is either stable or is monitored by other means, such as a GNSS receiver.

More recent algorithms, such as the SqueeSAR™ approach of Ferretti *et al.* (2011), allow one to increase the spatial density of measurement points over the area of interest by identifying two distinct families of points: permanent scatterers (PSs) and distributed

97

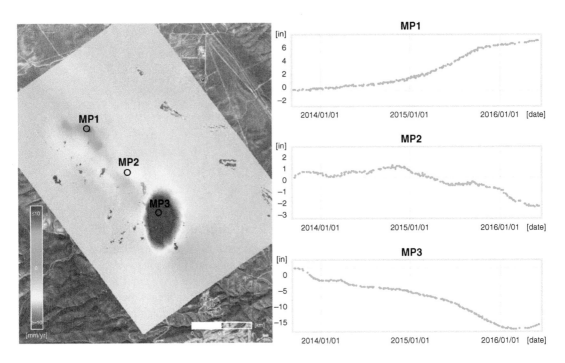

Figure 6.4 SqueeSAR™ analysis performed over an oil field. Red areas indicate measurement points (MP) affected by subsidence (negative displacement rates), while blue areas are affected by uplift. Some examples of MP displacement time series are displayed on the right side of the picture.

scatterers (DSs). PSs are radar targets characterized by high reflectivity values usually corresponding to buildings, metallic objects, pylons, antennae, outcrops, and such. PSs generally produce very bright, isolated, pixels in a SAR scene. Conversely, distributed scatterers are radar targets corresponding to a collection of pixels in the SAR image, all exhibiting very similar radar reflectivity. Phase decorrelation, though still affecting DSs, can be usually managed by means of spatial averaging procedures, allowing for a significant improvement of the quality of the interferometric fringes. These pixels usually correspond to rocky areas, detritus, and areas of short vegetation.

Provided that enough images are available, one can estimate a time series of range change with a precision dependent on the number of available data and the distance from the reference point, but typically better than 3–4 mm. This may be done regardless of the type of measurement point identified by the algorithm, that is, for both permanent and distributed scatterers. To perform a successful multi-interferogram analysis, a minimum number of satellite images, approximately 10–15, are required. This is necessary to create a reliable statistical analysis of the radar returns, making it possible to identify pixels exhibiting good SNR. The higher the number of images, the better the quality of the results. We reiterate that, for displacement data associated to both permanent and distributed scatterers, a key factor is the distance from the reference point. The relative accuracy can be better than a few millimeters for a distance less than the average correlation length of the atmospheric components (about 4 km at mid-latitudes). Average displacement rates can usually be estimated with a precision better than 1 mm/year, depending on the number of data available and the temporal span of the acquisitions (Ferretti, 2014). Figure 6.4 is an example of a SqueeSAR™ analysis performed over an area in the United States, using images acquired by the high-resolution, X-band, COSMO-SkyMed SAR sensors between 2013 and 2016. The measurement points are color-coded according to the average displacement rate. Subsidence patterns (negative displacement rates) are indicated by yellow to red) produced by oil production are clearly visible, as well as areas affected by uplift due to fluid injection (in blue). Over

Figure 6.5 Example of motion decomposition combining ascending and descending acquisition geometry.

nonvegetated areas, the spatial density of measurement points identified by multi-interferogram techniques can exceed 10,000 measurement points per square kilometer. For each measurement point identified, a displacement time series is available and it can be used successfully to run history matching algorithms and, in general, for reservoir monitoring purposes.

Two-Dimensional Displacement Decomposition

Satellite SAR interferometry measures only the projection of a three-dimensional displacement vector only along the satellite's line-of-sight. However, it is possible to combine data acquired from different acquisition geometries to estimate two-dimensional displacement fields (Fujiwara et al., 2000; Wright et al., 2004; Teatini et al., 2011; Rucci et al., 2013). In fact, all SAR sensors follow near-polar orbits and every point on the Earth can be imaged by two different acquisition geometries: one with the satellite flying from north to south (descending mode), looking westward (for right-looking sensors) and the other with the antenna moving from south to north (ascending mode), looking eastward. This is the reason why, by combining InSAR results from both acquisition modes, it is possible to estimate two components of displacement (Figure 6.5), the quasi-east–west and the quasi-vertical components. In fact,

because the satellite orbit is almost parallel to the Earth's meridians, the sensitivity to possible motion in the north–south direction is negligible.

An example of two-dimensional measurements is presented in Figure 6.6. The area of interest is a stacked carbonate reservoir in the Middle East (Klemm et al., 2010; Tamburini et al., 2010). For this case study, one ascending and one descending SAR data set have been acquired over a time period of four years. The ascending time series was acquired by the Envisat satellite, while the descending data set was acquired by RADARSAT-1. Over the 120-km^2 area, some 280,000 and 360,000 InSAR measurement points were identified in the ascending and the descending data set respectively.

Multidimensional Small Baseline Subset Methodology

The time series of one-dimensional line-of-sight deformation can be computed from a set of highly coherent interferograms acquired with small temporal and spatial baselines using the small baseline subset (SBAS) technique (Berardino et al., 2002; Lanari et al., 2004; Samsonov et al., 2011). The temporal resolution of the estimated time series is defined by satellite revisit time and their spatial resolution and coverage are similar to those of single-pair interferograms. The multidimensional small baseline subset (MSBAS) technique computes two-

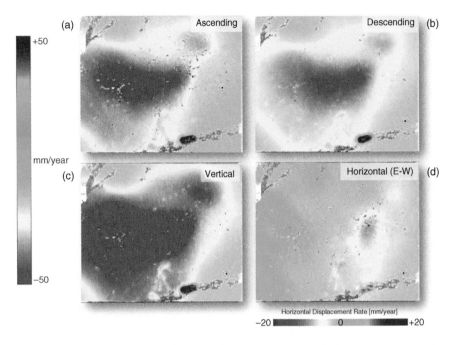

Figure 6.6 Example of estimation of vertical and horizontal (east–west) displacement components from ascending and descending radar data acquired by two SAR sensors over an oil field in the Middle East.

dimensional, horizontal east–west and vertical time series of ground deformation from ascending and descending DInSAR (differential InSAR) data acquired by various SAR sensors with different acquisition parameters, such as azimuth and incidence angles, spatial resolution, and wavelength:

$$\begin{pmatrix} \hat{\mathbf{A}} \\ \lambda \mathbf{L} \end{pmatrix} \begin{pmatrix} \mathbf{V}_e \\ \mathbf{V}_u \end{pmatrix} = \begin{pmatrix} \hat{\mathbf{\Phi}} \\ 0 \end{pmatrix} \qquad (6.4)$$

where the matrix $\hat{\mathbf{A}}$ consists of time intervals between consecutive SAR acquisitions and the east–west and vertical components of a line-of-sight vector, $\mathbf{V}_e$ and $\mathbf{V}_u$ represent unknown east–west and vertical velocities that are to be determined, $\hat{\mathbf{\Phi}}$ represents observed DInSAR data, geocoded, and resampled to a common grid, λ is a regularization parameter and $\mathbf{L}$ is a zero-, first-, or second-order difference operator. The unknown parameters $\mathbf{V}_e$ and $\mathbf{V}_u$ for each pixel are solved by applying the singular value decomposition (SVD) of the coefficient matrix (Press *et al.*, 2007) and the deformation time series are reconstructed from the computed deformation rates by numerical integration. Produced two-dimensional time series have combined temporal resolution, and near daily temporal resolution can be achieved by simultaneous processing data

from multiple SAR sensors. MSBAS methodology has been used successfully for measuring ground deformation due to carbon sequestration (Samsonov *et al.*, 2015; Czarnogorska *et al.*, 2016), mining (Samsonov *et al.*, 2013a,b, 2014a), urban development (Samsonov *et al.*, 2014b, 2016a), permafrost aggradation and pingo growth (Samsonov *et al.*, 2016b), and volcanic activity (Samsonov *et al.*, 2012, 2014c,d).

Interpretation Techniques

For oil and gas applications, InSAR analysis can provide data on (1) reservoir compaction/expansion and surface subsidence or uplift, (2) fault reactivation, and (3) areas of possible well failure. In some cases, surface deformation measurements can provide insight about where extracted fluids are drawn from and where injected fluids flow to, two of the most important questions in reservoir management. To link surface deformation to reservoir parameters, we need to use geomechanics and geophysical inversion methods. InSAR data are becoming more and more important in projects related to enhanced oil recovery, underground gas storage projects, and carbon capture and storage (CCS), where it is becoming a standard monitoring tool.

Both forward and inverse modeling can aid in the interpretation of geodetic observations related to the injection and storage of CO$_2$. In forward modeling one is given the properties of the medium, the injection rates, and the properties of the fluids and calculates the resulting fluid saturations, fluid pressures, reservoir volume changes, and deformation by solving the governing equations. Inverse modeling, or solving the inverse problem, involves using observed saturation, well pressure, and surface deformation to estimate the properties within the reservoir that are driving the deformation.

The first task is to develop a conceptual model of the reservoir and of the processes controlling flow within the reservoir. Such a model is usually based on well logs and core samples gathered during the drilling of injection and monitoring wells. The conceptual model may be based on additional geophysical data, such as seismic reflection surveys defining reservoir boundaries and imaging visible faults. The conceptual model is typically large scale and incomplete, with important features, such as the permeability distribution within the reservoir, poorly constrained. Therefore, one must be prepared to modify the conceptual model in the face of new information such as flow data and/or geophysical monitoring data. For example, at the In Salah site, an anomalous double-lobed pattern of InSAR range change suggested fault or fracture flow, rather than uniform and homogeneous fluid migration through the porous sandstone of the reservoir (Vasco *et al.*, 2010).

The conceptual model, along with reservoir properties from available well information, can be used in conjunction with a coupled hydromechanical reservoir simulator to calculate the fluid distribution and deformation due to the injection of CO$_2$. This is the essence of the forward problem. The resulting predictions can be compared with observed flow data and surface deformation. The flow and mechanical properties of the medium can be adjusted in order to produce better agreement between observed and calculated quantities. Such trial-and-error adjustments are the simplest approach for solving the inverse problem, where observations are used to estimate the properties of the medium. There are much more sophisticated techniques for solving the inverse problem, some of which are based on stochastic methods that rely on trial-and-error approaches. Other methods are based on sensitivity calculations in which the

rate of change of an observation with respect to a change in a model parameter is used to iteratively update a given model. We illustrate this approach in the text that follows.

To make things concrete, we shall consider a specific inverse problem. First, note that the inverse problem will be as complicated as the forward model on which it is based. In this example, based on an application from the In Salah storage project, we shall assume that the injected CO$_2$ migrates through a fracture zone intersecting the well (Vasco *et al.*, 2010, Rucci *et al.*, 2013). The injected CO$_2$ leads to aperture changes (fracture opening) distributed over the fracture zone. The aperture changes will be assumed to vary spatially within the fracture zone and also as a function of time. The fluid will be assumed to stay within the fault/fracture zone for the duration of the injection. The unknown model parameters for this inverse problem are the aperture changes within the fault/fracture zone during a given time interval. Note that we are not modeling the fluid flow within the fault/fracture zone and will not be relating aperture change to fluid pressure changes within the zone. That is another level of complexity that we shall not attempt to address, as we do not know the exact nature of the fracture zone because no core samples or log data were retrieved from it. Aperture changes in the fracture zone drive deformation within the surrounding medium, possibly leading to deformation at the Earth's surface. We will further assume that the material surrounding the fault/fracture zone behaves elastically during the weekly to monthly time intervals between observations. Therefore, the displacements are governed by the set of equations (Aki and Richards, 1980):

$$\frac{\partial}{\partial x_j}\left[\mu\left(\frac{\partial u_i}{\partial x_j} + \frac{\partial u_j}{\partial x_i}\right) + \lambda \frac{\partial u_k}{\partial x_k}\delta_{ij}\right] = \frac{\partial S}{\partial x_j} \quad (6.5)$$

where S is the source term. Using the representation theorem (Aki and Richards, 1980), we can write the solution to these equations in terms of an integral over the surface of the fault/fracture zone. A Green's function is the solution of the governing equations corresponding to a delta-function (impulsive) source. The expression for the i-th component of displacement at point $\mathbf{x}$ for time t is given by the integral:

$$u_i(\mathbf{x}, t) = \int_{\Sigma} a(\varsigma, t) g_i(\mathbf{x}, \varsigma) d\Sigma \quad (6.6)$$

where $a(\varsigma, t)$ is the aperture change at location ς at time t, Σ is the fault/fracture surface, and $g_i(\mathbf{x}, \varsigma)$ is the Green's function. The integral expression can be written as a discrete sum if we represent the aperture changes defined over the surface as piecewise constant functions defined over rectangular patches. Then the integral may be written as the sum:

$$u_i(\mathbf{x}_j, t) = \sum_{n=1}^{N} G_i^n(\mathbf{x}_j)a_n = \mathbf{G}_i(\mathbf{x}_j) \cdot \mathbf{a}(t) \quad (6.7)$$

where a_n is the average aperture change of the n-th fault/fracture patch and $\mathbf{G}_i(\mathbf{x}_j)$ is a vector with components that are the integral of the Green's function:

$$G_i^n(\mathbf{x}_j) = \int_{R_n} g_i(\mathbf{x}_j, \varsigma)d\Sigma \quad (6.8)$$

over the rectangular patch R_n. The expression for $u_i(\mathbf{x}_j, t)$ is suitable for vector displacement data, such as that obtained from the GNSS. However, a range change, as derived from InSAR observations, is a projection of the displacement vector onto the look vector pointing in the direction of the satellite, $\mathbf{l}$:

$$r(x_j, t) = \mathbf{l} \cdot \mathbf{u} = l_i \cdot u_i \quad (6.9)$$

where we are invoking the convention of summation over repeated indices. Substituting the representation of $u_i(\mathbf{x}_j, t)$ as a sum, or its vector matrix equivalent $\mathbf{G}_i(\mathbf{x}_j) \cdot \mathbf{a}(t)$, we can write the range change as

$$r(x_j, t) = \mathbf{m}_j \cdot \mathbf{a}(t) \quad (6.10)$$

where

$$\mathbf{m}_j = l_i \mathbf{G}_i(\mathbf{x}_j) \quad (6.11)$$

and $\mathbf{x}_j$ signifies an observation point, such as a permanent scatterer. Given a large collection of measurements we obtain a linear system of equations relating the aperture changes over the fracture zone during a particular time interval, $\mathbf{a}$, to get

$$\mathbf{r} = \mathbf{M} \cdot \mathbf{a} \quad (6.12)$$

where we have suppressed the explicit time dependence.

In the forward problem we use a model of aperture changes to calculate the resulting range change and this follows from simple matrix multiplication in Eq. (6.12). In the inverse problem we have to solve the linear system of equations for the vector $\mathbf{a}$ on the right-hand side. This is a much more difficult task,

especially because the matrix $\mathbf{M}$ is a smoothing operator that averages the aperture changes in order to produce calculated range changes. Because of the large number of InSAR observations, the linear system of equations may be formally overdetermined, with more equations than unknown parameters. However, owing to the averaging nature of the matrix $\mathbf{M}$ many of the equations are not independent and the system of equations will be effectively underdetermined, with too few independent equations to resolve each model parameter. Furthermore, because of errors in the observations and approximations in the modeling, the equations may be inconsistent and may not be directly solvable. To overcome this difficulty, one may adopt a least squares approach, whereby one finds the solution that minimizes the sum of the squares of the residuals, rather than trying to solve every equation exactly. Thus, we seek the aperture distribution g that minimizes:

$$\rho^2 = (\mathbf{r} - \mathbf{M} \cdot \mathbf{a})^t \cdot (\mathbf{r} - \mathbf{M} \cdot \mathbf{a}) \quad (6.13)$$

or some weighted version of the original equations. To treat the nonuniqueness that is introduced by the fact that the equations are effectively underdetermined, one may introduce additional terms, referred to as penalty terms biasing the solution in some fashion. This approach is known as regularization. For example, it may be desirable to have a solution that does not stray from a prior model of aperture changes $\mathbf{a}_p$; thus we would minimize:

$$\rho^2 = (\mathbf{r} - M \cdot \mathbf{a})^t \cdot (\mathbf{r} - M \cdot \mathbf{a}) + w_p(\mathbf{a} - \mathbf{a}_p)^t \cdot (\mathbf{a} - \mathbf{a}_p)I \quad (6.14)$$

where w_p is a weighting coefficient that controls the trade-off between fitting the data and remaining close to the prior model. The conditions for a minimum, or a maximum for that matter, are given by the vanishing of the gradient of the augmented functional with respect to $\mathbf{a}$:

$$\nabla_\mathbf{a}\rho^2 = -\mathbf{M}^t\mathbf{r} - w_p\mathbf{a}_p + [\mathbf{M}^t\mathbf{M} + w_p\mathbf{I}]\mathbf{a} = 0 \quad (6.15)$$

leading to a linear system of equations for the aperture changes. The estimated aperture changes are given by the formal solution of Eq. (6.16):

$$\hat{\mathbf{a}} = [\mathbf{M}^t\mathbf{M} + w_p\mathbf{I}][\mathbf{M}^t\mathbf{r} + w_p\mathbf{a}_p] \quad (6.16)$$

Other forms of regularization may be introduced in order to emphasize other aspects of the solution. For example, one may minimize the roughness of the solution by introducing a penalty term that represents one of the spatial derivatives of the model (Menke, 1989). Alternatively, it might be desirable to produce a model in which the aperture changes are largest near the injection well, where the fluid pressure changes are the greatest. Thus, one can introduce a penalty term for which aperture changes that are far from the well are penalized. Such terms can be written as quadratic forms in the model parameters, leading to a linear system of equations when differentiated. A critical component of the inverse problem is an assessment in which model parameter uncertainties and resolution are estimated (Menke, 1989). There is a well-established formalism for model assessment when the inverse problem is linear. We will not describe these techniques here. Rather, we refer the reader to in-depth discussions of inverse problems (Menke, 1989).

Applications

There are a number of sites worldwide where significant volumes of CO_2 are stored within the Earth, including Sleipner (Chadwick *et al.*, 2012), Weyburn (White *et al.*, 2011), and Cranfield (Hovorka *et al.*, 2013). Here we will discuss three additional sites where InSAR plays a significant role in the monitoring program.

In Salah, Algeria

At the In Salah storage project excess CO_2 from three natural gas fields is separated, processed, compressed, and reinjected into the flanks of one of the reservoirs. A total of three horizontal wells injected more than 3 million tons of CO_2 between 2004 and 2008. This made In Salah one of the largest storage projects in the world (Mathieson *et al.*, 2010, 2011). The reservoir chosen to contain the CO_2 was not ideal for the storage of a large amount of CO_2. While the sandstone reservoir was laterally extensive, it was only 20 m thick on average, and its porosity varied between 10% and 18%. Image logs and mud loss data suggested that the reservoir contained two or three northwest-trending fractures per meter, with apertures ranging between 0.1 and 1.0 mm, and fracture lengths varying between 6 and 25 m (Iding and Ringrose, 2010). The three horizontal wells were oriented in the direction of the least principle stress, perpendicular to the most likely

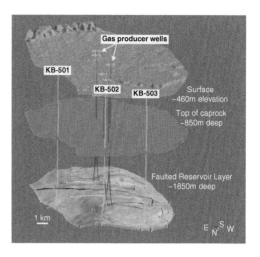

Figure 6.7 General structural setting and well placement at the In Salah gas storage site in Algeria.

fracture planes, in order to maximize the injectivity and the influence of the fractures.

The general structure of the reservoir is that of a gentle anticline with a few degrees of dip. The gas field and producing wells are located on the crest of the anticline and the injection wells are on the flanks of the anticline (Figure 6.7). The reservoir is overlain by more than 800 m of mudstone, forming an extensive caprock seal above the injection zone. Topping this is approx. 900 m of interbedded sands and shales that contains the regional Pan Saharan potable aquifer, an important water source for central and southern Algeria.

In 2005 the Joint Industry Project (JIP), a collaboration between various research institutions and the operating companies, was set up with the goal of monitoring the fate of the injected CO_2 using geophysical, geochemical, and reservoir production data (Mathieson *et al.*, 2011). As part of this effort, Lawrence Berkeley Laboratory contracted with TRE of Milan, Italy to provide estimates of range change from InSAR observations of the European Space Agency's (ESA) Envisat satellite. Fortunately, these data were available before the start of injection in 2004, allowing for the establishment of baseline observations with which to measure range changes due to the injection of carbon dioxide (Figure 6.8). In Salah's rough, boulder-strewn ground surface, with little movable sand, proved favorable for InSAR monitoring. The injected volume of carbon dioxide has produced more than 30 mm of observable range change,

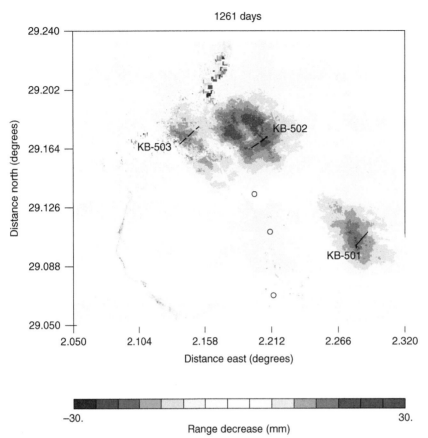

Figure 6.8 Range change after 1226 days of injection.

measured from roughly 300,000 permanent scatterers (Vasco *et al.*, 2010). Uplift at the Earth's surface causes reflectors on the ground to move closer to the reference point in space, resulting in a decrease in range. Significant uplift is observed over all three injector wells (KB-501, KB-502, and KB-503), with amplitudes that are much larger than the estimated 5-mm error associated with the measurements. A double-lobed pattern, first noted by collaborators from Pinacle Technologies, is visible in the uplift measured over injector KB-502. Such a pattern is suggestive of the opening over a near vertical planar surface, such as a fault or fracture (Davis, 1983). Based on this observation, Vasco *et al.* (2010) constructed a damage zone model with variable aperture change over a vertical planar feature. They conducted an inversion of the range changes with the goal of estimating the aperture change on the fracture zone and volume change in the reservoir itself. Subsequent work with two–component displacement estimates (quasi-east–west and quasi-vertical) suggested that subvertical damage

zones were required at all three injectors in order to fit the observations (Rucci *et al.*, 2013).

The Envisat C-band satellite was decommissioned and in October 2010 was no longer in service to monitor the range change over In Salah. Fortunately, two constellations of X-band satellites: TerraSAR-X and Cosmo-SkyMed, were available and the monitoring was transitioned to them in 2009 and 2010. The X-band satellites had better temporal sampling, with return times of 8 and 11 days, in comparison to the best possible 35-day repeat time of Envisat satellite. This allowed for better estimates of corrections to the phase estimates. TRE merged the three data sets – Envisat, TerraSAR-X, and Cosmo-SkyMed – into a common time series and provided it to Lawrence Berkeley National Laboratory for analysis.

The merged data set suggests a double lobed pattern over well KB-501 for selected time increments (Figure 6.9). This supports the findings of Rucci *et al.* (2013) that only a model containing aperture and

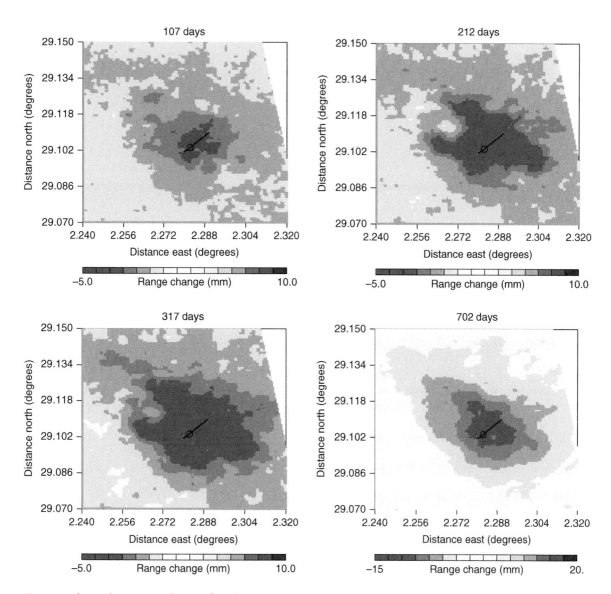

Figure 6.9 Range change over well KB-501 from the stiched Envisat, TERRASar-X, and Cosmo-SkyMet data set.

volume change on a subvertical damage zone could satisfy both the quasi-east–west and quasi-vertical data sets simultaneously (Rucci *et al.*, 2013). For time intervals of 702 days and greater, the second lobe disappears, perhaps because flow in the reservoir begins to dominate the source of the deformation. The generally higher quality corrections to the merged data set and some spatial averaging helped to bring out the subtle pattern in the measurements. Based on the suggestion of a double lobed pattern in Figure 6.9 and the results of Rucci *et al.* (2013), we

developed a conceptual model of flow and deformation associated with the injection at KB-501. A major component of the deformation appears to be due to aperture change over a vertical fault/fracture, or damage zone, intersecting the well. Note that this conclusion is also supported by the analysis of Vasco *et al.* (2008) and the linear high-permeability feature suggested by a sequence of inversions for flow in the reservoir layer. The concentrated flow was correlated with a hypothesized fault/fracture zone observed as a break in a seismic horizon.

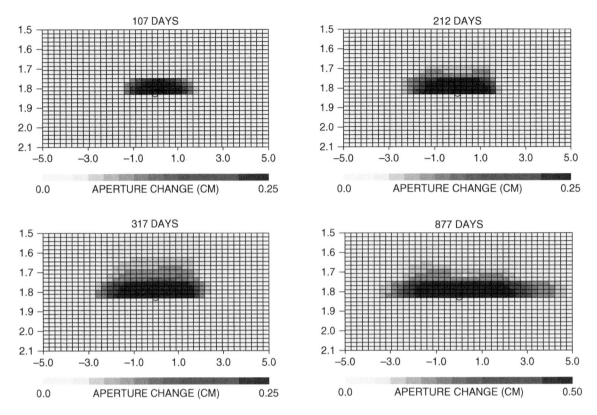

Figure 6.10 Distribution of aperture change over the damage zone for four time intervals, corresponding to the range change shown in Figure 6.9.

To study the evolution of the aperture change with time, due to the injection of CO_2, we can construct a model of the damage zone following the approach described in the section on interpretation techniques. Specifically, we postulated that aperture change generating the surface deformation is distributed over a vertical planar zone. As indicated in the previous section, we can subdivide such a zone into a grid of voxels encompassing the damage zone and set up a least squares problem for the aperture change within each element. The grid geometry is illustrated in Figure 6.10 and the solution to the least squares problem for the aperture change is also plotted in Figure 6.10 for four time intervals. That is, the solutions minimize the misfit to the range changes while simultaneously minimizing aperture changes that are far from the injection well (Vasco *et al.*, 2010).

Because we have range change estimates every month or so, we can calculate the time history of the fracture opening. Specifically, we can compute the time at which the aperture begins to change from its background value (Figure 6.11). From this figure one observes that the earliest aperture change is in a region surrounding the well, as might be expected due to the pressure changes associated with the injection of CO_2. With time the aperture changes extend along reservoir and into a weak zone overlying the reservoir. For a brief period the aperture change appears to have migrated upward before extending laterally outward.

The results of the inversions suggest that the activated or reactivated fracture has remained within the reservoir and the immediately overlying formations. However, owing to the nature of geodetic data, the solution to the inverse problem is nonunique and an infinite number of solutions are possible. Ramirez and Foxall (2014) used a stochastic approach to try to characterize the range of possible solutions associated

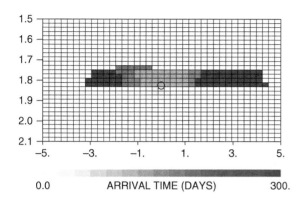

Figure 6.11 Onset of aperture change within the damage zone.

with the deformation observed over well KB-502. They found that the majority of solutions lie within at or below the hot shale, within 200 m of the reservoir.

Decatur Project, Illinois Basin, USA

The Illinois Basin Decatur project, started in November 2011, is a multiyear program managed by Archer Daniels Midland, the U.S. Department of Energy, the Illinois State Geological Survey, and Schlumberger Carbon Services (Falorni *et al.*, 2014). This is the only carbon capture and storage effort in the United States that is currently injecting large volumes of greenhouse gas emissions into a regional deep saline formation (Finley *et al.*, 2011, 2013). The injection, from an ethanol production facility, started at a rate of 1000 metric tons/day and is planned to increase to more than 2000 metric tons/day. The CO_2 is stored in the 550 m thick Mount Simon sandstone at a depth of 2.1 km. The overlying Eau Claire shale forms a 100–150 m thick seal. The Mount Simon formation was the site of a functioning natural gas storage facility and had performed well in that capacity. Several different geochemical, geophysical, and remote sensing technologies are employed at the site to monitor the evolution of the injected volume of CO_2. Preliminary results from microseismic and InSAR monitoring were reported by Kaven *et al.* (2014) and Falorni *et al.* (2014), respectively.

The injection site is situated in a mixture of industrial sites, farmland, forest, and residential areas and InSAR monitoring can be challenging in such a diverse environment. The SqueeSAR™ algorithm described in

the previous sections, with its combination of PSs and DSs, is a flexible method for treating such a wide variety of land surfaces. The distribution of the approx. 109,000 scatterers found in the region surrounding the injection well (Falorni *et al.*, 2014) is shown in Figure 6.12.

Another difficulty is introduced by the seasonal and atmospheric conditions in this region, including long periods of snow cover. Some PSs such as steep roofs and tall towers may not accumulate much snow, but large areas will be covered, reducing the coherence and introducing significant variations in the characteristics of the scattering surface. Artificial reflectors, manmade, stable targets designed to remain free of snow, are one remedy and are commonly used in areas prone to snowfall. To provide for year-round coverage, 21 artificial reflectors, spaced 75 m apart, were constructed and emplaced in an open area close to the injection well-pad (Figure 6.12). All 21 reflectors were found to have strong and stable reflectivity, both for ascending and descending satellite orbits.

The acquisition of COSMO-SkyMed X-band data began in July 2011, before the start of injection in November. Falorni *et al.* (2014) have reported on roughly two years of observations from July 2011 until June 2013. They note that there is little or no ground motion associated with the start of injection with the exception of two points. For example, a time series consisting of the average range change from all 21 reflectors is plotted in Figure 6.13. The two points showing some movement correlating with the start of injection were in the vicinity of the injection well and may reflect some movement related to the pressurization of the well bore itself (Figure 6.14).

The lack of injection-related surface deformation is an expected result. Owing to the high permeability and thickness of the reservoir sandstone, the bottom hole pressure has changed very little during the injection. Barring any unforeseen geological features, such as an unmapped fracture zone, the CO_2 is expected to migrate outward into the reservoir without generating large pressure changes or associated deformation.

Aquistore, Williston Basin, Canada

The Aquistore storage site acts as a buffer for the world's first commercial post-combustion CO_2 capture plant. The operation of the plant began in October 2014 at the Boundary Dam coal-fired power station in Saskatchewan, Canada (Worth *et al.*, 2014). The injection of CO_2 into the subsurface

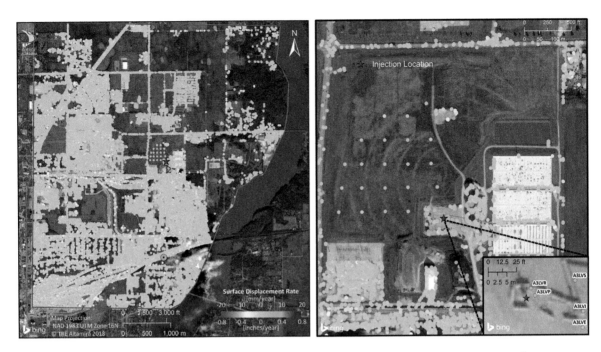

Figure 6.12 (Left) Distribution of scatterers in the region surrounding the injection well (magenta star), roughly 4000/km². (Right) Close-up view of the region surrounding the injection well-pad. The grid of artificial reflectors is visible to the northwest of the wellhead.

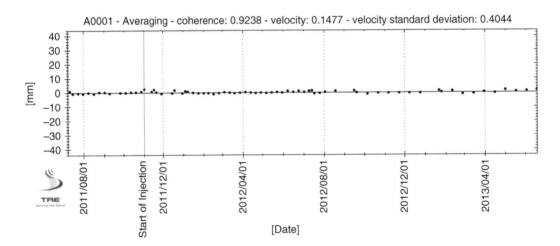

Figure 6.13 Average range change from the 21 artificial reflectors installed just to the northwest of the injection well site (Falorni et al., 2014). The start of injection is indicated by the vertical blue line.

began in April 2015 with maximum anticipated injection rates of up to 1500 tonnes/day to brine-filled clastic strata of the Deadwood and Winnipeg formations, in the depth range of 3150–3350 m (Norford et al., 1994). These two formations are the deepest sedimentary units in the Williston Basin, and are located below any oil-producing and potash-bearing formations (Norford et al., 1994). Similar deep saline formations are found elsewhere in western Canada and throughout North America and the world.

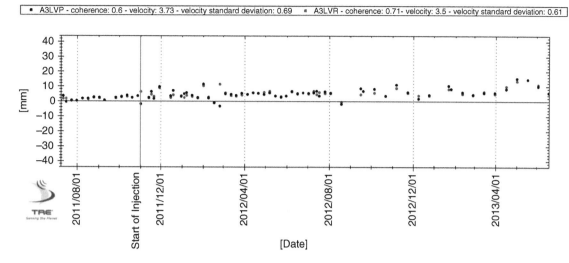

Figure 6.14 Time series of two measurement points in the vicinity of the injection well. The vertical blue line indicates the start of injection.

The rocks in these formations are well suited for the storage of CO_2 because they are porous, permeable, and thick, leading to huge storage volumes, and are overlain by substantial geological seals that will impede the upward migration of the injected fluid.

Various *in situ* and remote sensing data are collected at the Aquistore site for the determination of the fluid distribution in the subsurface, pressure changes, and ground deformation associated with the injection. For measuring surface deformation, high-resolution ascending and descending SAR RADARSAT-2 Spotlight and Wide Ultra-Fine SAR data were acquired starting in June 2012. The Spotlight data consist of ascending Spotlight 18 (SLA18) and descending Spotlight 12 (SLA12) images, acquired with the incidence angles of 44 and 40 degrees and the range-azimuth spatial resolution of 1.6 by 0.8 m. The Wide Ultra-Fine data consist of ascending and descending Wide 2 Ultra-Fine 7 (U7W2) images, acquired with an incidence angle of 37 degrees and the range-azimuth spatial resolution of 1.6 by 2.8 m. The overall temporal resolution is increased by a factor of 4, from 24 (i.e., standard RADARSAT-2 revisit cycle) to 6 days, by combining four independent data sets. A 1-m resolution LiDAR DEM is used to remove the topographic phase contribution.

Differential interferograms are computed from SAR data, geocoded, resampled to a common grid, and processed with the MSBAS technique. The result

is a two-dimensional, vertical and horizontal east–west deformation, time series referenced to a stable point (R in Figure 6.10). Although MSBAS is capable of measuring all three components of deformation, such processing requires data from other than the near-polar orbiting sensors. The availability of space-borne SAR data has been steadily improving, making this remote sensing technique a valuable, cost-effective alternative to GNSS receivers and leveling. In an effort to improve the precision of the deformation measurements, especially during winter when the ground is covered by snow, a network of corner reflectors was installed with reflectors at nine monitoring sites marked as NE1, NE2, SE1, SE2, SE3, SITE, SW1, NW1, and NW2 (Figure 6.15). These corner reflectors, designed at the Canada Centre for Mapping and Earth Observation, Natural Resources Canada, consist of two trihedral reflectors positioned for ascending and descending imaging geometry, mounted on a single monument. To more accurately capture ground deformation from a deep source such as the injected CO_2, and to reduce signals from the near-surface processes caused by seasonal changes and post-mining soil settlement, the corner reflector monuments were anchored at a depth of 20 m.

RADARSAT-2 SAR data have been collected since June 2012, prior to the injection that started in April 2015. The nearly three years of preinjection DInSAR observation allowed for the mapping of ground deformation due to background processes.

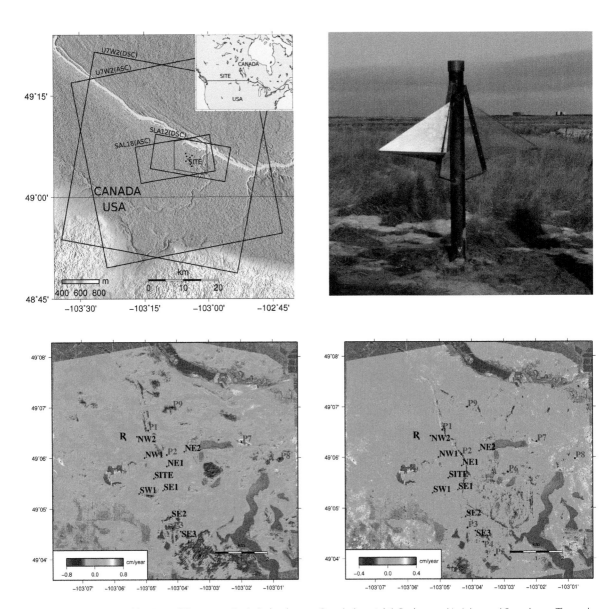

Figure 6.15 Location of Aquistore CO_2 storage site in Saskatchewan, Canada (top right). Background is Advanced Spaceborne Thermal Emission and Reflection Radiometer (ASTER) 30 m resolution Digital Elevation Model (DEM). RADARSAT-2 frames are outlined in black. Region of interest is outlined in brown. Extent in top right corner shows location of study region in North America. Photograph of corner reflector designed at Canada Centre for Mapping and Earth Observation, Natural Resources Canada by K. Murnaghan for ascending and descending imaging (top left). Observed vertical (bottom left) and east–west horizontal (bottom right) deformation rates computed by applying multidimensional small baseline (MSBAS) technique to RADARSAT-2 data spanning 20120615–20140923. Monitoring sites and reference region R are plotted in black. Points P1–P9 experiencing fast ground deformation are plotted in red. SAR intensity image is exposed in background in regions of low coherence.

The vertical and east–west horizontal deformation maps are shown in Figure 6.15. During the 2012–2015 time interval, active ground deformation was observed across the region. Vertical deformation exceeded 0.8 cm/year and subsidence was predominantly observed in areas of post-mining reclamation, where construction activities have affected the surface. Uplift was observed at manmade sites but also in other locations influenced by groundwater recharge and clay swelling. Horizontal motion of up to 0.4 cm/year was observed mainly along the steep slopes of the diversion canal. Analysis of the preinjection deformation maps documented that some monitoring sites are affected by near-surface deformation processes, in particular sites located south of the injection well (SW1, SE2, SE3), emphasizing the importance of having preinjection baseline observations. The InSAR observations serve to verify performance of the sequestration effort and to detect any anomalous behavior.

Conclusions

Geodetic methods, particularly interferometric synthetic aperture radar (InSAR), provide a cost-effective approach for monitoring the geological storage of CO_2. Satellite-based techniques are nonobtrusive and, in many cases, do not require the placement of instrumentation on the ground. Measurements of surface displacement are particularly well suited for detecting the upward migration of CO_2 under pressure, due to the sensitivity of ground motion to near-surface stress changes. The technology for geodetic monitoring is improving over time, with satellite revisit times decreasing and new processing algorithms under development. Inversion and interpretation approaches are also advancing, increasing in sophistication and efficiency.

There are a number of issues associated with the use of geodetic data for monitoring subsurface processes. Some techniques, such as InSAR, are sensitive to the nature of the ground surface and to the presence or absence of vegetation. Geodetic data gathered at the Earth's surface generally provide poor depth resolution of subsurface processes, and issues of uniqueness are often significant. The development of a conceptual model is not always straightforward and often requires additional data, such as seismic observations. Combining various data sets appears to be a promising approach for reducing the nonuniqueness of the inverse problem. However, such large-scale inversions and imaging can be a computational challenge, particularly when combining disparate data types, such as flow observations and geodetic measurements.

The geological storage of CO_2 is not yet a common occurrence and has been implemented only at a handful of sites. Therefore, our experience with monitoring techniques is rather limited. Still, InSAR monitoring has been documented in a few sites, including the three described here – In Salah, Algeria, Decatur, Illinois – and the Aquistore site in

Canada. The InSAR monitoring described in this chapter provided insight, and some unexpected results, regarding the migration of injected CO_2 at the In Salah gas development site. For example, the magnitude of surface deformation was significantly larger than anticipated. In addition, the geodetic monitoring suggested that fault/fracture zones controlled the flow around each of the injectors at In Salah. Such fault/fracture zones were not observed in the preinjection seismic survey and not detected during the drilling of the injection wells. The monitoring results from the Decatur and Aquistore sites appear to indicate much less pressure buildup and associated deformation due to the injected CO_2. This may be due to the fact that the target formations contain a much greater storage volume and the flow around the wells is much more uniform, leading to smoothly varying pressure changes and smaller pressure gradients.

Acknowledgments

Portions of this work were performed under the auspices of the GEOSEQ project for the Assistant Secretary for Fossil Energy, Office of Coal and Power Systems, through the National Energy Technology Laboratory, U.S. Department of Energy by Lawrence Berkeley National Laboratory under contract number DE-AC02-05CH11231 and by Lawrence Livermore National Laboratory under Contract DE-AC52-07NA27344. All authors would like to thank the members of the Joint Industry Project and the Joint Venture, in particular BP, Sonatrach, and Statoil for their cooperation and support at every stage of this work. Special thanks go out to Iain Wright, Allen Mathieson, Sue Raikes, John Midgley, Catherine Gibson-Poole, David Roberts, Stephen Cawley, Phil Ringrose, Lykke Gemmer, and the other members of the Joint Industry Project. We would like to thank the Canadian Space Agency for providing RADARSAT-2 data.

References

Aki, K., and Richards, P. G. (1980). *Quantitative seismology*. San Francisco: Freeman and Sons.

Bamler, R., and Hartl, P. (1998). Synthetic aperture radar interferometry. *Inverse Problems*, **14**: R1–R54.

Berardino, P., Fornaro, G., and Lanari, R. (2002). A new algorithm for surface deformation monitoring based on small baseline differential SAR interferograms. *IEEE Transactions on Geoscience and Remote Sensing*, **40**: 2375–2383.

Chadwick, R. A., Williams, G. A., Williams, J. D. O., and Noy, D. J. (2012). Measuring pressure performance of a large saline aquifer during industrial-scale CO_2 injection: The Utsira Sand, Nowegian North Sea. *International Journal of Greenhouse Gas Control*, **10**: 374–388.

Czarnogorska, M., Samsonov, S., and White, D. (2016). Airborne and spaceborne remote sensing characterization for Aquistore carbon capture and storage site. *Canadian Journal of Remote Sensing*, **42**: 274–291. DOI:10.1080/07038992.2016.1171131.

Davis, P. M. (1983). Surface deformation associated with a dipping hydrofracture. *Journal of Geophysical Research*, **88**: 5826–5834.

Dzurisin, D. (2007). *Volcano deformation: Geodetic monitoring techniques*. Chichester: Springer.

Falorni, G, Hsiao, V., Iannaconne, J., Morgan, J., and Michaud, J.-S. (2014). InSAR monitoring of ground deformation at the Illinois Basin Decatur Project. In *Carbon dioxide capture for storage in deep geological formations*, 4. Thatcham, Berks: CPL Press.

Ferretti, A. (2014). *Satellite InSAR data: Reservoir monitoring from space*. Houten, The Netherlands: EAGE Publications.

Ferretti, A., Prati, C., and Rocca, F. (2000). Nonlinear subsidence rate estimation using permanent scatterers in differential SAR interferometry. *IEEE Transactions on Geoscience and Remote Sensing*, **38**: 2202–2212.

Ferretti, A., Prati, C., and Rocca, F. (2001). Permanent scatterers in SAR inferometry. *IEEE Transactions on Geoscience and Remote Sensing*, **39**: 8–20.

Ferretti, A., Monti-Guarnieri, A., Prati, C., Rocca, F., and Massonnet, D. (2007). *InSAR Principles: Guidelines for SAR Interferometry Processing and Interpretation*. Noordwijk, The Netherlands: ESA Publications, TM-19.

Ferretti, A., Fumagalli, A., Novali, F., Prati, C., Rocca, F., and Rucci, A. (2011). A new algorithm for processing interferometric data-stacks: SqueeSAR. *IEEE Transactions on Geoscience and Remote Sensing*, **49**: 3460–3470.

Finley, R. J., Greenberg, S. D., Frailey, S. M., Krapac, I. G., Leetaru, H. E., and Marsteller, S. (2011). The path to a successful one-million tonne demonstration of geological sequestration: Characterization, cooperation, and collaboration. *Energy Procedia*, **4**: 4770–4776.

Finley, R. J., Frailey, S. M., Leetaru, H. E., Senel, O., Coueslan, M. L., and Marsteller, S. (2013). Early

operational experience at a one-million tonne CCS demonstration project, Decatur, Illinois. *Energy Procedia*, **37**: 6149–6155.

Fujiwara, S., Nishimura, T., Murakami, M., Nakagawa, H., Tobita, M., and Rosen, P. A. (2000). 2.5-D surface deformation of M 6.1 earthquake near Mt. Iwate detected by SAR interferometry. *Geophysical Research Letters*, **27**: 2049–2052.

Funning, G. J., Parsons, B., Wright, T. J., Jackson, J. A., and Fielding, E. J. (2005). Surface displacements and source parameters of the 2003 Bam (Iran) earthquake from Envisat advanced synthetic aperture radar imagery. *Journal of Geophysical Research*, **110**: B09406. DOI:10.1029/2004JB003338.

Hovorka, S. D., Meckel, T. A., and Trevino, R. H. (2013). Monitoring large-volume injection at Cranfield, Mississippi-Project design and recommendations. *International Journal of Greenhouse Gas Control*, **18**: 345–360. DOI:10.1016/j.ijggc.2013.03.021.

Iding, M., and Ringrose, P. (2010). Evaluating the impact of fractures on the performance of the In Salah CO_2 storage site. *International Journal of Greenhouse Gas Control*, **4**: 242–248. DOI:10.1016/j.ijggc.2009.10.016.

Kaven, J. O., Hickman, S. H., McGarr, A. F., Walter, S., and Ellsworth, W. L. (2014). Seismic monitoring at the Decatur, IL, CO_2 sequestration demonstration site. *Energy Procedia*, **63**: 4264–4272.

Klemm, H., Quseimi, I., Novali, F., Ferretti, A., and Tamburini, A. (2010). Monitoring horizontal and vertical surface deformation over a hydrocarbon reservoir by PSInSAR. *First Break*, **28**: 29–37.

Lanari, R., Mora, O., Manunta, M., Mallorqui, J. J., Berardino, P., and Sanosti, E. (2004). A small-baseline approach for investigating deformation on full-resolution differential SAR interferograms. *IEEE Transactions on Geoscience and Remote Sensing*, **42**: 1377–1386.

Massonnet, D., and Feigl, K. L. (1998). Radar interferometry and its application to changes in the Earth's surface. *Reviews of Geophysics*, **36**: 441–500.

Mathieson, A., Midgley, J., Dodds, K., Wright, I., Ringrose, P., and Saoul, N. (2010). CO_2 sequestration monitoring and verification technologies applied at Krechba, Algeria. *Leading Edge*, **29**(2): 216–222.

Mathieson, A. Midgley, J., Wright, I., Saoula, N., and Ringrose, P. (2011). In Salah CO_2 storage JIP: CO2 sequestration monitoring and verification technologies applied at Krechba, Algeria. *Energy Procedia*, **4**: 3596–3603.

McManamon, P. (2015). *Field fuide to Lidar*. Bellingham, WA: SPIE Press.

Menke, W. (1989). *Geophysical data analysis: Discrete inverse theory.* San Diego: Academic Press.

Misra, P., and Enge, P. (2001). *Global positioning system: Signals, measurements, and performance.* Lincoln, MA: Ganga-Jamuna Press.

Norford, B., Haidl, R., Bezys, F.M., Cecile, M., McCabe, H., and Paterson, D. (1994). Middle Ordovician to Lower Devonian strata of the Western Canada Sedimentary Basin. In G. Mossop and I. Shetsen (comp. eds.), *Geological Atlas of the Western Canada Sedimentary Basin.* Edmonton, Alberta: Canadian Society of Petroleum Geologists, Calgary, Alberta and Alberta Research Council, 109–127.

Petrovski, I. G., and Tsujii, T. (2012). *Digital satellite navigation and geophysics.* Cambridge: Cambridge University Press.

Press, W. H., Teukolsky, S. A., Vetterling, W. T., and Flannery, B. P. (2007). *Numerical recipes.* Cambridge: Cambridge University Press.

Ramirez, A., and Foxall, W. (2014). Stochastic inversion of InSAR data to assess the probability of pressure penetration into the lower caprock at In Salah. *International Journal of Greenhouse Gas Control*, **27**, 42–58.

Rosen, P. A., Hensley, S., Joughin, I. R., *et al.* (2000). Synthetic aperture radar interferometry. *Proceedings of the IEEE*, **88**: 333–382.

Rucci, A., Vasco, D. W., and Novali, F. (2013). Monitoring the geologic storage of carbon dioxide using multicomponent SAR interferometry. *Geophysical Journal International*, **193**(1): 197–208.

Rutqvist, J. (2011). Status of TOUGH-FLAC simulator and recent applications related to coupled fluid flow and crustal deformations. *Computational Geoscience*, **37**: 739–750.

Samsonov, S., and d'Oreye, N. (2012). Multidimensional time series analysis of ground deformation from multiple InSAR data sets applied to Virunga Volcanic Province: *Geophysical Journal International*, **191**: 1095–1108. DOI:10.1111/j.1365-246X.2012.05669.x.

Samsonov, S., van der Koij, M., and Tiampo, K. (2011). A simultaneous inversion for deformation rates and topographic errors of DInSAR data utilizing linear least square inversion technique. *Computers and Geosciences*, **37**: 1083–1091.

Samsonov, S., Gonzalez, P., Tiampo, K., and d'Oreye, N. (2013a). Methodology for spatio-temporal analysis of ground deformation occurring near Rice Lake (Saskatchewan) observed by RADARSAT-2 DInSAR during 2008–2011. *Canadian Journal of Remote Sensing*, **39**: 27–33.

Samsonov S., d'Oreye N., and Smets, B. (2013b). Ground deformation associated with post-mining activity at the French-German border revealed by novel InSAR time series method. *International Journal of Applied Earth Observation and Geoinformation*, **23**: 142–154.

Samsonov, S., Gonzalez, P., Tiampo, K., and d'Oreye, N. (2014a). Modeling of fast ground subsidence observed in southern Saskatchewan (Canada) during 2008–2011. *Natural Hazards and Earth System Sciences*, **14**: 247–257. DOI:doi:10.5194/nhess-14-247-2014.

Samsonov, S., d'Oreye, N., Gonzalez, P., Tiampo, K., Ertolahti, L., and Clague, J. (2014b). Rapidly accelerating subsidence in the Greater Vancouver region from two decades of ERS-ENVISAT-RADARSAT-2 DInSAR measurements. *Remote Sensing of Environment*, **143**: 180–191. DOI:10.1016/j.rse.2013.12.017.

Samsonov, S. V., Tiampo, K. F., Camacho, A. G., Fernandez, J., and Gonzalez, P. J. (2014c). Spatiotemporal analysis and interpretation of 1993–2013 ground deformation at Campi Flegrei, Italy, observed by advanced DInSAR. *Geophysical Research Letters*, **41**: 6101–6108. DOI:10.1002/2014GL060595.

Samsonov, S. V., Trishchenko, A. P., Tiampo, K. Tiampo, Gonzalez, P. J., Zhang, Y., and Fernandez, J. (2014d). Removal of systematic seasonal atmospheric signal from interferometric synthetic aperture radar ground deformation time series. *Geophysical Research Letters*, **41**: 6123–6130. DOI:10.1002/2014GL061307.

Samsonov, S., Czarnogorska, M., and White, D. (2015). Satellite interferometry for high-precision detection of ground deformation at a carbon dioxide storage site. *International Journal of Greenhouse Gas Control*, **42**: 188–199. DOI:10.1016/j.ijggc.2015.07.034.

Samsonov, S., Tiampo, K., and Feng, W. (2016a). Fast subsidence in downtown of Seattle observed with satellite radar. *Remote Sensing Applications: Society and Environment*, **4**: 179–187. DOI:10.1016/j.rsase.2016.10.001.

Samsonov, S. V., Lantz, T. C., Kokelj, S. V., and Zhang, Y. (2016b). Growth of a young pingo in the Canadian Arctic observed by RADARSAT-2 interferometric satellite radar. *The Cryosphere*, **10**: 799–810. DOI:10.5194/tc-10-799-2016.

Shi, J.-Q., Sinayuc, C., Durucan, S., and Korre, A. (2012). Assessment of carbon dioxide plume behavior within the storage reservoir and the lower caprock around the KB-502 injection well at In Salah. *International Journal of Greenhouse Gas Control*, **7**: 115–126.

Smith, J. R. (1997). *Introduction to geodesy: The history and concepts of modern geodesy.* New York: John Wiley & Sons.

Tamburini, A., Bianchi, M., Chiara, G., and Novali, F. (2010). Retrieving surface deformation by PSInSAR technology: A powerful tool in reservoir monitoring. *International Journal of Greenhouse Gas Control*, **4**: 928–937.

Teatini, P., Castelletto, N., Ferronato, M., *et al.* (2011). Geomechanical response to seasonal gas storage in depleted reservoirs: A case study in the Po River basin, Italy. *Journal of Geophysical Research*, **116**: 1–21.

Torge, W. and Muller, J. (2012). *Geodesy*. Berlin: Walter de Gruyter.

Vasco, D. W., Ferretti, A., and Novali, F. (2008). Estimating permeability from quasi-static deformation: Temporal variations and arrival time inversion. *Geophysics*, **73**: O37–O52. DOI:10.1190/1.2978164.

Vasco, D. W., Rucci, A., Ferretti, A., *et al.* (2010). Satellite-based measurements of surface deformation reveal fluid flow associated with the geological storage of carbon dioxide. *Geophysical Research Letters*, **37**: L03303, 1–5. DOI:10.1029/2009GL041544.

White, D. J., Meadows, M., Cole, S., *et al.* (2011). Geophysical monitoring of the Weyburn CO_2 flood: Results during 10 years of injection. *Energy Procedia*, **4**: 3628–3635.

Worth, K., White, D., Chalaturnyk, R., *et al.* (2014). Aquistore project measurement, monitoring and verification: From concept to CO_2 injection. *Energy Procedia*, **63**: 3202–3208. http://dx.doi.org/10.1016/j.egypro.2014.11.345.

Wright, C. A. (1998). Tiltmeter fracture mapping: From the surface and now downhole. *Petroleum Engineer International*, **71**: 50–63.

Wright, T. J., Parsons, B. E., and Lu, Z. (2004). Toward mapping surface deformation in three dimensions using InSAR. *Geophysical Research Letters*, **31**: 1–5.

Gravity
Surface and Borehole

Ola Eiken

Introduction

The technology of gravity monitoring has improved over the decades, with an increasing number and range of applications. Time-lapse gravity data can be more straightforward to interpret than most other geophysical data, because the signal is proportional to mass changes. This is a fundamental property to understand in a dynamic underground reservoir. The method, which also is termed microgravity, is particularly powerful when the geological framework is well known and fluids with large density differences flow. Gas–water and air–water are such favorable fluid contrasts, but also injected and stored carbon dioxide (CO_2) can provide significant negative density contrast when it pushes away water. Further, CO_2 injection into confined reservoirs will cause pore expansion, mass increase, and thus gravity increases around it. At the same time, pore and reservoir expansion will lift the surface up, which will cause gravity reductions at fixed observation platforms.

Gravity changes have been monitored for a variety of manmade and natural subsurface mass changes, including hydrologic, geothermal, and hydrocarbon gas extraction processes. The depth of the mass changes range from sinkholes a few meters below the surface (e.g., Debeglia and Dupoint, 2002) to gas reservoirs kilometers down (e.g., Alnes et al., 2010). Gravity monitoring of geothermal reservoirs has helped determine the natural and injected water inflow distribution and phase since the 1960s, with pioneering work in New Zealand (Allis and Hunt, 1986; Hunt and Kissling, 1994), Japan (Sugihara and Ishido, 2008), Philippines (Nordquist et al., 2004), and Indonesia (Sofyan, 2011). Development of magma chambers in volcanoes has been monitored (e.g., Battaglia et al., 2008; Carbone et al., 2017). Both campaign and continuous gravity observations have made valuable contributions, and volcanic eruptions have been correctly predicted. Jacob et al. (2010), Wilson et al. (2012), and Christiansen et al. (2011) all reported successful hydrology monitoring. Measurements on hydrocarbon fields with moving gas–fluid fronts started at the giant Groningen gas field in 1978 (Gelderen et al., 1999; Eiken et al., 2017). Seven gas fields in the North Sea have been monitored successfully (Alnes et al., 2008; Eiken et al., 2008; Vevatne et al., 2012; Van den Beukel et al., 2014). Successive microgravity surveys over the gas cap of the Prudhoe Bay oil field from 2003 to 2008 provided insights into the distribution of water injected into the gas cap (Brady et al., 2008; Ferguson et al., 2008; Hare et al., 2008; Yin et al., 2016).

Techniques of Gravity Surveying

A basic limitation is set by measurement accuracy, or noise level. While measurements with precision down to a few μGal (one μGal is 10^{-8} ms^{-2}) have been obtained at laboratory conditions for decades (e.g., Torge, 1989), recent advances have made it practically possible to carry out field surveys at such accuracy, and stationary measurement at sub-μGal precision. This increases the potential applications of gravity monitoring to smaller and deeper reservoirs, with lower density contrasts and at shorter time-lapse intervals. High-precision measurements of gravity may be done by (1) absolute free-fall gravimeters, (2) superconducting gravimeters, or (3) relative spring gravimeters. The various instruments all have their advantages and limitations. Some key accuracy properties are listed in Table 7.1.

Absolute free-fall gravimeters measure the drop time of a mass in a vacuum chamber. These are immune to sensor drift. Repeated measurements are thus not required during a survey, and reference stations outside the area of interest are not needed to see

Table 7.1 Summary of the most common gravity measurement techniques, and typical accuracies

Type of gravimeter	Accuracy (µGal)	Drift (µGal/yr)	Typical duration of a single measurement (hours)	Environment
Absolute free-fall	2–10	none	6–24	Land
Superconducting	0.01–1	5–50	>240	Land
Relative spring	1–5	>1000	0.05–0.5	Land/sea/borehole

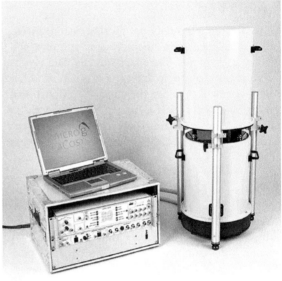

Figure 7.1 Absolute gravity instruments: FG5X (left) and A-10 (right). (From Micro-g LaCoste, http://www.microglacoste.com/absolutemeters .php)

time-lapse changes. Semiportable instruments, the FG5 series (Figure 7.1), require setup and demobilization times of a few hours at an indoor site, and may give 2 µGal accuracy and 1 µGal repeatability. The smaller and more easily portable absolute A-10 instrument can give 10 µGal accuracy and repeatability in a 10-minute reading on a quiet site. It can be operated outdoors, at temperatures ranging from –18°C to +38°C.

Superconducting gravimeters measure a spherical superconducting mass, levitated using a magnetic force that exactly balances the force of gravity. The iGRAV (Figure 7.2) requires a setup time of about five hours and a significantly longer sensor stabilization time, which makes the instrument best suited for months to years long continuous recording on a site. Resolution is better than 0.05 µGal after one-minute averaging, and it is therefore particularly

sensitive to measuring tides and atmospheric influence. Drift for state-of-the art instrumentation is given as less than 0.5 µGal/month, and scale factor will be stable within 10^{-2} for years (Calvo *et al.*, 2014). Even with these low values, for monitoring over several years, drift uncertainty can build up to several µGal, and calibration by, for example, an absolute meter is required. Sugihara *et al.* (2013) discussed the use of superconducting gravimeters in CO_2 storage monitoring, and considered it a promising, though yet unproved, technique. Kim *et al.* (2015) found that a CO_2 signal larger than about 0.5 µGal can be detected with an iGrav's continuous recordings, by superposing real time series from superconducting meters onto modeled CO_2 signals.

A relative spring meter utilizes a spring with a proof mass at the end, which extends proportional

Figure 7.2 iGrav instrument. (From GWR instruments, http://www
.gwrinstruments.com/pdf/principles-of-operation.pdf)

to the force of gravity. The position of the mass is accurately read off. The supplier Scintrex uses a quartz spring with a drift of some µGal/hour, and the company ZLS uses metal zero-length springs (Figure 7.3). Most of the drift is linear, and much of the residual

drift can be controlled by repeated measurements during a survey. The scale factor needs to be calibrated from time to time. Repeatability is given at a few µGal. A seafloor version of the CG5 gravimeter has been built for offshore monitoring (Zumberge *et al.*, 2008; Figure 7.3).

Relative gravimeters can be used in campaigns/surveys, which is when one instrument is doing sequential measurements at a grid of stations. Sensor drift is controlled by making repeat measurements at the same station. Using multiple measurements at a site also reduces other types of noise, such as random background noise. The uncertainty in final station values can be significantly lower than the reading accuracy of each measurement, and repeatabilities of 1–2 µGal have been reported from several recent surveys (Vevatne *et al.*, 2012; Eiken *et al.*, 2017; Lien *et al.*, 2017).

Gravity data need several corrections before station values and time-lapse changes can be estimated (Table 7.2). Land data need to be corrected for variations in groundwater and shallow aquifers (hydrology). Seafloor data need to be corrected for the varying gravity attraction from the ocean above (ocean tides and seawater density effects). All data need to be corrected for the varying gravity attraction from the atmosphere, earth tides, and ocean loading, either by a good model or by actual recordings by a stationary reference gravimeter.

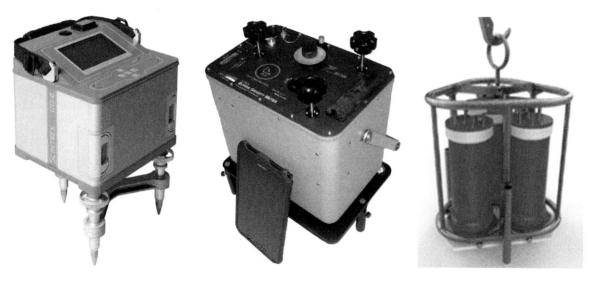

Figure 7.3 Spring-based instruments. (Left) Scintrex land meter (http://www.scintrexltd.com/dat/content/file/CG-6%20Brochure_R15(1)
.pdf). (Middle) ZLS land meter (http://zlscorp.com/?page_id=33). (Right) Quad Geometrics seafloor meters (http://quadgeo.com/services/
gravity-surveys/equipment/).

Table 7.2 Some gravity effects to be corrected for in time-lapse gravimetry processing

Effect	Typical size of signal (μGal)	Residual noise (μGal)
Sensor drift	10–1000/day	1–10
Earth tides	200	<1
Ocean loading	20	<1
Atmospheric variations	<10	<1
Microseism	50–5000 in the band >0.1 Hz	<1
Groundwater	<50	1–10

Stability of measurement platforms over years is required for high-quality time-lapse data. This can be achieved by installing benchmarks, made of, e.g., massive concrete or metal. In the North Sea, conical concrete benchmarks have been residing on the seafloor for this purpose for nearly 20 years (Stenvold et al., 2006). When vertical movements of the measurement platforms are measured independently, e.g., by optical land leveling, inteferometric synthetic aperture radar (InSAR), or GPS satellite measurements on land or by water pressure at the seafloor, the gravity data can be corrected for height changes, by using the appropriate vertical gravity gradient.

A surface gravity monitoring program may be designed as a survey of a grid with relative gravimeter-(s), or with portable absolute meters, or a sparser grid of stations with permanent superconducting gravimeters. Surveys using a combination of these types of instrumentation may be termed hybrid gravimetry.

By measuring gravity in boreholes, the sensors get closer to the storage reservoir and thus stronger time-lapse signals can be observed. Recent advances have made the instruments to work in smaller, more deviated boreholes (Ander and Chapin, 1997; Chapin and Ander, 2000; Seigel et al., 2007; Nind and MacQueen, 2013). Quoted measurement accuracy range from 1 to 5 μGal. Borehole gravity data will have a challenge with depth repeatability. The gravity gradient in a borehole is given by:

$$\frac{\Delta g}{\Delta z} = 3.086 - 0.838 \cdot \rho,$$

in μGal/cm, when ρ is in kg/m^3

Seigel et al. (2007) estimated ± 5 cm depth uncertainty, which transform to 5–7 μGal for rock densities

of 2000–2500 kg/m^3. Nind and MacQueen (2013) stated that depth-induced errors in Δg are typically comparable to the sensitivity of the gravity measurements themselves. Depth measurements may be obtained using high-resolution casing collar locator logs.

Borehole data are obviously limited by where the boreholes are, their geometry, and whether they can be accessed. CO_2 injection operations will aim at minimizing the number of injection wells, which may be relatively sparser than at many oil and gas fields.

Observation points may be above the reservoir, and time-lapse changes will be in the same directions as surface gravity but larger, due to closer proximity to the reservoir.

Borehole gravity can also be used as a density logging tool with deep penetration into the formation. When measurements are made at different depths, the gravity and height difference will tell the average formation density. A Shuttle Sonde (trademark of Microg LaCoste) can reduce the depth increment error to ± 1 mm over a range of 2.5 m or less. When done repeatedly, the density changes can be used to infer saturation changes in the depth interval. This has been done successfully for producing gas reservoirs, with an estimate of residual gas saturation (e.g., Adams, 1991). Such use of borehole gravity data "sees" much farther into the formation than most other logging tools, and is thus less vulnerable to hole breakouts, casing effects, etc.

Gravity gradiometry measures the spatial gravity gradients (in three spatial directions). If derived from gravity measurements, the gradients offer no additional information or reduced noise level in the interpretation stage. Some gradiometers can measure the gravity gradients directly. So far, no surveying capabilities have been reported that are comparable to or better than gravity sensors.

CO₂ Densities in the Subsurface

The phase and density of injected CO_2 in the storage reservoir will depend highly on temperature and pressure, as shown in the phase diagram (Figure 7.4). The critical point at 31.1°C and 7.39 MPa makes a supercritical state likely for depths exceeding 1 km. CO_2 density versus depth for three geothermal gradients are shown in Figure 7.5. The "cold case" gives sharp transition at about 500 m depth, while the warmer situations have more gradual density transitions.

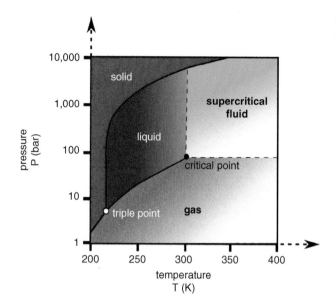

Figure 7.4 Phase diagram for CO_2 (http://en.wikpedia.org/wiki/Supercritical_fluid).

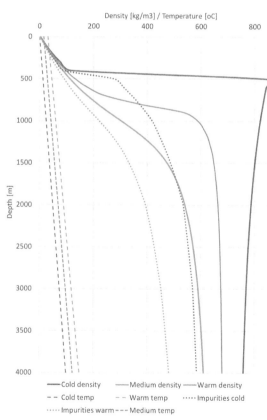

Figure 7.5 Density of CO_2 vs. depth shown as solid lines for three different geothermal gradients, shown as dashed lines. The gradients are (1) 0°C + 25°C/km, (2) 15°C + 27.5°C/km, and (3) 30°C + 30°C/km, and shown as dashed lines. The effects of 10% methane impurities are shown as fine dashed lines for the warmest and coldest cases.

CO_2 will typically have high pressure at the entrance of an injection well (wellhead). The temperature may be the ambient surface (air or water) temperature or higher, depending on the distance from the compression facility. As the CO_2 gets further compressed down the well by the weight of CO_2 above, the temperature will change (generally increase) due to compression and heat transfer with the surrounding rock (e.g., Singhe *et al.*, 2013). For instance, at Sleipner, the temperature has been estimated to increase from 25°C at the injection point to 48°C at the perforation (Alnes *et al.*, 2011). When flowing into the reservoir, the CO_2 will generally take the formation temperature quickly, due to the much greater heat capacity of the rock and formation water.

If CO_2 is pushing away and replacing formation water in the reservoir, this storage mechanism may involve only minor pressure changes. Neglecting pressure changes, the density contrast can be described as

$$\Delta\rho = \theta \cdot (S_{w,i} - S_{w,r}) \cdot (\rho_w - \rho_{CO2}) \qquad (7.1)$$

where

$\Delta\rho$ is rock density change
θ is porosity
$S_{w,i}$ is initial water saturation
$S_{w,r}$ is residual water saturation
ρ_w is water density
ρ_{CO_2} is CO_2 density

119

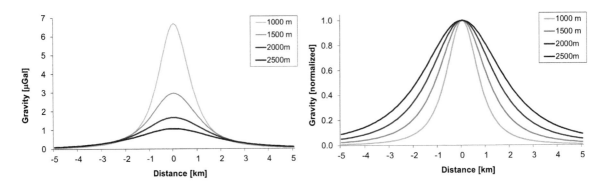

Figure 7.6 (Left) Point mass response to 1 Mtonne at various depths. (Right) Scaled to the same peak level.

In the case of CO_2 pushing away water, both the density and saturation changes may be significant, and vary significantly from case to case. Saturation changes will depend on the process of fluid replacement, and is likely to be higher for high porosity and permeability.

Pressure increase during injection will cause the storage formation to expand and create extra space. The pore volume increase may occur following linear elasticity:

$$V = V_0 \cdot (1 + c_p \cdot \Delta p) \qquad (7.2)$$

where

    V is rock volume
    V_0 is initial rock volume
    c_p is pore compressibility
    Δp is pressure change

Such expansion could occur as expansion of all pores, or more heterogeneously, e.g., by a much larger opening of a few fractures. In either case, the extra space will most often be filled with CO_2, causing mass increase and gravity changes, as well as deformation of the reservoir and the surrounding rocks.

The density of the formation water will generally be higher for higher salinity and pressure, and lower for higher temperature. Borehole data on these will usually provide sufficient information for the modeling.

Most often the injected CO_2 will have impurities, depending on the processes that have treated the CO_2 prior to entering the wellhead. If the CO_2 has been separated from flue gas in a combustion plant or separated from an initial mix with hydrocarbon gases, the impurities may be dominated by nitrogen or methane. The volumetric fractions of such impurities can dramatically increase from the surface to the reservoir. An example for a 10% mole fraction of methane is shown in Figure 7.5. This can reduce the density to less than half in the most extreme cases.

Overburden/Burial

The depth of the storage reservoir will not only influence pressure and temperature conditions and thus the *in situ* density of CO_2, but also the signal decay caused by the distance between the reservoir and the observation points on the surface. The scalar gravity change from a mass change Δm is given by Newton's law:

$$\Delta g = G \frac{\Delta m}{d^2}$$

where

    Δg is gravity change
    G is the gravitational constant; measured as 6.67408×10^{-11} m^3 kg^{-1}s^{-2}
    d is the distance between the mass change and the observation point

For a reservoir body of some extension, the surface gravity change can be found by volume integration of each mass change in Newton's formula. A gravimeter measures the scalar (vertical) gravity attraction. For small changes, therefore, only the vertical component of change can be inferred. The measured gravity change, Δg_z, then becomes

$$\Delta g_z = G \frac{\Delta m \cdot z}{d^3}$$

where z is depth.

An example of the point mass response of 1 Mtonne CO_2 at various depths is shown in Figure 7.6. The response will both have lower amplitude and longer wavelengths for deeper sources.

Gravity gradiometers represent another measurement principle. Gravity gradients decrease faster than gravity itself with increasing distance, by $\frac{1}{d^3}$, and therefore the strength of the change will decay more rapidly with increasing reservoir depth. However, the shorter wavelengths are relatively better preserved in a grid of surface observations, and can thus provide higher lateral resolution in data before inversion. After inversion, the signal-to-noise ratio will determine resolution, and current gradiometers are mostly inferior to gravimeters, except for very shallow reservoirs.

Surface Uplift/Height Changes

The reservoir pressure buildup associated with CO_2 injection will cause mechanical deformations, usually pore expansion. This will again cause deformations in the surrounding rocks, and likely uplift of the surface, which may be measurable or too small to detect. Some horizontal strain will also occur. Pore expansion will create additional storage space, which in some cases may be the primary storage mechanism, e.g., in tight reservoirs (Kabiradeh and Sideris, 2016). In most situations, except for very large and permeable formations with low injection rates, the pressure increase will be an important parameter for managing the injection.

Surface height changes may on land be monitored by optical leveling, InSAR, or GPS. This is discussed in Chapter 15 by Don White in this volume. At sea, water pressure surveys of stable seafloor benchmarks can provide height change data with relative accuracy of 2–4 mm (Stenvold et al., 2006), at water depths ranging from 80 to 1200 m and survey areas up to 1000 km^2.

Height change (uplift or subsidence) of the surface, and thereby the gravity stations themselves, cause a change in the observed gravity. The free-air gravity gradient, and uplift or subsidence gradient, is on average about 308 µGal/m, but may vary several tens of µGal/m across a survey area if it is influenced by the local topography (e.g., Hunt et al., 2002). When the seafloor is lifted up, there will be less water above the observation point. The influence on gravity will be a combination of the free-air gradient and the (opposite sign) gravity attraction of a Bouguer plate of water, $\frac{\Delta g}{z} = 2 \cdot \pi \cdot \rho_{water} \cdot G$, which on average reduces the vertical gradient to about 267 µGal/cm. Local variations in the gradient across a survey area are likely to

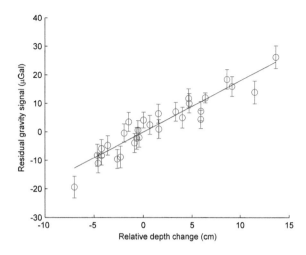

Figure 7.7 Correlation between depth changes and gravity changes for the outermost benchmarks of Sleipner 2005–2002. The slope of the best fitting line is a gravity gradient of 180 µGal/m. (From Alnes et al., 2008.)

be less significant offshore, as the seafloor topography generally is much flatter than terrestrial landforms.

Settlement of the benchmarks causes local height changes, which may have another vertical gradient, dependent on the mechanisms of deformation. As an example, scouring around and beneath the benchmarks at Sleipner caused them to sink, with a gradient of 160 mGal/m estimated from the data (Figure 7.7).

A significant uncertainty in the correction to the gravity data may arise from the subsidence measurements. Subsidence correction will often be a significant part of the total gravity error budget. For example, an uncertainty of 5 mm in height transforms to 1.3–1.5 µGal in gravity. Some uncertainty will also come from the vertical gradient, either the land topography or the seawater density.

Surface elevation change is not only needed for gravity monitoring, but can, in itself, be an important monitoring tool. The size of the uplift will provide information about the size of the reservoir expansion and thus the additional pore space available for storage. The aerial distribution of uplift can be used to infer the lateral distribution of pressure increase and how it decays away from the injection well, dependent on the reservoir geometry, barriers, etc. The In Salah injection project (Figure 7.8) is a good example of such an effect, and has been analyzed (Ringrose et al., 2013). Another example is the

Map of surface uplift

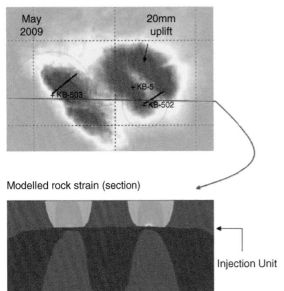

Modelled rock strain (section)

Figure 7.8 (Above) Surface uplift at the In Salah CO$_2$ injection sites as inferred from InSAR data. (Below) Modeled rock strain. (From Ringrose *et al.*, 2013.)

summer–winter cycling of a gas storage reservoir (Figure 7.9).

Such inferences from surface data to reservoir require a working model for the overburden response to deformations. Geomechanical modeling is a comprehensive topic of its own. Simple models as the Geertsma (1973) or van Opstal (1974) assuming homogeneous elastic properties are widely used and

appear to have worked well. More advanced models may reveal further detail.

Modeling Gravity Response for Various Storage Mechanisms

Time-lapse gravity changes are straightforward to calculate from flow models, based on the fluid saturations and pressures at each time step, as outputted from reservoir simulators such as Eclipse (www.software.slb.com/products/eclipse), TOUGH2 (Preuss, 1998; http://esd1.lbl.gov/research/projects/tough/) or STAR (Pritchett, 1995; www.cmgl.ca/stars). The temperature may be included as a model parameter, as CO$_2$ density alters significantly with temperature. This would differ from the common isothermal practice for modeling of production from or injection into oil and gas reservoirs. If flow models are not available, or not appropriate, less complex models, such as geometrical reservoir models with sketched fluid front and pressure development, may be used.

Forward modeling of scenarios spanning the uncertainties will be the core of gravity feasibility studies, which also should include sensor and survey setup, expected accuracy, and interpretive aspects. Different CO$_2$ storage mechanisms will cause gravity anomalies with different characteristics. Mechanisms may be divided into (1) aquifers, (2) depleted gas fields, (3) overpressure space, and (4) enhanced oil recovery (EOR) projects.

When injecting CO$_2$ into large, usually saline, aquifers, pressure increases are likely to be modest. The main gravity effect will be replacement of water

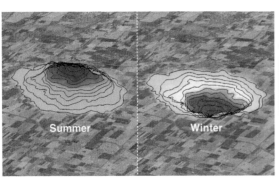

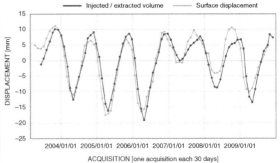

Figure 7.9 Summer injection and winter extraction of gas produce seasonal surface displacement. In the right graph, the displacement time series obtained with SqueeSAR™ shows a strong correlation with the injected/extracted gas volume. (From www.tre-altamira.com/oil-and-gas/)

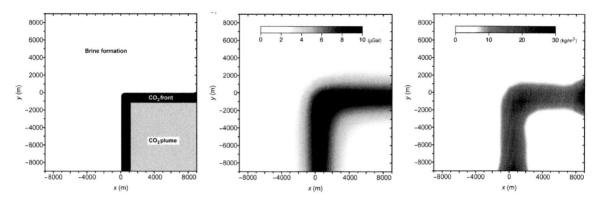

Figure 7.10 (Left) Map view of model of 1-km-side CO_2 fronts. (Middle) Surface gravity response from the model. (Right) Density change by inversion of the gravity data in the middle. (From Gasperikova and Hoversten, 2008.)

with CO_2. The largest density changes will be at the CO_2 front, and the dynamics of the front will be the dominating effect to model and capture. Thermal effects – heating or cooling of the injected CO_2 when filtering through the formation – may have a secondary effect on the CO_2 density. Dissolution of CO_2 in the formation water may be a slow process, but ongoing for a long time after site closure. It will increase the density of the water and decrease the volume of the CO_2 plume, and thus cause a slight increase in gravity with time. Such information may be valuable in a postinjection, site closure, phase.

Three model studies of aquifer storage have been published. Gasperikova and Hoversten (2008) modeled the time-lapse gravity response for an advancing 1 km wide CO_2 front in a 20 m thick brine formation with 20% porosity at a depth of 1900 m (Figure 7.10). With a maximum of 30% CO_2 saturation in the generated plume, the maximum surface change was 10 μGal. The modeled data were inverted after 2.5 μGal random noise was added. Inversion with the density variations constrained to occur between the top and bottom reservoir surfaces recovered the density changes within 30% of the true value. Gasperikova and Hoversten (2008) concluded that it is possible to recover the general position of density changes caused by advancement of the CO_2 front, but not the absolute value of the change. They further advocated gravity monitoring as a low-cost supplement to or replacement of seismic monitoring.

Sugihara et al. (2013) modeled the injection of 10 Mtonnes of CO_2 into an aquifer at a depth of 2050 m,

causing a 4 μGal gravity change. To discriminate between models, they concluded that a precision better than 0.1 μGal/year would be needed. Sugihara et al. (2013) further discussed the feasibility of such detection using superconducting gravimeters (iGrav) together with free-fall absolute calibrations, and concluded it would be challenging with currently available instrumentation.

Jacob et al. (2016) modeled the injection of 18 Mtonnes at 1800 m depth, developing a plume of 2600 m radius (Figure 7.11) and causing an up to 8 μGal gravity reduction. They inverted the plume outline after 5 μGal random noise was added. They found generally that the shape could not be reliably mapped, but the total mass estimate was more reliable (Figure 7.12).

For CO_2 injection into depleted gas fields with pressure much below initial or hydrostatic pressure, the CO_2 is likely to mix with the remaining gas. If there is insignificant water inflow from aquifers, the mass change will be equal to the injected CO_2, and consequently the gravity change will be positive (increase) and of a size roughly proportional to the injected mass. The point mass response shown in Figure 7.6 is relevant for such situations. However, the response must be scaled for the injected mass and adjusted for the reservoir geometry having a finite extension. There might be less uncertainty and interest in the detailed spatial distribution of CO_2 in a depleted gas field, as long as it is contained within the depleted reservoir, and time-lapse gravity may have more limited value than for some of the other storage mechanisms.

123

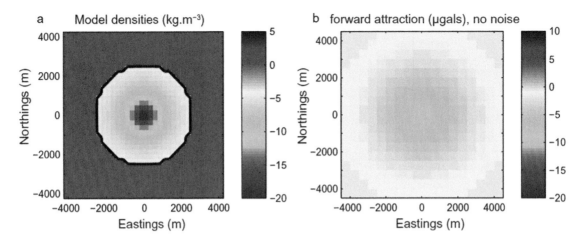

Figure 7.11 (a) Model rock density changes after injection of 18 Mtonnes of CO_2 at 1800 m. The black line represents the saturation front, color scale is clipped. (b) Gravity attraction at the surface from the model densities. (From Jacob *et al.*, 2016.)

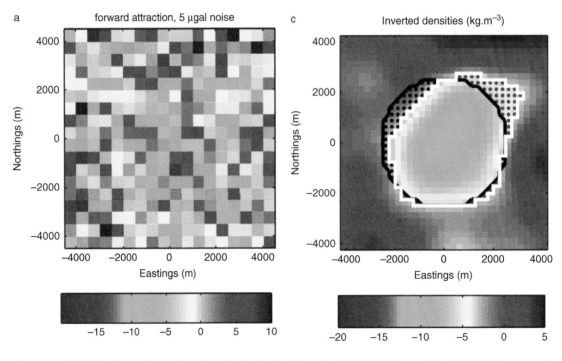

Figure 7.12 (Left) Gravity attraction at the surface with 5 µGal Gaussian noise added. (Right) Inverted densities for the L-curve method. Thick black line: position of the plume in the forward model; white line: inverted plume position. (From Jacob *et al.*, 2016.)

Sherlock *et al.* (2006) modeled the gravity change above a depleted gas reservoir that was 29 m thick and at a depth of about 2 km, with an active aquifer beneath. At the site, 100 000 tonnes were injected over 18 months, and pressure was assumed to be hydrostatic throughout the injection period. The injected CO_2 partly mixed with the remaining methane cap, partly created a plume with CO_2

saturation >50% within the initial gas reservoir, and pushed the gas–water contact downwards. The first two effects increase reservoir density, while the latter reduces the density. The modeled surface gravity changes summed up to a change of about 0.1 μGal, much less than any realistic detection levels. Gravimeters in boreholes closely above the reservoir could detect signals of several μGals, but this would likely be a cost-prohibitive survey (Sherlock *et al.*, 2006).

Gravity monitoring in the late gas extraction phase and after gas depletion (postproduction) could give unique and valuable information about the natural water inflow into the reservoir when in a low-pressure state. This can provide information about regional aquifer strength and connectivity, which can be important for basin-wide CO_2 storage capacity. Such understanding would otherwise require costly acquired well pressure data, and be rare.

Petroleum reservoirs subject to CO_2 injection for EOR have potentially more complex flow patterns than brine formations because they typically have reduced vertical extent and multiple *in situ* fluids (oil, hydrocarbon gas, brine, and CO_2). Gasperikova and Hoversten (2006, 2008) modeled the gravity change of a 25–30 m thick reservoir at 1150–1350 m depth after 20 years when injecting in 22 wells. Maximum gravity change, at the surface was on the order of 3 μGal, not much above current state-of-the-art noise levels, they concluded.

In confined storage reservoirs, pore pressure will increase during injection, and expansion of the formation will create extra storage space. The extra injected CO_2 will cause a positive gravity change at the surface, while the surface uplift will cause a negative change at the observation platforms. Kabirzadeh *et al.* (2017) calculated the uplift effect to easily dominate gravity for typical reservoir parameters. When uplift is measured with sufficient precision (as discussed earlier) this part of the gravity signal can be corrected, and the reservoir changes can be extracted. Real cases may be in between the end cases of confined and unconfined reservoirs, and thus have less uplift and less gravity change. In such situations, it will be important to monitor both gravity and height changes.

Coal formations subject to methane production and CO_2 storage tend to be thin and shallow. Gasperikova and Hoversten (2008) modeled a 3 m thick zone at depths between 400 m and 700 m, with water in the fractures replaced by CO_2, causing a maximum response of 4 μGal – on the edge of detection. Thicker or shallower coal seams would clearly increase the signal.

History Matching

A straightforward analysis of gravity monitor data is done by comparing with the modeled (predicted) time-lapse changes. "History matching" is often the term used when flow models are adjusted to get a better match to measured data. The traditional data used in the petroleum industry are well pressures and production rates/volumes. History matching may also use gravity data, as has been demonstrated by Glegola *et al.* (2012). They found that pressure data combined with gravity gave history matched models with less uncertainty in the initial gas mass and production prognosis than models that had been matched to pressure only.

Mass Change Estimates

The time-lapse version of Gauss's theorem says that the gravity change can be area-integrated to get the total mass change. This is irrespective of shape or depth of the volume undergoing change (e.g., Goetz, 1958; Campos-Enriquez *et al.*, 1998), and the mass change is

$$\Delta M = \frac{1}{4\pi G} \int_A \Delta g_z dA$$

where

ΔM is the anomalous mass
G is the gravitational constant
Δg_z is the change in attraction of gravity
A is a surface that covers the surface gravity anomaly

Signal-to-noise ratios and spatial sampling will limit the accuracy. For estimating mass changes from increasingly deeper reservoirs, an increasing survey area will be required.

A mass change estimate can be a valuable constraint in situations of CO_2 injection and storage. This may be of interest, for instance, when the injected mass is known, and the mass flowing out of the reservoir (area) can be calculated.

Leakage Detection

An important purpose of gravity monitoring can be the detection of any leakage of CO_2 upward from the storage reservoir, toward the surface. The density of

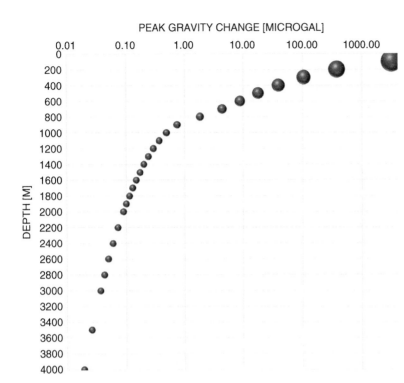

Figure 7.13 Peak gravity change for a spherical CO_2 plume (bubble) replacing saline formation water at various depths. Assumptions are hydrostatic pressure, geothermal gradient of 15°C + 27.5°C/km, water salinity 3.5%. The relative size of the CO_2 bubble is shown for each depth.

CO_2 will decrease as pressure is lowered on the way upwards (Figure 7.5). This will be a particularly strong effect for depths shallower than 500–1000 m, dependent on the temperature distribution. The density contrast and mass change when CO_2 expels water will increase correspondingly. Greater proximity to the observation points on the surface or seafloor will further increase the sensitivity of gravity to small amounts of leaking CO_2.

The peak gravity response of a bubble (plume) of 100 ktonnes of CO_2 that has replaced formation water is shown in Figure 7.13. The size of each bubble is illustrated, with a radius of 33 m at 4000 m depths, expanding to >100 m at 100 m depth. With a detection level of 1–2 µGal, a 100 ktonne concentrated leakage could be detected at about 800 m depth, while a superconducting meter with sensitivity 50 nGal could detect a sudden leak at 2.5 km depth.

A leakage could occur along vertical cracks or fissures with high permeability. Flow rates could be high while only small amounts of CO_2 are resident in the overburden at a given time. This could be challenging to detect. On the other hand, some of the leaking CO_2 may accumulate in pockets or strings of porous rock, e.g., sand or silt, as is commonly found with shallow gas pockets. Such shallow CO_2 accumulations may have easily detectable time-lapse gravity signals.

Shallower sources of mass changes will require denser spatial sampling than the reservoir, in order to keep an aerial coverage without holes. If predefined areas with a higher risk of leakage, such as a mapped zone of fractures, can be identified, permanent gravimeters could be placed above for continuous monitoring. For instance, superconducting gravimeters could detect smaller and more abrupt changes than survey campaigns can.

Sleipner

At the Sleipner CO_2 storage site, seafloor microgravity surveys have so far been carried out in 2002, 2005, 2009, and 2013, using the ROVDOG technique with relative spring meters (Sasagawa et al., 2003; Zumberge et al., 2008). As described in Chapter 13 by Eiken (this volume), nearly 1 Mtonne of CO_2 has been injected yearly since 1996. The Utsira Fm. storage reservoir is situated 700–1000 m below the seafloor, at water depths of 82–83 m. The initial purpose of the gravity monitoring was to test the feasibility of observing a time-lapse signal from the evolving CO_2 plume. In 2005, when the first gravity repeat measurements were made, five time-lapse seismic surveys (Chadwick

Figure 7.14 Photograph of one of the seafloor benchmarks at Sleipner, in 2013, 11 years after the deployment.

et al., 2004, 2005) had provided rich information on the outline and internal stratification of the CO_2 plume. However, the seismic data are limited at quantifying CO_2 saturations and resolving the vertical distribution of low- and high-saturation zones of CO_2 at scales below the seismic wavelengths.

Benchmarks were deployed on the seafloor prior to the first gravity survey, and serve as permanent measurement platforms for measurements (Figure 7.14). Stations were distributed in a main profile across the CO_2 plume, and with additional stations giving an aerial cover of the plume (Figure 7.15). In each survey, which lasted about one week, three separate station visits were made with three gravity sensors recording in parallel. After data processing, estimated station time-lapse accuracies of about 3 µGal were estimated (Alnes et al., 2011).

The seafloor benchmarks show relative height changes exceeding 20 cm between surveys (Figure 7.16). The movements vary greatly between neighboring stations, and can hardly be explained by uplift caused by expansion of the storage reservoir or subsidence caused by compaction of the gas reservoirs beneath. More likely, the benchmarks have been subject to scouring. This is supported by changes in measured tilt of the top of the platform between

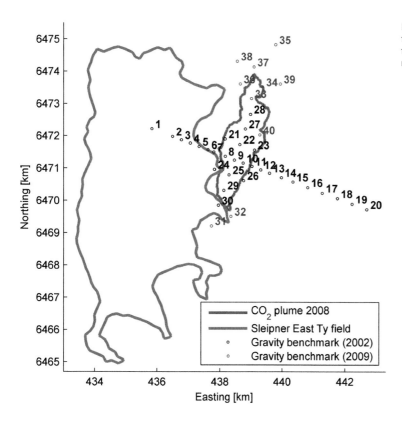

Figure 7.15 The layout of gravity stations, with the rim of the CO_2 plume as of 2008 in red, and the rim of the Sleipner Øst Ty gas-condensate reservoir in green. (From Alnes et al., 2011.)

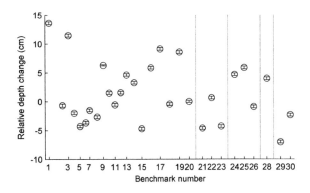

Figure 7.16 Depth changes 2002–2005. (From Alnes *et al.*, 2008.)

surveys, by water-filled gaps between the bottom of the benchmarks and the seafloor (Figure 7.14), and by shell fragments, possibly exposed after local erosion of the uppermost silt layer around the benchmarks.

After correcting the gravity for height changes, using a vertical gravity gradient derived from the data as shown in Figure 7.17, a time-lapse gravity increase of nearly 40 µGal was calculated at the westernmost station, gradually decreasing to no change over 2–3 km eastward. Alnes *et al.* (2008) explained this strong increase at the end of the profile, well aside from the CO_2 plume, as caused by water inflow to the Ty Fm. gas reservoir. The required water volumes agreed with later acquired saturation logs in a nearby well and with 4D seismic data.

After corrections were made for both the time-lapse height changes and water flow into Ty Fm,

a gravity reduction could be identified above the central part of the CO_2 plume (Nooner *et al.*, 2007; Alnes *et al.*, 2008, 2011, 2015; Arts *et al.*, 2008). The estimated maximum gravity reduction increased from about 10 µGal over 2002–2005 to more than 20 µGal over the time interval 2002–2013. The gravity data have in various studies been used to invert for the average CO_2 density, dissolution rate of CO_2 in the brine and lately also the thickness of CO_2. For such inversions, the geometry of the CO_2 plume as imaged in 4D seismic data, and the injected amount of CO_2, are further constraining the models.

Both Nooner *et al.* (2007) and Alnes *et al.* (2008) estimated the average CO_2 density, in a joint inversion with (1) the gravity height gradient and (2) a scale factor for the Ty Fm gas reservoir flow model (to adjust the water influx), as free parameters. Nooner *et al.* (2007) estimated an average CO_2 density of 530 ± 65 kg/m³ (95% confidence interval). Based on this, they argued that a relatively high geothermal gradient was required in the area to have most of the injected CO_2 at such low densities. Alnes *et al.* (2008) estimated a significantly higher average CO_2 density: 760 kg/m³, corresponding to a temperature of 29°C at 870 m depth for pure CO_2. This change in the estimate based on the same data was explained by a series of small changes: (1) reprocessing of the raw gravity data, (2) updated reservoir model, (3) new 4D seismic data of the CO_2 plume, and (4) a different treatment of (unknown) dissolved CO_2. Alnes *et al.* (2011) used the longer time span of 2002–2009 and reprocessed all gravity and depth data, and arrived at 720 ± 80 kg/m³ for average CO_2 density. They used an additional

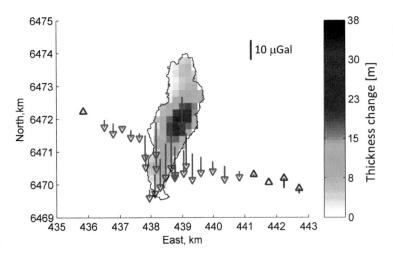

Figure 7.17 Estimated gravity changes caused by CO_2 injection, from 2002 to 2013 (from Alnes, 2015). Inverted CO_2 thickness is shown as blue colors.

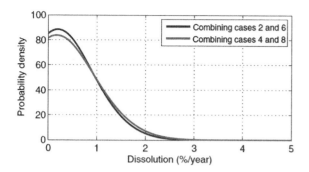

Figure 7.18 Probability distribution of dissolution rate, based on the 2002–2009 time span. (From Hauge and Kolbjørnsen, 2015.)

constraint on rock temperature from a well producing water from the same formation about 10 km away, giving a geothermal gradient of

$$T(z) = 31.7°C/km\ z\ +\ 3.4°C(\pm\ 0.5°C/km)$$

From this they estimated a dissolution rate of CO_2 in the formation water between zero and 1.8% per year. Hauge and Kolbjørnsen (2015) performed a Bayesian analysis of the *in situ* CO_2 density and dissolution, using the same 2002–2005–2009 data set, with the Ty Fm and height change effects removed. They arrived at

similar conclusions as Alnes *et al.* (2011), as shown in Figure 7.18. After the 2013 data were acquired, Alnes (2015) updated the dissolution estimate to <2.7% per year, based on the longer 2002–2013 time span.

Landrø and Zumberge (2017) combined time-lapse seismic AVO data with gravity data to estimate saturation changes also for values exceeding 0.3. They claimed that a simple inversion procedure can be used.

Gravity Monitoring at Other Storage Sites

The In Salah gas field started CO_2 injection in 2004. With depths of 1700 m below the surface, temperatures of 80–100°C, and pressures between about 180 bar initially and up to 280 bar during injection (Eiken *et al.*, 2011), the CO_2 is supercritical, with densities in the range of 500–700 kg/m³. However, porosities of 10–15% and reservoir thicknesses of only about 20 m cause the signal to hardly exceed a few μGal. Elevation monitoring by InSAR has successfully measured uplift to a few mm's accuracy. No microgravity data have been acquired at the site.

Hydrocarbon gas has been injected in the Izaute gas storage reservoir during summer, and produced in the winter (Bate, 2005). Gravity measurements were made over a period exceeding a year, with one

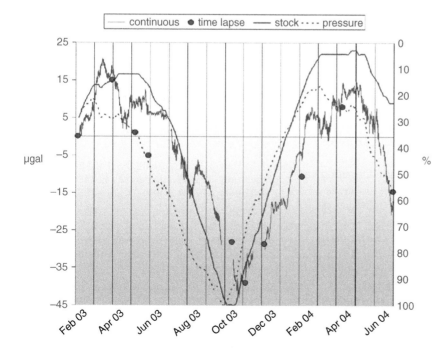

Figure 7.19 Diagram showing the time-lapse gravity (blue dots), continuous gravity (blue trace), gas stock levels (red trace), and reservoir pressure (red dashed trace) at a monitoring well over a 16-month period (from Bate, 2005). The polarity of the gravity data has been inverted to facilitate easier comparison.

Table 7.3 Benefits and limitations of gravity monitor data compared with other data

Benefit	Limitation
Direct and linear relation to mass change (better than the complicated rock physics relations that applies to 4D seismic data)	Detectability (need a minimum mass change, which is depth dependent)
Areal coverage (better than wells)	Limited lateral resolution (depth and signal-to-noise ratio dependent)

permanent gravimeter installed near the expected peak signal, as well as a grid that was surveyed about once a month. The permanent station shows a clear seasonal gravity signal that correlated well with gas injection/production and pressure fluctuations (Figure 7.19). As the reservoir is depleted, the gas/water contact rises, resulting in an increase in the time-varying gravity field.

Discussion and Summary

Gravity's unique strength compared to other monitoring techniques is the ability to measure mass changes for a larger volume, directly with no transformation except for Newton's gravity constant. Such data are thus straightforward to use in quantitative estimates and flow model matching. This could be of particular interest for CO_2 storage, as the mass balance will be a critical aspect of the injection and storage management. Gravity may be regarded as a complementary rather than competing technique to well time series (pressure and flow rates) and seismic or electromagnetic techniques, giving different information. Some benefits and limitations, as compared with other reservoir monitoring data, are listed in Table 7.3.

The rate of dissolution of CO_2 into formation water will be of high interest for decades- to centuries-long storage performance, and can be difficult to measure. Precise gravity changes offer unique data on this process. Microgravity data may also provide early warning for leakage to the surface, as the time-lapse gravity signal would quickly increase if CO_2 should rise to shallower levels than the storage reservoir.

Mass changes will, regardless, be key to understanding the subsurface flow.

References

Adams, S. J. (1991). Gas saturation monitoring in North Oman Reservoir using a borehole gravimeter. *SPE Journal*, **21414**: 669–678.

Allis, R. G., and Hunt, T. M. (1986). Analysis of exploitation-induced gravity changes at Wairakei Geothermal Field. *Geophysics*, **51**: 1647–1660.

Alnes, H. (2015). Gravity surveys over time at Sleipner. Presentation at the 10th IEAGHG Monitoring Network Meeting, June 10–12, 2015. http://ieaghg.org/docs/General_Docs/8_Mon/6_Gravity_surveys_over_time_at_SleipnerSEC.pdf

Alnes, H., Eiken, O., and Stenvold, T. (2008). Monitoring gas production and CO_2 injection at the Sleipner field using time-lapse gravimetry. *Geophysics*, **73**: WA155–WA161.

Alnes, H., Stenvold, T., and Eiken, O. (2010). Experiences on seafloor gravimetrics and subsidence monitoring above producing reservoirs. In Extended Abstract, *72nd EAGE Conference*.

Alnes, H., Eiken, O., Nooner, S., Stenvold, T., and Zumberge, M. A. (2011). Results from Sleipner gravity monitoring: Updated density and temperature distribution of the CO_2 plume. *Energy Procedia*, **4**: 5504–5511 (*10th International Conference on Greenhouse Gas Control Technologies*).

Ander, M. E., and Chapin, D. A. (1997). Borehole gravimetry: A review. In Extended Abstracts, *67th Annual Society of Exploration Geophysicists Meeting*, 531–534.

Arts, R., Chadwick, A., Eiken, O., Thibeau, S., and Nooner, S. (2008). Ten years' experience of monitoring CO_2 injection in the Utsira Sand at Sleipner, offshore Norway. *First Break*, **26**: 65–72.

Bate, D. (2005). 4D reservoir volumetrics: A case study over the Izaute gas storage facility. *First Break*, **23**: 69–71.

Battaglia, M., Gottsmann, J., Carbone, D., and Fernández, J. (2008). 4D volcano gravimetry. *Geophysics*, **73**(6): WA3–WA18.

Brady, J. L., Hare, J. L., Ferguson, J. F., *et al.* (2008). Results of the world's first 4D microgravity surveillance of a Waterfloor – Prudhoe Bay, Alaska. *SPE Journal*, 101762.

Calvo, M., Hinderer, J., Rosat, S., *et al.* (2014). Time stability of spring and superconducting gravimeters through the analysis of very long gravity records. *Journal of Geodynamics*, **80**: 20–33.

Campos-Enríquez, J. O., Morales-Rodrigues, H. F., Domínguez-Mendez, F., and Birch, F. S. (1998). Gauss's theorem, mass deficiency at Chicxulub crater (Yucatan,

Mexico), and the extinction of the dinosaurs. *Geophysics*, **63**(5): 1585–1594.

Carbone, D., Poland, M. P., Diament, M., and Greco, F. (2017). The added value of time-variable microgravimetry to the understanding of how volcanoes work. *Earth-Science Reviews*, **169**: 146–179.

Chadwick R. A., Arts, R., Eiken, O., Kirby, G. A., Lindeberg, E., and Zweigel, P. (2004). 4D seismic imaging of an injected CO_2 plume at the Sleipner Field, central North Sea. In R. J. Cartwright, S. A. Stewart, M. Lappin, and J. R. Underhill, (eds.), *3D seismic technology: Application to the exploration of sedimentary basins*. Geological Society, London, Memoirs, **29**: 311–320. The Geological Society of London.

Chadwick, R. A., Arts, R., and Eiken, O. (2005). 4D seismic quantification of a growing CO_2 plume at Sleipner, North Sea. In A. G. Doré and B. A. Vining (eds.), *Petroleum geology: North-west Europe and global perspectives: Proceedings of the 6th Petroleum Geology Conference*, 1385–1399.

Chadwick, R.A., Arts, R., Bentham, M., *et al.* (2009). Review of monitoring issues and technologies associated with the long-term underground storage of carbon dioxide. London: Geological Society, Special Publications, 313: 257–275.

Chapin, D. A., and Ander, M. E. (2000). Advances in deep-penetration density logging. Society of Petroleum Engineers Conference Papers, 59698.

Christiansen, L., Lund, S., Andersen, O. B., Binning, P. J., Rosbjerg, D., and Bauer-Gottwein, P. (2011). Measuring gravity change caused by water storage variations: Performance assessment under controlled conditions. *Journal of Hydrology*, **402**: 60–70.

Debeglia, N., and Dupont F. (2002). Some critical factors for engineering and environmental microgravity investigations. *Journal of Applied Geophysics*, **50**: 435–454.

Dodds, K., Krahenbuhl, R., Reitz, A., Li, Y., and Hovorka, S. (2013). Evaluating time-lapse borehole gravity for CO_2 plume detection at SECARB Cranfield. *International Journal of Greenhouse Gas Control*, **18**: 421–429.

Eiken, O., Stenvold, T., Zumberge, M., Alnes, H., and Sasagawa, G. (2008). Gravimetric monitoring of gas production from the Troll field. *Geophysics*, **73**: WA149–WA154.

Eiken, O., Ringrose, P., Hermanrud, C., Nazarian, B., Torp, T. A., and Høier, L. (2011). Lessons learned from 14 years of CCS Operations: Sleipner, In Salah and Snøhvit. *Energy Procedia*, **4**: 5541–5548.

Eiken, O., Glegola, M., Liu, S., and Zumberge, M. A. (2017). Four decades of gravity monitoring of the Groningen Gas Field. Extended Abstract, First EAGE Workshop on Practical Reservoir Monitoring.

Ferguson, J. F., Klopping, F. J., Chen, T., Seibert, J. E., Hare, J. L., and Brady, J. L. (2008). The 4D microgravity method for waterflood surveillance: Part 3–4D absolute microgravity surveys at Prudhoe Bay, Alaska. *Geophysics*, **73**(6): WA163–WA171.

Furre, A.-K., Eiken, O., Alnes, H., Vevatne, J. N., and Kiær, A. F. (2017). 20 years of monitoring CO_2 injection at Sleipner. *Energy Procedia*, **4**: 5541–5548.

Gasperikova, E., and Hoversten, G. M. (2006). A feasibility study of nonseismic geophysical methods for monitoring geologic CO_2 sequestration. *Leading Edge*, October: 1282–1288.

Gasperikova E., and Hoversten, G. M. (2008). Gravity monitoring of CO_2 movement during sequestration: Model studies. *Geophysics*, **73**(6): WA105–WA112.

Geertsma, J. (1973). Land subsidence above compacting oil and gas reservoirs. *Journal of Petroleum Technology*, **59** (6): 734–744.

Gelderen, M. v, Haagmans, R., and Bilker, M. (1999). Gravity changes and natural gas extraction in Groningen. *Geophysical Prospecting*, **47**: 979–993.

Glegola, M., Didmar, P., Hanea, R. G., *et al.* (2012). History matching time-lapse surface-gravity and well-pressure data with ensemble smoother for estimating gas field aquifer support: A 3D numerical study. *SPE Journal*, 161483.

Goetz, J. F. (1958). A gravity investigation of a sulphide deposit. *Geophysics*, **23**(6): 606–623.

Hare, J. L., Ferguson, J. F., and Brady, J. L. (2008). The 4D microgravity method for waterflood surveillance: Part IV – Modeling and interpretation of early epoch 4D gravity surveys at Prudhoe Bay, Alaska. *Geophysics*, **73** (6): WA173–WA180.

Hauge, V. L., and Kobjørnsen, O. (2015). Bayesian inversion of gravimetric data and assessment of CO_2 dissolution in the Utsira Formation. *Interpretation*, sp1–sp10.

Hunt, T. M., and Kissling, W. M. (1994). Determination of reservoir properties at Wairakei Geothermal Field using gravity change measurements. *Journal of Volcanology and Geothermal Research*, **63**: 129–143.

Hunt, T., Sugihara, M., Sato, T., and Takemura, T. (2002). Measurement and use of the vertical gravity gradient in correcting repeat microgravity measurements for the effects of ground subsidence in geothermal systems. *Geothermics*, **31**: 524–543.

Jacob, T., Bayer, R., Chery, J., and Le Moigne, N. (2010). Time-lapse microgravity surveys reveal water storage heterogeneity of a karst aquifer. *Journal of Geophysical Research*, **115**: B06402.

Jacob, T., Rohmer, J., and Manceau, J.-C. (2016). Using surface and borehole time-lapse gravity to monitor CO_2

in saline aquifers: A numerical feasibility study. *Greenhouse Gas Science and Technology*, **6**: 34–54.

Kabirzadeh, H., Kim, J. W., and Sideris, M. G. (2017). Micro-gravimetric monitoring of geological CO_2 reservoirs. *International Journal of Greenhouse Gas Control*, **56**: 187–193.

Kabirzadeh, H., Sideris, M. G., Shin, Y. J., and Kim, J. W. (2017). Gravimetric monitoring of confined and unconfined geological CO_2 reservoirs. *Energy Procedia*, **114**: 3961–3968.

Kim, J. W., Neumeyer, J., Kao, R., and Kabirzadeh, H. (2015). Mass balance monitoring of geological CO_2 storage with a superconducting gravimeter: A case study. *Journal of Applied Geophysics*, **114**: 244–250.

Landrø, M., and Zumberge, M. (2017). Estimating saturation and density changes caused by CO_2 injection at Sleipner: Using time-lapse seismic amplitude-variation-with-offset and time-lapse gravity. *Interpretation*, T243–T257.

Lien, M., Agersborg, R., Hille, L. T., Lindgård, J. E., Ruiz, H., and Vatshelle, M. (2017). How 4D gravity and subsidence monitoring provide improved decision making at a lower cost. Extended Abstract, *First EAGE Workshop on Practical Reservoir Monitoring*.

Nind, C. J. M., and MacQueen, J. D. (2013). The borehole gravity meter: *Development and Results: 10th Biennial International Conference & Exhibition.*

Nooner, S. L. (2005). *Gravity changes associated with underground injection of CO2 at the Sleipner storage reservoir in the North Sea, and other marine geodetic studies.* PhD thesis, University of California, San Diego.

Nooner, S. L., Eiken, O., Hermanrud, C., Sasagawa, G. S., Stenvold, T., and Zumberge, M. A. (2007). Constraints on the *in situ* density of CO_2 within the Utsira formation from time-lapse seafloor gravity measurements. *International Journal of Greenhouse Gas Control*, **1**: 198–214.

Nordquist, G., Protacio, J. A., and Acuna, J. A. (2004). Precision gravity monitoring of the Bulalo geothermal field, Philippines: Independent checks and constraints on numerical simulation. *Geothermics*, **33**: 37–56.

Preuss, K. ed. (1998). *Proceedings of the TOUGH Workshop '98*, Berkeley, California, May 4 –6, 1998. Lawrence Berkeley National Laboratory report LBNL-41995.

Pritchett, J. W., and Garg, S. K. (1995). *STAR: A geothermal reservoir simulation system: Proceedings of the World Geothermal Congress 1995*, Florence, Italy, May 18–31, International Geothermal Association, 2959–2963.

Ringrose, P. S., Mathieson, A. S., Wright, I. W., *et al*. (2013). The In Salah CO_2 storage project: Lessons learned and knowledge transfer. *Energy Procedia*, **37**: 6226–6236.

Sasagawa, G., Crawford, W., Eiken, O., Nooner, S. L., Stenvold, T., and Zumberge, M. A. (2003). A new sea-floor gravimeter. *Geophysics*, **68**(2): 544–553.

Seigel, H. O., Nind, C., Lachapelle, R., Choteau, M., and Giroux, B. (2007). Development of a borehole gravity meter for mining applications. In B. Milkereit (ed.), *Proceedings of Exploration 07: Fifth Decennial International Conference on Mineral Exploration*, 21143–21147.

Sherlock, D., Toomey, A., Hoversten, M., Gasperikova, E., and Dodds, K. (2006). Gravity monitoring of CO_2 storage in a depleted gas field: A sensitivity study. *Exploration Geophysics*, **37**: 37–43.

Singhe, A. T., Ursin, J. R., Pusch, G., and Ganzer, L. (2013). Modeling of temperature effects in CO_2 injection wells. *Energy Procedia*, **37**: 3927–3935.

Sofyan, Y., Kamah, Y., Fujimitsy, Y., Ehara, S., Fukuda, Y., and Taniguchi, M. (2011). Mass variation in outcome to high productionactivity in Kamojang Geothermal Field, Indonesia: A reservoir monitoring with relative and absolute gravimetry. *Earth Planets and Space*, **63**: 1157–1167.

Stenvold, T., Eiken, O., Zumberge, M. A., Sasagawa, G. S., and Nooner, S. L. (2006). High-precision relative depth and subsidence mapping from seafloor water-pressure measurements. *SPE Journal*, **11**(3): 380–389.

Sugihara, M., and Ishido, T. (2008). Geothermal reservoir monitoring with a combination of absolute and relative gravimetry. *Geophysics*, **73**(6): WA37–WA47.

Sugihara, M., Nawa, K., Nishi, Y., Ishido, T., and Soma, N. (2013). Continuous gravity monitoring for CO_2 geo-sequestration. *Energy Procedia*, **37**: 4302–4309.

Torge, W. (1989). *Gravimetry*. Berlin: Walter de Gruyter.

Van den Beukel, A. (2014). Integrated reservoir monitoring of the Ormen Lange field: Time lapse seismic, time lapse gravity and seafloor deformation monitoring. The Biennial Geophysical Seminar, NPF, Kristiansand.

Van Opstal, G. H. C. (1974). The effect of base-rock rigidity on subsidence due to reservoir compaction. In *Proceedings of the 3rd Congress of the International Society for Rock Mechanics*, Denver, II, Part B, 1102–1111.

Vevatne, J. N., Alnes, H., Eiken, O., Stenvold, T., and Vassenden, F. (2012). Use of field-wide seafloor time-lapse gravity in history matching the Mikkel gas condensate field. Extended Abstract, *74th EAGE Conference.*

Wilson, C. R., Scanlon, B., Sharp, J., Longuevergne, L., and Wu, H. (2012). Field test of the superconducting gravimeter as a hydrologic sensor. *Ground Water*, **50** (3): 442–449.

Yin, Q., Krahenbuhl, R., Li, Y., Wagner, S., and Brady, J. (2016). Time-lapse gravity data at Prudhoe Bay: New understanding through integration with reservoir simulation models. Expanded Abstract, *Society of Exploration Geophysicists Annual Meeting.*

Zumberge, M., Alnes, H., Eiken, O., Sasagawa, G., and Stenvold, T. (2008). Precision of seafloor gravity and pressure measurements for reservoir monitoring. *Geophysics*, **73**(6): WA133–WA141.

Estimating Saturation and Density Changes Caused by CO₂ Injection at Sleipner

Martin Landrø and Mark Zumberge

Introduction

To estimate rock density directly from seismic data is both challenging and associated with high uncertainties. Prestack inversion methods exploiting amplitude versus offset (AVO) information have been tested to estimate P-wave velocity, S-wave velocity, and density simultaneously by several researchers. Helgesen and Landrø (1993) formulated an AVO inversion scheme where P- and S-wave velocities, densities, and layer thicknesses were inverted for. Buland et al. (1996) used a nonlinear AVO inversion in the tau-p domain to estimate these three parameters at the Troll Field, offshore Norway. Leiceaga et al. (2010) used multicomponent data to reduce the uncertainty related to inversion of density. Bai and Yingst (2014) tested full waveform inversion (FWI) on synthetic data and found that it is challenging to avoid cross-talk between velocity and density. Roy et al. (2006) explored the possibility of using wide-angle data for density inversion and found that by incorporating incidence angles from 40 to 55 degrees, more stable and reliable density estimates can be achieved. However, such data must be processed with care and often involve anisotropic imaging and wavelet stretch corrections. Despite the fact that there have been significant research efforts on how to estimate density reliably from seismic data, it is still considered difficult (see, e.g., Roy et al., 2008) and much more unstable compared to impedance inversion.

For time-lapse seismic data we find more or less the same challenges; however, the possibility of extracting density information might be somewhat better in cases in which high-quality repeated surveys are undertaken. For time-lapse seismic data, when oil or gas is replaced with water, for example, there is a linear relationship between the saturation change and the density change. Therefore, for 4D seismic data one can say that density estimation is to some extent equivalent to estimating saturation changes. There are several examples of saturation estimation from time-lapse seismic data (see, e.g., Landrø, 1999, 2001; Tura and Lumley, 1999; Trani et al., 2011; Grude et al., 2013; Bhakta and Landrø, 2014). If the saturation is not uniform (patchy), then the rock physics relations between saturation and seismic parameters should be changed (see, e.g., Grude et al., 2013).

In the work presented here our purpose is to combine saturation estimates from time-lapse seismic data with repeated gravity measurements in order to reduce the large uncertainty related to time-lapse estimated density changes. For this purpose we will use data acquired at the Sleipner carbon dioxide (CO₂) storage site. Since Statoil launched this project in 1996, more than 150 papers have been published on various topics attached to it. Several seismic surveys and complementary data have been made available to researchers, and therefore it is expected that a variety of analyses will continue to grow in the future. Arts et al. (2008) published an overview of the seismic results achieved after 10 years of CO₂ injection, where they discuss combined use of time-lapse seismic and gravity. One major difference between our paper and Arts et al. (2008) is that we derive explicit expressions for saturation changes using time-lapse seismic AVO data as input. Furthermore, we constrain these estimates by using gravity measurements in the inversion procedure. Queisser and Singh (2013) used a 2D full waveform inversion scheme to invert for P-wave velocity using prestack time-lapse seismic data from 1994 and 1999. Rabben and Ursin (2011) used the 2001 seismic data and performed an amplitude variation with angle (AVA) inversion for the top

This chapter was published in *Interpretation*, 5(2) (May 2017), 1–15. DOI: 10.1190/INT-2016–0120.1. It is reprinted here with the permission of The Society of Exploration Geophysicists and American Association of Petroleum Geologists.

Utsira formation and estimated P- and S-wave impedances as well as density. They clearly show the potential for using amplitude information to distinguish between acoustic impedance and density at Sleipner. Evensen and Landrø (2010) used a time-lapse tomographic inversion method and seismic data sets from 1994 and 2001 to estimate P-wave velocity in a thin CO_2 layer.

In 1996 Statoil and Scripps Institute of Oceanography embarked on a project to develop high-precision seafloor gravimeters focusing on accurate measurements of density changes caused by hydrocarbon production in a reservoir or storage of CO_2 in the subsurface (Sasagawa et al., 2003; Zumberge et al., 2008). Here we combine the gravimetric results published in Alnes et al. (2011) with seismic estimation of saturation changes between 2001 and 2008.

Another way to constrain and help the interpretation of 4D seismic data acquired above CO_2 storage sites is to use fluid flow simulation techniques. One of the early published simulation results was published by Lindeberg and Bergmo (2003), and a more recent example can be found in Cavanagh and Haszeldine (2014).

This chapter is organized as follows. First we present the input data, both time-lapse seismic and gravity data. An overview of the basic assumptions made in the chapter is given in the next section. A simple rock physics model based on earlier work is presented as a basis for the analysis work. Then we derive a simple formula relating near and far offset time lapse changes directly to saturation changes, based on the rock physics model. The calibration procedure used to couple seismic amplitude changes directly to saturation changes is described in a separate section. A key section in the chapter is the formulation of a simple inversion problem to combine measured time-lapse seismic data with the measured gravity anomalies. Before we conclude we discuss various limitations and precautions for the presented work.

The Input Data

The input seismic data used in this project are listed in Table 8.1. Note that what we refer to here as the far offset stack (450–1050 m) is not the stack containing the largest offsets. Offsets from 1200 to 1650 m were collected but are not used owing to significantly lower 4D repeatability for this stack. In this project our aim is to study 4D changes in AVO, and hence we need to use data that are highly repeatable. The average incidence angle using

Table 8.1 An overview of available offset stacks

Data type	150–450 m offset stack NEAR	600–1050 m offset stack FAR	1200–1650 m offset stack ULTRA FAR
2001 3D seismic	Used	Used	Not used
2008 3D seismic	Used	Used	Not used

Those that were actually used in this study are marked by green. Note that what we refer to as far offset here (corresponding to an average reflection angle of approx. 30 degrees) does not represent the offset stack including the largest offsets.

simple ray tracing for the near offset stack (150–450 m) is approx. 11 degrees, and the corresponding angle for our far offset data (600–1050 m) is approx. 30 degrees. A moderate Q-filter assuming a constant Q-value of 300 was used in the processing of the seismic data. The remaining amplitude compensation was done consistently for the 2001 and 2008 seismic data sets. We will therefore assume that apart from an additional Q-compensation that we introduce, there should be only one global scalar necessary to convert the seismic data into "true" reflectivity. As we use a root mean square (RMS) window technique to extract reflection amplitudes, this scalar will vary with the length of the window we are using, and hence the global scalar will increase as the length of the RMS window increases.

The injection of CO_2 at Sleipner started in 1999. This means that the 2001 data set that we are using as a baseline is not a true baseline survey because some changes had already occurred when the 2001 survey was acquired. Hence, we have to estimate roughly the extent and saturation in 2001 based on the 2001 data only, and this is done by assuming that the amplitude anomaly observed close to the injection well on the 2001 seismic data is caused mainly by the CO_2 injection. A better choice would of course be to use data from 1994 (which unfortunately were not available to us during this project), and use the time-lapse seismic difference between 2001 and 1994 to estimate the saturation distribution in 2001.

Time-lapse gravity data from 2002 and 2009 are also used, and here we have simply used the same data as those presented by Alnes et al. (2011). The seismic data and the gravity data cover approximately the same number of years (seven); however, the gravity data have a delay of one year compared to the seismic data. We will assume that this delay is zero, and that the seismic and gravity time-lapse data were acquired simultaneously. It is hard to assess or quantify the error caused by this assumption.

Some Basic Assumptions

As this is a combined methodology and case study, we have to make several assumptions. Some of these assumptions are based on scientific considerations and others are used in order to simplify the case study. In the following we list such assumptions and add some comments to each item:

- We assume that the empirical relations given by Span and Wagner for CO_2 at various pressures and temperatures are valid. The initial reservoir temperature at the injection point is measured to 35.5°C (Alnes *et al.*, 2011). The initial pore pressure is hydrostatic, which means that it is 80 bar at 800 m depth, and hence it is reasonable to assume that the injected CO_2 is most likely supercritical, as the critical point is at 31°C and 74 bar. (For more discussion on the temperature of injectant see Chapter 13 by Eiken in this volume.)
- We assume that when supercritical CO_2 is injected into the brine-filled sandstone rock, the two liquids are immiscible, and that the CO_2 pushes the water away from the injection point.
- A calibration procedure is needed to convert seismic amplitudes into reflection coefficients that are used for the AVO inversion. Our choice is to use one global scalar for this purpose. Ghaderi and Landrø (2009) found a near offset reflection coefficient of – 0.06 for the top sand layer outside the plume. This value was used to determine this global scalar (again using seismic data outside the CO_2 plume) to 0.02.
- We use a simple constant Q-model to account for amplitude variations between near and far offset stacks (we found that $Q = 80$ was a good choice). This is explained in Appendix A.
- Because we are using seismic data from 2001 and 2008, we need an estimate for the saturation change from 1996 (injection start) to 2001. We have assumed that the average CO_2 saturation was 0.1 in 2001. 3 Mtonnes of CO_2 has been injected into the reservoir in the period between 1996 and 2001.
- We assume that the amplitude versus offset response of many thin CO_2 layers can be approximated by using one thick layer and Backus

averaging. This is a rough approximation and we regard it as the lowest order approximation to the time-lapse seismic AVO tuning problem. This issue is discussed in more detail in Appendix B.
- We assume that the average reservoir thickness, or the thickness of the Utsira sand layer, is approx. 200 m. From the seismic data we estimate that the top reservoir varies by approx. 15 m, and that the base has somewhat larger variations over the area where the CO_2 plume occurs.
- We assume that saturation effects dominate over temperature and pore pressure on the time-lapse seismic data. This is motivated mainly by the dramatic decrease in P-wave velocity caused by a relatively small change in CO_2 saturation, and the fact that the pressure and temperature changes are moderate in the Sleipner CO_2 project.

A Simple Rock Physics Model for CO_2 Injection at Sleipner

We will assume that the CO_2 that is injected into the Utsira sand layer at Sleipner is supercritical and does not mix with the brine water that occupies the pore space prior to injection. In a P–T phase diagram, the typical pore pressure in the Utsira formation is above 80 bar and the temperature is most likely above 30°C, which leads to a supercritical state of CO_2. This is discussed by Alnes *et al.* (2011), who mentioned that the injected CO_2 at Sleipner is close to the critical point. Furthermore, we assume that the porosity is constant and equal to 37%. Ghaderi and Landrø (2009) used the empirical relations obtained by Span and Wagner (1996) to estimate the bulk modulus and density of the injected CO_2. It is important to notice that the density is strongly dependent on both temperature and pressure at the storage site. For example, a temperature increase from 30° to 40° C reduces the CO_2 density from 680 kg/m^3 to 300 kg/m^3, assuming that the pore pressure is 80 bar. Ghaderi and Landrø (2009) therefore present two curves (Figure 8.4 in their paper) representing the P-wave velocity change as a function of CO_2 saturation. In this chapter we use a simple exponential decay curve to model how the P-wave velocity (α) change with CO_2 saturation (S) within the Utsira sand (using a porosity value of 37%):

$$\alpha = \alpha_1 + \alpha_2 e^{-\kappa S} \qquad (8.1)$$

Here $\alpha_1 = 1437 \text{m/s}$, $\alpha_2 = 613 \text{m/s}$ and $\kappa = 10$ was found by simple trial-and-error curve fitting

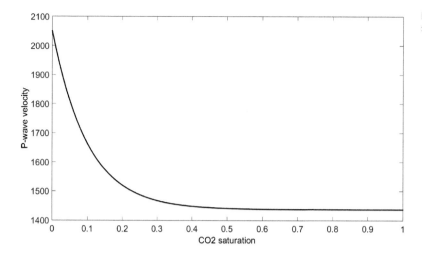

Figure 8.1 P-wave velocity versus CO_2 saturation using Eq. (8.1).

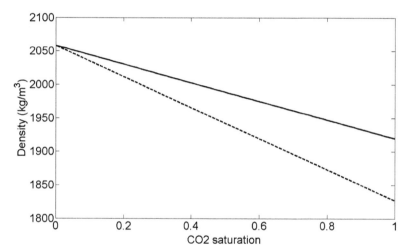

Figure 8.2 Rock density versus CO_2 saturation assuming that the density of CO_2 is 675 kg/m³ (solid line) and 425 kg/m³ (dashed line).

compared to Figure 8.4 in Ghaderi and Landrø (2009). This empirical formula represents an average between the two curves used in Ghaderi and Landrø (2009). The important feature of Eq. (8.1) is that it captures the steep decrease in velocity when the CO_2 saturation increases from 0 to 0.1, as shown in Figure 8.1. Assuming nonmiscible mixture of CO_2 and water in the pore space the fluid density (ρ_F) is a linear combination of the two phases:

$$\rho_F = S\rho_{CO_2} + (1 - S)\rho_W \qquad (8.2)$$

where W denotes water, which we assume has a density equal to 1050 kg/m³. Alnes et al. (2011) discussed the ranges for densities within the Utsira formation. The initial reservoir temperature is 35.5°C at the injection point at 1050 m (the corresponding fluid pressure is then 105 bar). Alnes et al. (2011) estimated the well-bottom CO_2 temperature to be 48°C and using a fluid pressure of 105 bar they found a density of 485 kg/m³. At top Utsira the corresponding density value is as low as 425 kg/m³. However, they concluded that the average CO_2 density within the Utsira formation is 675 kg/m³. We will therefore use this value in most of our calculations and the 425-value as an extreme low-density value.

The density of the rock including the fluid in the pore space is given as

$$\rho = \phi\rho_F + (1 - \phi)\rho_S \qquad (8.3)$$

where $\phi = 0.37$ is porosity, and $\rho_S = 2650$ kg/m³ is the density of the rock matrix or in this case the density of the quartz minerals. Density versus CO_2

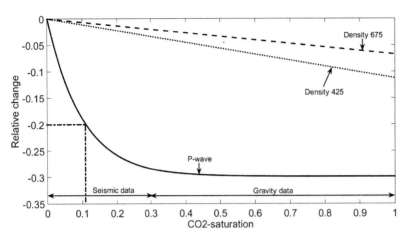

Figure 8.3 Relative change (change divided by original value) in P-wave velocity (solid line) and density assuming CO_2 density of 675 kg/m^3 (dashed line) and density assuming CO_2 density of 425 kg/m^3 (dotted line) versus CO_2 saturation. A relative velocity change of −0.2 corresponds to a saturation change of 0.11, as shown by the dashed-dotted line. The two double arrows indicate which saturation bands we use for time-lapse seismic and gravity data, respectively, in this chapter.

saturation is shown in Figure 8.2 for the two end members of 675 and 435 kg/m^3 of CO_2 density. The expected relative changes in P-wave velocity and density are summarized in Figure 8.3. We notice that the relative velocity changes are large for saturation changes between 0 and 0.4. The relative density changes are small (less than 0.05) in most cases, and hence it is more appropriate to estimate velocity changes for the Sleipner CO_2 case. We will therefore estimate density changes caused by CO_2 injection by first estimating the saturation changes, and then estimate the density change by combining Eqs. (8.2) and (8.3):

$$\Delta\rho = \phi\Delta S(\rho_{CO_2} - \rho_W) \qquad (8.4)$$

Assuming that the density difference between CO_2 and water is not varying spatially, we see that the density change is directly proportional to the saturation change. However, it is very likely (see discussion by Alnes et al., 2011) that the density of CO_2 increases away from the injection point because the injected CO_2 is gradually cooled by the surrounding rock as it propagates away from the injection point. For simplicity, we will assume that this effect is second order and assume that observed density changes from time-lapse gravity can be directly compared to estimated saturation changes from time-lapse seismic data.

A Simple Method to Estimate Density Changes

It is inherently difficult to estimate density directly from seismic data. In seismic inversion, it is

commonly accepted that it is robust and stable to estimate seismic impedance. The same is the case for time-lapse seismic data: There are few examples where density changes have been estimated, and in most cases this means estimating saturation changes and then deriving density directly from the estimated saturation changes is used. Landrø (2001) formulated a direct inversion method using time-lapse AVO data to estimate pressure and saturation changes in a producing hydrocarbon reservoir, or alternatively an injection site for CO_2. Using Eq. (8.7) in Landrø (2001) we find that the change in P-wave reflection coefficient (ΔR) when the pore fluid saturation changes, is given as

$$\Delta R = \frac{1}{2}\left(\frac{\Delta\rho}{\rho} + \frac{\Delta\alpha}{\alpha}\right) + \frac{\Delta\alpha}{2\alpha}\tan^2\theta \qquad (8.5)$$

where θ is the incidence angle, α the P-wave velocity, ρ the density, and Δ represents time-lapse changes in the parameters. We will use this to directly invert for density changes. In the following sections we will assume that the top reservoir has a clear amplitude increase when shale is overlaying CO_2-filled sandstone rock compared to when the same rock is water filled. Next, we will assume that the time-lapse Δ in Eq. (8.5) represents the difference between water-filled and CO_2-filled rock. This means that it is straightforward to estimate changes in the near and far offset stacks (ΔN and ΔF) between the two regions. We will further assume that these stacks have been calibrated to modeled reflection coefficients (as described in the section "Calibrating time lapse AVO data") so that they represent reflection

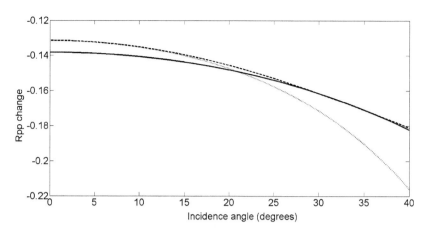

Figure 8.4 Modeling reflectivity changes using Zoeppritz equation (solid line), Eq. (8.5) (dotted line) and Eq. (8.10) (dashed line). Input parameters are listed in the first column in Table 8.2.

coefficients. The time-lapse amplitude changes at near and far offset are given (where θ_N and θ_F are the near and far offset angles):

$$\Delta N = \frac{1}{2}\left(\frac{\Delta\rho}{\rho} + \frac{\Delta\alpha}{\alpha}\right) + \frac{\Delta\alpha}{2\alpha}\tan^2\theta_N \qquad (8.6)$$

and $$\Delta F = \frac{1}{2}\left(\frac{\Delta\rho}{\rho} + \frac{\Delta\alpha}{\alpha}\right) + \frac{\Delta\alpha}{2\alpha}\tan^2\theta_F. \qquad (8.7)$$

Subtracting Eqs. (8.7) and (8.6) we find an explicit expression for the change in P-wave velocity:

$$\frac{\Delta\alpha}{\alpha} = \frac{2(\Delta F - \Delta N)}{\tan^2\theta_F - \tan^2\theta_N}. \qquad (8.8)$$

Substituting this back into Eq. (8.6) again we find an expression for the relative density change:

$$\frac{\Delta\rho}{\rho} = 2\Delta N - 2\frac{(\Delta F - \Delta N)}{\tan^2\theta_F - \tan^2\theta_N}(1 + \tan^2\theta_N). \qquad (8.9)$$

The nice feature of this equation is that it is simple and represents a direct seismic estimate of the density contrast between water-filled and CO_2-filled reservoir rock. One obvious problem is, however, that the assumption behind Eqs. (8.8) and (8.9) is that all relative contrasts should be much less than 1. In many cases these assumptions hold true, and in such cases these equations are valid. It is also possible to include higher order terms. However, for our CO_2-injection case at Sleipner, the above equations are not sufficiently accurate. The relative P-wave velocity change is expected to be larger than 20% for the Sleipner CO_2 example. Therefore, we will use a slightly different version of Eq. (8.5), where we assume that because the

P-wave velocity effect is the dominant effect, an acoustic approximation will be sufficient. Assuming that we can use an acoustic approximation for the Zoeppritz equation we get

$$\Delta R = \frac{1}{2}\left(\frac{\Delta\rho}{\rho} + \frac{\Delta\alpha}{\alpha}\right) + \frac{\Delta\alpha}{2\alpha}\sin^2\theta. \qquad (8.10)$$

As Eqs. (8.10) and (8.5) differ only in the angle-dependent term, we simply replace tangent by sine in Eqs. (8.8) and (8.9) to obtain the alternative and more accurate equations for the Sleipner CO_2 case. A comparison between Eqs. (8.5) and (8.10) to the exact Zoeppritz equation is shown in Figure 8.4, using realistic values from the Sleipner CO_2 case. We see that Eq. (8.10) is more accurate in predicting the AVO behavior for the far offsets. Both approximations show a minor discrepancy of approx. 6–7% at zero offset, which is caused by the simplification of replacing the zero-offset reflection coefficient by the first term on the right-hand side in Eq. (8.10) (or 8.5).

Estimating CO_2 Saturation Changes

If we assume that the CO_2 saturation of the time for the first seismic survey (A) is S_A and the corresponding saturation for the second survey (B) is S_B, we find from Eq. (8.1) that the change in P-wave velocity is

$$\Delta\alpha = \alpha_1 + \alpha_2 e^{-\kappa S_B} - \left(\alpha_1 + \alpha_2 e^{-\kappa S_A}\right) \qquad (8.11)$$

and hence the relative velocity change is given by

$$\frac{\Delta\alpha}{\alpha} = \frac{\alpha_2\left(e^{-\kappa S_B} - e^{-\kappa S_A}\right)}{\alpha_1 + \alpha_2 e^{-\kappa S_A}}. \qquad (8.12)$$

Combining this with Eq. (8.8) we find that the CO_2 saturation at time B is given as

139

$$S_B = -\frac{1}{\kappa}\ln\left[e^{-\kappa S_A} + 2\frac{(\Delta F - \Delta N)(\alpha_1 + \alpha_2 e^{-\kappa S_A})}{\alpha_2(\sin^2\theta_F - \sin^2\theta_N)}\right]$$

$$(8.13)$$

The difference between the near and far offset seismic stacks is measured, and the angle span is known, so the most critical issue is to determine the CO_2 saturation for the first survey. Estimating CO_2 saturation from time-lapse seismic data becomes unstable for saturation values above 0.3–0.4 because there are practically no velocity changes when the saturation is increased from 0.3 to 1 (Figure 8.3). Therefore, this equation must be used with great care, and it definitely must be limited to saturation values below 0.3. Furthermore, we observe that the saturation change is directly dependent on the difference between the time-lapse far and near offset differences. This means that it is critical to calibrate the near and far offset differences prior to using Eq. (8.13). For example, if the argument to the logarithmic function in Eq. (8.13) is negative, it means that the calibration of the near and far offset reflectivity changes is poor. In practice, to avoid problems related to negative arguments for the logarithm in Eq. (8.13), we simply set S_B equal to 0.4 if this occurs. The value of 0.4 is chosen based on trial and error and is also related to the point where the P-wave variation with CO_2 saturation flattens. Furthermore, we observe from Figure 8.4 that there is a minor discrepancy between Zoeppritz modeling and Eq. (8.10) (which is used to derive Eq. 8.13) for near offsets. This discrepancy can be reduced by introducing higher-order terms in Eq. (8.10) (which then will lead to a corresponding modification of Eq. 8.13).

Calibrating Time-Lapse AVO Data

In Eq. (8.13) it is assumed that both near and far offset seismic data have been calibrated so that they both represent true reflectivity changes. In our calibration procedure we will first scale both the near offset and far offset seismic data by one constant scalar. From rock physics modeling and well information, Ghaderi and Landrø (2009) found a near offset reflection coefficient of –0.06 in a water-filled part of the top sand layer (outside the CO_2 plume). We found that by applying a global scalar of 0.02, the estimated RMS amplitudes using a 26 ms time window gives a zero-offset reflection strength of approx. 0.06 for the

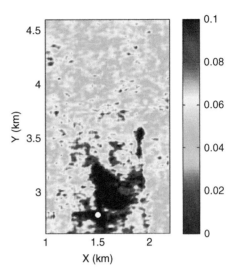

Figure 8.5 RMS amplitudes for near offset stack from 2001 using a 26 ms time window. The injection point is shown as a white circle and the dark red anomaly extending both north and south to the injection point is caused by the CO_2 plume. North of approx. 3.5 km we assume that there is no CO_2 in 2001, and we will therefore assume that the reflectivity above this line should represent the untouched reservoir. The color scale (0–0.1) represents absolute values of the reflection coefficient for the top Utsira interface.

untouched part of the reservoir, as shown in Figure 8.5.

Near and far offset stacks of the data for two cross-lines are shown in Figures 8.6 and 8.7, where the first cross-line intersects the CO_2 injection point. We notice that the CO_2 effect on the seismic data is significant, both on near and far offset data. This is consistent with the rock physics model and corresponds well with the modeled velocity and density changes displayed in Figures 8.1–8.3. The top of the upper sand layer has a depth variation of the order of 15 ms within the plume mapped in 2008 (Figure 8.8). In our analysis this depth variation has been included in a very simple way by adding a stepwise start time for the 26-ms time window used from north to south, so that the added time for the southern areas is approximately 15 ms. In this way we ensure that the RMS signal captures the main seismic reflection associated with the top sand layer. We also tried to use the minimum and maximum values within the time window, and found similar spatial maps when using the RMS window, and we therefore decided to use RMS as a way to estimate reflection coefficients from the near and far offset data.

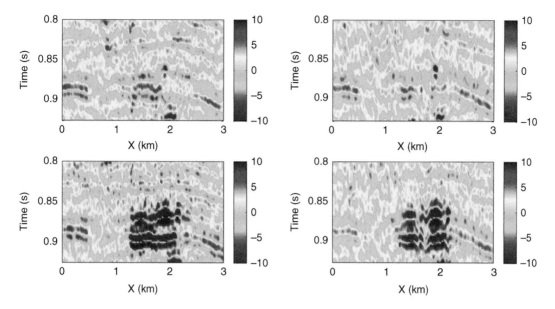

Figure 8.6 Near offset stacks (left) from 2001 (top) and 2008 (bottom), and corresponding far offset stacks (right) for xline 1120 (see Figure 8.10). This cross-line intersects the injection point at approx. 1.5 km. The color bar represents the original seismic amplitudes after processing into near and far offset stacks. The location of xline 1120 is shown in Figure 8.10, and the direction is from west to east.

The next step is to find the relative scaling between the near and far offset stacks; we use a constant Q-model (see Appendix A) for the overburden to correct for the extra geometric damping of the far offset data. Prior to this calibration step we modeled four different scenarios based on the simple rock physics model discussed earlier. The four scenarios are given in Table 8.2, which include CO_2 saturation changes from 0 to 0.2, 0 to 0.5, 0.2 to 0.5, and finally 0.5 to 1.0. The reflection coefficients versus angle for the four cases are shown in Figure 8.9. We notice that for the two latter scenarios, both the overall differences and the AVO differences are small, as expected. The seismic data are not very sensitive to saturation changes above 0.4. The two first scenarios on the contrary show significant reflectivity differences at zero offset as well as a slight AVO increase with offset. In 2001, the extent of the plume for the uppermost CO_2 layer was limited to the area close to the injection point (Figure 8.11). In 2008 the extent of the seismic anomaly caused by CO_2 injection has increased significantly (Figure 8.10) and followed the structurally higher areas toward the north. We find that the largest changes in reflectivity (both for near and far

offset data) occur approx. 1.4 km northeast of the injection point. At this point (or area) we observe an RMS difference change of approx. −0.14 (Figure 8.10). This near offset value (shown by a black star in Figure 8.9) corresponds very well with the modeled black solid line in Figure 8.9. This means that the global scalar of 0.02 is consistent with the observed RMS near offset differences, *given that the saturation has changed from zero to 0.2 in this area*. The next step is to calibrate the far offset difference data. From the modeled curve we observe that the far offset RMS amplitude difference should be somewhat larger in magnitude and close to −0.16. By using a constant Q-model and assuming a Q-value of 80 we obtain an RMS difference value that is in perfect agreement with the modeled curve in Figure 8.9. From Figure 8.10 we see the effect of applying this Q-compensation to the far offset data: The Q-compensated far offset difference data (to the right) are somewhat stronger in amplitude compared to the uncompensated data (in the middle). In summary, we have now used one global scalar (equal to 0.02) to convert the data from seismic amplitude values into reflection coefficient values, and another calibration step to calibrate the far offset

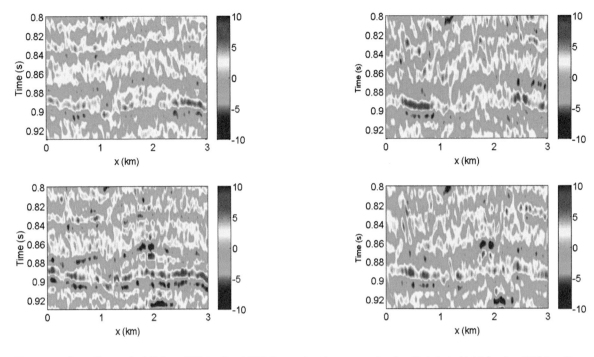

Figure 8.7 Near offset stacks (left) from 2001 (top) and 2008 (bottom), and corresponding far offset stacks (right) for xline 1258 (see Figure 8.10). This is approx. 1 km north of the injection point. The CO_2 anomaly is observed for 2008 at $x = 1.9$ km. The color bar represents the original seismic amplitudes after processed into near and far offset stacks. The location of xline 1258 is shown in Figure 8.10, and the direction is from west to east.

data by using a constant Q-model. For the calibration area (1.4 km northeast of the injection well) the observed near and far offset amplitude changes corresponds to a saturation change from 0 to 0.2.

It should be noted that if we assume that the saturation change is 0.5 instead of 0.2, a similar correction (using $Q = 80$) would still give a reasonable fit to this curve (shown by the red solid line in Figure 8.9). This is because the two curves for the two first scenarios are close to parallel. The estimated near offset time lapse difference after application of the 0.02 correction is shown in Figure 8.10 together with uncorrected and Q-compensated far offset time-lapse difference. A time window of 26 ms has been used to estimate RMS amplitudes prior to the calibration steps. After these two calibration steps we are ready to use Eq. (8.13) to estimate saturation changes between 2001 and 2008. It should be noted that if the analysis is extended by using longer time windows for the RMS calculation, the scaling factor

(which is 0.02 for a 26 ms long window) should be changed accordingly.

Estimating the Saturation Change from 1996 to 2001

In Eq. (8.13) we need the initial saturation distribution within the plume (S_A). As we do not have the 1996 offset stacks, we used a simplified method to get an estimate of this, by using the near offset stack from 2001, and simply assuming that the strong anomaly observed close to the injection point is caused by the CO_2 injection. This is a reasonable assumption, as we know that the pressure changes are close to zero. One remaining cause is temperature changes, because we are injecting relatively warm CO_2 into a somewhat cooler reservoir (35.5°C). We will assume that the seismic impacts of such temperature changes are minor.

Figure 8.11 shows near offset seismic data from 2001, and we notice a clear anomaly close to the

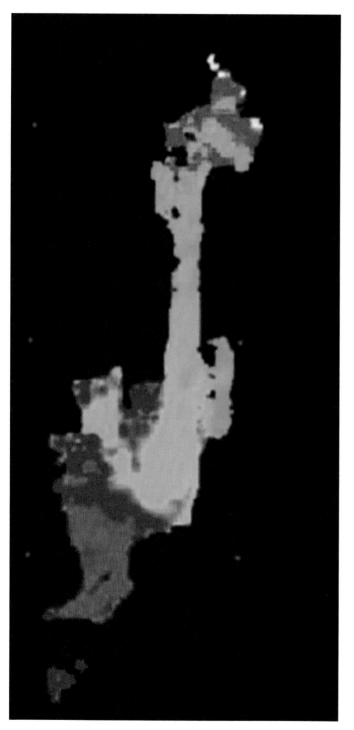

Figure 8.8 Top of Utsira time surface. Deep purple colors in the south correspond to 872 ms and red colors in the north correspond to 857 ms. (Figure provided by Statoil.) The size of the figure is approx. 1.8 km (horizontal) by 5 km (vertical).

injection well. We interpret this as being caused by the CO_2 injection, and more precisely by the CO_2 saturation change. If we subtract the background average amplitude RMS values, we obtain the RMS map shown to the right in Figure 8.11. We observe that the average RMS level is close to 0.1 after subtracting this background amplitude value. In our case the zero-offset reflection coefficient is approximately equal to $\frac{\Delta\alpha}{2\alpha}$, where α is the P-wave velocity. This means that a reflection coefficient of –0.1 corresponds to $\frac{\Delta\alpha}{\alpha} = -0.2$. From Figure 8.3 we observe that this value corresponds to a CO_2 saturation of approximately 0.1 (which means that we use $S_A = 0.1$).

An alternative way to estimate the 2001 CO_2 saturation is to use earlier publications to obtain a rough estimate. Using Arts *et al.* (2008), we observe from their Figure 8.6 that the seismic response in 2001 is distributed over approximately the same area also in depth, and that the reflectivity strength is fairly constant with depth, approximately down to the injection point (which is at 1050 m). From Kiær *et al.* (2015) we observe from Figure 8.1 in this chapter that the areal extent of the uppermost layer is approx. 0.23 km^2 in 2001. Assuming a homogeneous distribution of CO_2 from the top layer to the injection point, that is over a depth range of 250 m, yields an available volume $V_a = 0.058$ km^3. We know that approximately 3 Mtonnes of CO_2 has been injected at Sleipner between 1996 and 2001, corresponding to a compressed volume $V_C = 0.0044$ km^3 assuming that the CO_2 density is 675 kg/m^3. Dividing these two volumes, yields a very rough estimate for the average CO_2 saturation of 0.08, which is not too far from our estimate of 0.1.

Estimating the Saturation Change from 2001 to 2008

Using the estimated saturation change in 2001 as input to Eq. (8.13), and using the near and far offset differences between 2001 and 2008 as input to Eq. (8.13), we obtain the saturation in 2008 (S_B). If we subtract S_A (equal to 0.1) from this, we get the saturation change between 2001 and 2008 as shown in Figure 8.12. Again, this estimate is meant to represent the uppermost sand layer (often referred to as layer 9; see, e.g., Furre and Eiken, 2014). A smoothed version of the saturation change is also shown in this figure. We notice saturation changes to the north of the injection point, but also fairly close to the injection point at the east and south side.

In order to combine seismic and gravity data, we need to find a way to estimate saturation changes for

Table 8.2 Relative changes in P-wave velocity and density for the four time-lapse scenarios

Change in CO$_2$ saturation	Relative P-wave velocity change	Relative density change assuming rhoCO$_2$ = 675 kg/m^3	Relative density change assuming rhoCO$_2$ = 425 kg/m^3
0 to 0.2	–0.24	–0.014	–0.023
0 to 0.5	–0.28	–0.034	–0.056
0.2 to 0.5	–0.06	–0.021	–0.034
0.5 to 1.0	–0.02	–0.035	–0.060

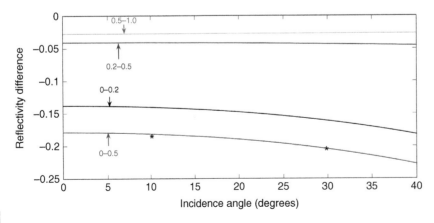

Figure 8.9 Zoeppritz AVO curves for four CO$_2$ saturation change scenarios: 0–0.2 (black line); 0–0.5 (red line); 0.2–0.5 (blue line), and 0.5–1.0 (green line). Calibration of far offset stack data was obtained by using the black curve: First the near and far offset data were multiplied with 0.02 based on the near offset reflectivity outside the plume. Notice that the near offset difference data point shown by a star at 10 degrees is very close to the first scenario. Then the far offset data (shown by the second star at 30 degrees) was scaled by using a constant Q-value of 80, so that the far offset difference follow the modeled trend predicted by the rock physics curve.

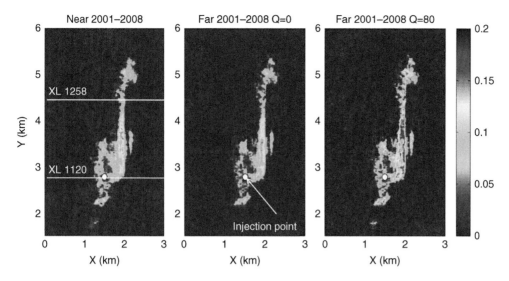

Figure 8.10 RMS amplitude (26 ms time window) for the 2001–2008 near offset stack difference (left), corresponding far offset amplitudes (middle, no Q-compensation), and after application of a constant Q-model using $Q = 80$ (right). The white circle at approx. (1.5, 2.8) km is the CO_2 injection point. The location of the cross-lines shown in Figures 8.6 and 8.7 are shown as white horizontal lines on the left plot in this figure.

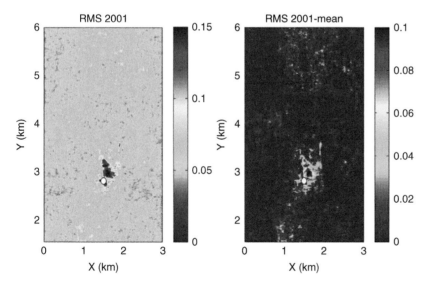

Figure 8.11 Near offset reflection amplitude (RMS) from 2001 (left) and same data after subtracting the average value (right). The maximum value for this difference is 0.1, which corresponds to a relative velocity change of –0.2. The corresponding saturation change (see dashed line in Figure 8.3) is approximately 0.1. This means that the right figure can also serve as an initial guess for the saturation changes between 1996 and 2001, without further scaling.

the entire reservoir thickness. If we use a very long window instead of the short window used so far, the algorithm should pick up some saturation changes also from deeper layers. We apply the same procedure as described earlier for a time window ranging from 850 ms to 1100 ms (which should include all layers above the injection point). A new calibration was done for this purpose and a single scalar of 0.08 was

applied to the data, and the initial saturation (Figure 8.13) was estimated by multiplying the near offset amplitude change by 0.4. In this procedure we assume that the time-lapse AVO behavior of a stack of thin layers follows the same AVO trend as one thick layer represented by the RMS amplitude over the entire layer. The accuracy of this assumption is discussed in Appendix B. The estimated saturation

145

changes are displayed in Figure 8.13, showing both an unsmoothed and a smoothed version. We observe that the algorithm suggests that there are practically no saturation changes in an area close to the injection point equal to approximately 1 square kilometer. This area corresponds more or less exactly to the estimated initial saturation change (left image in Figure 8.13). It is very likely that the 4D seismic estimates are inaccurate and underestimate the saturation changes caused by both the shadow effect from the CO_2 that was occupying this volume in 2001, and the fact that saturation changes above 0.3 are hard to detect from seismic measurements. One way to circumvent this problem is to use the time-lapse gravity data as a complementary source of information in this area.

Time-Lapse Gravity

Offshore time-lapse gravity measurements have a relatively short history. This monitoring technique was possible by the development of high-precision seafloor gravimeters (Sasagawa et al., 2003). By gradual improvements of the seabed gravimeters the repeatability between two gravity surveys was pushed from approx. 4 µgal to 2 µgal. For the Sleipner CO_2 project, we have gravity data from 2002 and 2009 available. We will use these data to constrain our saturation inversion, and especially put weight on the gravity data in areas where we assume that the CO_2 saturation is larger than 0.3. Our gravity signal is caused by the density difference between supercritical

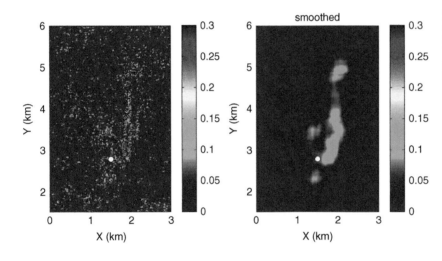

Figure 8.12 Estimated saturation changes (left) between 2001 and 2008 for the upper layer at Sleipner. The white circle shows the injection point. A smoothed version of the same plot is shown to the right.

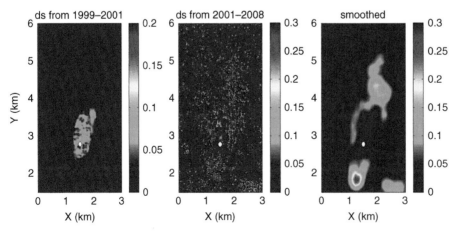

Figure 8.13 (Left) Near off-set RMS amplitude for a time window from 850 to 1100 ms, after subtracting the average value, and scaled by 0.4. This is used as an initial estimate of the average CO_2 saturation in 2001. (Middle) Estimated saturation changes between 2001 and 2008 from the 4D seismic data. (Right) Smoothed version of the estimated saturation changes.

CO_2 and the brine water originally trapped in the pore space within the Utsira sand layer.

Following for instance Keary *et al.* (2002; Eq. 8.9) we can write down the expression for the gravity anomaly caused by the CO_2 injection. We find that the gravity difference response modeled at the seabed in (x, y) is simply estimated by computing the following 2D integral:

$$\Delta g_{mod}(x,y) = G \int dx' dy' \varphi \Delta\rho \Delta S(x',y') \frac{d \cdot z}{\left((x-x')^2 + (y-y')^2 + z^2\right)^{\frac{3}{2}}}. \quad (8.14)$$

Here G is the gravity constant, φ the porosity, $\Delta\rho$ the density difference between CO_2 and water, z the depth to the reservoir, and d the thickness of the reservoir. ΔS is the CO_2 saturation change between 2002 and 2009. Figure 8.14 (left) shows the estimated gravity signal based on the smoothed saturation change estimate (Figure 8.13, left). As our main concern here is the spatial distribution, we have simply scaled the final estimated gravity change to match the maximum measured gravity change. We clearly observe from Figure 8.14 (left) that there is a mismatch between the measured gravity data (shown by circles) and the modeled response based on the estimated saturation changes from 4D seismic. The white circle in Figure 8.14 is the injection point.

Now we will use a simple inversion strategy: Let the estimated saturation changes in the areas where the estimated saturation changes are below a critical threshold (S_c) be scaled by a factor (k) in order to enhance the estimated saturation changes. This critical threshold is estimated by trial and error and it is a practical number that identifies areas where the seismic shadow effect has been strong. This means that we change the modeled gravity difference given in Eq. (8.14) to

$$\Delta g_{mod}(k,x,y) = k \cdot G \int dx' dy' \varphi \Delta\rho \Delta S(x',y') \frac{d \cdot z}{\left((x-x')^2 + (y-y')^2 + z^2\right)^{\frac{3}{2}}} \quad (8.15)$$

where $k = 1$ (meaning no change) if $\Delta S > S_C$. In the present example we used $S_c = 0.03$. The risk of introducing noise in the final saturation image increases as the critical threshold decreases, so this parameter

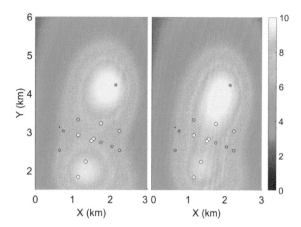

Figure 8.14 Modeled gravity signal change based on the 4D seismic estimate of saturation change (left). The color bar is in µGal, and 10 µGal corresponds to a reduction of 10 µGal between 2001 and 2008. The circles show the measured gravity anomaly data. The circles are color-coded, so a huge color difference between the circle and the surrounding modeled signal means discrepancy between modeling and observations. The size of the circles is proportional to the gravity change signal. (Right) Corresponding modeled gravity change, after least square error minimization. The injection point is shown as a white circle.

has to be determined on a trial-and-error basis. In this way we allow the estimated saturation changes based on time-lapse seismic to be scaled by k for instance in the shadow zone. This means that we keep the spatial distribution of the saturation changes suggested by the time-lapse seismic inversion, but let the measured gravity data determine how much the estimated saturation change should be *up-scaled*. In effect, we let the gravity data "talk" in the shadow zone, and simply assume that the saturation change suggested by the time lapse seismic inversion should be strengthened by this scalar in order to compensate for the seismic shadow effect and the fact that there is low seismic sensitivity for saturation changes from 0.3 and above. To determine an optimal value for k we use the following least squares norm:

$$LS(\kappa) = \sum_{i=1}^{N} \left(\Delta g_{mod}(\kappa, x_i, y_i) - \Delta g_{obs}(x_i, y_i)\right)^2 \quad (8.16)$$

where N is the number of gravity anomaly measurements (denoted Δg_{obs}) made at locations (x_i, y_i). By letting this factor (κ) vary between 1 and 7, we

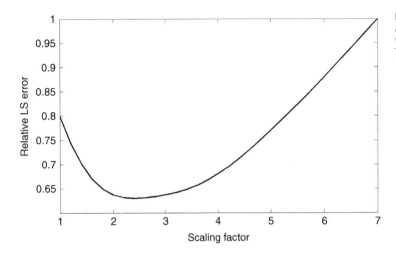

Figure 8.15 Relative least squares error as a function of the scaling factor used to enhance the saturation changes in the "shadow zone." The optimal scaling factor is 2.4.

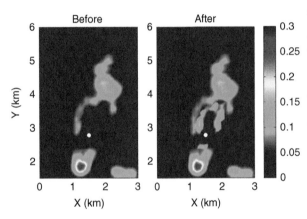

Figure 8.16 Initial saturation change estimate between 2001 and 2008 based on 4D seismic (left) and inverted saturation changes constrained by a least squares inversion using the gravity measurements.

find a global least squares error minimum between the gravity data and the modeled gravity response from the estimated saturation changes at approx. 2.4, as shown in Figure 8.15. The optimal modeled gravity change is shown in Figure 8.14 (right), and we see that the match between the measured gravity data (shown as circles) and the modeled gravity data has improved. This is particularly true for the saddle area between the two main peaks in the modeled gravity anomaly. For other measurements we can actually find areas where there is no improvement. However, the overall result is that the least squares error is reduced by approximately 20%, as shown in Figure 8.15.

The final estimated saturation changes are simply given as $\Delta S_{final} = k\Delta S$, where $\Delta S(x, y)$ represents the estimated saturation changes based on seismic data only. In this inversion we used only gravity measurements that were larger than 4 µGal. The 4 µGal threshold represents the accuracy of the time-lapse gravity measurements (Alnes *et al.*, 2011). A comparison between the initial saturation change estimate (between 2001 and 2008) and the inverted one is shown in Figure 8.16. We clearly see the role of the gravity data: to increase the saturation changes close to the injection point.

On the lower right corners of the estimated saturation changes shown in Figure 8.16, there is an anomaly that we interpret as noise. If we study the RMS differences for near and far offset (Figure 8.17) we notice that there is a weak difference signal in the lower right corner on the far offset data. As we have not implemented a cutoff related to the amplitude of the 4D differences, and we are using the difference between the near and far offset differences, this noise will show up as signal on the estimated saturation changes.

However, the uncertainties and pitfalls associated by using Eq. (8.13) for a long time window are numerous. First, there will be shadow effects and focusing/defocusing effects, causing the deeper reflections to be less accurate and actually misleading to be used by a simple AVO interface approach as proposed here. Second, remaining multiples caused by the strong reflections associated with the top layer will cause misleading results. Third, additional filling of a lower CO_2 layer will be observed only if the CO_2

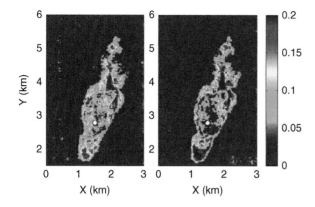

Figure 8.17 Near (left) and far (right) offset RMS-differences (between 2001 and 2008) using a time window from 850 to 1100 ms. Notice the weak noise in the lower right corner on the far offset differences. These differences (although they are weak) create a noisy saturation signal (see Figure 8.16).

saturation increases to a level that is less than 0.3. For instance, an increase in saturation from 0.3 to 0.6 will not give a large reflectivity change in the time-lapse seismic data.

Discussion

In the constrained inversion we used only gravity data that showed a decrease in gravity, and where the absolute value of the gravity change is larger than 4 µGal. If we include all available measurements, some of the weaker gravity signals and especially those measurements resulting in increased gravity change will give a very strong signal close to the injection point. In addition to this, gravity measurements that are farther away from the Sleipner CO_2 site and closer to the area of gas production will be more influenced by the gas production. The Sleipner gas field is situated to the west of the Sleipner CO_2 area. We think this is misleading for the following reasons: Between years 2002 and 2009, 5.88 Mtonnes of CO_2 had been injected, and therefore a decreased gravity should be expected, especially for those measurements that are taken outside the plume (the farther away from the plume the closer the gravity signal will be to a point source). There are some scenarios that might create increased gravity, for instance, if the CO_2 dissolves in water (after a time) and hence water goes back into the plume again. In our opinion this is not very likely to happen at a large scale. However, in a detailed mass balance computation such an effect might influence

the final result. A downward migration of the CO_2 that was in the plume in 2001 will also cause a positive gravity change, but this is also very unlikely to happen. It is much more likely that injection of 5.88 Mtonnes of CO_2 will cause decreased gravity, both when measured right above the plume and when measured at long distances away from the plume. Therefore, we used a threshold of 4 µGal (corresponding to the uncertainty of the time lapse gravity measurements) for the constrained inversion. Figure 8.18 shows a simple comparison between the measured gravity data plotted against the radial distance from the injection point. In this figure we compare the experimental data to simple gravity modeling assuming that the 5.88 Mtonnes of CO_2 has been injected as a point mass at the injection point (shown by solid lines), and assuming that the same mass is distributed as two point masses separated by 1.6 km (dashed lines). Black curves represent a CO_2 density of 675 kg/m^3 while red curves correspond to a density of 720 kg/m^3. We clearly see that the simple monopole distribution does not fit the data, and that the two-point distribution is closer to the measurements. It should be noted that the dashed lines correspond to an in-line two-point distribution. For other directions, the gravity signal will be weaker, and even closer to the measurements. Compared to the best estimate of CO_2 density changes obtained by the joint inversion method (Figure 8.14), we observe that this estimate is far from such simple distributions: It is more like an elongated structure with a weak concentration structure in the north–south direction. This might indicate that the uppermost layers are being filled from several sources that are spread, and not only from the area close to the injection point.

How sensitive is the time-lapse AVO method to errors in the estimated calibration parameters? A simple test where the global scaling parameter is changed from 0.06 to 0.08 is shown in Figure 8.19. The major effect is to increase the saturation by a constant factor. Maybe the increase is slightly less in the south compared to the north. A similar effect (but opposite in sign) is observed if the Q-factor is increased from 80 to 100: The estimated saturation changes have approximately the same spatial distribution, but the overall saturation changes are less, as shown in Figure 8.20. From these simple tests, we conclude that the calibration procedure does not influence the spatial distribution of the estimated

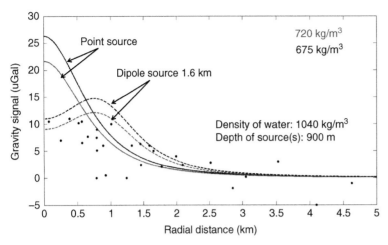

Figure 8.18 Change in observed gravity anomaly signals (black dots) between 2002 and 2009 plotted as function of the radial distance from the injection point. Solid lines show point source (assuming that 5.88 Mtonnes of CO_2 has been injected) gravity signals assuming that the density of CO_2 is constant and equal to 675 kg/m³ (black line) and 720 kg/m³ (red line). Density of water is assumed to be 1040 kg/m³. A depth of 900 m has been used for the point source gravity modeling. Dashed lines show corresponding (inline) result assuming two point sources separated by 1.6 km.

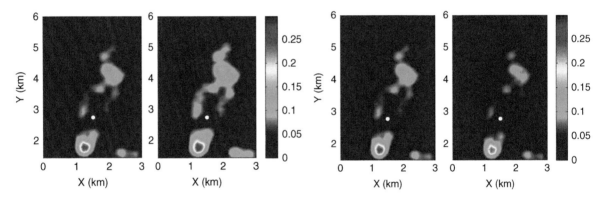

Figure 8.19 The effect of increasing the global scaling factor from 0.06 (left) to 0.08 (right) on the estimated saturation changes based on time-lapse seismic data only.

Figure 8.20 The effect of increasing the Q-factor from 80 (left) to 100 (right) on the estimated saturation changes based on time-lapse seismic data (2001 and 2008) only.

saturation changes significantly, but it changes the DC component or the absolute value of the estimated changes.

We have chosen to use Eq. (8.13) to estimate the saturation changes and then use the direct relationship between saturation and density to get the density changes. This choice is related to the fact that the sensitivity for density changes on the seismic data is significantly less than that for velocity (Figure 8.3). However, if we use Eq. (8.9) to estimate density changes directly from the seismic data we get a result (Figure 8.21) that is not too far from the saturation estimate (Figure 8.16, left). It should be noted, however, that the estimated size of the density change – 30% – is not very realistic.

The total amount of CO_2 injected between the 2002 and 2009 seismic surveys is 5.88 Mtonnes. If we use the estimated saturation change values as shown in Figure 8.16b, we find that the total mass of CO_2 injected in this period is equal to 7.91 Mtonnes. In this simple calculation we have assumed an average reservoir thickness of 200 m, and that the density difference between CO_2 and water is 230 kg/m³. Furthermore, we have assumed that the average porosity is 0.37. The injection point is not located at the base of the Utsira sandstone unit, but approx. 50 m above. Using 150 m for the reservoir thickness will therefore reduce the total injected CO_2 from 7.91 to 5.93 Mtonnes, which is much closer to the actual value. However, the uncertainties coupled to these

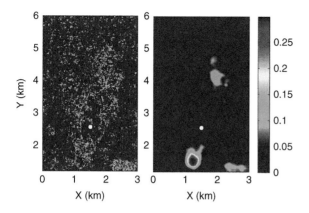

Figure 8.21 Estimated density changes using the time window from 850–1100 ms (left) using seismic data from 2001 and 2008. The smoothed version is shown to the right.

estimates are huge, and therefore this exercise should not be taken too seriously.

Conclusion

A calibrated time-lapse seismic method using near and far offset differences as input to estimate CO_2 saturation changes, has been tested on field data from the Sleipner CO_2 injection site, offshore Norway. We find that this method is of limited value if the CO_2 is stored in several layers on top of each other, as the time-lapse seismic data have less sensitivity to detect saturation changes in such a multilayered medium. By combining the seismic method with time-lapse gravity measurements we demonstrate that a simple inversion procedure can be used to estimate saturation changes also in areas where multiple CO_2 layers are stacked on top of each other.

Acknowledgments

We thank Statoil and the Sleipner license partners ExxonMobil and Total for permission to use the data. Anne Kari Furre is acknowledged for assistance, discussions, and providing the seismic data to us. Thanks also go to Alistair Harding for helping us to load and read the data. Matthew Dzieciuch is acknowledged for assistance in writing Matlab scripts. ML wants to thank Scripps for hosting him during his sabbatical stay, and the Norwegian Research Council for financial support.

Appendix A: Using a Simple Q-Model to Calibrate Near versus Far Offset Stacks

From Eq. (8.13) it is evident that it is critical to determine the ratio between the near and far offset changes as precisely as possible. The time-lapse processing of the data has been done as accurately as possible, but the Q-compensation has been very moderate, as a relatively high Q-value of 300 has been applied to the data. Typical Q-values for overburden sediments in the North Sea is more likely to be less than 100 (Reid *et al.*, 2001), and therefore we introduce a simple Q-model correction term:

$$A_Q = e^{2\pi f\left(\frac{z}{\cos\theta_F} - \frac{z}{\cos\theta_N}\right)/(QV)} \qquad (8.A\text{-}1)$$

where Q is the quality factor, z is the depth to the Utsira formation, f is the average frequency, and V is the average P-wave velocity for the overburden layers. The far offset data should be divided by A_Q to correct for this absorption effect. One way to implement Eq. (8.A-1) is to simply test various Q-values and see how robust Eq. (8.13) is to such variations. For calibration purposes we used a z-value of 800 m, an average frequency of 50 Hz and an average P-wave velocity for the overburden (V) of 1800 m/s.

Appendix B: Using Backus Averaging to Estimate Time-Lapse Response of a Multilayered CO_2 Storage Site

Backus (1962) introduced a systematic way of averaging finely layered media. Using this averaging technique Stovas *et al.* (2006) showed that it is possible to replace a sequence of finely layered medium, with one thick layer being described by average parameters and the net to gross (N/G) ratio, which is a number describing the relative thickness of all CO_2 layers within the Utsira sand divided by the total thickness. In Stovas *et al.* (2006) the N/G ratio is defined as the thickness of the sand layer divided by the total thickness of the reservoir layer. In this work we have adapted the N/G ratio for our purpose, and therefore we replace the sand thickness by thickness of CO_2 layers. For the Sleipner case, the N/G ratio is relatively small, as the thickness of each layer is of the order of 5–15 m. To the lowest order, the reflection coefficient versus offset can be written as

$$R(\theta) = R_0 + G\sin^2\theta \qquad (8.B\text{-}1)$$

151

where θ is the incidence angle, R_0 is the zero angle reflection coefficient, and G is the gradient.

If we assume that the contrasts in all elastic constants are small, Eq. (8.11) from Stovas *et al.* (2006) reads (note that in Stovas *et al.* [2006] $\Delta\rho$ and ΔV_P are dimensionless entities in contrast to this chapter where $\Delta\rho/\rho$ is the corresponding dimensionless quantity):

$$R_0 = \frac{N/G}{2}\left(\frac{\Delta\alpha}{\alpha} + \frac{\Delta\rho}{\rho}\right)$$
$$G = \frac{N/G}{2}\left(\frac{\Delta\alpha}{\alpha} - 2\frac{\beta^2}{\alpha^2}\left(\frac{\Delta\rho}{\rho} + \frac{2\Delta\beta}{\beta}\right)\right) \quad (8.B\text{-}2)$$

where α and β denote P- and S-wave velocities, respectively. This means that to the lowest order, both R_0 and G are directly proportional to the expressions for one thick layer (if we let N/G = 1 in Eq. 8.B-2 we are back to the conventional AVO formula for PP reflections). For medium contrasts in all elastic parameters Stovas *et al.* (2006) found expressions that include higher order terms in N/G ratios. These expressions will alter the analysis done in this chapter somewhat. However, if the N/G ratio is small, these terms will not change our estimates significantly. For N/G ratios close to 0.5 these corrections will be more significant and that requires that we use Eq. 8.17 in Stovas *et al.* (2006) instead of Eq. (8.B-2).

References

Alnes, H., Eiken, O., Nooner, S., Sasagawa, G., Stenvold, T., and Zumberge, M. (2011). Results from Sleipner gravity monitoring: Updated density and temperature distribution of the CO_2 plume. *Energy Procedia*, **4**: 5504–5551.

Arts, R., Chadwick, A., Eiken, O., Thibeau S., and Nooner, S. (2008). Ten years' experience of monitoring CO_2 injection in the Utsira sand at Sleipner, offshore Norway. *First Break*, **26**: 65–72.

Backus, G. E. (1962). Long-wave elastic anisotropy produced by horizontal layering. *Journal of Geophysical Research*, **67**: 4427–4440.

Bai, J., and Yingst, D. (2014). Simultaneous inversion of velocity and density in time-domain full waveform inversion. Expanded Abstracts, *Society of Exploration Geophysicists Technical Program*, 922–927.

Bhakta, T., and Landrø, M. (2014). Estimation of pressure-saturation changes for unconsolidated reservoir rocks with high V_p/V_s ratio. *Geophysics*, **79**: M35–M54.

Boait, F. C., White, N. J., Bickle, M. J., Chadwick, R. A., Neufeld J. A., and Huppert, H. E. (2012). Spatial and temporal evolution of injected CO_2 at the Sleipner Field, North Sea. *Journal of Geophysical Research*, **117**: B03309.

Buland, A., Landrø, M., Andersen M., and Dahl, T. (1996). AVO inversion of Troll Field data. *Geophysics*, **61**: 1589–1602.

Cavanagh, A. J., and Haszeldine, R. S. (2014). The Sleipner storage site: Capillary flow modeling of a layered CO_2 plume requires fractured shale barriers within the Utsira Formation. *International Journal of Greenhouse Gas Control*, **21**: 101–112.

Evensen, A. K., and Landrø, M. (2010). Time-lapse tomographic inversion using a Gaussian parameterization of the velocity changes. *Geophysics*, **75**: U29–U38.

Furre, A. K., and Eiken, O. (2014). Dual sensor streamer technology used in Sleipner CO_2 injection monitoring. *Geophysical Prospecting*, **62**: 1075–1088.

Garcia Leiceaga, G., Silva, J., Artola, F., Marquez E., and Vanzeler, J. (2010). Enhanced density estimation from prestack inversion of multicomponent seismic data. *Leading Edge*, **29**: 1220–1226.

Ghaderi, A., and Landrø, M. (2009). Estimation of thickness and velocity changes of injected carbon dioxide layers from prestack time-lapse seismic data. *Geophysics*, **74**: O17–O28.

Grude, S., Landrø M., and Osdal, B. (2013). Time-lapse pressure saturation discrimination for CO_2 storage at the Snøhvit field. *International Journal of Greenhouse Gas Control*, **19**: 369–378.

Helgesen, J., and Landrø, M. (1993). Estimation of elastic parameters from AVO effects in the tau-p domain. *Geophysical Prospecting*, **41**: 341–366.

Keary, P., Brooks M., and Hill, I. (2002). *An introduction to geophysical exploration*, 3rd edn. Oxford: Blackwell.

Kiær, A. F., Eiken O., and Landrø, M. (2015). Calendar time interpolation of amplitude maps from 4D seismic data. *Geophysical Prospecting*, **64**: 421–430.

Landrø, M. (1999). Discrimination between pressure and fluid saturation changes from time lapse seismic data. Expanded Abstracts, *Society of Exploration Geophysicists Technical Program*, 1651–1654.

Landrø, M. (2001). Discrimination between pressure and fluid saturation changes from time lapse seismic data. *Geophysics*, **66**: 836–844.

Lindeberg, E., and Bergmo, P. (2003). The long-term fate of CO_2 injected into an aquifer. In J. Gale and Y. Kaya (eds.), *Proceedings of the 6th International Conference on Greenhouse Gas Control Technologies (GHGT-6)*, Kyoto, Japan, October 1–4, 2002. Oxford: Pergamon, 489–494.

Queißer, M., and Singh, S. C. (2013). Localizing CO_2 at Sleipner: Seismic images versus P-wave velocities from waveform inversion. *Geophysics*, **78**: B131–B146.

Rabben, T. E., and Ursin, B. (2011). AVA inversion of the top Utsira Sand reflection at the Sleipner field. *Geophysics*, **76**: C53–C63.

Reid, F. J. L., Nguyen, P. H., MacBeth, C., Clark R. A., and Magnus, I. (2001). Q estimates from North Sea VSPs. Expanded Abstracts, *Society of Exploration Geophysicists Technical Program*, 440–443.

Roy, B., Anno, P., and Gurch, M. (2006). Wide-angle inversion for density: Tests for heavy-oil reservoir characterization. Expanded Abstracts, *Society of Exploration Geophysicists Technical Program*, 1660–1664.

Roy, B., Anno, P., and Gurch, M. (2008). Imaging oil-sand reservoir heterogeneities using wide-angle prestack seismic inversion. *Leading Edge*, **27**: 1192–1201.

Sasagawa, G., Crawford, W., Eiken, O., Nooner, S., Stenvold T., and Zumberge, M. (2003). A new seafloor gravimeter. *Geophysics*, **68**: 544–553.

Span, R., and Wagner, W. (1996). A new equation of state for carbon dioxide covering the fluid region from the triple-point temperature to 1100 K at pressures up to 800 MPa. *Journal of Physical and Chemical Reference Data*, **25**: 1509–1596.

Stovas, A., Landrø M., and Avseth, P. (2006). AVO attribute inversion for finely layered reservoirs. *Geophysics*, **71**: C25–C36.

Trani, M., Arts, R., Leeuwenburgh O., and Brouwer J. (2011). Estimation of changes in saturation and pressure from 4D seismic AVO and time-shift analysis. *Geophysics*, **76**: C1–C17.

Tura, A., and Lumley, D. E. (1999). Estimating pressure and saturation changes from time-lapse AVO data. In *69th Annual International Meeting, The Society of Exploration Geophysicists*. Expanded Abstracts, *Society of Exploration Geophysicists Technical Program*, 1655–1658.

Zumberge, M., Alnes, H., Eiken, O., Sasagawa G., and Stenvold, T. (2008). Precision of seafloor gravity and pressure measurements for reservoir monitoring. *Geophysics*, **73**: WA133–WA141.

Electrical and Electromagnetic Methods

Erika Gasperikova and Michael Commer

Introduction

Monitoring approaches for geological carbon dioxide (CO_2) storage (GCS) or geosequestration depend on objectives, subsurface reservoir/site dimensions, and the stage of a CO_2 storage site. Different monitoring techniques would be selected for a site characterization prior to CO_2 injection, for monitoring while injecting CO_2, or for a postinjection stage. Electromagnetic (EM) and electrical techniques are used to map subsurface electrical resistivity. They are complementary to seismic monitoring because they are sensitive to fluid properties and hence able to detect elevated CO_2 saturations.

Electrical and EM techniques measure electric (E) and magnetic (B or H) fields caused by currents that are injected into the ground by contacting electrodes or a time-varying magnetic field induced to flow into the ground by inductive sources. Electrical techniques, also called resistivity techniques, use only current and voltage measurements at frequencies low enough at which EM induction effects are negligible. EM techniques require frequency-dependent sources to induce currents in the ground. Magnetic fields are produced from currents created from both types of sources. The basic concept in both techniques is to measure these electric and magnetic fields and to infer from these measurements the configuration and amplitudes of the current in the subsurface and hence the distribution of electrical resistivity.

Injected CO_2 is expected to form lenticular lenses or plumes of a finite size and change the subsurface resistivity resulting in a resistivity contrast with the enclosing formation. The goal of the survey is to identify a local variation in resistivity relative to the background geology. The changes in resistivity and the associated perturbations in the measured electric and magnetic fields are referred to as *anomalies*. The process of continuously measuring field variations due to the anomalies is referred to as *monitoring*.

Given a complete description of the physical system, that is, with the subsurface structure in terms of its physical property distribution precisely known, the field values could be predicted uniquely. This process is called the simulation or modeling problem. In geophysical surveying, the problem is the opposite: estimate the properties of the subsurface structure on the basis of actual field measurements. This procedure is analogous to medical imaging. In medicine, radiologic or ultrasound methods are routinely employed to make pictures of organs and structures inside the human body. Similarly, geophysical measurements have the ultimate goal of producing structural pictures of the Earth's subsurface. Both instances involve the simulation of fields that propagate through the body/subsurface and field anomalies due to anomalous material properties along the travel path. Compared to medical imaging, the obvious difference is the size of the investigated body. This has the major implication that geophysical imaging problems are harder to solve, because the measurements at hand are usually sparse in relation to the subsurface volume of interest. Because of the sparsity issue, geophysical data analysis in the pioneering days was typically restricted to a trial-and-error repetition of the following procedure:

1. Assume a certain subsurface structure by establishing and quantifying the relevant parameters that make up this structure.
2. Simulate the underlying physical processes that are sensitive to these structure parameters.
3. Predict the measurements that would be produced by these simulations.
4. Compare these predictions to actual real-world measurements.

Depending on the number of structure parameters, this procedure may need to be repeated many times in the hope of closing in on a parameter set that leads to a good match in step 4. The final step is to graphically visualize the spatial distribution of the parameters and obtain an interpretable subsurface image. The procedure of carrying out steps 1–3 is also known as forward modeling because a geophysical survey scenario is modeled, usually by means of computer simulations.

The aforementioned trial-and-error modeling procedure of a complex subsurface would become rather tedious because it would have to be repeated many times. Therefore, a reversed approach, inverse modeling or simply inversion, is more appropriate. It takes given measurements as input with the goal of predicting the "right" numbers for the structural (modeling) parameters such that step 4 gives a match. A graphically enhanced version of such an algorithm would finally produce an easy-to-interpret layman's image of the "true" model parameters. However, the forward modeling routine is an essential ingredient of inversion codes. Internally, inversion algorithms carry out many forward modeling instances in order to figure out the best search direction in the parameter space. An overarching optimization framework controls this search by modifying model parameter guesses in an iterative way until data predictions match observations.

Rock Properties and Resistivity

The form and behavior of CO_2 depend on temperature and pressure. CO_2 behaves as a gas in air at standard pressure and temperature, as a solid (called dry ice) when frozen, or as a supercritical fluid at temperatures and pressures at or above the critical point (temperature of 31°C and pressure of 7.4 MPa) (Figure 9.1). Under supercritical conditions, which appear at depths greater than 800 m, CO_2 takes a much smaller volume than in the gas phase. The resistivity of CO_2 is high, similar to gas or air, independent of its state. Brine-bearing formations that are below and hydrologically separated from drinking water reservoirs have been widely recognized as having high potential for GCS. The resistivity of brine depends on the amount of total dissolved solids (TDS), but in general is low.

The electrical resistivity of the subsurface is highly sensitive to changes in key formation properties such as porosity, pore fluid resistivity, and fluid saturation. A wide range of empirical relations exists for linking formation and electrical

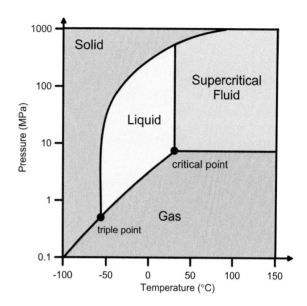

Figure 9.1 Phase diagram of CO_2.

properties. Commonly used is Archie's Law (Archie, 1942), which describes the electrical resistivity (ρ_b) of sedimentary rocks as a function of water saturation (S_w), porosity (ϕ), and pore fluid resistivity (ρ_w):

$$\rho_b = a \, \phi^{-m} \frac{\rho_w}{S_w^n}$$

where a is tortuosity, and m and n are constants with $1.8 < m \leq 2$ and $n \cong 2$.

During GCS, CO_2 may be injected into a formation originally filled with brine. Replacing brine with CO_2 results in a CO_2 saturation $S_{CO2} = 1 - S_w$. Figure 9.2 shows the rock bulk resistivity (ρ_b) as a function of CO_2 saturation (S_{CO2}) for the formation with brine resistivity of 0.3 Ohm-m and 25% porosity. CO_2, as well as all petroleum fluids (e.g., oil, condensate, and hydrocarbon gas), is electrically resistive; hence the relation shown in Figure 9.2 is appropriate not only for CO_2 but also for any combination of oil, hydrocarbon gas, or condensate. The replacement of highly conductive saline fluids (TDS = 10 000 ppm and up; parts per million; 1 ppm = 1 mg of salt in 1 liter of water) with resistive CO_2 results in a resistivity increase in the storage reservoir. When CO_2 is present at shallow depths, dissolution of CO_2 causes an increase in TDS and results in resistivity decrease.

155

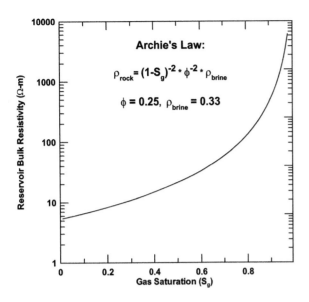

Figure 9.2 Reservoir bulk resistivity (ρ_{rock}) as a function of gas saturation ($S_g = 1 - S_w$) for a reservoir with brine resistivity equivalent to sea water ($\rho_{brine} = 0.33$ Ω-m) with 25% porosity. (Figure 1 from Gasperikova and Hoversten, 2016.)

Electrical resistivity can be used to determine CO_2 saturation:

$$S_{CO_2} = 1 - S_w = 1 - \frac{a\,\rho_w}{\varphi^m \rho_b}.$$

Complex mineral composition may affect bulk rock parameters, and estimates of CO_2 saturations using Archie's equation might not be accurate. In such situations, another useful and simple relationship between resistivity and brine saturation, the resistivity index (Gueguen, 1994), can be used:

$$RI = \frac{\rho}{\rho_0} = (S_w)^{-n},$$

where ρ is the resistivity of the rock partially saturated with brine, ρ_0 is the resistivity of fully saturated rock with brine, and n is the saturation exponent. Again, in the case of CO_2 injection, it is possible to estimate CO_2 saturation from the initial resistivity of the fully saturated rock with brine and partial brine saturation during CO_2 injection using:

$$S_{CO_2} = 1 - \left(\frac{1}{RI}\right)^{1/n}$$

When a formation contains a substantial amount of clay, an additional parameter – the ratio of volume of sand to volume of clay – is necessary (e.g., Nakatsuka et al., 2010).

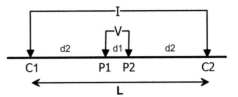

Figure 9.3 Schlumberger electrode array for a surface electrical survey. C1 and C2 are current electrodes, P1 and P2 are voltage electrodes, d1 is spacing between potential electrodes, d2 is spacing between current and potential electrodes, L is total array length, I is current, V is voltage.

Basic Principles of Electrical and EM Techniques

Resistivity Techniques

Resistivity is a material property that describes the resistance to current flow when a voltage is applied across a sample of the material. The resistance R (in Ohms) between the current (I) flowing through a sample of length L and cross section A and the applied voltage (V) across the sample is given by

$$R = V/I$$

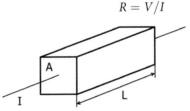

When expressed in terms of the electric field $E = V/L$, in volts per meter (V/m), and current density $J = I/A$, in amperes per square meter (A/m^2), this becomes the resistivity ρ (Ohm-m):

$$\rho = \frac{E}{J} = R\frac{A}{L}$$

The relation $J = \frac{E}{\rho}$ is Ohm's law and it defines the resistivity. It is often written as $J = \sigma E$, where $\sigma = \frac{1}{\rho}$ is the conductivity in Siemens per meter (S/m). For the sample cross sectional area of 1.0 m^2 and the length 1.0 m $\rho = R$.

The resistivity of the ground is measured by injecting electrical currents into the ground and measuring the resulting potential differences. The current source is typically a wire and two current electrodes (C1, C2) connected to a battery or a generator, called an electric bipole (Figure 9.3). The voltage difference is measured

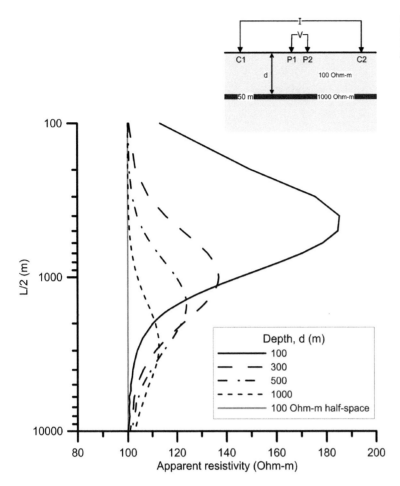

Figure 9.4 Schlumberger apparent resistivity response for a layer 50 m thick and resistivity of 1000 Ohm-m in a 100 Ohm-m half-space as a function of layer depth.

between two potential electrodes (P1, P2 in Figure 9.3). The electrodes can be on the surface as well as in boreholes. The measured electric field, E, is given by Ohm's law. If the ground is of uniform resistivity, and the injected current (I) and the spacing of the electrodes (d1, d2) are known, then the measured voltage (V) is linearly related to the ground resistivity through a simple formula. If the ground is not uniform, the currents will be distorted by the inhomogeneities, and voltage measurements become indirect measures of the resistivity distribution.

The array shown in Figure 9.3 is called the Schlumberger array. As the array length L increases, the injected currents flow deeper into the earth, and the measured V becomes sensitive to deeper layers or features. In any resistivity survey, the value of V is greatly influenced by the geometry of the source–receiver configuration as well as by the resistivity distribution in the subsurface. To normalize the

geometric effect, the observed voltage V for any source–receiver array can be interpreted in terms of the resistivity of a uniform half-space that would give rise to the measured value of V. This is called the apparent resistivity ρ_a and it is the standard form of representing the results of resistivity surveys.

Figure 9.4 shows the Schlumberger apparent resistivity response for a layer 50 m thick and resistivity of 1000 Ohm-m in a 100 Ohm-m half-space as a function of layer depth. The gray vertical line is a response of 100 Ohm-m half-space. This figure shows that with increasing layer depth, the peak response gets smaller and necessitates an increasing current electrode spacing. The figure is illustrative of a practical deployment problem with the technique; the current electrode spacing has to be greater than 8 km ($L/2 = 4$ km) to define the response curve for the layer at 1.0 km depth. In addition to the logistical difficulty of deploying an 8 km long current-carrying

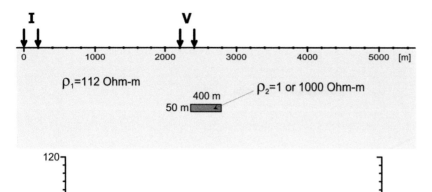

Figure 9.5 Dipole–dipole response for deep conductive and resistive bodies. (Figure 10 from Gasperikova and Morrison, 2019.)

cable, it is very likely that on such a scale, the local shallow geological variations would produce resistivity inhomogeneities that would distort the sounding response.

If the resistive layer at 1 km depth in Figure 9.4 is not continuous, but instead creates a thin slab of finite lateral extent (400 m), the response is significantly less (Figure 9.5). Figure 9.5 shows a model with resistive (1000 Ohm-m) or conductive (1 Ohm-m) finite size body at the depth of 1 km in 112 Ohm-m background. The response in Figure 9.5 is for a traversing dipole–dipole array with 200 m current (I) and potential (V) dipoles and the separation of 2 km; the array moves in 200 m steps over the body. The dipole–dipole array combines sounding and lateral resolution, and has a significant deployment advantage because the cable needs to be laid only between relatively closely spaced electrodes. The depth of detection is controlled by the dipole separation – the depth of the detection increases with the dipole separation. The apparent resistivities are plotted at the midpoint of the array for the conductive and resistive bodies. The conductive target causes 7% apparent resistivity decrease, while the presence of the resistive target causes only 1.6% apparent resistivity increase. Thus, the small resistive body (or CO_2 plume) would be very difficult to detect or to monitor.

Electrodes can be configured in many different patterns where the distance between current and potential electrodes and their ratios vary. The reader is referred to general geophysics textbooks (e.g., Telford *et al.*, 1990) for details about these arrays. To achieve the best results in CO_2 reservoir or plume monitoring, arrays are selected based on forward modeling and analysis of a model that captures the geological complexity at the monitoring site (e.g., Zhou and Greenhalgh, 2000; Christensen *et al.*, 2006; Hagrey, 2012).

EM Techniques

The current loops carrying an alternating current or grounded current dipoles injecting alternating current into the ground have an associated changing magnetic field that, through Faraday's law, induces currents to flow in the subsurface. A time-varying magnetic field passing through a circuit produces an electromotive force (emf), which is proportional to the time rate of change of the magnetic flux threading the circuit. The total flux, Φ, through the circuit is defined as the integral of the component of **B** normal to the surface contained by the circuit:

$$\Phi = \int \mathbf{B} \cdot \mathbf{ds}$$

Faraday's law then states that

$$\text{emf} = -\frac{d}{dt}\int \mathbf{B}\cdot \mathbf{ds} = -\frac{d\Phi}{dt}$$

In the circuit, the emf is the integral of the electric field around the circuit:

$$emf = \oint \mathbf{E}\cdot \mathbf{dl} = -\frac{d\Phi}{dt}$$

Applying Stokes' theorem we get the differential form of Faraday's law:

$$\nabla \times \mathbf{E} = -\frac{d\mathbf{B}}{dt}$$

and Ampere's law:

$$\nabla \times \mathbf{B} = \mu \mathbf{J}.$$

Ampere's and Faraday's laws, the constitutive relations, $\mathbf{J} = \sigma \mathbf{E}$, $\mathbf{B} = \mu \mathbf{H}$, and the fact that $\nabla \cdot \mathbf{B} = 0$ are all the equations and relationships that are needed to solve any problem in low-frequency electromagnetic induction.

Both the current source (transmitter) and receivers are usually multiturn loops of wire. A field produced by a multiturn loop is proportional to its dipole moment, M, which is equal to the product of the current, I, the area of the loop, A, and the number of turns, N. The moment is a vector whose direction is normal to the plane of the loop (along the axis of the loop). The currents induced in the subsurface are a function of the time rate of change of the primary field at the inhomogeneity and of its size, shape, resistivity (ρ), and magnetic permeability (μ), and the resistivity and permeability of the surrounding medium. The response of the ground or an anomalous structure (target) is defined as the measured field for a given configuration of transmitter and receiver. The configuration of the sources and receivers, and the frequency of operation, need to be optimized for the target. CO_2 is usually injected into relatively flat lying formations and is expected to result in lenticular lenses or plumes of a finite size and of increased resistivity, so this study will focus on the responses of resistive targets.

Figure 9.6 illustrates that a conductive (low resistivity) target produces a larger signal and therefore it is easier to detect than the resistive target. Figure 9.6a shows two models: the resistor, which is a 100 m thick resistive layer (30 Ohm-m) at a depth of 1000 m in a conductive background (2 Ohm-m), and the conductor, which is a 100 m thick conductive layer

(2 Ohm-m) at a depth of 1000 m in the resistive background (30 Ohm). The blue star indicates the position of the source and inverted triangles indicate receiver positions. The curves in Figures 9.6b–i are plotted as a function of offset, which is the distance between the source and receiver. The response of the resistor is shown in blue, while the response of the conductor is shown in red. The fields are plotted as a ratio with and without the anomalous layer present. Two types of sources are considered: a horizontal electric dipole (HED) (Figures 9.6d–i), and a vertical magnetic dipole (VMD) (Figures 9.6b, c). Depending on the orientation of the source, EM field ratios are plotted for receiver components for which the ratios are larger than 10%. For the conductor, all configurations produce a signal above this limit and therefore could be used for imaging this target. In contrast, only Figures 9.6d and e show a significant anomaly for the resistive target. The poor detectability of resistors to most source–receiver geometries is related to currents preferably flowing within more conductive bodies. The relatively good detectability of resistors to the HED–Ex and HED-Ez configurations is associated with the guided-wave mode developing and propagating within the resistive layer (Streich, 2016).

Measured signal levels depend on the volume affected by CO_2 or the target size. Responses calculated using layered models give the upper limit of the response to be expected. However, in many practical applications the storage reservoir or the target is of a finite size, which needs to be considered when designing a monitoring survey. Figure 9.7 illustrates a design of marine controlled source EM (CSEM) survey for a sensitivity analysis of a finite size CO_2 plume with uniform or patchy saturations. The plume dimensions are 2 km diameter and 40 m thickness, with its upper edge located 800 m below seafloor. The plume has either uniform CO_2 saturation or layered saturation (Figure 9.7b). The background model is 1 Ohm-m. The resistivities of the CO_2 plume were estimated using the relationships of the section "Rock Properties and Resistivity" and Figure 9.2, with 30% porosity, brine resistivity of 0.3 Ohm-m, and 37% CO_2 saturation for the uniform distribution, and 26%, 37%, and 85% CO_2 saturations for the layered distribution. The sources are positioned on five north-heading tow-lines (Tx01–Tx05). The receivers (Rx001–Rx030) are 1 km apart and cover the entire study area. Additional parameters in

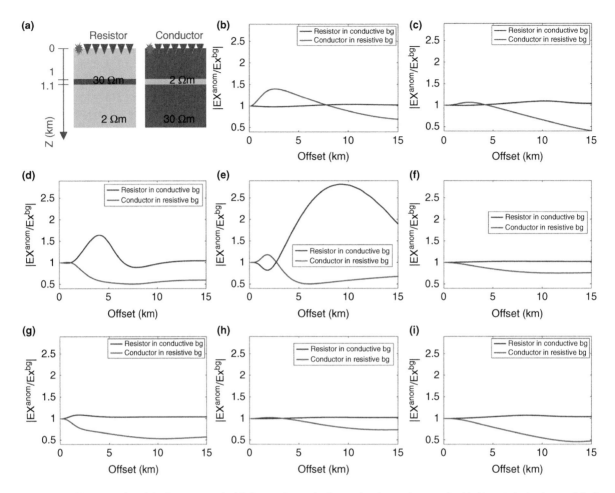

Figure 9.6 (a) Canonical models of a resistor embedded in conductive background and a conductor embedded in resistive background. (b–i) Ratios of fields for the models with the anomalous (anom) layers to the fields for the respective background (bg) models. (b) Vertical magnetic dipole (VMD) source and field component Hx, (c) VMD and Hz, (d) x-directed horizontal electric dipole (HED) and inline Ex, (e) x-directed HED and Ez, (f) x-directed HED and Hy, (g) y-directed HED and Ey, (h) y-directed HED and Hx, (i) y-directed HED and Hz. (Figure 1 from Streich, 2016.)

CSEM monitoring design are the frequency of the operation and the configuration of the sources and receivers or the source–receiver offset, which depend on the depth and size of the target. In this example, the frequency is 1 Hz and results in Figure 9.8 are for the source–receiver offset of 2.5 km. The normalized electric field responses shown in Figure 9.8 are normalized to background responses at the upper left corner (Rx005). The results show that the CO_2 plume with patchy and uniform saturations for the layered distribution (Figures 9.8a, b) causes a 50% increase in the electric field magnitude, while the plume with uniform CO_2 saturation (Figures 9.8c, d) causes a 25% increase of the electric field magnitude, and locates the plume with good lateral resolution.

Electrical Resistance Tomography

Electrical resistance tomography (ERT) is an indirect method for visualizing the movement of fluids in porous media requiring the intermediate application of inversion algorithms that convert raw measurements of electrical resistance to a tomographic image (resistivity or concentration) of a fluid plume. Figure 9.9 conceptualizes the monitoring of a CO_2 reservoir using the ERT method. The placement of the current (C1, C2) and potential (P1, P2) electrode dipoles on either the surface or in boreholes leads to a variety of possible ERT configuration types: surface-to-surface, surface-to-borehole, borehole-to-surface, and borehole-to-borehole.

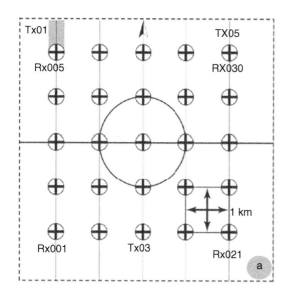

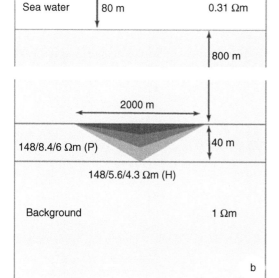

Figure 9.7 (a) CSEM survey configuration. Receiver's (Rx001–Rx030) locations are shown by black circled cross markers. Gray lines show five north-heading towlines (Tx01–Tx05). Footprint of the CO_2 plume is shown by a large black circle. (b) CO_2 plume model. Resistivities of different gray shades of the CO_2 plume are given on the left, the background resistivity is 1 Ohm-m, and sea water resistivity is 0.31 Ohm-m. (Figure 1 from Bhuyian et al., 2011.)

The placement of the dipoles depends on survey economics, logistics, and sensitivity to the target. While forward modeling studies are typically conducted to find the optimal layout, GCS scenarios

usually favor the borehole-to-borehole method given the large application depths.

Within a monitoring context, we consider two types of electric field measurements. The first one responds to the background geology, also called background field. It is recorded before the CO_2 injection starts in order to have a reference to any subsequent reservoir state change. The second field type arises when the CO_2 enters the reservoir and changes the resistivity of the background geology, thus called the anomalous field. Given that CO_2 behaves like an electrical insulator, the reservoir's bulk electrical resistivity will increase. The borehole-to-borehole configuration in Figure 9.9 is designed to image the CO_2-saturated region between two wells. On CO_2 intrusion, anomalous voltage measurements will translate to an increase of apparent resistivity, ρ_a. Monitoring measurements, also called time-lapse measurements, taken at regular intervals after injection show how the evolution of the measured voltages over time is linked to CO_2 plume growth and structural change over time.

The underlying physical processes that lead to the spatial CO_2 distribution over time are flow and transport processes. They govern how CO_2 accumulates and travels in the subsurface and are primarily controlled by two reservoir rock properties: porosity and hydraulic conductivity. Basically, porosity governs how much CO_2 can be stored, whereas hydraulic conductivity describes how easily the gas can travel through a rock's pore space. While these parameters are hard to measure right in the reservoir, electrical resistivity is an indirect indicator for their combined effect, as it correlates with CO_2 saturation changes. Therefore, ERT time-lapse observations also provide some capability of predicting preferential flow paths, which may be helpful for risk assessment purposes.

Inverse Modeling

Mathematical Principles of Inverse Modeling

Recall the following modeling steps:

1. Assume a certain subsurface structure by establishing and quantifying the relevant parameters that make up this structure.
2. Simulate the underlying physical processes that are sensitive to these structure parameters.

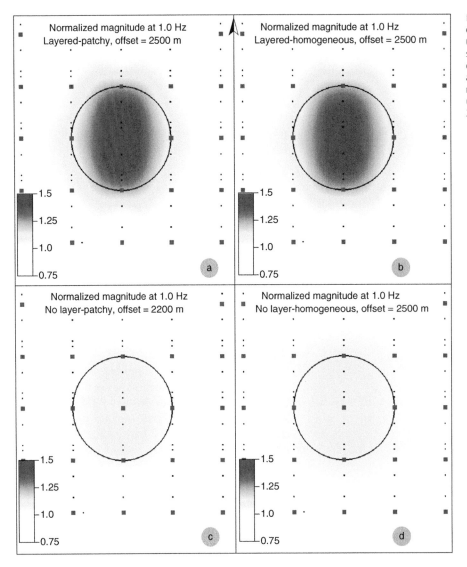

Figure 9.8 Normalized electric field responses for (a) patchy and (b) uniform saturations for layered distribution, and (c) patchy and (d) uniform saturations in nonlayered distribution. (Figure 3 from Bhuyian et al., 2011.)

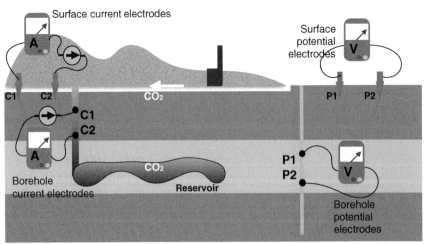

Figure 9.9 Illustration of ERT survey configurations within a CO_2 injection monitoring scenario. The electrode pairs on the left indicate current dipoles (C1, C2), which can be placed either at the surface or in a borehole. The electrode pairs on the right indicate potential electrodes (P1, P2), which can also be located either at the surface or in a borehole. Electric fields are generated by the current dipole and are measured at the potential dipole. Increasing CO_2 saturations in a reservoir cause anomalous electric field measurements with respect to the preinjection state.

3. Predict the measurements that would be produced by these simulations.
4. Compare these predictions to actual real-world measurements.

We will use them to outline the mathematical basics of geophysical inverse modeling problems. The modeling step (1) involves the definition of a model parameter vector $\mathbf{m}$, to be fed into a modeling operator F in step (2), which in practice is a simulation algorithm in form of a computer code. The simulation step (3) delivers the final modeling product $F(\mathbf{m})$, the data predictions at a number of predefined receiver locations. To assess the quality of fit to an actually measured data vector $\mathbf{d}^{obs}$ of N observations in step (4), the misfit measure ϕ_d is commonly used:

$$\phi_d = \frac{1}{2} \sum_{i=1}^{N} \left(\frac{d_i^{obs} - F_i[\mathbf{m}]}{\varepsilon_i} \right)^2 = \frac{1}{2} \| \mathbf{W}_d(\mathbf{d}^{obs} - F[\mathbf{m}]) \|^2$$

(9.1)

where the diagonal matrix $\mathbf{W}_d$ contains each datum's standard deviation, ε_i.

Two major obstacles may hinder the process of finding an adequate solution $\mathbf{m}$ that minimizes the misfit ϕ_d. The first is related to a model's level of detail. To honor true nature, that is, to completely model spatial property variations on even the smallest scale, the size of $\mathbf{m}$ would be infinite. In contrast, geophysical surveys can only provide observations that have a finite number of data points. This dilemma is referred to as the nonuniqueness problem, because one may find many different vectors $\mathbf{m}$ with high granularity that provide a fit to a data set $\mathbf{d}^{obs}$ with insufficient spatial coverage. The second obstacle is related to solution stability. A solution may not depend continuously on the data. In other words, $\mathbf{m}$ can exhibit large property contrasts caused by only small variations in the data. Other data deficiencies given by excessive noise or poor sensitivity to the physical property of interest aggravate these difficulties.

These two issues, the nonuniqueness and solution instability, render geophysical inverse problems as generally ill-posed. A common remedy is the incorporation of additional model knowledge into the solution process by means of a regularizing function, ϕ_m. Prior model information can be cast into mathematical constraints in various ways. Here, we augment ϕ by a so-called penalty term, arriving at the Tikhonov functional (Tikhonov and Arsenin, 1977),

$$\phi(\mathbf{m}) = \frac{1}{2} \phi_d + \frac{\lambda}{2} \phi_m(\mathbf{m}) = \frac{1}{2} \| \mathbf{W}_d(\mathbf{d}^{obs} - F[\mathbf{m}]) \|^2$$
$$+ \frac{\lambda}{2} \| \mathbf{W}_m(\mathbf{m} - \mathbf{m}^{ref}) \|^2.$$

(9.2)

The second measure ϕ_m penalizes deviations from a given reference model $\mathbf{m}^{ref}$, thus encouraging model solutions that stay close to $\mathbf{m}^{ref}$. Further, the matrix $\mathbf{W}_m$ is constructed in a way that penalizes model roughness, that is, the product $\mathbf{W}_m$ becomes large in the presence of large property contrasts between spatially adjacent model parameters. The variable λ is called the regularization parameter, providing a means of finding a balance between the influence of ϕ_d and ϕ_m on the inverse solution.

Given a sufficiently simplified model parameterization $\mathbf{m}$ that still honors the expected spatial property variation in the subsurface to a sufficient degree, we can now solve the inverse problem in a stable manner by minimizing ϕ subject to a predefined target misfit. In practice, this can be accomplished by setting the derivative of Eq. (9.2) to zero. The resulting expression is then solved for the solution vector $\mathbf{m}$. This process varies depending on the nature of the forward modeling operator F, specifically whether F is a linear or nonlinear function. Limiting this section to only an introduction, we highly recommend the tutorial by Oldenburg and Li (2005) for both a more in-depth study and for a review of further readings on the broad subject of geophysical inverse modeling.

Field Data Time-Lapse Inversion Example

The inversion outcome of time-lapse borehole-to-borehole electrical field data from a pilot CO_2 sequestration experiment is summarized in Figure 9.10. The experiment took place at the Cranfield site near Natchez, Mississippi, and involved the injection of more than 1 million metric tons of CO_2 into a deep subsurface reservoir (approx. 3000 m). The recovered change in resistivity is plotted as a function of distance from the injector (x-axis) and depth (y-axis) for eight different days over a period of 103 days. Two monitoring boreholes, marked as F2 and F3, were separated by 30 m, with their ERT electrode locations indicated by black dots. A string of 14 electrodes with 4.6 m vertical spacing and 61 m total length was deployed in F2. Instrument failure led to only

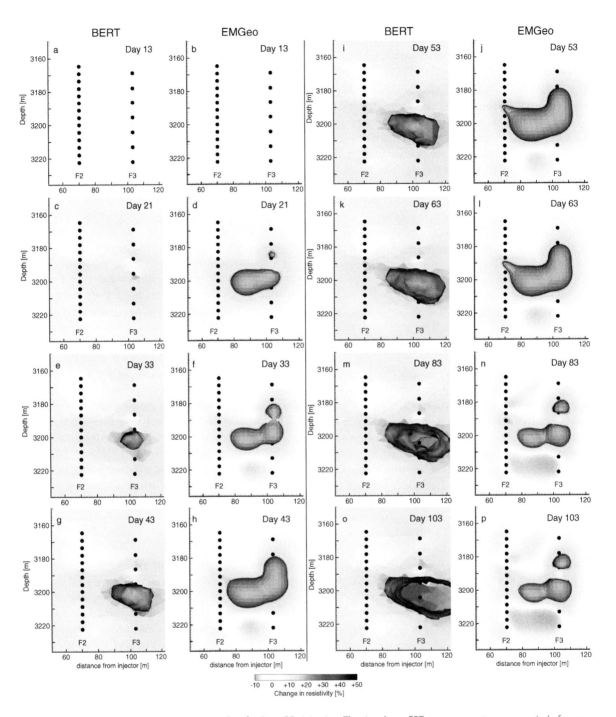

Figure 9.10 Comparative 3D ERT inversion results of a deep CO_2 injection. The time-lapse ERT measurements span a period of over 100 days and involve a borehole-to-borehole configuration. The two different inversion algorithms BERT (images in columns 1 and 3) and EMGeo (images in columns 2 and 4) indicate a similar plume evolution over time, ultimately reducing ambiguity in assessing preferential flow paths.

seven deployable electrodes, hence the doubled vertical spacing in F3. The CO_2 injector (F1) was located in another well 70 m to the left of F2 (not shown). The dipole–dipole switching was scheduled in a way that current (transmitting) and potential (receiving) electrode pairs moved along both boreholes, leading to data with both pairs in one of either well, and data with the electrode pairs across the two wells.

The field measurements were inverted using two different algorithms. Each inversion algorithm features a different kind of forward modeling engine as well as a different optimization driver. In this particular case, the usage of different inversion methods was motivated by the prospect of reduced ambiguity: a similar model parameter estimation produced by two different algorithms might indicate a higher degree of geological meaningfulness, or less ambiguity. The employed methods are the inversion code BERT (Boundless Electrical Resistivity Tomography) (Ruecker *et al.*, 2006) and EMGeo (ElectroMagnetic Geological mapping software) (Commer *et al.*, 2011, 2016). For each of the eight selected days, BERT results are shown on the left and EMGeo results are shown on the right. The principal difference in their forward modeling methods is the volume discretization, i.e., the process of transferring a continuous model into discrete elements. BERT uses the finite-element method, here characterized by the mesh elements with tetrahedral geometry, while EMGeo is a finite-difference method, where typical volume discretizations are given by regular meshes of cubic cells. Each finite-element or finite-difference mesh cell is assigned a Δ-parameter, representing the change of electrical resistivity with respect to the preinjection state. Recall that increased CO_2 saturation causes a resistivity increase that further translates into positive Δ-values, shown as a percent change in Figure 9.10.

Regarding the main features of the Δ-parameter distributions, the image sequences of the two difference methods show a similar CO_2 plume development over time. These inversion results further confirmed a curiosity that emerged in an earlier ERT data interpretation (Doetsch *et al.*, 2013) and also showed in gas and tracer concentration measurements near the F3 well (Lu *et al.*, 2012): CO_2 transport between F1 and F3 occurred faster than between F1 and F2, which is likely due to differential flow paths in this reservoir region. The early onset (starting at day 21) of elevated resistivities near F3 in Figure 9.10 also supports this

finding. This experiment demonstrated that even in the situation when the reservoir is relatively deep and geophysical measurements become logistically demanding, this monitoring approach is feasible and provides useful information about CO_2 movement.

Advantages and Limitations of Electrical and EM Techniques

Electrical and EM techniques are complementary to seismic monitoring because resistivity has a different dependence on CO_2 saturation than seismic velocity. There is the potential in CSEM for providing important complementary information on its spatial distribution in the subsurface (e.g., Eiken *et al.*, 2000; Hoversten *et al.*, 2003; Gasperikova and Chen, 2009). 3D images of subsurface resistivity can be produced routinely thanks to simultaneous development of acquisition hardware, computers, processing, and modeling and inversion algorithms (e.g., Grayver *et al.*, 2014 and references within).

CO_2 and all petroleum fluids (e.g., oil, condensate, and hydrocarbon gas) are electrically resistive; therefore, many approaches developed for oil or gas reservoir monitoring are applicable to CO_2 monitoring. However, because the resistive targets produce much smaller signals than the conductive targets and are more difficult to detect, the choice of the technique will depend on the CO_2 plume size/volume and its depth. Gasperikova and Morrison (2019) present detectability analyses using a thin resistive layer and sphere models for a range of techniques. Although unlikely, the increased reservoir pressure can lead to leakage of reservoir brine (high-salinity water) or CO_2 into shallow aquifers (Birkholzer *et al.*, 2009). High-salinity reservoir brine can therefore increase salinity levels in groundwater, which would be characterized by a strong decrease in electrical resistivity, caused by the dissolved salt. In such a scenario, electrical techniques could be used to locate these conductive anomalies. The electrical signature of dissolved CO_2 in groundwater can be either resistive or conductive, which creates challenges for detecting dissolved CO_2 in freshwater aquifers. Fluid salinity and the complex geochemical reactions of CO_2 with these fluids would have to be considered in the electrical monitoring design.

Crucial requirements for monitoring are sufficient accuracy, repeatability of the measurements, and

165

sufficient sensitivity to the subsurface changes. Data errors have to be significantly smaller than the EM field anomalies resulting from changes within the target structure. Repeatability errors may accumulate through repositioning errors of the acquisition equipment, hardware changes, or temperature effects influencing hardware performance over time.

Data acquisition and monitoring on land face different challenges compared to marine environments. Increasing levels of cultural noise pose a growing challenge to EM surveys. The future success of EM applications may thus depend on improving the techniques for handling the various types of noise encountered. The influence of noise may be reduced by recording field components less affected by noise or by defining source geometries and signals such that they are optimally separable from the noise. Further noise reduction may be achieved by developing advanced processing schemes that exploit some a priori knowledge of the noise at hand. Certain types of signal undesired at the outset, such as the effects of well casings, may have to be explicitly accounted for in data interpretation.

Acronyms and Abbreviations

B	magnetic field
BERT	Boundless Electrical Resistivity Tomography
CSEM	controlled source electromagnetics
CO_2	carbon dioxide
DC	direct current
E	electric field
EM	electromagnetic
EMGeo	ElectroMagnetic Geological mapping software
ERT	electrical resistance tomography
H	magnetic field ($B = \mu H$)
HED	horizontal electric dipole
TDS	total dissolved solids
VMD	vertical magnetic dipole

References

Archie, G. E. (1942). The electrical resistivity log as an aid in determining some reservoir characteristics. *Transactions of the AIME*, **146**: 54–62.

Bergmann, P., Schmidt-Hattenberger, C., Kiessling, D., *et al.* (2012). Surface-downhole electrical resistivity tomography applied to monitoring of CO_2 storage at Ketzin, Germany. *Geophysics*, **77**: B253–B267.

Bhuyian, A. H., Ghaderi, A., and Landro, M. (2011). CSEM sensitivity study of CO_2 layers with uniform versus patchy saturation distributions. Expanded Abstracts, *Society of Exploration Geohysicists Technical Program*, 655–659.

Birkholzer, J. T., Zhou, Q., and Tsang, C.-F. (2009). Large-scale impact of CO_2 storage in deep saline aquifers: A sensitivity study on pressure response in stratified systems. *International Journal of Greenhouse Gas Control*, **3**: 181–194.

Christensen, N. B., Sherlock, D., and Dodds, K. (2006). Monitoring CO_2 injection with cross-hole electrical resistivity tomography. *Exploration Geophysics*, **37**: 44–49.

Commer, M., Newman, G. A., Williams, K. H., and Hubbard, S. S. (2011). Three-dimensional induced polarization data inversion for complex resistivity. *Geophysics*, **76**: F157–F171.

Commer, M., Doetsch, J., Dafflon, B., Wu, Y., Daley, T. M., and Hubbard, S. S. (2016). Time-lapse 3-D electrical resistance tomography inversion for crosswell monitoring of dissolved and supercritical CO_2 flow at two field sites: Escatawpa and Cranfield, Mississippi, USA. *International Journal of Greenhouse Gas Control*, **49**: 297–311.

Doetsch, J., Kowalsky, M. B., Doughty, C., *et al.* (2013). Constraining CO_2 simulations by coupled modeling and inversion of electrical resistance and gas composition data. *International Journal of Greenhouse Gas Control*, **18**: 510–522.

Eiken, O., Brevik, I., Arts, R., Lindeberg, E., and Fagervik, K. (2000). Seismic monitoring of CO_2 injected into a marine aquifer. Expanded Abstracts, *Society of Exploration Geophysicists Technical Program*, 1623–1626.

Gasperikova, E., and Chen, J. (2009). A resolution study of non-seismic geophysical monitoring tools for monitoring of CO_2 injection into coal beds. In L. I. Eide (ed.), *Carbon dioxide capture for storage in deep geologic formations. Results from the CO_2 Capture Project, Vol. 3: Advances in CO_2 capture and storage technology.* Berkshire: CPL Press, 403–420.

Gasperikova, E., and Hoversten, G. M. (2006). A feasibility study of nonseismic geophysical methods for monitoring geologic CO_2 sequestration. *Leading Edge*, **25**: 1282–1288.

Gasperikova, E., and Morrison, H. F. (2019). Electrical and electromagnetic techniques for CO_2 monitoring. In L. Huang (ed.), *Geophysical monitoring for geologic carbon sequestration*. Hoboken, NJ: AGU Wiley.

Grayver, A. V., Streich, R., and Ritter, O. (2014). 3D inversion and resolution analysis of land-based CSEM data from the Ketzin CO_2 storage formation. *Geophysics*, **79**: E101–E114.

Gueguen, Y. (1994). *Introduction to the physics of rocks*. Princeton, NJ: Princeton University Press.

Hagrey, S. A. (2012). 2D optimized electrode arrays for borehole resistivity tomography and CO_2 sequestration modelling. *Pure and Applied Geophysics*, **169**: 1283–1292.

Hoversten, G. M., Gritto, R., Washbourne, J., and Daley, T. (2003). Pressure and fluid saturation prediction in a multicomponent reservoir using combined seismic and electromagnetic imaging. *Geophysics*, **68**: 1580–1591.

Lu, J., Kharaka, Y. K., Thordsen, J. J., *et al.* (2012). CO_2 rock-brine interactions in Lower Tuscaloosa Formation at Cranfield CO_2 sequestration site, Mississippi, U.S.A. *Chemical Geology*, **291**: 269–277.

Morrison, H. F., and Gasperikova, E. The Berkeley course in applied geophysics (interactive textbook). http://appliedgeophysics.berkeley.edu

Nakatsuka, Y., Xue, Z., Garcia, H., and Matsuoka, T. (2010). Experimental study on CO_2 monitoring and quantification of stored CO_2 in saline formations using resistivity measurements. *International Journal of Greenhouse Gas Control*, **4**: 209–216.

Oldenburg, D. W., and Li, Y. (2005). Inversion for applied geophysics: A tutorial. In D. K. Butler (ed.), *Investigations in geophysics, No. 13: Near-surface geophysics*. Society for Exploration in Geophysics, 89–150.

Rucker, C., Gunther, T., and Spitzer, K. (2006). Three-dimensional modelling and inversion of dc resistivity data incorporating topography. I. Modelling. *Geophysical Journal International*, **166**: 495–505.

Schmidt-Hattenberger, C., Bergmann, P., Bosing, D., *et al.* (2013). Electrical resistivity tomography (ERT) for monitoring of CO_2 migration: From tool development to reservoir surveillance at the Ketzin pilot site. *Energy Procedia*, **37**: 4268–4275.

Streich, R. (2016). Controlled-source electromagnetic approaches for hydrocarbon exploration and monitoring on land. *Surveys in Geophysics*, **37**: 47–80.

Telford, W. M., Geldart, L. P., and Sheriff, R. F. (1990). *Applied geophysics*. Cambridge: Cambridge University Press.

Tikhonov, A. V., and Arsenin, V. Y. (1977). *Solution of ill-posed problems*. Hoboken, NJ: John Wiley & Sons.

Zhou, B., and Greenhalgh, S. A. (2000). Cross-hole resistivity tomography using different electrode configurations. *Geophysical Prospecting*, **48**: 887–912.

Chapter 10

Microseismic Imaging of CO_2 Injection

Shawn Maxwell

Passive microseismic monitoring is commonly used to image the geomechanical impact from a variety of industrial activities, including fluid or gas injection (e.g., Maxwell *et al.*, 2010). Stress or pressure changes during injection can result in either activation of preexisting fractures or potentially creation of new fractures, both of which can result in induced seismicity associated with inelastic deformation of existing fractures or the rock itself. Passive seismic monitoring can be used to detect the transient seismic waves of these sources, which can then be processed for location and source intensity, or magnitude, along with other characteristics. Induced seismicity follows the same Gutenberg–Richter law as tectonic earthquakes, a power-law relationship between number of events and magnitude such that there is approximately an order of magnitude increase in the number of sources for one magnitude unit decrease in magnitude. Therefore, there will be a relatively large number of small magnitude sources, which are typically referred to as "microseismicity" generally with negative magnitudes (below zero) on the seismic magnitude scale. The spatial and temporal characteristics of microseismicity can then be used to image the geomechanical deformation and hence the subsurface characteristics of an injection.

Geomechanical processes associated with microseismicity during an injection operation include both pressure changes and poroelastic stress changes. Pore pressure increases resulting from injection can release the clamping stress on fractures and induce shear slip (Hubbert and Rubey, 1959). Increased pore pressure can also lead to volumetric expansion and resulting local stress changes that can also lead to fracture movement (Segall and Lu, 2015). During injection, including compressible gas as in the case of carbon dioxide (CO_2) sequestration, direct pressure induced fracture slip is typically considered the main mechanism. Microseismic events are normally associated with

fractures with length dimensions of the order of 10s of meters. Larger faults could potentially also be associated with slip which leads to potential questions around induced seismicity and possible seismic hazard (Zoback and Gorelick, 2012).

There are three main motivations for microseismic monitoring of gas injection and geosequestration:

1. Imaging the injection by detecting pressure-induced fracture movements
2. Monitoring for seal integrity
3. Monitoring for induced seismicity

These three objectives represent similar geomechanical processes, but with different scales and risk implications to the injection.

In this chapter, microseismic details will be described, including geomechanics and source theory, acquisition, processing, and interpretation. Several case studies will also be presented, highlighting field example of microseismic monitoring of geosequestration and other gas injections.

Geomechanics and Microseismic Sources

The classic investigation of brittle fracture stability relies on a Mohr circle depiction of the stress state (Hubbert and Rubey, 1959). Mohr circles represent the stress state on arbitrarily oriented fractures, decomposed into shear and normal stress components (Figure 10.1). Principal stress magnitudes correspond to where the circle crosses the normal stress axis (horizontal) such that the center of the circle is positioned around the average stress and the radius of the circle is proportional to the deviatoric stress difference between maximum and minimum stress. Conditions for fracture stability resulting in brittle failure, frictional sliding, or slip of preexisting fractures, can be established using a specific failure

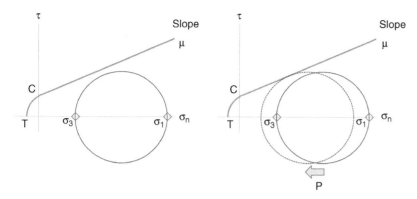

criterion, commonly the Mohr–Coulomb criteria for frictional stability using frictional and cohesive strength. When the Mohr circle intersects the failure criteria, the combination of normal clamping force and shear slipping force will be such that a fracture in the appropriate orientation will undergo frictional sliding or slip.

During injection, increased pressure will reduce the effective normal stress and cause the Mohr circle to translate to the left toward the failure condition (Figure 10.1). Unless in a tectonically active region, initial stresses should be stable with the Mohr circle to the right of the stability condition. During injection, elevated pressures could lead to induced slip on previously stable fractures. Alternatively, poroelastic stress changes can change the Mohr circle radius and increase the slip potential. Slip conditions would depend on the initial stress state and *in situ* pressure changes, and orientations of preexisting fractures and their strength characteristics. It is important to note that fractures in a certain orientation will tend to slip under shear, a preferred direction being at an angle of approx. 30° to the maximum principal stress direction. Fracture slip can then be associated with the occurrence of a microseismic source, although the slip can potentially be aseismic (without detectable seismicity), particularly if the slip is relatively slow such as cases with a small clamping force where aseismic, stable slip can occur instead of stick-slip behavior associated with seismicity.

Geosequestration and other fluid disposal or storage operations typically involve a relatively low injection pressure. In contrast, however, hydraulic fracturing operations involve a high injection pressure slightly higher than the minimum principal stress and above the pressure to create new tensile

fractures. At these elevated pressures where the Mohr circle is shifted to the left into negative normal stress conditions for tensional fractures, a significantly higher rate of microseismic events would be expected, with a faster growing pressure plume and more fractures encountered in orientations susceptible to slip. Indeed, microseismic imaging is a common technology to map hydraulic fracture networks for stimulation treatments of geothermal and oil and gas wells, where the abundance of small magnitude microseismicity can provide a detailed image of the resulting fracture network (e.g., Maxwell *et al.*, 2010). In contrast, where injection operations are for enhanced oil recovery (EOR) to either sweep pore-space hydrocarbon through flow of injected fluids, injection pressure would be locally offset by pressure reductions associated with oil production. Therefore, a relatively low microseismicity rate would be expected during EOR injections.

The most fundamental microseismic source characteristic is origin time and hypocentral location, and indeed are important to microseismic imaging. The so-called Brune source model (Brune, 1970) is often used to characterize source deformation attributes and can be related to the physical slip geometry (Figure 10.2). The Brune model primarily uses two signal attributes to quantify the source deformation: dominant frequency of ground velocity is inversely proportional to the source radius of the slip (*r*) and amplitude is proportional to the source strength normally measured as the seismic moment (M_0 in units of Nm), defined as

$$M_0 = \mu A d$$

where μ is the shear modulus, A is area of slip, and d is the slip displacement. The logarithm of seismic

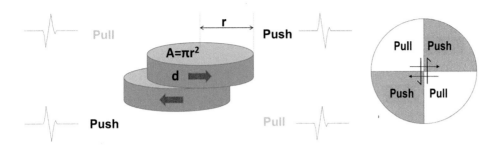

Figure 10.2 Schematic representation of shear slip associated with a microseismic event shown as two blue disks, with a displacement (*d*) and area (*A*). The displacement creates a particular radiation pattern depicted as the four pulses in each surrounding quadrant where the P-wave first motion changes polarity based on whether the rock experiences a compression ("push") or dilation ("pull"). The radiation pattern is represented by a "beach-ball" diagram representing a stereonet plot where the P-wave first motion is appropriately shaded.

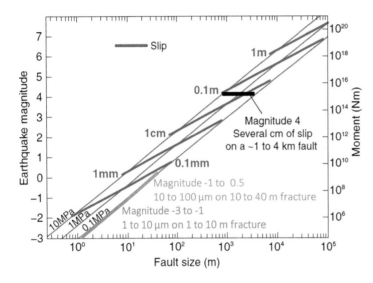

Figure 10.3 Plot of magnitude versus fault size, and corresponding amounts of slip (blue line) for different stress release estimates. Fault characteristics of a typical range of microseismicity are shown in green and orange. A magnitude 4 earthquake is shown for reference. (Modified after Zoback and Gorelick, 2012, Figure 10.2.)

moment is used to compute a "moment magnitude" (M_w), which is calibrated to be similar to the well-known "Richter magnitude" scale:

$$M = 2/3\log(M_0) - 6$$

The relationship between small magnitude microseismicity and larger earthquakes from both induced and tectonic origins is of particular interest and part of a seismology topic of "source scaling" (e.g., Abercrombie, 1995). Small and larger magnitude events generally follow similar scaling, such that the larger the magnitude, the larger the source radius (Figure 10.3). Slip on a larger fault could then be expected to have the potential of a larger magnitude induced seismic event (e.g., Zoback and Gorelick, 2012). Induced seismicity is a growing topic of concern associated with fluid injections, following an increase in earthquake occurrences in the central United States attributed to salt water disposal.

Investigating the relative number of microseismic events at different magnitudes is a common analysis technique (Aki, 1965). The frequency–magnitude relationship typically follows a power law as described through the Gutenberg–Richter law:

$$\log(N) = a - bM$$

where N is the number of events. The slope is referred to as the "*b*-value" and is typically approximately 1 for tectonic earthquakes but may be closer to 2 for fluid related induced seismicity. Equally important, but less commonly discussed is the "*a*-value" proportional to the intrinsic microseismic activity rate. The *a*-value or activity rate varies significantly from site to site and with injection type. From an imaging perspective, the lower the *a*-value the greater the challenge of recording sufficient microseismicity to image the injection. However, from an induced seismicity perspective a lower *a*-value relates to lower seismic hazard. Higher *b*-values will also tend to relate to lower seismic hazard where the total seismic deformation takes place with an abundance of smaller magnitude microseismicity and proportionally fewer large-magnitude events.

A seismic source associated with a shear slip will also result in a specific radiation pattern of seismic waves (Jost and Hermann, 1989). Two types of seismic body waves will result from a seismic source: a compressional P-wave and a slower traveling shear s-wave. The P-wave first motion and relative amplitude of each wave radiated by the source in different directions depend on the seismic wave propagation angle relative to the direction of slip and can be used to construct a "beach-ball diagram" or fault-plane solution (Figure 10.2). Such a fault-plane analysis can be used to infer the orientation of a slip plane (also a secondary auxiliary plane orthogonal to the primary slip plane results in an identical radiation pattern, creating a nonunique identification of which plane actually slipped) associated with individual microseismic sources. In some instances, multiple events are grouped together in order to construct a composite fault plane solution. In addition to constraining the fault plane, the classic beach-ball diagram is also useful to define the stress regime as normal/extensional, strike-slip or thrust/compressional. The beach-ball construction fundamentally assumes a shear slip source, but can be generalized using a "moment tensor inversion" to include other potential source types including fracture dilation.

Microseismic Acquisition

Microseismic data acquisition typically relies on a digitizer connected to some type of seismic transducer, most commonly a geophone, seismometer, or accelerometer. Seismometers are useful for surface networks for monitoring induced seismicity, having the sensitivity and frequency response to accurately characterize larger seismic events. Accelerometers are useful to quantify the ground motion associated with induced seismicity for seismic hazard applications. However, for monitoring small-magnitude microseismicity, arrays of three component (3C) geophones are most commonly used consisting of geophones in three orthogonal orientations to record the complete seismic wavefield. Geophones are a common and inexpensive sensor that typically respond to seismic waves in the frequency range of 10's of Hz and greater and therefore are a convenient transducer.

Geophones can be deployed in deep or shallow boreholes or on surface in 2D patches, lines, or a grid of single sensors (see Maxwell, 2014, for specifics). In some instances, geophones permanently deployed for time-lapse seismic surveys are used for passive monitoring between active surveys (see Chapter 15 by White, this volume). The number of sensors deployed on surface is variable, but in some cases very large channel counts (1000+ sensors) are used to stack and detect low-amplitude signals that would otherwise be buried in background noise (Duncan and Eisner, 2010). However, these stacked surface arrays can have false triggers and misidentify possible microseismic sources. Downhole monitoring is also common in observation wells, either cemented in place or temporarily clamped to casing. Geophones can also be deployed behind casing but requires careful planning during drilling, or in the annular space behind tubing. These configurations allow deployment in an injection well without the need for dedicated observation well, but often result in elevated background noise and loss of sensitivity to record small-amplitude events. Regardless of the specifics of the well deployment, a number of 3C geophones will typically be used so that there is sufficient array aperture to accurately estimate microseismic source locations. There is potentially benefit monitoring microseismic from fiber optic, but the technology is in its infancy.

While selecting a sensor type and corresponding array geometry for a specific application, the first question should be related to the ability to address the motivation for the monitoring. For example, will passive microseismic imaging be used to image the injection and potential loss of containment, or is it for seismic hazard? Once defined, subsequent seismic array design questions commonly involve

1. Should a surface or a borehole array be used?
2. How many sensors to use and at what spacing?
3. How closely to the injection point should borehole sensors be deployed and over what distance can microseismicity be detected?

Answers to these specific questions can be gained through an analysis of the array performance (Maxwell *et al.*, 2003a), including comparisons of logistics and costs of each monitoring option along with estimates of the technical monitoring capabilities. Given a specific background noise level and the intrinsic seismic attenuation characteristics, the array sensitivity or smallest detectable seismic signal amplitude and associated moment or magnitude can be computed. If these parameters are not available, a sensitivity study of potential characteristics can be performed. Seismic amplitudes decrease with distance according to the seismic attenuation characteristics, such that the smallest detectable magnitude will increase with distance. The probable magnitude range of recorded microseismicity is another unknown factor that ultimately will control the number of events of the sensitivity limits and equally how far away from the array events will be detected. Magnitudes from injection in analogue scenarios can be used as an estimate, or a geomechanical assessment can be made to compute fracture slip conditions and likely microseismic magnitudes.

Similarly, microseismic location uncertainties associated with a particular array can be estimated, which is useful in setting expectations for the robustness of eventual microseismic data sets and ensuring that the microseismicity will be sufficiently accurate to resolve the intended targets. For example, depth accuracy will be needed to a sufficient level to identify if microseismicity is occurring out of the injection target interval for seal integrity applications. Lateral and depth accuracy requirements will be scaled by the expected size of the injected CO_2 plume, where microseismicity is being used to image the injection.

Monitoring array specifics will therefore be project dependent; however, some general rules of thumb can be considered. The question of surface or downhole monitoring will ultimately dictate the sensitivity of the recording, where typically only larger magnitude events are detectable from a surface compared to a proximal downhole monitoring array. Depth location uncertainties are also typically higher from a surface, although tend to increase with distance from a downhole array and may eventually become larger than surface values. Horizontal uncertainties are typically smaller for surface arrays. Surface arrays tend to be ideal for monitoring induced seismicity during injection, where larger potentially felt events are of interest.

Microseismic Processing

Microseismic processing typically involves three steps:

1. Detection of a microseismic signal
2. Estimation of the hypocentral location
3. Computation of source attributes including magnitude

There are two general workflows that can be used for either surface or downhole data (see Maxwell, 2014, and references therein for additional details). The classic method was adopted from earthquake methods, involving detection of a coherent signal above background noise followed by estimating the arrival times of the seismic wave at each sensor and performing an arrival-time inversion. Arrival-time inversion involves finding the location that minimizes the time difference between the observed arrival times and computed arrival times by ray tracing through a velocity model. Associated location uncertainties can be determined through the arrival-time inversion, resulting in uncertainties in different directions often represented by an error ellipsoid. Arrival times can be automatically determined, or manually picked through visual inspection. For single monitoring wells, signal polarization of the 3C sensors is also used to constrain a unique location. An alternative is a source scanning or "migration" method, which generally combines detection and location. A series of locations are assessed by stacking the recorded signals with time delays equivalent to differences in the computed arrival times for each location. The signals are also sequentially scanned in time, with the stacked amplitude reaching a peak at a microseismic source location and onset time. Generally, these methods are fully automated, but can be refined using arrival time picking constraints ideally once the data are manually reviewed.

Both workflows require a seismic velocity model to compute theoretical arrival times. Most methods use both p- and s-wave arrival times and so a velocity model is required for each. As injection is typically in sedimentary rocks, depth varying velocity models are needed. Velocities in each layer can be either isotropic or, better, anisotropic to account for layering effects. Lateral velocity variations can also be included in these models, although often 1D models are commonplace. Initial velocity models are often constructed from dipole sonic logs, but are ideally "calibrated" using travel time measurements from one or more checkshots. Checkshots are most effective if located near the expected microseismic locations and help verify the sonic logs, quantify anisotropy, and mitigate any lateral velocity variability. The velocity model construction and calibration are important steps in microseismic processing, as inaccuracies in the velocity model can result in mislocation of the microseismicity. Sensitivity studies can be made with different velocity models to quantify the potential mislocation.

Once detected and located, other microseismic source characteristics can be estimated. The most important is seismic moment, which can be estimated from a displacement spectrum of the recorded signal. Beach-balls or source mechanisms can also be determined, although often single monitoring wells are used that are unable to uniquely constrain the source mechanism for a single event. Although useful information, source mechanisms are not routinely determined.

After an initial processing, advanced techniques can also be used to improve attributes such as location. One of these methods is a relative location technique, where arrival time differences between events are used to estimate a location difference between sets of events. These techniques tend to improve precision although not absolute location and sometimes allow underlying patterns in microseismic location to emerge.

The final processing product is a "catalog" of microseismicity consisting of data, time, 3D location, magnitude, additional source characteristics, and quality control attributes. Useful QC parameters include location uncertainty ellipsoids, which will vary between sources, and signal-to-noise ratio (SNR). SNR is useful to define the signal quality and hence a proxy for the confidence in arrival time and associated location confidence. A comprehensive documentation of suggested deliverables has been defined for hydraulic fracturing applications (Maxwell and Reynolds, 2012), the majority of which are also suitable to CO_2 sequestration applications.

Microseismic Interpretation

Microseismic catalogs consist of a series of events that require an important interpretation step to address the original monitoring objectives. Prior to interpretation, it is important to assess the location uncertainty and potential recording biases associated with the sensitivity of the monitoring (see Maxwell, 2014, and references therein for additional details). Location uncertainties should be a specific component of the reported data set and, provided these estimates

are realistic, can be used to quantify the location confidence. Often these uncertainties are associated with random errors associated with arrival time uncertainties input to the location algorithm. Systematic location uncertainty associated with the velocity model is equally important and can be quantified through a velocity model sensitivity study, but is often not undertaken. It is critical to properly interpret the microseismic locations to quantify the entire location uncertainty, particularly when interpreting the depth of the microseismic sources in relation to stratigraphic layers. Other QC attributes such as SNR can also be useful to potentially identify a high confidence subset of the data and mitigate possible false triggered or misidentified signals. Recording biases where the minimum detected magnitude varies in space and time are important to quantify and if possible to normalize. For downhole monitoring, a plot of magnitude versus distance between the monitoring stations and microseismicity is useful to define a distance sensitivity bias and how far away events can be detected. A minimum magnitude can be established, for instance, so that only events above that magnitude are considered. For events above the minimum detectable magnitude, an analysis of the magnitude of completeness is important for accurate comparison of activity rates between and within specific projects.

The starting point of the interpretation is to examine the spatial and temporal locations for specific patterns. Microseismicity often forms clusters of events in space and time. The temporal evolution should be compared against the injection schedule, for correlations with changes in injection rate or pressures. Spatial clustering is also important, specifically in relation to the injection well and geological structure. The depth of the microseismicity in relation to the injection depth in a stratigraphic profile is a fundamental consideration and vital for characterizing geological containment of the CO_2. Lateral clustering of the microseismicity is also informative, providing insights into preferred CO_2 plume directions. Microseismicity often illuminates interactions with preexisting fractures and faults and so integration with a geological model and 3D seismic volumes can assist with understanding heterogeneities. Time-lapse interpretation of the microseismic sources is also an important aspect and can be more completely interpreted when integrated with time-lapse seismic such as 4D seismic.

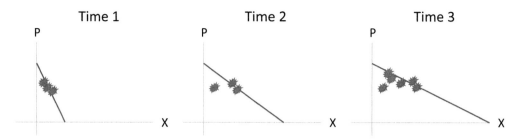

Figure 10.4 Schematic depiction of pressure profiles and associated microseismicity (orange) at increasing injection time steps. Microseismicity would be expected behind the point of furthest pressure increase that moves with time relative to the injection point as the pressure extends. Remnant microseismicity can be expected after the pressure is first increased to the critical level. Green colored microseismicity shows the possibility of slip on less favorably oriented fractures, associated with higher pressure and therefore would occur closer to the injection point at later times. The edge of the injected fluid will likely be somewhere between the extreme pressure change and the microseismicity.

It is important while interpreting microseismicity to avoid forming a direct association between the CO_2 plume and any recorded activity. As discussed in the geomechanics section, microseismic sources occur when localized subsurface pressures increase to a level sufficient to cause slip on preexisting fractures optimally oriented in relation to the stress field (Figure 10.4). The spatial position of the CO_2 plume relative to the critical pressure increase to generate microseismicity depends on the initial geomechanical conditions. The microseismicity will also be concentrated in areas where optimally oriented fractures exist, resulting in the possibility of spatially distinct microseismic clusters. A coupled simulation of flow and geomechanical response to injection can be used to interpret the microseismicity and predict the CO_2 saturations to identify the plume extents. Such a simulation requires specific input parameters, particularly the permeability characteristics to model flow and pressure and the stress state and fracture characteristics to model strains and microseismic sources. These parameters are often challenging to robustly characterize, but the microseismicity along with pressure recording and other plume images can be used to validate the modeling and potentially calibrate uncertain input parameters.

Case Studies

In this section, specific case studies are reviewed. In addition to pure CO_2 sequestration examples, a few relevant microseismic examples of monitoring gas injections for other types of wellbore operations are also included.

CO_2 Hydraulic Fracturing

During a hydraulic fracture treatment of a gas well, one depth interval was treated by pumping supercritical CO_2 instead of the typical water-based gel fracturing fluid used at other depths (Verdon et al, 2010). For these injections above fracturing pressure, similar microseismicity rates and magnitudes (below magnitude −3) were found for both fluid types (Maxwell et al., 2008). Injection volumes, rates, and viscosity of the injected fluids were similar. Compressible gases could be expected to generate less pressure increase than an incompressible fluid, but in this example the CO_2 would be expected to stay in a supercritical state during pumping at least within the high-pressure hydraulic fracture system. Nevertheless, the similarity of the microseismic responses is particularly relevant for qualifying microseismicity as an imaging tool for gas injection.

Talco Oilfield Natural Gas EOR Injection

A microseismic monitoring project was performed while injecting natural gas for EOR in the Talco field in Texas. As existing array, composed of three shallow (extending down approximately 425 m from surface) string of geophones originally intended for time-lapse monitoring of the reservoir injection at a depth of 1200 m below surface, was utilized. Seventy-seven microseismic events were detected during 6 months of monitoring (Maxwell et al., 2003a) with moment magnitudes between −1.6 and 0. More microseismicity could have been recorded if a deeper monitoring array had been used to get better sensitivity closer to the microseismically active region. The microseismicity was confined within the oil reservoir and generally proximal to a bounding fault.

Aneth Oilfield CO_2 EOR Injection

A vertical string of 3C geophones was deployed in a monitoring well to monitor microseismicity during EOR operations in the Aneth field in Utah (Soma and Rutledge, 2013). Microseismicity was detected at distances between 1 and 5 km from the geophones, and locations were computed using arrivals from refracted raypaths to supplement direct arrivals and enhance the accuracy. Specific microseismic clusters were identified, one spatially correlating with the stratigraphic depth of the CO_2 injection, although the majority of the activity was found at a depth of a simultaneous injection for salt water disposal.

Weyburn CO_2/Water EOR Injection

An EOR injection in the mature Weyburn oilfield in Saskatchewan, Canada was monitored using a vertical string of 3C geophones in an abandoned well (White, 2009). Very few microseismic events have been recorded, with the exception of limited time periods where injection switched between CO_2 and water (Maxwell et al., 2004). Although limited in number, the microseismic cluster between an injection and production well where a CO_2 production increase was observed following the start of injection. The lack of microseismicity at Weyburn can be attributed to the depleted nature of the field, where production has occurred since 1955.

Lacq-Rousse CO_2 Capture and Storage Pilot

Several shallow monitoring arrays, a surface seismometer, and downhole fiber optic sensors were used to monitor CO_2 injection into a depleted gas field in France (Prinet et al., 2013). During the first two and half years of injection, three microseismic events with magnitudes below 0 were recorded. The microseismic quiescence was attributed to the depleted nature of the field.

CO2CRC Otway Project

CO_2 injection at the Otway project in Australia was monitored using downhole sensors (Daley et al., 2009). A small number of microseismic events were detected and, although not sufficient to image the plume, did provide an indication that a fault proximal to the injection was not associated with induced seismicity.

Longyearbyen CO_2 Lab

In preparation for CO_2 injection, water injection tests have been monitored with limited microseismicity detected (Albaric et al., 2014). Characteristics of the recorded seismicity during the injection test were used to evaluate the initial monitoring array, leading to the addition of a downhole array closer to the injection depth to improve sensitivity and coverage to detect any microseismicity associated with future CO_2 injections.

In Salah CO_2 Project

CO_2 injection at the In Salah project in Algeria was monitored with various methods. Surface deformation was detected from inteferometric synthetic aperture radar (InSAR) (Ramirez and Foxall, 2014), a remote sensing technique using repeated radar satellite measurements to potentially detect millimeter scale surface movements. Surface deformation is a useful compliment to microseismic monitoring, and can also be used to invert for reservoir geomechanical strains (Maxwell et al., 2008). A coupled simulation of flow and geomechanics was performed and indicated a probable penetration into the caprock above the injection interval.

Microseismic monitoring was also performed by passively recording 6 of 48-3C geophones deployed in a downhole monitoring well (Oye et al., 2013). Instrumentation problems resulted in reliable recording of signals from only two sensors and limited directional information. The impact of the limited recording meant that reliable locations could not be determined, but the data have been analyzed for common signal characteristics to identify distinct spatial clusters with possible locations based on S- to P-wave arrival times. Although specific locations are uncertain, the analysis was able to define a proximal distance to the sensors and associate some of the microseismicity with the distance to the injection well.

Despite the quality of the microseismic data, the quantity of the events is significant. Approximately 1500 microseismic events were initially detected over a 22-month period, increasing to about 5000 events using a signal matching technique to detect weak signals based on signal characteristics consistent with the larger originally detected events (Oye et al., 2013). The

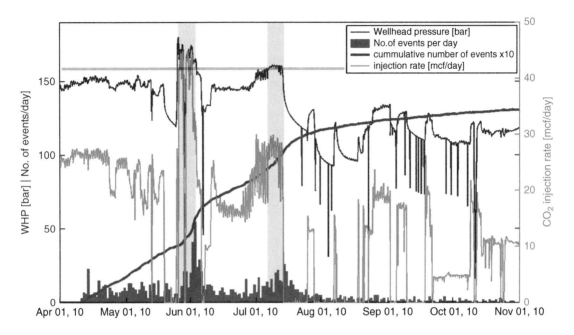

Figure 10.5 Comparison of injection rate and microseismicity rate for In Salah. Note the increased microseismicity rate during the initial period with a higher injection rate prior to August 2010. Also note the two shaded time periods where the injection pressure is interpreted to exceed the fracture pressure (pink horizontal line). (After Oye *et al.*, 2013.)

largest moment magnitude was 1. Over and above a background microseismicity rate, greater activity was found during periods of increased injection rate and associated pressure increases (Figure 10.5). The injection pressure was interpreted to be above the hydraulic fracturing pressure, suggesting that hydraulic fracturing was occurring during these periods of increased activity. Notwithstanding the processing issues related to acquisition problems, the In Salah project appears to be the first to demonstrate that an adequate amount of microseismicity could be detected to image the injection.

Illinois Basin–Decatur Project

A comprehensive characterization and monitoring project was performed to investigate CO_2 injection (see also Chapter 19 by Bauer *et al.*, this volume). Microseismicity was initially monitored using three geophones deployed in the injection well along with a string of the 3C geophones used for both active and passive imaging (Will *et al.*, 2014). Baseline monitoring of the background microseismicity was performed over a one–and-half-year period before injection (Smith and Jacques, 2016). Three years of CO_2

injection was monitored over a period between 2011 and 2014, during which time approximately a million tonnes of CO_2 was injected into the Mount Simon Sandstone located above Pre-Cambrian crystalline basement. A surface seismometer array consisting of 13 stations was later deployed during the last half of the injection period and able to record the larger magnitude events (Kaven *et al.*, 2015).

The preinjection monitoring resulted in identification of eight microseismic events with magnitudes between -2.16 and -1.52, suggesting little background activity near the injection point (Smith and Jacques, 2016). During the injection, a total of 4747 microseismic events were detected by the downhole arrays with moment magnitudes between -2.12 and 1.17 (Bauer *et al.*, 2016). The microseismic was found to be distributed in depth between the upper Pre-Cambrian and lower Mount Simon. No significant activity was found above a clay layer above the injection point, which is believed to act as a flow baffle and contain the injection. Laterally from the injection point, the microseismic formed several discrete clusters primarily to the north of the injection (Figure 10.6). Each cluster tended to start at a specific time with clusters both at greater and lesser offsets beginning at different

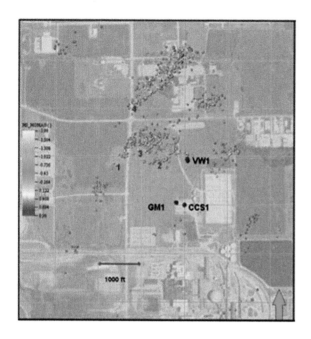

Figure 10.6 Map view of downhole detected microseismicity resulting from injection into well CCS1 at the Illinois Basin–Decatur Project showing clusters proximal to the injection. The timing of the onset of the microseismic clusters is labeled sequentially. (From Will et al., 2014.) (Courtesy Illinois Geological Survey.)

times. Within each cluster, the number of microseismic tended to increase to a peak and then decline.

The surface monitoring of the second half of the injection detected 179 locatable events (Figure 10.7) in the moment magnitude range between –1.13 and 1.26 (Kaven et al., 2015). The surface seismometer network was not able to record relatively small magnitude events, consistent with increased noise levels and greater offset from the activity that is expected for a surface compared to a downhole array. The depth distribution and spatial clustering were consistent with the results of the contemporaneous downhole monitoring, although the hypocentral depths were spread over a larger depth interval consistent with larger depth uncertainty expected for the surface array geometry. Focal mechanisms could be computed for individual events and indicated an approximate NE–SW strike-slip mechanism (Figure 10.8). A similar mechanism was found with the downhole data as a composite solution of events within specific clusters.

3D seismic volumes were analyzed for preexisting fractures using "edge detection" methods, which indicated lineation aligned with the microseismic source

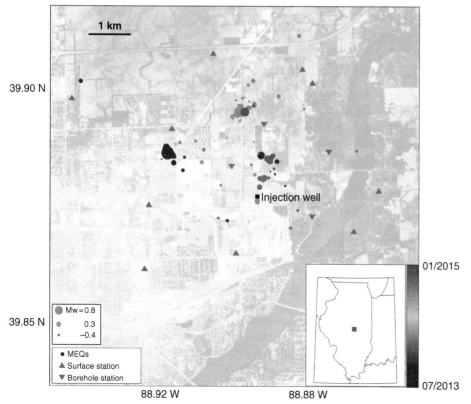

Figure 10.7 Map view of surface detected microseismicity at the Illinois Basin–Decatur Project depicting additional clusters detected more distal to the injection. (From Kaven et al., 2015.)

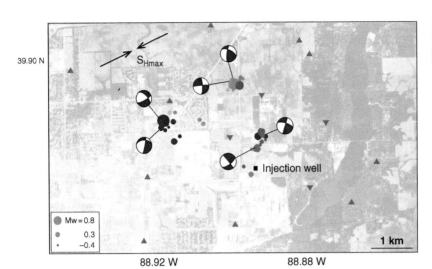

Figure 10.8 Fault-plane solution for select events detected by the surface array at the Illinois Basin–Decatur Project. (From Kaven *et al.*, 2015.)

mechanisms (Will *et al.*, 2014). A flow simulation was performed to estimate pore pressure changes resulting in an approximately circular pressure increase centered on the injection well (Figure 10.9), which was then used in a geomechanical model to examine fault slip conditions (Lee *et al.*, 2014). Preexisting faults were assumed based on the microseismic cluster geometry, location, and orientation. Relatively low friction had to be assumed on these faults in order for them to slip, with frictional angles of 10°. Based on the modeled slip, equivalent seismic moments were estimated, but were approximately two orders of magnitude larger than the observed moments. Differences between modeled and observed seismic moments could be attributed to energy partitioning between seismic and aseismic deformation, and dynamic aspects of the slip associated with individual sources. The modeling also did not attempt to simulate seismically quiescent regions, which is an important potential aspect to understand the spatial clustering. Nevertheless, the Decatur project is the most thorough to date from a microseismic perspective, with significant numbers of high-quality microseismic sources leading into an integrated interpretation.

Conclusion

In conclusion, passive microseismic monitoring is an important tool for characterizing hazard of induced seismicity and potentially to image the injection. The various published examples of microseismic monitoring of gas injection show highly variable data quality

and microseismic activity rates. Microseismic data quality can be partially addressed using reliable recording equipment, and using a sensitive monitoring array to improve the sensitivity of the recording. However, microseismic rates are clearly highly variable and depend on the injection rates and associated pressures, but also the geomechanical reservoir conditions. At sufficiently high injection pressures, CO_2 can result in detectable microseismicity, particularly if hydraulic fractures are created. Even below the fracture pressure, microseismic can occur if *in situ* pressures are sufficient to cause slip on preferentially oriented fractures. For some depleted reservoirs, the injection pressure does not appear to be sufficient to generate significant levels of microseismicity, although this can be seen as favorable from an induced seismic hazard perspective. Discernible microseismicity have been reported during some gas EOR injections, although the reservoirs are also depleted and the injection is being balanced by hydrocarbon production. Potential for microseismicity appears to be highest during injection into saline aquifers, although still dependent on the site specifics.

Although no felt seismicity has been attributed to CO_2 injection to date, injection induced seismicity concerns are increasing with experiences in geothermal and salt water disposal. As sequestration projects grow in scale with larger injection volumes, the seismic hazard question will likely be under increased scrutiny. With the challenges associated with sufficiently characterizing the injection target to confidently estimate seismic potential, the motivation for passive monitoring can be expected to continue. Increased monitoring

Figure 10.9 Perspective view showing modeled pressure increase contours in relation to microseismicity (black). Injection well (CCS1) is shown in blue and the VW1 monitoring well in green. (From Lee *et al.*, 2014.)

sensitivity can then provide details of the seismic response to injection to better forecast future injection scenarios at a particular site. Information about gas leaks and mapping the pressure plume for integration with complementary monitoring technologies further adds to the value proposition of microseismic monitoring.

References

Abercrombie, R. E. (1995). Earthquake source scaling relationships from –1 to 5 ML using seismograms recorded at 2.5-km depth. *Journal of Geophysical Research*, **100**(B12): 24, 015–24, 036.

Aki, K. (1965). Maximum likelihood estimate of b in the formula log N = a – bM and its confidence limits. *Bulletin of the Earthquake Research Institute*, **43**: 237–239.

Albaric, J., Oye, V., and D. Kühn. (2014). Microseismic monitoring in carbon capture and storage projects: Fourth EAGE CO₂ geological storage workshop, Stavanger, Norway.

Bauer, R. A., Carney, M., and Finley, R. J. (2016). Surface monitoring of microseismicity at the Decatur, Illinois, CO2 Sequestration Demonstration Site. *International Journal of Greenhouse Gas Control*, **54**: 378–388.

Brune, J. (1970). Tectonic stress and the spectra of shear waves from earthquakes. *Journal of Geophysical Research*, **75**: 4997–5009.

Daley, T. M., Sharma, S., Dzunic, A., Urosevic, M., Kepic, A., and Sherlock, D. (2009). Borehole seismic monitoring at Otway using the Naylor-1 instrument String. LBNL-2337E report. Berkeley, CA: Lawrence Berkeley National Laboratory.

Duncan, P., and Eisner, L. (2010). Reservoir characterization using surface microseismic monitoring. *Geophysics*, **75**. DOI:10.1190/1.3467760.

Hubbert, M. K., and Rubey, W. W. (1959). Role of fluid pressure in mechanics of overthrust faulting: 1. Mechanics of fluid-filled porous solids and its application to over-thrust faulting. *Bulletin of the Geological Society of America*, **70**: 115–166.

Jost, M. L., and Hermann, R. B. (1989). A student's guide to and review of moment tensors. *Seismological Research Letters*, **60**: 37–57.

Kaven, O., Hickman, S. H., McGarr, A. F., and Ellsworth, W. L. (2015). Surface monitoring of microseismicity at the Decatur, Illinois, CO_2 Sequestration Demonstration Site. *Seismological Research Letters*, **86**. DOI:10.1785/0220150062.

Lee, D.W., Mohamed, F., Will, R., Bauer, R., and Shelander, D. (2014). Integrating mechanical earth models, surface seismic, and microseismic field observations at the Illinois Basin – Decatur Project. *Energy Procedia*, **63**: 3347–3356.

Maxwell, S. C. (2014). Microseismic imaging of hydraulic fracturing: Improved engineering of unconventional shale reservoirs. SEG Distinguished Instructor Short Course. DOI:10.1190/1.9781560803164.

Maxwell, S. C., and Reynolds, F. (2012). Guidelines for standard deliverables from microseismic monitoring of hydraulic fracturing. *CSEG Recorder*. https://csegrecorder.com/articles/view/guidelines-for-standard-deliverables

Maxwell, S. C., Urbancic, T., and McClellan, P. (2003a). Assessing the feasibility of reservoir monitoring using induced seismicity. *EAGE Abstracts*.

Maxwell, S. C., Urbancic, T. I., Prince, M., and Demerling, C. (2003b). Passive imaging of seismic deformation associated with steam injection in Western Canada. *SPE 84572*.

Maxwell, S. C., White, D. J., and Fabriol, H. (2004). Passive seismic imaging of CO_2 sequestration at Weyburn. Expanded Abstracts, *Society of Exploration Geophysicists*.

Maxwell, S. C., Du, J., and Shemeta, J. (2008). Passive seismic and surface monitoring of geomechanical deformation associated with steam injection. *Leading Edge*, **27**: 260–266.

Maxwell, S. C., Rutledge, J., Jones, R., and Fehler, M. (2010). Petroleum reservoir characterization using downhole microseismic monitoring. *Geophysics*, **75**. DOI:10.1190/1.3477966.

Oye, V., Aker, E., Daley, T. M., Kuhn, D., Bohloli, B., and Korneev, V. (2013). Microseismic monitoring and interpretation of injection data from the In Salah CO2 storage site (Krechba), Algeria. *Energy Procedia*, **37**: 4191–4198.

Prinet, C., Thibeau, S., Lescanne, M., and Monne, J. (2013). Lacq-Rousse CO_2 capture and storage demonstration pilot: Lessons learnt from two and a half years monitoring. *Energy Procedia*, **37**: 3610–3620.

Ramirez, A., and Foxall, W. (2014). Stochastic inversion of InSAR data to assess the probability of pressurepenetration into the lower caprock at In Salah. *International Journal of Greenhouse Gas Control*, **27**: 42–48.

Segall, P., and Lu, S. (2015). Injection-induced seismicity: Poroelastic and earthquake nucleation effects. *Journal of Geophysical Research: Solid Earth*, **120**: 5082–5103. DOI:10.1002/2015JB012060.

Smith, V., and Jaques, P. (2016). Illinois Basin-Decatur Project pre-injection microseismic analysis. *International Journal of Greenhouse Gas Control*, **54**: 362–377.

Soma, N., and Rutledge, J. T. (2013). Relocation of microseismicity using reflected waves from singlewell, three-component array observations: Application to CO_2 injection at the Aneth oil field. *International Journal of Greenhouse Gas Control*, **19**: 74–91.

Verdon, J. P., Kendall, J. M., and Maxwell, S. C. (2010). A comparison of passive seismic monitoring of fracture stimulation from water and CO_2 injection. *Geophysics*, **75**. DOI:abs/10.1190/1.3304825.

White, D. (2009). Monitoring CO2 storage during EOR at the Weyburn-Midale Field. *Leading Edge*, July: 838–842.

Will, R., Smith, V., Leetaru, H. E., Freiburg, J. T., and Lee, D. W. (2014). Microseismic monitoring, event occurrence, and the relationship to subsurface geology. *Energy Procedia*, **63**: 4424–4436.

Zoback, M. D., and Gorelick, S. M. (2012). Earthquake triggering and large-scale geologic storage of carbon dioxide. *Proceedings of the National Academy of Sciences of the USA*. DOI:10.1073/pnas.1202473109.

Chapter

Well Logging

11

Zaki Bassiouni

Introduction

Carbon dioxide (CO$_2$) has been injected into underground geological reservoirs essentially for two main purposes: enhanced oil recovery (EOR) and the storage of this greenhouse gas (CGS, CO$_2$ geological storage or geosequestration). In both applications, characterization of movement of CO$_2$ within the reservoir is critical to ensuring efficient oil recovery in EOR projects and effective sequestration in CGS projects. Monitoring the progress of injected CO$_2$ provides insight into the extent of gravity segregation and the degree of stratification and other permeability distributions and their effect on project management. In EOR applications, changes of quantitative parameters such as volumetric sweep efficiency and oil saturation are critical to recovery optimization.

Several monitoring approaches may be used to judge the progress of CO$_2$ movement in the reservoir. These approaches encompass surface measurements, as in seismic time-lapse techniques, and measurements performed in boreholes, such as well logs, pressure surveys, and the use of core samples. This chapter is devoted to well logging methods using tools sensitive to the presence of CO$_2$. In general, monitoring using well logs consists of log data acquired in cased holes. The presence of the steel tubular string and the cement sheath behind it favors the use of pulsed neutron and gamma ray tools. In newly drilled wells, such as infill and observation wells, open-hole logs, namely formation resistivity and formation density tools, are incorporated in the logging program.

The interpretation of the acquired logs can be qualitative, i.e., limited to the detection of the presence of CO$_2$, or quantitative, which involves the estimation of changes in the saturation of fluids present in the formation. At least two fluids will always be present, formation water and CO$_2$, as in the case of aquifer carbon storage (CS). In some cases,

hydrocarbons may also be present, as in EOR projects or CS in depleted hydrocarbon reservoirs.

CO$_2$ properties that can be used for detection and quantification are its relatively low density, its nonconductive nature, and its chemical composition characterized by the presence of carbon and oxygen and absence of hydrogen.

Concepts of Open-Hole Logs

All subsurface injection projects start with the drilling of boreholes, which are initially uncased, referred to as "open-hole." Well logs are acquired in open-hole wells to evaluate the geological environment and estimate the reservoir potential. These logs usually represent the base case in monitoring operations schemes.

Once the well is cased, open-hole logs cannot be repeated for monitoring purposes and so cased-hole logs are substituted. Understanding the measurement and analysis of open-hole logs is critical to understanding all subsequent measurements taken in cased-holes. Figure 11.1 is a common presentation of an open-hole well log usually referred to as "resistivity log." In addition to resistivity curves traced in the middle track, the log displays several other useful curves. The solid curve on the first track is the spontaneous potential (SP), measured between an electrode traveling in the borehole and a grounded electrode. The SP log is used to distinguish between shales and permeable formations, in this case sandstones. The SP curve displays maximum voltage readings in shales forming a shale baseline. Deflection away from the shale baseline usually occurs at strata boundaries. The deflection reaches a maximum opposite clean or shale-free sand, in this example –100 mV. The top and bottom boundaries of the thick sand displayed by the log are at 9975 ft. and 10,035 ft. respectively. The gross thickness of the sand, i.e.,

181

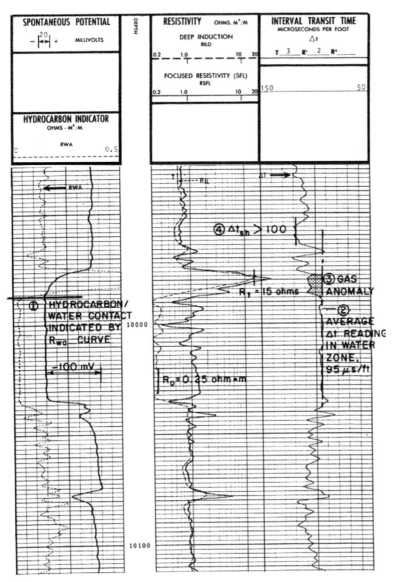

Figure 11.1 Example of a "resistivity" log, courtesy of Schlumberger.

the interval between top and bottom geological boundaries is approx. 60 ft. The net, i.e., permeable interval, thickness is estimated by subtracting the thickness of shale streaks within the sand, if any exist, from the gross thickness.

Theoretical and experimental work demonstrated that the SP deflection in clean sand is determined by the contrast of the resistivity of the formation water R_w, and the resistivity of the water phase of the drilling fluid or mud filtrate R_{mf}. Using available algorithms, R_w, and in turn the formation water salinity, can be estimated from the SP reading. Formation water

salinity determines what well logs can be run and plays an important role in the quantitative interpretation of resistivity logs.

Because hydrocarbons do not conduct electric current, the formation resistivity is a key hydrocarbon indicator. However, low-porosity formations and formations saturated with relatively low-salinity water also display high resistivity values. To avoid mistaking the latter formations for a hydrocarbon bearing one, a hydrocarbon indicator value R_{wa} is estimated using formation resistivity, R_t, and porosity, ϕ, values, (Bassiouni, 1994):

$$R_{wa} = R_t \phi^2 \qquad (11.1)$$

R_{wa} is also traced on the first track. In a section of a well where formation water salinity and lithology remain the same, the minimum value $(R_{wa})_{min}$ of the hydrocarbon indicator approximates the true R_w, in this example 0.025 ohm-m at formation temperature. In permeable zones, R_{wa} values, higher than $(R_{wa})_{min}$ indicate hydrocarbon presence even in low porosity formations. This is because the calculation of R_{wa} makes use of porosity, thus screening out the effect of low porosity.

Based on the response of the resistivity curves displayed on the second track, and the calculated R_{wa} curve, the few feet at the top of the sand display a zone partially saturated with hydrocarbons. Using similar reasoning, the bottom zone of the sand is water bearing. In this analysis, the dashed resistivity curve is used as it represents deep measurement beyond the zone affected by the borehole and the mud filtrate invaded zone of the formation.

The contrast between the resistivity, R_t, in a hydrocarbon-bearing interval and the resistivity, R_o, in a water-bearing interval is inversely proportional to the saturation of formation water, S_w. According to Archie's model (Archie, 1942):

$$S_w = \left(\frac{R_o}{R_t}\right)^{1/2}, \text{ and } S_{HC} = 1 - S_w \qquad (11.2)$$

where S_{HC} is the hydrocarbon saturation.

Using the R_t value displayed within the hydrocarbon zone (15 ohm-meter or ohm-m) and the R_o value displayed in the water-bearing zone (0.25 ohm-m), the water saturation in the top of the sand is 13% and subsequently the hydrocarbon saturation is 87%.

The third track displays the interval transit time Δt, which is the inverse of the acoustic velocity. Δt is a weighted average of the interval transit times of the rock matrix, Δt_{ma}, and the formation fluids, Δt_f:

$$\Delta t = \phi \Delta t_f + (1 - \phi)\Delta t_{ma} \qquad (11.3)$$

where ϕ is the formation porosity. Solving for the porosity yields

$$\phi = (\Delta t - \Delta t_{ma})/(\Delta t_f - \Delta t_{ma}) \qquad (11.4)$$

The model represented by Eqs. 11.3 and 11.4 was developed for water-filled formations; however, in

this particular example of an unconsolidated sand, the sonic curve indicates the hydrocarbon to be gas. The low acoustic velocity of gas results in the relatively high Δt. Yet, because the response of the SP and sonic logs support the premise of a uniform porosity throughout the sand, the porosity of the hydrocarbon zone is better estimated in the water-bearing zone. Using a Δt value of 95 µs/ft. together with the empirical values $\Delta t_{ma} = 55.5 \mu s/ft.$, and $\Delta t_f = 189 \mu s/ft.$ yields a porosity value of 30% which, in this case, is representative of both the hydrocarbon and water zones.

From the log of Figure 11.1 the presence of a relatively thick permeable sand containing a gas-bearing interval at its top was detected. It was also possible to quantify the interval's thickness, average porosity, and average hydrocarbon saturation. The presence of CO_2 will result in a similar response of the resistivity and sonic logs and can be analyzed in similar fashion.

Formation-fluid types are usually determined using the type of log displayed on Figure 11.2. This type of log presentation is usually referred to as "porosity log." The dashed curve on the first track of the log is a caliper log approximating the diameter of the borehole. The caliper indicates a uniform borehole diameter averaging 8.5 inches in the clean sand intervals. Borehole enlargement is shown in shale intervals, which is the result of shale sloughing. The solid curve on the first track is the gamma ray log, which is a measure of the natural radioactivity of the formations. Studies of the origin of radioactivity in sedimentary rocks indicate that shale is characterized by a relatively high level of gamma ray radioactivity in contrast to clean sands. In general, the gamma ray curve correlates reasonably well with the SP curve.

When formations are irradiated with gamma rays or neutrons, the gamma ray and neutron energies are moderated as they travel from the source of energy to the detectors of the moderated energy. The degree of moderation depends on the rock type, e.g., sandstone, limestone, etc., the fluid type, e.g., gas or liquid, and the amount of fluid that is related to the formation porosity.

The two curves displayed on the second track are the calculated density porosity (ϕ_D) and neutron porosity (ϕ_N) values assuming sandstone matrix and fresh formation water. The calculated porosity is a true value only in water-bearing

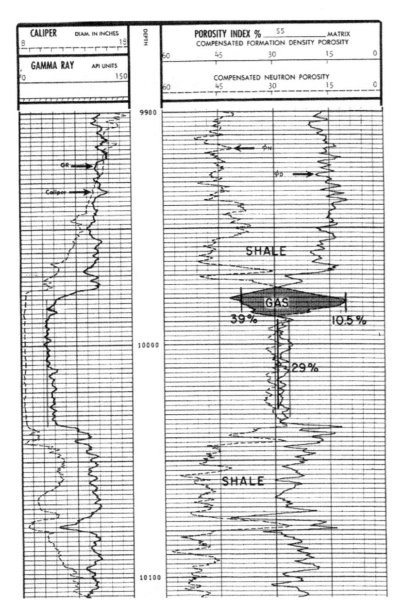

Figure 11.2 Example of a "porosity" log, courtesy of Schlumberger.

sandstones. Hence, the density porosity (ϕ_D) and neutron porosity (ϕ_N) values are close in value in liquid-filled formations and are quite different in gas-bearing formations where $(\phi_D) \gg (\phi_N)$, and in shales, where $(\phi_D) \ll (\phi_N)$.

These concepts confirm that the hydrocarbon present in the upper part of the sand is gas. The porosity curves of Figure 11.2 display average apparent porosity values of $\phi_D = 39\%$ and $\phi_N = 10.5\%$ respectively. To determine the true porosity value, ϕ, of the gas-bearing section, the following

empirical transform may be used (Bassiouni, 1994):

$$\phi = [(\phi_D^2 + \phi_N^2)/2]^{1/2} \qquad (11.5)$$

This empirical transform yields an estimated true porosity value of 29%, which is in agreement with the porosity value of the underlying liquid-bearing zone where both curves indicate similar porosity values. Such agreement is expected in uniform sands, which is the case in this example log. This value is slightly different from that estimated using the sonic log because of the assumed parameters used in the interpretation.

Analysis of the logs of Figures 11.1 and 11.2 is relatively straightforward, as the sand of interest is a clean sand. The log responses in shaly sands are more complex as they reflect the presence of clay-minerals. Because clean formations are usually selected for CS projects to ensure sufficient permeability and in turn adequate injection rates, shaly sand analysis is not covered here. Readers interested in the topic are referred to the references, e.g., Archie (1942) for the coverage of shaly sand analysis.

Concepts of Cased-Hole Resistivity Logs

Open-hole conventional resistivity tools emit a relatively low frequency electromagnetic field. In cased holes, the presence of a highly conductive steel casing shunts the emitted field through the casing thereby rendering the conventional measurement impossible. However, a relatively new technology led to the design of a cased-hole resistivity tool. The new tool is basically a focused current device. A current electrode in contact with the casing injects current into the borehole environment. Most of the current will flow up and down the casing steel, the path of least resistance. A high-frequency current will stay almost entirely within the steel. The use of a low-frequency current allows a tiny fraction of the current to leak into the formation. The higher the leaking current, the lower the resistivity of the formation. The result is a cased-hole resistivity log that is used to monitor the advance of injected fluids through the reservoir.

The tool measures the current leaking into the formation I_f as the difference of axial currents over two adjacent casing segments. The formation resistivity R_t is expressed as (Harold *et al.*, 2004)

$$R_t = K(V/I_f) \tag{11.6}$$

where V is the casing voltage and K is the tool geometric factor, which depends on the tool design. The leakage current is typically in the order of several milliamperes, V is in 10s of nanovolts, which can be measured using new technology.

Cased-hole resistivity tools have, however, several limitations related to the measurement environment. The presence of high-resistivity cement might require environmental corrections. Scale accumulation on the casing surface may inhibit efficient

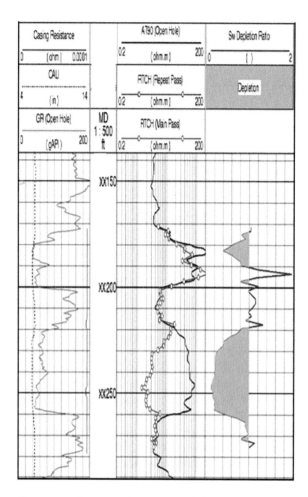

Figure 11.3 Comparison between open-hole and cased hole formation resistivity. (From Jiang *et al.*, 2009.)

contact between the tool electrodes and the casing. The tool may not be able to get good contact over casing collars. In addition, the measurement itself is impaired by the presence of corrosion-inhibitor nonconductive coating.

Figure 11.3 illustrates the potential application of cased-hole formation resistivity (CHFR) measurement in monitoring reservoir performance. This example shows significant reduction in resistivity in the lower part of the displayed hydrocarbon formation, which indicates depletion, i.e., reduction in the saturation of the resistive fluid. Equation (11.1) can be used to quantify saturation changes. The depletion can be quantified using the ratio of existing saturation to initial saturation, as seen on the third track of Figure 11.3.

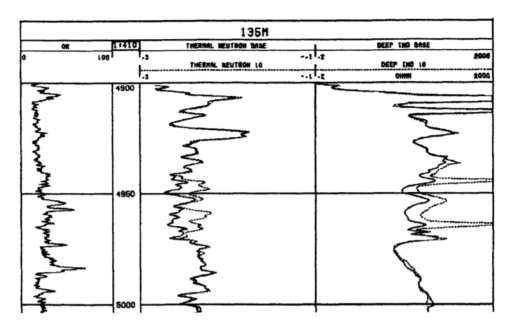

Figure 11.4 Time-lapse induction and sigma logs recorded in Fiberglass casing. (From Gould *et al.*, 1991.)

If observation or monitoring wells are utilized, the wells can be cased using high-resistivity fiberglass tubulars. Such casing material allows the propagation of the electromagnetic field through the casing and into the formation. Resistivity type logs can be recorded throughout the life of the reservoir, providing means to monitor changes in water saturation. Figure 11.4 shows time-lapse induction logs recorded in fiberglass casing compared to time-lapse pulsed neutron logs recorded in the same well. Induction logs recorded through fiberglass casing provides information that correlates well with that obtained using pulsed neutron tools. Resistivity logs thus acquired can be integrated with neutron logs for quantitative analysis.

Concepts of Neutron Logs

In its basic design, neutron tools consist of a source periodically emitting high-energy neutrons. As they propagate, these neutrons interact with the individual nuclei of the atoms of the formation. The results of these interactions are measured at two detectors within the tool placed at two different distances from the source.

There are two types of interactions, moderating and absorptive. Moderating interactions, which can be inelastic or elastic, result in the slowing down of the neutron, i.e., the loss of kinetic energy. On the other hand, absorptive reactions result in the disappearance of the slow neutron in its free form.

Inelastic scattering is a moderating interaction of fast neutrons possessing relatively high energy. In this moderating reaction, the neutron collides with the nucleus. A fraction of the neutron's kinetic energy transfers to the nucleus in the form of internal energy. This in turn results in the excitation of the nucleus. The excited nucleus is immediately stabilized by emitting a gamma ray photon. Deexcitation gamma rays, commonly referred to as scattering gamma rays, are characteristic of the nucleus involved in the inelastic interaction. Spectral analysis of the scattering gamma rays indicates the type and quantity of nuclei present in the formation.

Inelastic scattering requires that the neutron energy exceeds the threshold energy needed to excite the nucleus. When the neutron energy is moderated to a level below that threshold, elastic scattering takes place. In this latter interaction, the neutron collides with a nucleus in a manner similar to that of two colliding billiard balls. The neutron scatters off the nucleus with less kinetic energy. The energy balance is transferred to the nucleus in the form of kinetic energy; its internal energy remains unchanged. Therefore, no radiation is associated with elastic scattering.

After sufficient elastic collisions, the neutron is thermalized, i.e., reaches a low kinetic energy level in the vicinity of 0.025 eV. The number of collisions resulting in the thermalization of the neutron depends on the mass of the nuclei present in the formation. Because hydrogen nuclei have almost the same mass as the neutron, fewer collisions with hydrogen atoms are needed to thermalize a neutron.

Thermal neutrons are captured by a nucleus. Neutron capture by a nucleus results in a heavier isotope in an excited state. The excited nucleus releases the excess energy in the form of gamma photons, referred to as "gamma ray of capture." Energy levels of capture gamma rays are characteristic of the capturing nucleus. Spectral analysis of capture gamma rays indicates the types and quantity of nuclei present in the formation.

Physical parameters associated with the phenomena occurring during the moderation and capture of neutrons are measured and empirically related to formation properties. Because hydrogen is responsible for most of the slowing down of the epithermal neutrons, i.e., neutrons with an energy level higher than that of the thermal neutrons. Measuring the concentration of epithermal neutrons indicates the hydrogen concentration in the formation, which in turn is empirically related to porosity because, in clean formations, hydrogen exists only in the fluids saturating the pore space.

The rate of decay of thermal neutrons is related to the capture cross sections, Σ, of the formation constituents. Chlorine has the highest capture cross section, Σ, of common formation elements; hence measuring Σ reflects chlorine content and, in turn, water content and, by complement, gas, or oil content. Figure 11.5 is an example of an earlier sigma log. In addition to the Σ curve, the raw readings of the near and far detectors are also displayed. The scales used to display these two curves are different and are selected in a way that the readings of the near and far detectors will track each other in liquid-bearing formations, and separate in gas-bearing formations.

The energy of scattering gamma rays and capture gamma rays are, respectively, characteristic of the nucleus involved in the scattering or the capture. This principle leads to several types of measurements, e.g., the spectral contribution of carbon and oxygen allows the determination of carbon/oxygen (C/O) ratio. The C/O ratio can be determined in both low

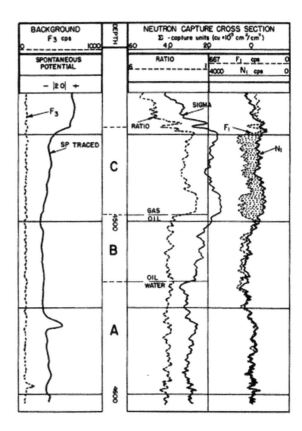

Figure 11.5 Example of an early sigma log. (From McGhee *et al.*, 1974.)

and high water salinity environments, unlike Σ, which requires relatively high salinity. In addition, the spectral analysis of the captured gamma ray can be used to determine the contents of iron, silicon, and calcium needed for lithology determination. It is also used to determine chlorine and hydrogen contents, which allows us to investigate salinity.

A schematic of a C/O ratio log is displayed in Figure 11.6. The C/O log measurements may be obtained by moving the tool at an optimal speed continuously or by stationary measurements for enhanced log quality. To assist with the analysis, a "C/O envelop" is computed, using the theoretical model presented in the section on water saturation and Figures 11.7 and 11.8. The C/O envelop consists of two curves $(C/O)_{min}$ computed by setting $S_w = 1$ and $(C/O)_{max}$ computed by setting $S_w = 0$. To perform such computation, the lithology and porosity are usually obtained from open-hole data. The

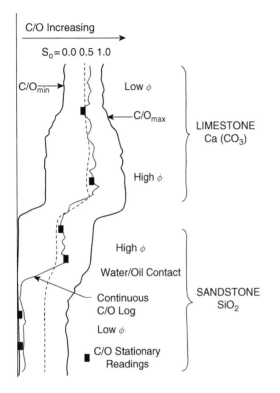

Figure 11.6 Schematic illustrating the C/O envelop interpretation technique. (From Smolen, 1996.)

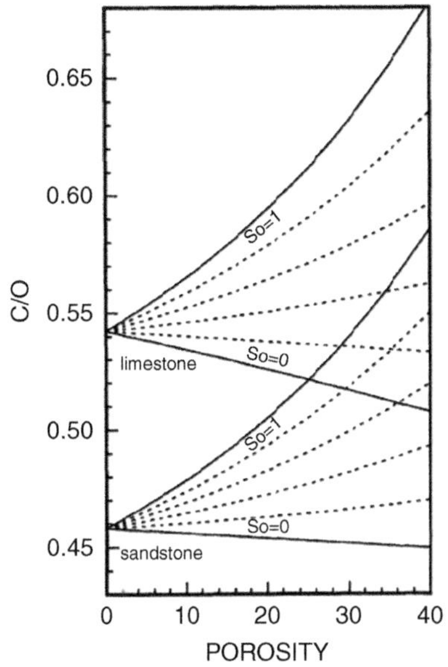

Figure 11.7 C/O interpretation chart constructed for a specific matrix. (From North, 1987.)

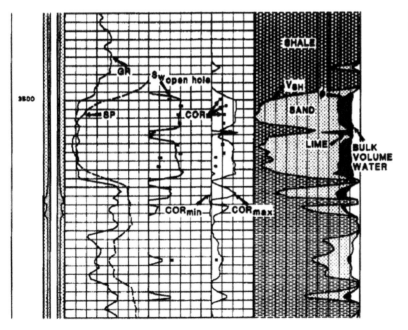

Figure 11.8 Water-flood monitoring case using C/O log. (From Gilchrist *et al.*, 1983.)

variations of the envelop curves reflect lithology and porosity variations. The separation between the envelope's two curves at any level scales linearly in terms of water saturation.

Quantitative Analysis Methods

Water Saturation from Sigma Logs

The tool measured Σ_{log} is related to the constituents of the formation. For a clean, porous formation containing water and hydrocarbon, Σ_{log} is expressed as (Clavier et al., 1971):

$$\Sigma_{log} = (1 - \phi)\Sigma_{ma} + \phi S_w \Sigma_w + \phi(1 - S_w)\Sigma_h \quad (11.7)$$

where Σ_{ma}, Σ_w, and Σ_h are the capture cross section of rock matrix, water, and hydrocarbon, respectively. Solving Eq. (11.6) for water saturation, S_w, gives

$$S_w = \frac{(\Sigma_{log} - \Sigma_{ma}) - \phi(\Sigma_h - \Sigma_{ma})}{\phi(\Sigma_w - \Sigma_h)} \quad (11.8)$$

The water, rock matrix, and hydrocarbon cross sections of capture are determined experimentally or deduced from graphical interpretation techniques using actual field data. Porosity values are derived from other log measurements. In shaly formations, Eq. (11.6) is modified to include the shale contribution as warranted.

The time-lapse technique allows monitoring of water saturation. If only one other fluid is present in addition to the water, e.g., oil in an undersaturated oil reservoir and CO_2 in the case of aquifer CS, changes of the saturation of oil or CO_2, the complement of the water saturation, can in turn be monitored.

Water Saturation from C/O Log

The C/O interpretation model can be written in terms of the matrix, pore space, and borehole contributions, which can be expressed as the product of their respective volume and atomic densities, yielding the following expression (North, 1987):

$$C/O = Af(C))/f(O)$$
$$f(C)) = \alpha(1 - \varphi) + \beta S_o \phi + B_c \quad (11.9)$$
$$f(o)) = \gamma(1 - \varphi) + \delta S_w \phi + B_o$$

where

A = Ratio of average C and O fast neutron (scattering gamma ray producing) cross sections

$f(C)$ and $f(O)$ are respectively the total carbon and oxygen contribution from borehole, matrix, and formation fluids

B_c, B_o = C and O contributions from borehole

S_w, S_o = Water and oil saturation

α = Atomic concentration of C in matrix

β = Atomic concentration of C in formation fluid

γ = Atomic concentration of O in matrix

δ = Atomic concentration of O in formation fluid

ϕ = Porosity

Because the model represented by Eq. (11.8) is independent of water salinity, the C/O log can be applied in any salinity environment including unknown salinity or mixed salinity, as in the case of CO_2 chased by water where the injected water salinity is different than that of the formation water.

The C/O model can be solved in several ways. For a specific known matrix, S_w can be estimated from a prepared chart similar to that of Figure 11.7. Families of similar graphs exist for various borehole environment and equipment type used.

The log of Figure 11.8 is from an observation well drilled for reservoir characterization and monitoring purposes. The reservoir is in a water-flood program where the injected water is fresher than the formation water. Water is injected in the main sand body, which lies between 3499 and 3518 ft. It consists of two lobes separated by a thin, shaly limestone section. Open-hole logs and cores indicated that the flood front had not yet reached the well. Cased-hole logs were run more than 6 months after the well had been drilled. Owing to the injection of water of different salinity than the formation water, the salinity at the flood front is unknown. The C/O log was hence selected for monitoring purposes.

Figure 11.8 displays open-hole logs (SP and gamma ray). Bulk volume analysis and water saturation are calculated using the resistivity log. It also displays the C/O envelop and C/O stationary measurements. Water saturation values calculated from C/O data are shown superimposed on open-hole water saturation.

The monitor log clearly indicated water breakthrough in the bottom of the lower sand lobe as the current water saturation is higher than the open-hole water saturation. This information was integrated with other petrophysical and production data to evaluate the flood performance and take necessary steps to optimize its performance. Also, should the operator elect to follow water flooding with CO_2 injection, a similar monitoring approach can be implemented.

CO_2 Saturation from Neutron Porosity Logs

The neutron porosity log measurement reflects the bulk hydrogen index of the formation. The hydrogen index of a material is defined as the ratio of the concentration of hydrogen atoms per unit volume in the material to that of pure water at 75°F. The hydrogen index of water is thus equal to 1. To simplify the analysis, the hydrogen index of oil is usually assumed equal to 1 as well.

If the lithology is devoid of hydrogen atoms, e.g., clean sandstone or limestone, neutron porosity measurement allows the estimation of the amount of liquid-filled porosity, ϕ_t. In a formation partially saturated with CO_2 the log displays an apparent porosity, ϕ_a, lower than the total porosity because the hydrogen index of CO_2 is almost nil, actually 0.03. It follows that the CO_2 saturation, S_{CO2}, can / be expressed as

$$S_{CO2} = 1 - (\phi_a / \phi_t) \qquad (11.10)$$

The decrease of neutron porosity with an increase of gas saturation is illustrated in the section of the log of Figure 11.9 showing a series of neutron porosity logs run to monitor the movement of a gas cap. Open-hole logs indicated that initially the gas/oil contact (GOC) was at 9310 ft. The open-hole log recorded on July 11, 1979 displays the true porosity of the thick uniform sand. The log of November 5, 1983 shows a marked decrease in porosity between 9310 ft. and 9475 ft., the dark shaded area, due to gas effect. The increase in the gas saturation above the GOC situated at 9475 ft. can be calculated using Eq. (11.9).

Another monitoring log was run on June 7, 1986. The new log indicates that the GOC is now at a depth of 9545 ft. Geological analysis of the structure indicated that other anomalies at the top of the sand and between 9410 and 9420 ft. are the result of gas encroachment under shale layers. The example of Figure 11.9 mimics the growth of the reservoir volume occupied by injected CO_2.

Residual Oil Saturation from Cased- Hole Resistivity and Pulsed Neutron Logs

The estimation of residual oil saturation is desired in CO_2 EOR projects and in cases of CGS in depleted oil

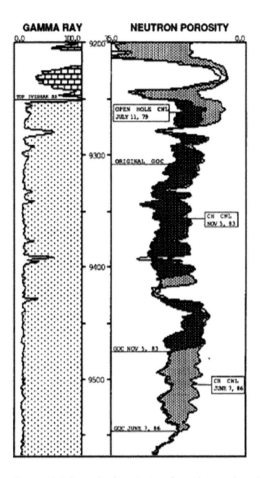

Figure 11.9 Example of monitoring of gas oil contact based on the change of gas effect on the neutron porosity. (From Dupree, 1989.)

reservoirs. The estimation is possible where both time-lapse cased-hole resistivity and pulsed neutron tools are available. Current water saturation, S_w, is determined from resistivity measurement and the CO_2 saturation, S_{CO2}, can be determined from the neutron porosity log. Hence, the saturation of the residual oil, S_o, can be monitored, since $S_o = 1 - S_w - S_{CO2}$.

CO_2 Flood Monitoring Examples

Monitoring of Timbalier Bay Miscible CO_2 Injection Project

This example's monitor well was drilled in the Gulf of Mexico as part of the Timbalier Bay Miscible CO_2 Injection Project (Moore, 1986; Ruhovets and Wyatt,

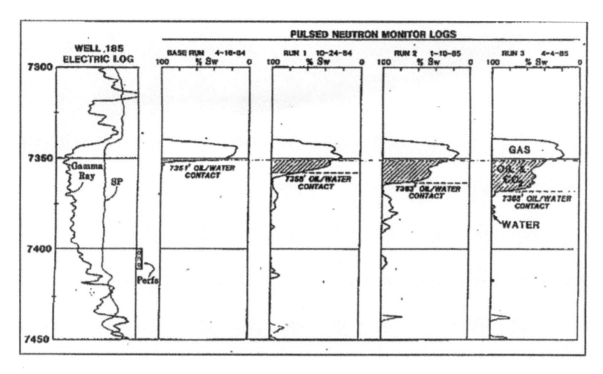

Figure 11.10 Sequential monitor logs showing reservoir displacement. (From Moore, 1986.)

1995). The formation is a clean, continuous, high-porosity sandstone. The reservoir was a nearly depleted saturated-oil accumulation produced by a strong water drive.

After the base logs were run in April, 1984, CO_2 injection was initiated in an up-dip injection well at the reservoir gas/oil contact. Approximately 6 months later, October, 1984, the first monitor logs were run. Additional monitor logs were run in January, 1985, April, 1985, and August, 1985. Figure 11.10 displays water saturation calculated from the base and the three monitor logs. The first monitor log indicates the arrival of the injected CO_2 at the monitoring well. The consecutive monitor logs show that the water contact is moving downward as additional CO_2 is injected resulting in oil displacement. The analysis made use of available log and core data. Saturation values of water, oil, and CO_2 were estimated using the techniques described in the section "Quantitative Analysis Methods" and Ruhovets and Wyatt (1995).

This example demonstrates that the remaining oil saturation in the CO_2 flooded interval can be estimated and a decision as to the economic

feasibility of continued CO_2 injection can be reached based on *in situ* values. The accuracy of this technique depends on the quality of the data and the representativeness of the models used in the quantitative evaluation. Nevertheless, reasonable qualitative and quantitative monitoring of CO_2 flooding can be achieved.

Monitoring of Saturation Changes for CO_2 Sequestration in Aquifers

The Frio Brine Pilot Experiment began in 2002 in South Liberty field, Dayton, Texas is discussed in Sakurai *et al.* (2006). The target of the CGS test is a fluvial sandstone of the upper Frio Formation at a depth of 5000 ft. An existing well was used as the observation well and a new injection well was drilled 100 ft. away from the observation well, as illustrated in Figure 11.11. A significant amount of data was obtained, including cores and open-hole logs. The formation water was sampled using a wireline formation tester. Results of core and open-hole log analysis are shown in Figure 11.12. Core analyses indicated porosity in the range of 20–35%, and the analysis of water samples

191

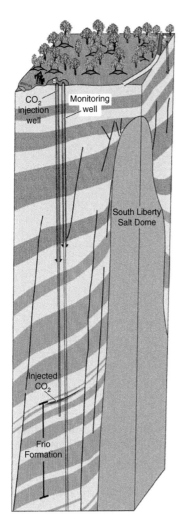

Figure 11.11 A schematic approximating the subsurface geology surrounding the injection and observation wells. (From Sakurai *et al.*, 2006.)

indicated a high salinity of 93 000 ppm. Because of the high porosity and formation water salinity, the pulsed neutron tool is optimal for monitoring saturation changes. The perforated interval, shown by the shaded red area, was selected in the interval displaying best reservoir quality.

Injection of CO_2 started on October 4, 2004 and lasted for 11 days. Base logs were recorded in the observation well before the start of injection. Monitor logs were run at preselected dates afterward to detect the CO_2 breakthrough and its subsequent

distribution over the aquifer interval. Figure 11.13 shows the presence of CO_2 indicated by the decrease in Σ. The change of CO_2 saturation is displayed in Figure 11.14. At the end of injection on October 14, 2004 the CO_2 saturation exceeded 80% in the vicinity of the observation well, although these results display CO_2 accumulation and migration only around the observation well.

References

Archie, G. E. (1942). Electrical resistivity log as an aid in determining some reservoir characteristics. *Transactions of the AIME*, **146**: 54–61.

Bassiouni, Z. (1994). *Theory, measurement, and interpretation of well logs*. SPE Textbook Series, 4. Richardson, TX: Society of Petroleum Engineers.

Clavier, C., Hoyle, W. R., and Meunier, D. (1971). Quantitative interpretation of TDT Logs. *Journal of Petroleum Technology*, **23**: 743–763.

Dupree, J. H. (1989). Cased-hole nuclear logging interpretation, Prudhoe Bay, Alaska. *Log Analyst*, 162–177.

Gilchrist, W. A., Jr., Roger, L. T., and Watson, J. T. (1983). Carbon/oxygen interpretation, a theoretical model. In *SPWLA* 24th Symposium, paper FF.

Gould, J., Wackler, J., Quiren, J., and Watson, J. (1991). CO_2 monitor logging: East Mallet Unit, Slaughter Field, Hockley County, Texas. In SPWLA 32nd Annual Symposium, paper PP.

Harold, B. H., Benimeli, D., Leveques, C., Bebourg, I., and Cadenhead, J. (2004). Combinable through-tubing cased hole formation resistivity tool. *SPE* 90018.

Jiang L., Tokar, T., and Zyweck, M. (2009). Innovation to enhanced oil recovery from slim cased hole resistivity measurement in Gulf of Thailand. In *SPWLA* 50th Annual Symposium, paper YY.

McGhee, B. F., McGuire, J. A., and Vacca, H. I. (1974). Examples of dual spacing thermal neutron decay time login Texas coast oil & gas reservoirs. *SPWLA* 15th Annual Symposium, paper R.

Moore, J. S. (1986). Design, installation, and early operation of the Tambalier Bay Gravity–Stable Miscible CO2 Injection Project. SPE Production Engineering, 369–378, also *SPE* 14287.

North, R. J. (1987). Through-casing reservoir evaluation using gamma ray spectroscopy. *SPE* 16356.

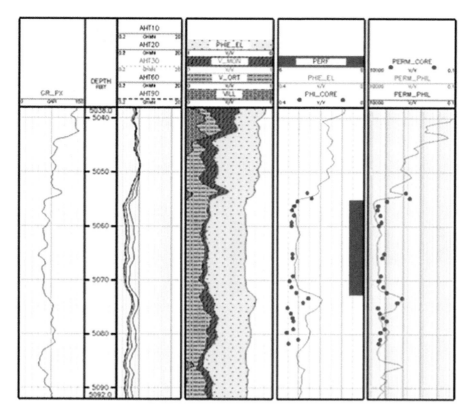

Figure 11.12 Results of core- and open-hole log-analysis. (From Sakurai *et al.*, 2006.)

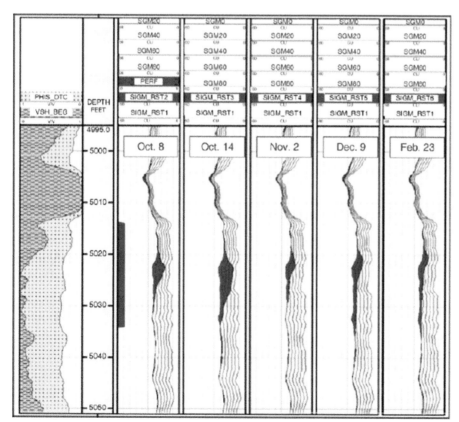

Figure 11.13 Time lapse decrease of Σ due to presence of CO_2 (shaded red area). (From Sakurai *et al.*, 2006.)

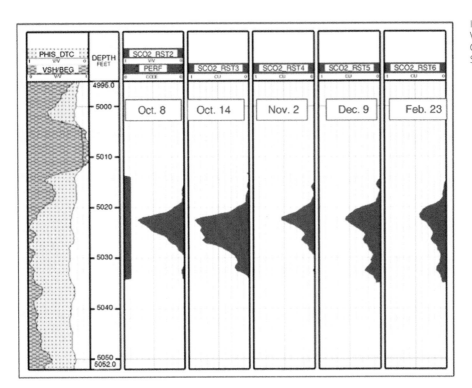

Figure 11.14 Time Lapse Variations of the saturation of CO_2 (shaded red area). (From Sakurai et al., 2006.)

Ruhovets, N., and Wyatt, D. F. (1995). Quantitative monitoring of gas flooding in oil-bearing reservoirs by use of a pulsed neutron tool. *SPE* 21411.

Sakurai, S., Ramakrishnan, T. S., Boyd, A., Mueller, N., and Hovorka, S. (2006). Monitoring saturation changes for CO_2 CSS; petrophysical support of the Frio Brine pilot experiment. *Petrophysics*, **47**(6): 483–496.

Smolen, J. J. (1996). Cased hole and production log evaluation. Tulsa, OK: PennWell, 108.

12

Offshore Storage of CO₂ in Norway

Eva K. Halland

Introduction

Three projects for offshore storage of carbon dioxide (CO_2) are ongoing in Norway. Chapters 13 by Eiken and 18 by Grude and Landrø in this book provide additional detail. This provides extensive experience with injection and storage of CO_2 in geological formations. Since 1996, approx. 1 Mtonne of CO_2 has, each year, been separated from gas production at the Sleipner Vest Field, in the North Sea. The CO_2 has been stored in the Utsira Formation and, from 2014, a further 0.1–0.2 Mtonne/year of CO_2 from the Gudrun Field has been injected into the same formation.

From the treatment of the production stream from the subsea developed on Snøhvit Field and the liquefied natural gas production on Melkøya in the Barents Sea, about 0.7 Mtonne of CO_2 is annually safely injected and stored in geological formations. The start of injection was in 2008 and the CO_2 was injected into the Tubåen Formation, about 2300 m beneath the seabed. In early 2010 the field operator realized that there was less storage capacity than first expected at the injection site. In 2011, the site of injection was relocated to the Stø Formation and a new geological structure is now under evaluation for further CO_2 storage in the area. The structures are identified with seismic data. High-quality 3D seismic data that properly image the storage site, storage complex, and the connecting area are required to identify geological formations and structures that can be suitable for CO_2 storage.

With the recognition of an increased climate challenge and based on the experience from these storage projects, the Norwegian Ministry of Petroleum and Energy (MPE) requested a CO_2 storage Atlas of Norway. The main objective was to identify storage capacity for safe and effective long-term storage of CO_2, and to facilitate selection of sites that could be suitable for future CO_2 sequestration projects, without having any negative effects on ongoing or future petroleum production.

The Norwegian CO_2 Storage Atlas aims to document the distribution and properties of the aquifers having a relevant storage potential for CO_2 offshore Norway (Halland et al., 2014). Several storage options have been evaluated such as structured aquifers, structural closures, dipping aquifers, abandoned fields, and the possibility of storing CO_2 in combination with enhanced oil recovery (EOR). In 2015, the Norwegian government decided to look at the possibility of establishing a full-chain carbon capture and storage (CCS) project in Norway, including capture, ship transport, and storage of CO_2. Based on the Norwegian CO_2 Storage Atlas, Statoil was appointed to further mature three storage sites. The aim was an injection capacity of 1 Mtonne of CO_2 annually with an opportunity for third-party access at a later stage. One geological structure in the Utsira Formation, one saline aquifer in the area east of the Troll gas field and the third site in the Heimdal depleted gas field have now been studied in more detail and all show good characteristics as CO_2 storage sites.

The Norwegian Continental Shelf

Petroleum activity on the Norwegian Continental Shelf (NCS) has been ongoing for about 50 years and today takes place in the North Sea, the Norwegian Sea, and the southern part of the Barents Sea (NPD fact pages). These regions also contain several saline aquifers (all aquifers referred to in this chapter are saline in nature) at a suitable depth for CO_2 storage, and are considered favorable for large-scale CO_2 sequestration (Fig. 12.1).

The NCS belongs to three different geological provinces. The northern North Sea is a subsiding basin developed above the continental Paleozoic and Mesozoic rift between the UK and Norway. The Basal

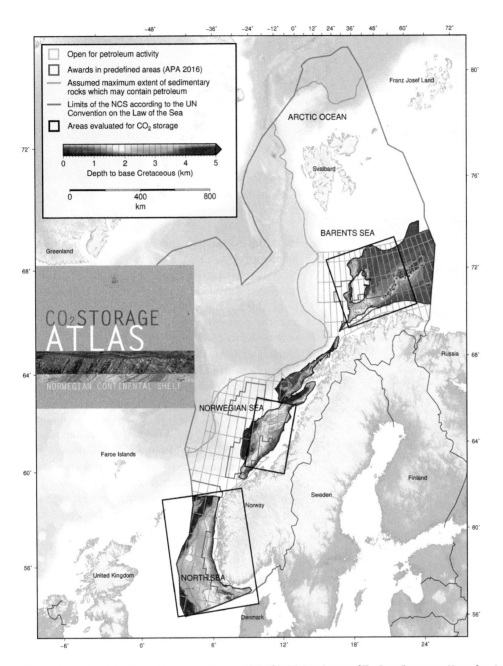

Figure 12.1 Area status for the Norwegian Continental Shelf (NCS). Depth map of The Base Cretaceous Unconformity and areas evaluated for CO₂ storage. The boundaries to oceanic crust and main volcanic provinces are indicated. (Source: Norwegian Petroleum Directorate, NPD Facts 2017.)

Cretaceous Unconformity (BCU) marks the end of the last major rifting event and the top of the cap rock of the Jurassic aquifers. Younger sediments in the North Sea Basin were sourced from all landmasses surrounding the basin and contain several important aquifers (Deegan and Skull, 1977; Vollset and Doré, 1984; Isaksen and Tonstad, 1989).

The Norwegian Sea shelf faces the oceanic crust in the Norwegian-Greenland Sea. The BCU is deeply buried under the deep waters beyond the shelf slope

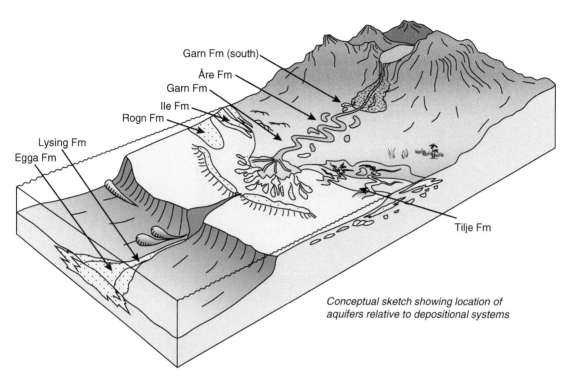

Garn Fm (south)
Åre Fm
Garn Fm
Ile Fm
Rogn Fm
Lysing Fm
Egga Fm
Tilje Fm

Conceptual sketch showing location of aquifers relative to depositional systems

Figure 12.2 Conceptual sketch showing location of aquifers relative to depositional systems. (Source: Norwegian Petroleum Directorate.)

(NPD, 1995). The map in Figure 12.1 does not display the BCU at greater burial depths than 5000 m, as these depths are not relevant for CO_2 injection. The main aquifers suitable for CO_2 injection in the Norwegian Sea belong to the Jurassic section.

The Barents Sea shelf was filled-in by thick sequences of Upper Paleozoic, Mesozoic, and younger rocks following a major carboniferous rifting phase (NPD, 1996). In the Norwegian sector, petroleum activity has been concentrated to the southwestern part, which is made up of platforms, shallow basins, and highs. The main aquifers consist of Jurassic and Triassic sediments. In the western margin, these sediments are too deeply buried to be suitable for CO_2 storage. The studied area was exposed to deep erosion in the Cenozoic and in the Quaternary. Because reservoir properties reflect maximum burial and not the present burial, aquifers with good storage potential are located at relatively shallow depths.

Geological Storage

Depending on their specific geological properties, several types of geological formations can be used to store CO_2. In the North Sea Basin, the greatest potential capacity for CO_2 storage will be in deep saline-water saturated formations (saline aquifers), or in depleted oil and gas fields.

Aquifers may consist of several sedimentary formations and cover large areas. A good understanding of the geology and sedimentologic development of the area is of upmost importance (Figure 12.2). An aquifer is a body of porous and permeable sedimentary rocks where the water in the pore space is in communication throughout (Figure 12.3). All the identified aquifers on the NCS are saline; most of them have salinities in the order of seawater or higher. To be suitable for CO_2 storage, aquifers need to be at a depth where CO_2 can be stored in a supercritical state and overlain by impermeable cap rocks acting as a seal to prevent CO_2 migration into other formations or to sea (Figure 12.4) (NPD, 1996). An extensive seismic data acquisition is used to provide the best possible basis for mapping the extent of the potential storage site for CO_2. The seismic data is an important part in order to plan the placement of injection wells for injecting CO_2 in an optimal manner. Finding an

197

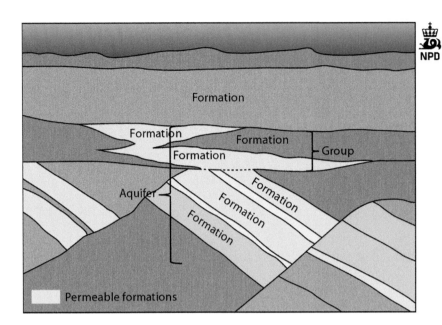

Figure 12.3 Relation between geological formations and aquifers. (Source: Norwegian Petroleum Directorate.)

optimal location of injection wells is important to ensure that the CO_2 can be stored and that it remains in the reservoir in the future.

CO_2 is held in place in a storage reservoir through one or more of five basic trapping mechanisms: stratigraphic, structural, residual, solubility, and mineral trapping. Generally, the initial dominant trapping mechanisms are stratigraphic trapping or structural trapping, or a combination of the two. The lighter CO_2 will move upwards through the porous rocks until it reaches the top of the formation, where it is trapped by an impermeable layer of cap-rock. In residual trapping, the CO_2 is trapped in the tiny pores in rocks by the capillary pressure of water. Once injection stops, water from the surrounding rocks begins to move back into the pore spaces that contain CO_2. As this happens, the pressure of the added water immobilizes the CO_2. This process, which further traps the CO_2, includes solubility (or dissolution) trapping (discussed further in Chapter 13 by Eiken, this volume). Dissolution of the CO_2 in the saline water forms a denser fluid that may sink to the bottom of the storage formation. Depending on the rock formation, the dissolved CO_2 may react chemically with the rock matrix to form stable minerals. Known as mineral trapping, this provides the most secure form of storage for the CO_2, but it is a slow process and may take thousands of years.

Data Availability

The Norwegian Petroleum Directorate (NPD) has access to all data collected on the NCS and has a national responsibility for the data. The NPD's data, overviews, and analyses make up an important information base for the oil and gas activities, which is also available for the evaluation of CO_2 storage.

The main objective of these Reporting Requirements is to support the efficient exploitation of Norway's hydrocarbon reserves. Around 50 years of petroleum activity, has generated a large quantity of data. These data were available for the CO_2 storage-mapping project. This includes 2D and 3D data and data from exploration and production wells such as logs, cuttings, and cores (Figure 12.5). Production data and pore pressure data from hydrocarbon fields and years of dedicated work to establish geological play models have all been utilized to define the main aquifers.

Workflow and Characterization

An approach for assessing the suitability of the geological formations for CO_2 storage is summed up in a flowchart (Figure 12.6). The intention has been to develop a systematic way to identify all the aquifers and to find which aquifers are prospective in terms of large-scale storage of CO_2.

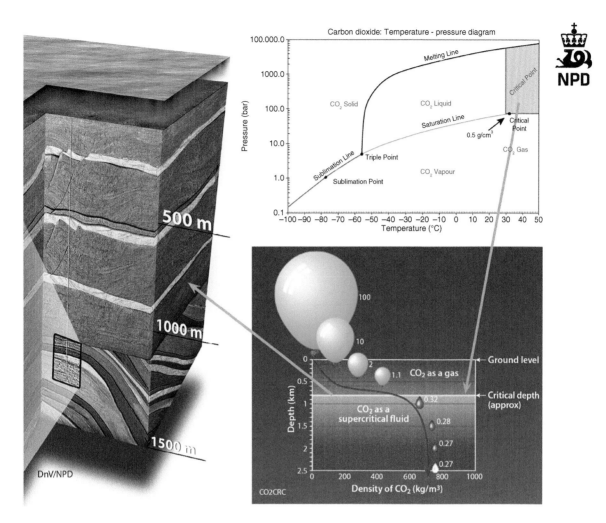

Figure 12.4 Supercritical fluids behave like gases, in that they can diffuse readily through the pore spaces of solids. However, like liquids, they take up much less space than gases. Supercritical conditions for CO_2 occur at 31.1°C and 7.38 megapascals (MPa), which occur approx. 800 m below surface level. This is where the CO_2 has both gas and liquid properties and is 500–600 times denser (up to a density of about 700 kg/m³) than at surface conditions, while remaining more buoyant than formation brine. (Source: Norwegian Petroleum Directorate, CO2CRC.)

In subsequent steps in the workflow, each potential reservoir and seal we identified was evaluated and characterized for its CO_2 storage prospectivity. Based on this, the potential storage sites are mapped and the storage capacity calculated. The evaluation is based on available data in the given areas. This evaluation does not provide an economic assessment of the storage sites.

Aquifers and structures are characterized in terms of capacity, injectivity, and safe storage of CO_2. To complete the characterization, the aquifers are also evaluated according to the data coverage and

their technical maturity. Some guidelines are developed to facilitate the characterization. A summary chart for characterization of reservoir, seal, and data coverage (Figure 12.7) with more detailed checklists is used for both reservoir and sealing properties.

Parameters used in the characterization process are based on data and experience from the petroleum activity and the CO_2 storage projects on the NCS, and the fact that CO_2 should be stored in the supercritical phase to obtain the most efficient and safest storage.

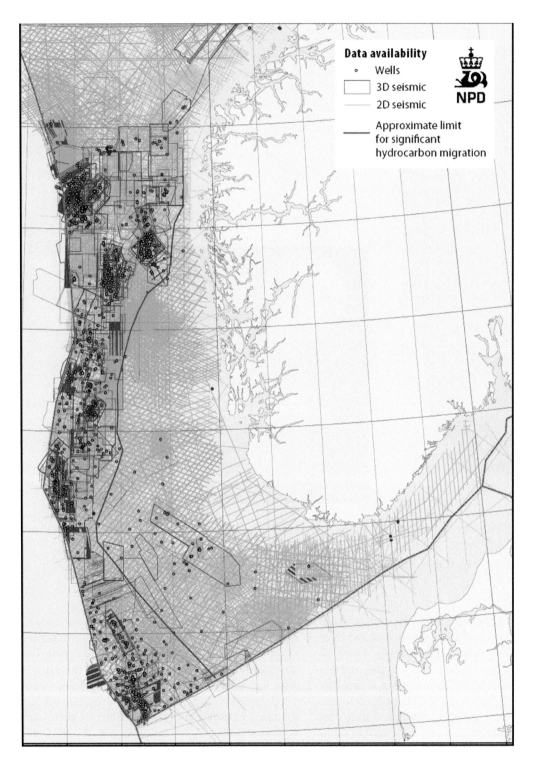

Figure 12.5 Data availability, seismic, and wells in the North Sea. (Source: Norwegian Petroleum Directorate.)

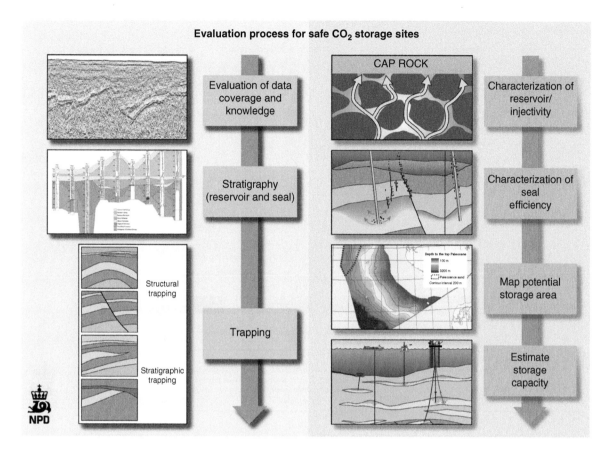

Figure 12.6 The workflow for characterization of geological formation for safe CO₂ storage. (Source: Norwegian Petroleum Directorate.)

The scores for capacity, injectivity, and seal quality are based on evaluation of each aquifer/structure.

For evaluation of regional aquifers in CO₂ storage studies, the mineralogical composition and the petrophysical properties of the cap rocks are rarely well known. In order to characterize the sealing capacity, we have relied mainly on regional pore pressure distributions and data from leak-off tests combined with observations of natural gas seeps.

Capacity Estimation

Storage capacity depends on several factors, primarily the reservoir pore volume and the fracturing pressure. The relation between pressure and injected volume depends on the compressibility of the rock and the fluids in the reservoir. The solubility of CO₂ in the different phases will also play a part. It is important to know if there is communication between multiple reservoirs, or if the reservoirs are in communication with larger aquifers. The pressure will increase when injecting fluid into a closed or half-open aquifer. Pressure increase can, however, be mitigated by production of formation water. The fracture pressure depends on the state of stress in the bedrock and is typically 10–30% lower than the lithostatic pressure (Figure 12.8). Fracture gradients were estimated by comparing pore pressures in overpressure reservoirs with data from leak-off tests.

The storage efficiency factor (S_{eff}) is assessed individually for each aquifer based on simplified reservoir simulation cases. For saline aquifers, the amount of CO₂ that can be stored can be determined using the following formula:

$$M_{CO_2} = V_b \times \emptyset \times n/g \times \rho_{CO_2} \times S_{eff}$$

where

M_{CO_2} = mass of CO₂
V_b = bulk volume

201

CHARACTERIZATION OF AQUIFERS AND STRUCTURES

	Criteria		Definitions, comments
Reservoir quality	Capacity, communicating volumes	3	Large calculated volume, dominant high scores in checklist
		2	Medium - low estimated volume, or low score in some factors
		1	Dominant low values, or at least one score close to unacceptable
	Injectivity	3	High value for permeability *thickness (k*h)
		2	Medium k*h
		1	Low k*h
Sealing quality	Seal	3	Good sealing shale, dominant high scores in checklist
		2	At least one sealing layer with acceptable properties
		1	Sealing layer with uncertain properties, low scores in checklist
	Fracture of seal	3	Dominant high scores in checklist
		2	Insignificant fractures (natural / wells)
		1	Low scores in checklist
Other leak risk	Wells	3	No previous drilling in the reservoir / safe plugging of wells
		2	Wells penetrating seal, no leakage documented
		1	Possible leaking wells / needs evaluation
Data coverage	Good data coverage	Limited data coverage	Poor data coverage

Other factors:
How easy / difficult to prepare for monitoring and intervention. The needs for pressure relief. Possible support for EOR projects. Potential for conflicts with future petroleum activity.

Data coverage
Good : 3D seismic, wells through the actual aquifer/structure
Limited : 2D seismic, 3D seismic in some ares, wells through equivalent geological formations
Poor : 2D seismic or sparse data

Figure 12.7 The summary chart for characterization of reservoir, seal, and data coverage. (Source: Norwegian Petroleum Directorate.)

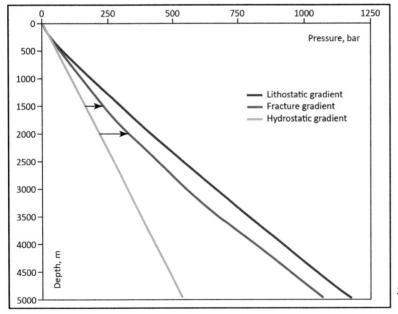

Figure 12.8 Pressure gradients obtained from pore pressure data and leak-off tests in wells from the Norwegian Sea Shelf and North Sea at water depths between 250 and 400 m. The fracturing gradient marks the lower boundary of measured leak-off pressures and the upper boundary of measured pore pressures. The lithostatic gradient was calculated from general compaction curves for shale and sand with a 300-m water column. The hydrostatic gradient assumes seawater salinity. The arrows show how much pressure can be increased from hydrostatic pressure before it reaches the fracture gradient. (Source: Norwegian Petroleum Directorate.)

$\varnothing$ = porosity

n/g = net to gross ratio

ρ_{CO_2} = density of CO_2 at reservoir conditions

S_{eff} = storage efficiency factor (source: Geocapacity, 2009)

S_{eff} is calculated as the fraction of stored CO_2 relative to the pore volume.

For a homogeneous reservoir with a permeability of 200 mD and reservoir thickness of 100 m, the storage efficiency in a closed system is simulated to be 0.4–0.8%, with a pressure increase of 50–100 bar. Discussion of the storage efficiency in the Utsira is undertaken in Chapter 13 by Eiken in this volume.

The storage potential is estimated for the different types of storage. The storage capacities are presented in a pyramid diagram, where the highest level in the pyramid represents the capacity of sites that are already used for CO_2 storage, while the lowest level represents theoretical capacity in lesser-known aquifers (Figure 12.9). An extensive database has been available for this evaluation. Nevertheless, evaluation of some areas is more uncertain due to limited seismic coverage and a lack of well information. The data coverage is color-coded to illustrate the data available for each aquifer/structure. Characterization and capacity estimates will obviously be more uncertain where data coverage is poor.

In summary, the capacity of each aquifer is given in the tables as a deterministic volume. Scores from 1 to 3 indicate the injectivity and sealing properties (3 is the best mark). The characterization is based on a best estimate of each parameter. Uncertainty is not quantified, but is indicated by the color coding for data availability and maturity. An example from the Utsira/Skade aquifer is shown in Figure 12.10.

Monitoring

An important matter when injecting CO_2 is to ensure that there is no leakage to other geological formations or to the sea bottom. Evaluation of the sealing capacity, faults, and fractures through the seal, in addition to old wells penetrating the seal, can provide important information on the sealing quality and the monitoring challenges.

Monitoring of injected CO_2 in a storage site is important for two main reasons: first, to see that the CO_2 is contained in the reservoir according to plans and predictions, and second, if there are deviations, to provide data that can update the reservoir models and support eventual mitigation measures. Seismic surveys are most widely used for monitoring of CO_2 geological storage. Careful monitoring of the behavior of the storage facility is required to establish its safety, that the injected CO_2 is behaving as predicted and is not migrating out of the intended storage site. An extensive program to monitor and model the distribution of injected CO_2 in the Utsira Formation has been undertaken by a number of organizations (and has been partly funded by the European Union). The Utsira monitoring program includes a baseline 3D seismic survey and several time-lapse (4D) seismic surveys. These surveys provide a good picture of how CO_2 moves horizontally and vertically (see Eiken, Chapter 13 in this volume).

In exploration wells on the NCS, pressure differences across faults and between reservoir formations and reservoir segments are commonly observed. Such pressure differences give indications of the sealing properties of cap rocks and faults. Based on such observations in the hydrocarbon provinces, combined with a general geological understanding, one can use the sealing properties in explored areas to predict the properties in less explored or undrilled areas.

Natural seepage of gas is observed in the hydrocarbon provinces in the NCS. Such seepage is expected from structures and hydrocarbon source rocks where the pore pressure is close to or exceeds the fracture gradient. Seepage at the sea floor can be recognized by changes in biological activity and by free gas bubbles. Seismically, gas chimneys or pipe structures indicate seepage. The seepage rates at the surface show that the volumes of escaped gas through a shale or clay dominated overburden are small in a time scale of a few thousand years. Rapid leakage can take place only if open conduits are established to the sea floor. Such conduits could be created along wellbores or by reactivation of faults or fractures. Established natural seepage systems are also regarded as a risk factor for CO_2 injection.

A wide range of monitoring technologies have been used by the oil and gas industry to track fluid movement in the subsurface. These techniques can be adapted to CO_2 storage. For example, repeated

The maturation pyramid

The evaluation of geological volumes suitable for injecting and storing CO_2 can be viewed as a step-wise approximation, as shown in the maturation pyramid. Data and experience from over 40 years in the petroleum industry will contribute in the process of finding storage volumes as high up as possible in the pyramid.

Step 4 is the phase when CO_2 is injected in the reservoir. Throughout the injection period, the injection history is closely evaluated and the experience gained provides further guidance on the reservoir's ability and capacity to store CO_2.

Step 3 refers to storage volumes where trap, reservoir and seal have been mapped and evaluated in terms of regulatory and technical criteria to ensure safe and effective storage.

Step 2 is the storage volume calculated when areas with possible conflicts of interest with the petroleum industry have been removed. Only aquifers and prospects of reasonable size and quality are evaluated. Evaluation is based on relevant available data.

Step 1 is the volume calculated on average porosity and thickness. This is done in a screening phase that identifies possible aquifers suitable for storage of CO_2. The theoretical volume is based on depositional environment, diagenesis, bulk volume from area and thickness, average porosity, permeability, and net/gross values.

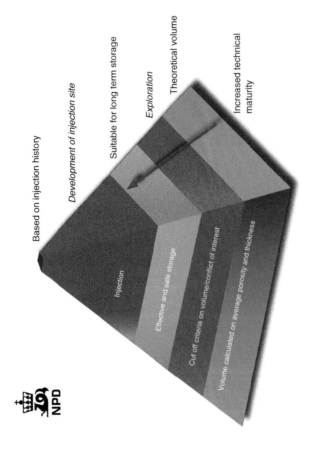

Based on injection history

Development of injection site

Suitable for long term storage

Exploration

Theoretical volume

Increased technical maturity

Injection

Effective and safe storage

Cut off criteria on volume/conflict of interest

Volume calculated on average porosity and thickness

NPD

1 tonne = one metric tonne = 100 kg
1 Mt = one megatonne = 10^6 tonnes
1 Gt = one gigatonne = 1000 Mt = 10^9 tonnes

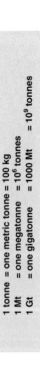

Figure 12.9 The maturation pyramid. (Source: Norwegian Petroleum Directorate.)

Utsira and Skade Fms		
Storage system	half open to fully open	
Rock volume		2500 Gm^3
Pore volume		526 Gm^3
Average depth		900 m
Average permeability		>1000 mD
Storage efficiency		4 %
Storage capacity aquifer		16 Gt
Storage capacity prospectivity		0,5-1,5 Gt
Reservoir quality		
	capacity	3
	injectivity	3
Seal quality		
	seal	2
	fractured seal	3
	wells	2
Data quality		
Maturation		

Figure 12.10 Evaluation of the Utsira/Skade aquifer. Showing capacities in aquifer and in prospects, reservoir and seal quality, data quality and maturation. (Source: Norwegian Petroleum Directorate.)

seismic surveying provides images of the subsurface, allowing the behavior of the stored CO$_2$ to be mapped and predicted. Other techniques include pressure and temperature monitoring, down-hole sensors as well as seabed monitoring. Surface sensors and satellite imaging will be available for onshore storage sites.

Conclusion/Summary

An overview of the results of this study are displayed in the table and illustrated by maturation pyramids for the North Sea, Norwegian Sea, and southern Barents Sea (Figure 12.11a, b). All the geological formations have been individually assessed, and grouped into saline aquifers. The aquifers are evaluated with regard to reservoir quality and presence of relevant sealing formations. Those aquifers that may have a relevant storage potential in terms of depth, capacity, and injectivity have been considered. Structural maps and thickness maps of the geological formations were used to calculate pore volumes.

All areas have a significant potential for CO$_2$ storage, but the three pyramids show that the regions are quite different. Most of the estimated capacities are in the green part of the maturation pyramid, defined as exploration phase, with an estimated capacity around 70 Gtonne. In the more mature aquifers and structures, a capacity of 1.3 Gtonne is more or less ready for CO$_2$ injection. The injectivity of the studied aquifers and the sealing properties of their cap rocks are considered acceptable or good. The studied areas are located in parts of the Norwegian Continental Shelf (NCS) that are now open for petroleum activity and appears to be the most promising areas for CO$_2$ storage.

We hope that this study will provide useful information for future exploration for CO$_2$ storage sites.

A team at the Norwegian Petroleum Directorate has developed the CO$_2$ Storage Atlas: Eva Halland, Fridtjof Riis, Andreas Bjørnestad, Christian Magnus, Rita Sande Rød, Van T. H. Pham, Inge M.Tappel, Ine Tørneng Gjeldvik, Jasminka Mujezinović, Maren Bjørheim, and Ida Margrete Meling.

NPD

Aquifer	Capacity Gt	Injectivity	Seal	Maturity	Data quality
North Sea aquifers					
Utsira and Skade Formations	15,8	3	2		
Bryne and Sandnes Formations	13,6	2	2/3		
Sognefjord Delta East	4,1	3	2/3		
Statfjord Group East	3,6	2	3		
Gassum Formation	2,9	3	2/3		
Farsund Basin	2,3	2	2/3		
Johansen and Cook Formations	1,8	2	3		
Fiskebank Formation	1	3	3		
Norwegian Sea aquifers					
Garn and Ile Formations	0,4	3	3		
Tilje and Åre Formations	4	2	2/3		
Barents Sea aquifers					
Realgrunnen Subgroup, Bjarmeland Platform	4,8	3	2		
Realgrunnen Subgroup, Hammerfest Basin	2,5	3	2		
Evaluated prospects					
North Sea	0,44				
Norwegian Sea	0,17				
Barents Sea	0,52				
Abandoned fields					
North Sea	3				
Producing Fields_2050					
North Sea 2050	10				
North Sea _Troll aquifer	14				
Norwegian Sea	1,1				
Barents Sea	0,2				

Figure 12.11 (a) Storage capacities, maturation, and data quality in the evaluated geological formations, prospects, and hydrocarbon fields on the NCS. (Source: Norwegian Petroleum Directorate.) (b) Total storage capacities in the different regions of NCS and the maturity of storage sites. (Source: Norwegian Petroleum Directorate.)

Figure 12.11b (Cont.)

References

CO2CRC. www.co2crc.au/

Deegan, C. E., and Skull, B. J. (1977). A standard lithostratigraphic nomenclature for the central and northern North Sea. *NPD Bulletin*, 1, www.npd.no/en/Publications/NPD-bulletins/251-Bulletin-1/

Halland, E., Bjørnestad, A., Magnus, C., *et al.* (2014). CO$_2$ Storage Atlas- Norwegian Continental Shelf. www.npd.no

Isaksen, D., and Tonstad, K. (1989). A revised Cretaceous and Tertiary lithostratigraphic nomenclature for the Norwegian North Sea. *NPD Bulletin*, 5. www.npd.no/en/Publications/NPDbulletins/255-Bulletin-5/.

NPD (Norwegian Petroleum Directorate). Factpages. http://factpages.npd.no

NPD (Norwegian Petroleum Directorate). (1995). Structural elements of the Norwegian continental shelf. Part II: The Norwegian Sea Region. *NPD Bulletin*, 8. www.npd.no/no/Publikasjoner/NPD-bulletin/258-Bulletin8/

NPD (Norwegian Petroleum Directorate). (1996). Geology and petroleum resources in the Barents Sea: *Norwegian Petroleum Directorate (NPD)*, Stavanger.

Vollset, J., and Doré, A. G., eds. (1984). A revised Triassic and Jurassic lithostratigraphic nomenclature for the Norwegian North Sea. *NPD Bulletin*, 3. www.npd.no/en/Publications/NPDbulletins/253-Bulletin-3/

Chapter 13

Twenty Years of Monitoring CO_2 Injection at Sleipner

Ola Eiken

Introduction

Sleipner is the world's largest and longest-running carbon dioxide (CO_2) storage project, with injection since September 1996. CO_2 is separated from natural gas at the production platform (Hansen et al., 2005) and reinjected into highly porous sands containing saline water of the Utsira Fm., through a single near-horizontal injection well at a depth of 1012 m below sea level. Water depth is 80 m, and the top of the storage reservoir is 800 m below sea level. About 17 Mtonnes of CO_2 had been injected and stored by the end of 2017. The field is operated by Equinor (former name, Statoil).

The Sleipner project, incentivized by the Norwegian state tax on offshore CO_2 emissions, was developed in a petroleum engineering setting (Baklid et al., 1996), with none of today's carbon capture and storage (CCS) legislation in place. Geophysical monitoring was not part of the original plan, but spectacular four-dimensional (4D) seismic images since 1999 have been a driver for understanding the injection and storage process, as well as for gaining public and authority acceptance. The data have been vital for improving reservoir characterization, developing new analysis techniques, cross-disciplinary work, verification, and prediction. Much of the early monitoring work was undertaken through EU- and industry-supported research projects, summarized in, e.g., Arts et al. (2008), Chadwick et al. (2008), and Chadwick and Eiken (2013). More than 300 scientific papers are testament to the wealth of knowledge that has evolved and been shared, and the great influence of the project. Significant contributions have also been shared from the operator's internal work of managing the site. This chapter aims to summarize the history and current knowledge.

Reservoir

The storage reservoir is a saline aquifer of regional extent (e.g., Chadwick et al., 2004a; Zweigel et al., 2004); stretching for more than 400 km north to south and 50–100 km east to west, locally exceeding thicknesses of 300 m in the Sleipner area. The CO_2 injection point is located beneath a domal feature that rises about 10–15 m above the surrounding seal topography (Figure 13.1). Significant faulting with a structural origin is absent. The sand bodies are commonly separated by thin intrareservoir mudstone beds, seen on logs, but barely visible in three-dimensional (3D) seismic images (Figure 13.1; Bitrus et al., 2016). The depositional environment is uncertain, possibly a turbiditic sand (Gregersen et al., 1997).

The Utsira Sand characteristically shows a sharp top and base in both logs and seismic data (Figure 13.1), with the proportion of clean sand in the reservoir unit typically above 70%. The nonsand fraction corresponds mostly to the thin mudstones (typically about 1 m thick), which show as peaks on the gamma-ray and resistivity logs. Except for one thicker shale near the top of the storage reservoir, it is not possible to correlate these shale layers between wells, or to identify them in the baseline seismic survey.

Core and cuttings samples show the Utsira Sand to be mostly fine-grained and largely uncemented. Porosity estimates from core, based on microscopy and laboratory experiments, are in the range of 27–42% and regional porosity estimates from geophysical logs are in the range of 35–40%. Permeabilities are correspondingly high, with measured values (from both cores and water-production testing) ranging from 1 to 8 Darcy.

A significant spread in temperatures is used in the literature, with early work based on nearby wells and gravity data (Lindeberg et al., 2000; Nooner et al.,

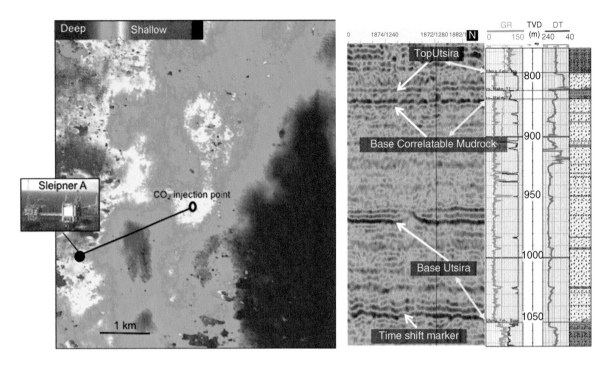

Figure 13.1 (Left) Map showing traveltime to top Utsira Fm. Red is shallow, blue is deep. (Right) Seismic section and typical well log from the Utsira Fm. (From Furre and Eiken, 2014.)

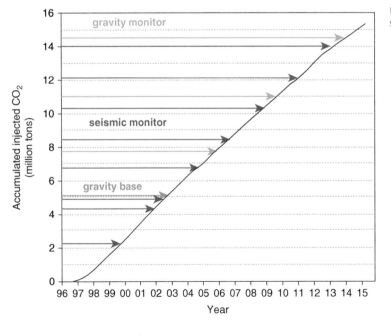

Figure 13.2 Injected mass and times of seismic and gravity surveys. (From Alnes, 2015.)

Table 13.1 Injection gas composition for three samples, in mol%

Component	1998 Sample	2002 Sample	2006 Sample
CO_2	98.07	98.38	96.53
CH_4	1.23	1.29	1.85
C_2H_6	0.19		0.17
N_2	0.10		1.15
BTX	0.40	0.33	0.30

Source: Equinor.

2007). Water production from the Utsira Sand at the nearby Volve field (approx. 8 km distant) since 2007 yields robust reservoir temperatures, summarized by Alnes *et al.* (2011). A water temperature of 32.2°C was measured during flow, from a perforation interval of 822–1009 m with unknown inflow profile. Projecting these values vertically gives a linearized relationship $T(z) = 31.7z + 3.4$ ($\pm$ 0.5) °C, where z is in kilometers below the sea surface. Applying this to the Sleipner injection area gives virgin temperatures of about 29°C at the reservoir top and 35.5°C at the depth of injection (1012 m). CO_2 at these temperatures and hydrostatic pressure would be supercritical and dense.

There is no published formation water composition from Utsira Fm. at Sleipner. Properties have been assumed to be close to seawater (about 35 g/L). Gregersen (1998) reported salinity of 30–43 g/L at the Grane and Oseberg fields 100 and 240 km north of Sleipner. Under these conditions, water density is within 1020–1031 kg/m³. Modeling studies have used values of 1020–1021 kg/m³ (Zhu *et al.*, 2015; Williams *et al.*, 2018), which is at the lower end of the range, and would require salinity of about 30 g/L.

The 700 m thick overburden comprises a primary caprock comprising a 50–100 m thick mudstone (Zweigel *et al.*, 2001; Harrington *et al.*, 2010). Above this, prograding sediment wedges are dominantly muddy, but coarsen into a sandier facies upward. The shallower overburden is mostly glaciomarine clays and glacial tills.

Injection

CO_2 is separated from the Sleipner Vest gas, since 2014 also from the Gudrun field, and has been injected at relatively uniform rates of about 0.8–0.9 Mtonnes per year (Figure 13.2). There are no commercial rate measurements, which results in greater than 10% uncertainty in the volumes injected. The injection fluid contains 1–2% methane, and smaller amounts of nitrogen and heavier hydrocarbons (Table 13.1).

The injection well 15/9-A-16 is highly deviated, with a sail angle of 83° at the injection point. The wellbore lies beneath the buoyant CO_2 plume, with no direct contact in the upper part of Utsira Fm, and hence no increased containment risk. Injectivity challenges caused by sand influx during the first months were remediated by a reperforation and installation of a 38 m long gravel pack in August 1997 (Hansen *et al.*, 2005).

The temperature of the CO_2 injectant is set at the last stage of compression, and has mostly been 25°C. Wellhead pressure has been around 63–66 bar (Figure 13.3). The state of CO_2 at the wellhead is at the gas/liquid phase boundary, but the ratio has not been measured (Hansen *et al.*, 2005). The CO_2 passes into supercritical state down the wellbore (Figure 13.4). The 1–2% methane content decreases the required gas/liquid ratio slightly compared to pure CO_2, but otherwise has little effect on the injection. The bottom hole pressure could vary from 90 to 140 bar for the same wellhead pressure and temperature, dependent on the gas/liquid ratio at the wellhead (Figure 13.5; Lindeberg, 2010; Kiær, 2011). The calculated bottom hole temperature is 49–51°C for gas/liquid ratios below 0.8 and 43°C if all the CO_2 was in liquid phase at the wellhead. In any case, the injected CO_2 will be warmer than the formation, and become cooled down in the reservoir (Alnes *et al.*, 2011).

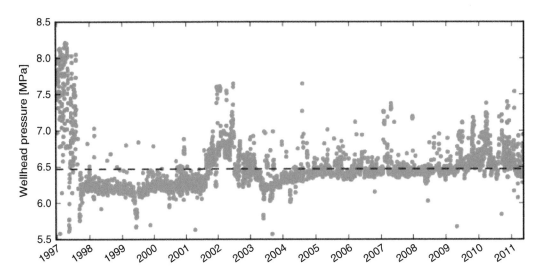

Figure 13.3 Measured wellhead pressure up until 2011. The increase in 2002–2003 was related to the temperature setting being increased above 25°C. (From Kiær, 2011.)

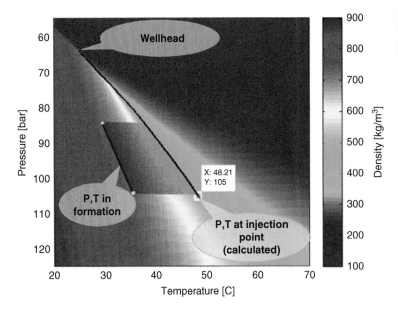

Figure 13.4 Density of pure CO_2 (kg/m^3) as function of pressure and temperature. An estimate of the conditions down the injection well is shown in black. Source: Equinor.

The injection point was placed 2.5 km NE of the Sleipner A platform (Figure 13.1), to prevent or delay backflow to the production wells, where they penetrate the storage formation. Exposure to CO_2 would increase the risk of casing corrosion or CO_2 leakage outside the casing. Engineering assessments, based on buoyant flow beneath the cap rock, assumed CO_2 would flow in a northerly direction. Baklid *et al.* (1996) envisioned an upward coning plume, as they drew in Figure 13.6, with lateral spreading underneath the top seal. Detailed flow modeling was not used to assist the decision of well placement.

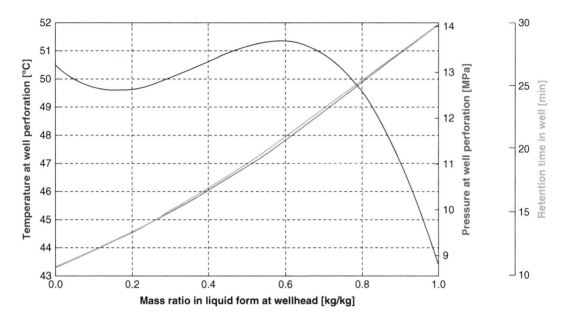

Figure 13.5 Calculated pressure and temperature at the perforation, as function of the gas/liquid ratio at the wellhead, assuming adiabatic conditions in the well. Retention time for CO_2 in the well is shown in green, and varies from 10 to 30 minutes. (From Kiær, 2011.)

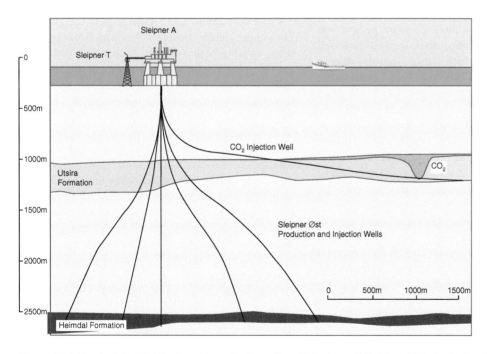

Figure 13.6 Sketch of the CO_2 injection with production wells and injection well. Envisioned CO_2 distribution after 20 years of injection is marked in orange. (From Baklid *et al.*, 1996.)

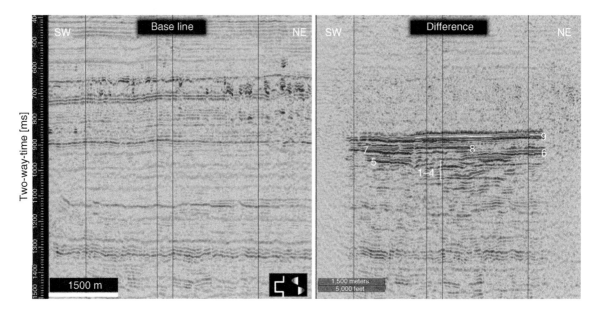

Figure 13.7 Vertical sections of (left) the 1994 baseline. (Right) The difference between the 2013 and 1994 seismic data. (From Furre *et al.*, 2017.)

Monitoring

The technology for downhole pressure gauges was not well established in 1996, when the injection well was drilled. Retrofitting such sensors imposes risks to the well integrity, and has not been undertaken. The overpressure during injection has likely been a few bars only, because of the high permeability and the 200 m thick sandy formation above the perforation. CO_2 is pushed in all directions, including downward several tens of meters. Increasing relative permeability as a result of higher CO_2 saturations in the pathway above are thought to have reduced the overpressure near the perforation after some time. No injection pressure data are available to test such a model.

At Sleipner, seismic monitoring is by far the most efficient containment monitoring tool, owing to the ability to remotely cover a large area, the strong time-lapse reflections (Figure 13.7) even for small CO_2 saturations, and consequently the ability to detect small accumulations of CO_2 filled pore space (of order 1 m vertically and 10 m laterally). The overwhelming seismic response has for many become a symbol of the Sleipner storage project. Already in 1999, after three years of injection, the

CO_2 was distributed in nine reflective levels, and a small amount of CO_2 had reached the top seal. The upper levels (layer 5–9) have been growing since, causing both a densification and expansion of the CO_2 plume. The seismic survey interval has been two to three years (Figure 13.2; Table 13.2), and no big surprises have emerged from one survey to the next. In addition, four gravity surveys, one controlled source electromagnetic (CSEM) survey, and one seafloor imaging survey have provided complementary data.

The first data acquisitions were research funded, and from 2008 on, the operator took over responsibility. The seismic monitoring results are now required by, and reported to, the Norwegian pollution authorities. Equinor has made all 4D seismic data up to 2010 publicly available, together with well logs and depth surfaces.

Seismic

All surveys have been acquired using streamers (Table 13.2), covering roughly the same 4 km × 7 km area. The six first had the combined purpose of monitoring deeper hydrocarbon targets, which reduced the cost and made some of them possible

Table 13.2 Seismic acquisition parameters

Year/date	Comment	Shooting direction	Source tow depth (m)	Flip-flop x-line separation (m)	Shot-point interval (m)	No. of cables	Cable separation (m)	Swath separation (m)	Cable length (m)	Cable tow depth
8/6–9/10, 1994	Preinjection, baseline	N–S	6	50	18.75	5	100	250	3000	8
10/8–10,1999	First image of CO_2	N–S	6	50	12.5	4	100	200	3600	8
9/27–10/1, 2001		N–S	6	50	12.5	6	100	200	3000	8
5/26–6/1, 2002		N–S	6	50	18.75	6	100	300	3600	8
6/13–8/13, 2004	Orthogonal line direction	E–W	6	one	18.75	10	37.5	250	4500	8
6/2–20, 2006		N–S	6	50		8	100	300	3600	8
	High-resolution lines	(2D)	3			1	—			3
5/4–6/15, 2008		N–S	6	50	18.75	9	50	200	3000	8
10/15–17, 2010	Dual sensor streamer	N–S	5	37.5	12.5	12	75	375	6000	15
1/2013	Slanted cable	N–S	7	37.5	12.5	8	100		5100	9
2016		N–S								

to realize at all. Tow depths of 6 m and 8 m are optimized for 2–3 km targets, and cause ghost notches that attenuate frequencies above 60 Hz. The dominant frequency in these images is about 35 Hz, and temporal thicknesses can be derived with an accuracy of 1–2 ms (Williams and Chadwick, 2012; Furre and Eiken, 2014). High-resolution two-dimensional (2D) lines with 3-m tow depths, as well as broadband technologies (Furre and Eiken, 2014), have demonstrated the ability to get frequencies above 100 Hz reflected back from the CO_2 plume.

The number of cables has increased with technology developments, from 4–6 in the first surveys to 8–12 in the most recent ones (Table 13.2). Overlapping swaths have reduced some of the azimuth and coverage variations at the outer streamers. The datasets have been processed in a time-lapse manner to enhance repeatability, including global matching and stacking/migration velocity reduction in the central part of the plume in subsequent surveys. Much of the analysis and interpretation of the plume development has been done on time-migrated full-offset cubes. Normalized root mean square (NRMS) repeatability values are 50–60% for the full frequency range (Furre and Eiken, 2014), and time shift noise is generally below 1 ms. The orthogonal shooting direction in 2004 caused variations that cannot be compensated in the processing, causing higher NRMS values of 80% when this survey is involved. All surveys and differences have high signal-to-noise ratios due to the large contrast in acoustic properties between the *in situ* saline aquifer and the injected CO_2, and have been valuable for understanding the plume development.

Figure 13.8 shows the time-lapse development in reflection response, in a section along the crest of the NNE–SSW trending anticline. The bright layers 1–9 (Chadwick *et al.*, 2004b) can be easily recognized in all the surveys. These are interpreted as thin layers of CO_2-filled sand capped by shale strings. Injected CO_2 thus acts to illuminate the internal reservoir architecture.

The visible plume area has been growing in all directions (Figure 13.9), and fastest to the NNE, giving an increasingly elongated shape. The area was 3.5 km^2 in 2010, which means that 5–6% of the pore volume under this area is filled with CO_2,

assuming 200 m thickness of the Utsira sand and 37% porosity. The 5–6% number on storage efficiency has increased slightly over the years. Reflection amplitudes have generally been increasing over the years, giving the plume growth a character of simultaneous expansion and densifying. The growth of each layer, measured in area-integrated amplitude, is shown in Figure 13.10. In the early years the shallower layers (6–9) were small, and have grown faster over the last 10 years. Layers 1–4 have, on the contrary, stopped growing, and some of them have apparently shrunk. They have become more challenging to interpret, which could be caused by inelastic attenuation, transmission loss, CO_2 migration/dissemination or all of these (Boait *et al.*, 2011). The area-integrated sum of seismic amplitudes has been remarkably proportional with injected mass (Eiken *et al.*, 2011).

Layers 5 and 9 were by 2010 the largest ones. The uppermost layer 9 does not suffer from the "shadow" effects of the lower layers, and is well imaged, as the amplitude maps in Figure 13.11 show. Its growth clearly follows topographic highs. After some years of accumulation above the injection point, the CO_2 has spread rapidly northward in a narrow corridor or channel, and started accumulating below a structural high approx. 3 km north/northeast of the injection point. The highest amplitudes are found in the corridor (Figure 13.11; Chadwick *et al.*, 2009, 2016).

Pushdown can be observed as an increasing time delay for a horizon in later vintages, and is caused by lower velocities and longer travel times for the seismic waves passing through CO_2-saturated rock. The amount of pushdown is increasing down through the plume, with a maximum at the marker horizons below. It has become increasingly difficult to reliably estimate the pushdown, owing to poorer image quality in the lower part and below the plume in later years. The strong and rapid lateral velocity variations in the CO_2 plume complicate the processing and distort the poststack time-migrated cubes. Seismic tomography and prestack depth migration could potentially handle these variations better, but attempts have so far shown limited improvements (Evensen and Landrø, 2010; Rossi *et al.*, 2011; Romdhane and Querendez, 2014). Broto *et al.* (2011) were able to extract horizontal velocity slices from tomographic analysis of the 2006 data that resemble

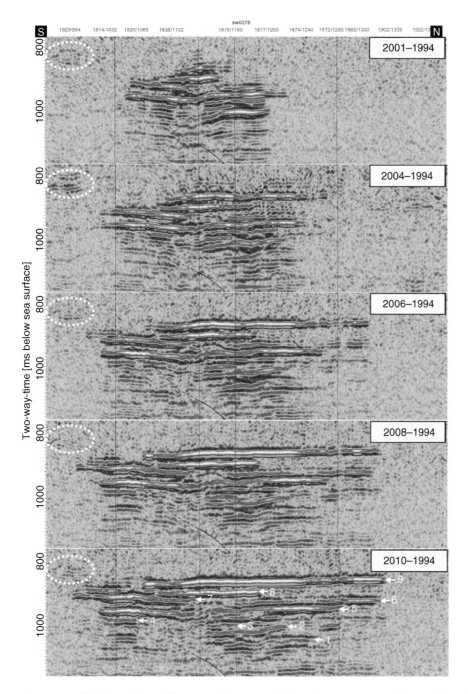

Figure 13.8 Seismic time-lapse difference with base survey for five subsequent repeats. White numbers refer to CO$_2$ layers. The white dashed ellipses show an area of strong amplitudes in the baseline, which serves as a reference for the time-lapse repeatability. (From Furre and Eiken, 2014.)

the pushdown maps, with most of the low-velocity area north and east of the injection point. Pushdown measure can be the time of either peak amplitude of a horizon or of cross-correlation between vintages. The observed pushdown was up to about 50 ms already in 1999 (Chadwick *et al.*, 2004b), in a narrow

217

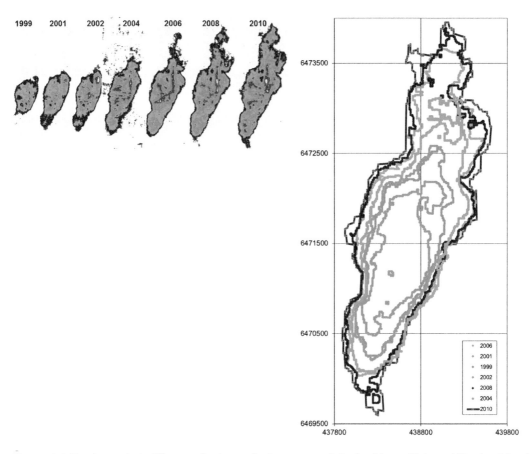

Figure 13.9 Time-lapse seismic difference reflection amplitude maps, cumulative for all layers. (Rightmost) The rim of the plume for each year. (Courtesy of Equinor.)

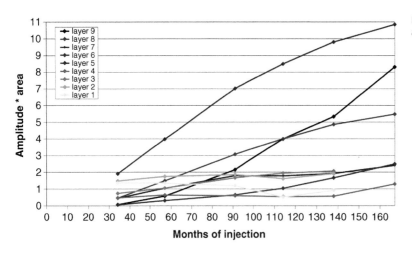

Figure 13.10 Layer growth quantified as amplitude * area, up until 2010.

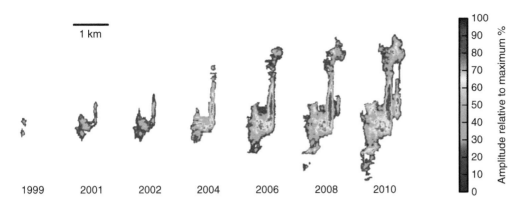

Figure 13.11 Layer 9 amplitude differences with the baseline survey. (From Kiær *et al.*, 2015b.)

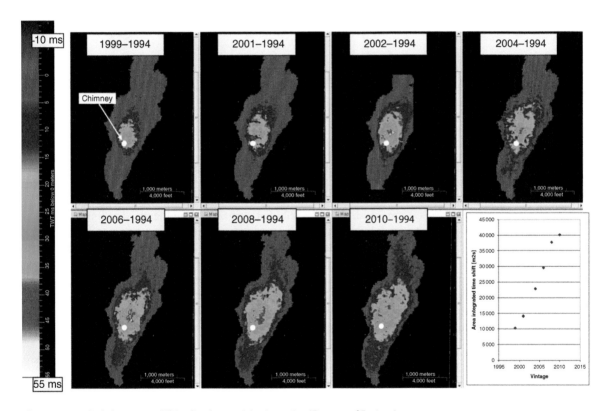

Figure 13.12 Pushdown maps. White disc denotes injection point. (Courtesy of Equinor.)

area (Figure 13.12). The area-integrated pushdown has increased regularly from about 11 to 40 × 10^3 m^{2}s between 1999 and 2010 (Chadwick *et al.*, 2004b; Boait *et al.*, 2012).

A seismic "chimney," circular with diameter 60–90 m and with disturbances of the flat reflections at all levels, is observed in the center of the plume (Figure 13.13). It can be observed only after 3D migration. Arts *et al.* (2009) demonstrated that a vertical cylinder completely filled with low-velocity material such as CO$_2$-saturated rock can cause such seismic features. The center is 90 m to one side of the

219

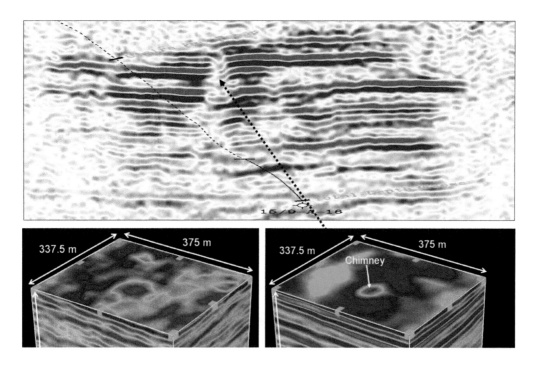

Figure 13.13 Seismic chimney. (Bottom left) 1994. (Bottom right) 2006. (Courtesy of Equinor).

perforation, which is 2σ on the sideway well position uncertainty, and it is thus possible that the chimney is directly above the perforation. It could have been created by the injection process (Hermanrud *et al.*, 2009), possibly by the flow of CO_2 breaking up the sealing capacity of the thin shales. On the other hand, a circular feature was seen at the top Utsira Fm. level in the preinjection data (Figure 13.13, lower left), indicating a relation with preexisting geology. In either case, the chimney seems important for vertical migration of CO_2, as both layer amplitudes and pushdown increase toward it.

Quantitative Interpretation

From rock physics modeling, Arts *et al.* (2004a) expected initial P-wave velocities in the Utsira Fm. of 2050 m/s and decreases of 600 m/s or more after CO_2 flooding. Queisser and Singh (2013a) used similar models, but with lower velocities for fully CO_2 saturated rocks, and a range of curves between the saturation end points, depending on the way of mixing of CO_2 and water (Figure 13.14). Falcon-Suarez *et al.* (2018) measured acoustic properties of an Utsira

sand core sample flooded with CO_2/brine, and this can be used to calibrate a rock physics model. Whether the CO_2 is homogeneously or patchily distributed, remains an ambiguity that limits the interpretation of saturations from measured or inferred seismic velocity reductions. Chadwick *et al.* (2016) estimated velocities in the topmost CO_2 filled layer in the range of 1350–1430 m/s, consistent with values derived independently from rock physics.

Reflection amplitudes generally increase from the rim and to the center, as seen for layer 9 in Figure 13.11. The amplitude variations have been explained by thin-layer tuning of two strong acoustic contrasts of opposite sign at the top and bottom of a layer filled with CO_2 (Figure 13.15; Arts *et al.*, 2004b; Chadwick *et al.*, 2004b, 2005). Average layer thickness of a few meters is to be expected: By 2010, when 12 Mtonnes had been injected, the cumulative area of the nine layers was 12 km^2, giving an average CO_2 column height of 1.3 m in each layer, corresponding to a rock column of about 5 m if 70% saturated. Amplitudes arising from thicknesses below maximum tuning can be converted to time thickness, when the wavelet and the maximum amplitudes are known. Further, layer

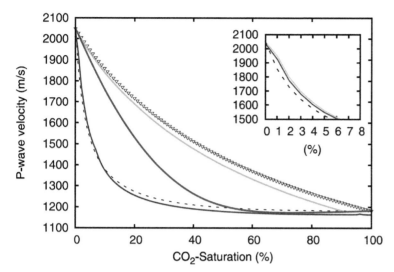

Figure 13.14 P-wave velocity versus CO$_2$ saturation, for various rock physics models. Upper patchy bound (triangles), modified upper patchy bound (green), Brie's curve (red), modeled lower bound (blue and gray), lower homogeneous bound (dotted). The inset shows the zoomed view of the homogeneous and modeled lower limits. (From Queisser and Singh, 2013a.)

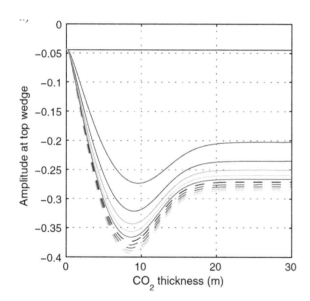

Figure 13.15 Amplitude from a CO$_2$-saturated wedge as a function of CO$_2$ layer thickness. Different colors represent increasing CO$_2$ saturation downward. (From Furre et al., 2015.)

thicknesses can be calculated from time thickness and the velocity of CO$_2$ saturated rock. Various estimates (Chadwick et al., 2005; Chadwick and Noy, 2010; Furre et al., 2015; Kiær, 2015) have given similar results, with minimum thin-layer detection level of about 1 m, and maximum tuning at about 8 m. For layer 9, the thicknesses derived from amplitude tuning fit with a flat CO$_2$–water contact (CWC) and infill

of topography (Chadwick and Noy, 2010). After about 10 years of injection, layer 5 showed decreasing amplitudes in the central part, interpreted as the temporal thicknesses exceeding maximum seismic tuning. With increasing layer thicknesses, time separation of top and bottom reflections becomes more reliable for thickness estimates (Furre et al., 2015). Spectral decomposition analysis can further enhance the lower resolution limit of the seismic data, utilizing the properties of the tuning wavelet and spectral decomposition (Williams and Chadwick, 2012; White et al., 2013).

Arts et al. (2004a) compared pushdown below a single layer at the edge of the plume with the reflection amplitude from that layer, and found on average a relation resembling the tuning curve. The pushdown/amplitude ratio is highest at the center of the plume, where a <500 m wide area has strong pushdown (Figures 13.9 and 13.12). Some of this could be caused by layers having exceeded the maximum tuning thickness, and, probably more important, additional pushdown is caused by non-reflective CO$_2$ (Chadwick et al., 2005), for example, in the migration pathways. Ghaderi and Landrø (2009) developed a method for estimation of thickness and velocity changes of CO$_2$ layers for prestack time-lapse data, and applied the method to a small part of the plume with one layer only.

Reflective layers with inferred thicknesses can account for most of the injected mass, if they are

221

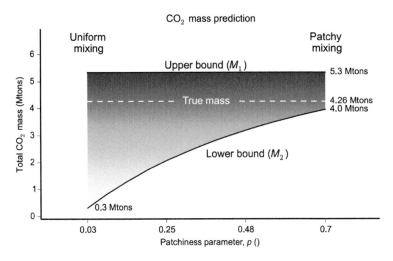

Figure 13.16 The inferred total mass bounds for 2006 plotted against the patchiness parameter. (From Bergmann and Chadwick, 2015.)

highly saturated (Chadwick *et al.*, 2004b). The additional CO_2 needed to explain the pushdown in the center of the plume must be at lower saturations to comply with the known injected mass, dependent on the velocity–saturation relation. Bergmann and Chadwick (2015) discussed this ambiguity in mass calculations from pushdown from a methodological point of view. They introduced the patchiness parameter, and found that the injected mass is within the upper and lower bound (Figure 13.16). Boait *et al.* (2011) suggested that the CO_2 is tending toward uniform mixing within the highly saturated layers and patchy mixing elsewhere, which could provide another constraint on the calculation of the amount of CO_2 at low saturations. The linear growth of the area-integrated pushdown may indicate that the principal characteristics of the velocity saturation in the CO_2 swept regions have remained rather constant through the years (Bergmann and Chadwick, 2015). The pushdown has, to a limited extent, been brought further from saturation distributions to flow models. Boait *et al.* (2012) analyzed the evolution of both reflectivity and pushdown for all layers and noted that pushdown has increased in the central part while amplitudes have dimmed, suggesting a continued growth of interlayer CO_2.

Chadwick *et al.* (2012) could not discern time shifts outside the CO_2 plume from noise, which indicates that pressure buildup likely does not exceed 1 bar.

Borgos *et al.* (2003) relaxed the thin-layer tuning assumption and inverted for a pair of opposite polarity reflections of variable size. They found the lower

spike of layer 5 to have lower reflectivity than the upper. Kiær *et al.* (2015a) investigated the amplitudes close to the rim, and found the rapid decline in amplitude when approaching the rim to be inconsistent with the topography, tuning relationship, and assumption of gravity dominated flow. Introducing capillary fringe effect alone did not help, but including macroscopic fluid flow for low frequencies can give a better match. The highly porous and permeable sandstones can cause wave-induced fluid flow (WIFF) at seismic frequencies, which may reduce the acoustic velocity for the lowest frequencies (Hofmann, 2006). For a thin layer of CO_2 capped by a string of shale, this could reduce the reflectivity at the lower interface (CWC) and thus alter the assumed tuning relation. WIFF is especially significant in the case of patchy saturation, and will also make the pushdown factor frequency dependent (Rubino and Velis, 2011). Rubino *et al.* (2011) suggested the effect can modify the seismic attributes significantly, but this has not yet been thoroughly investigated in the quantitative interpretations at Sleipner.

Several attempts at seismic inversion have been made. Stratigraphic inversion (Clochard *et al.*, 2010; Delépine *et al.*, 2011; Labat *et al.*, 2012) showed layers of low acoustic impedance. It is not clear how this compares to layer thicknesses found from amplitude inversion by the thin-layer tuning assumption. The inverted layers are possibly thicker and with less velocity reduction. Jullum and Kolbjørnsen (2016) developed a framework for Bayesian inversion, inverted a 2D line for saturation, and found the

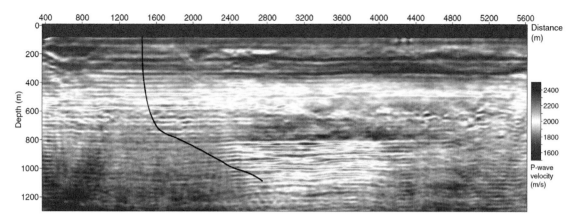

Figure 13.17 The P-wave velocity model derived from FWI. The black line corresponds to the projection of the injection well into the seismic section. The 2D section is located 533 m east of the injection point. (From Debuy *et al.*, 2017b.)

same five CO$_2$ layers as could be seen in qualitative interpretation of the seismic data. Queisser and Singh (2013a, 2013b) tested full waveform inversion (FWI) on a 2D line, and related the velocity changes to CO$_2$ saturation using a rock physics model. They found that applying FWI to real seismic data was challenging. Possibly they could identify places where the CO$_2$ saturation had reached maximum. Raknes *et al.* (2015) carried through a 3D elastic FWI study. Ghosh *et al.* (2015) inverted poststack data for saturation, and found large uncertainties attached to the rock physics modeling. Romdhane and Querendez (2014) inverted, and then prestack depth migrated, a 2D test line and Dupuy *et al.* (2017b) used FWI (Figure 13.17) and rock physics models to arrive at saturations in the thin layers of 30–35% for homogeneous mix and 75% for patchy mix, with true mixing between these. Owing to uncertainty in the choice of saturation model, velocity sensitivity to gas saturation is uncertain, meaning that without additional constraints, CO$_2$ saturation estimates are not conclusive from seismic inversions alone. The Sleipner inversion studies have been method developments, but have not been vital to reservoir understanding or estimating CO$_2$ plume properties. More extensive comparisons and testing of inversion results against other techniques such as amplitude tuning analysis might add further insight.

It has not been easy to retrieve additional information from amplitude variations with offset (AVO). Tuning effects need to be disentangled (Neele and Arts, 2010). Chadwick *et al.* (2010) focused on the

discrimination of layer thickness effects, where conventional Zoeppritz-based methods usually fail. Rabben and Ursin (2011) found from AVA inversion of the top sand reflection good results for the P-wave impedance, while S-wave and density were overestimated. Landrø and Zumberge (2017) combined AVO and gravity data to better constrain the density changes when the multilayered CO$_2$ accumulations limit the AVO technique. Haffinger *et al.* (2017) used wave-equation–based AVO to calculate injected mass, without uncertainty bounds or comparison with other work, however. Dupuy *et al.* (2017a) included rock physics models in their AVO workflow, and were able to derive high-resolution impedance contrasts, but not elastic contrasts.

No significant changes have been observed in the seismic response from the overburden, in any survey (Chadwick *et al.*, 2017). How sure can we be that no CO$_2$ has leaked? Chadwick *et al.* (2014) estimated detection limits based on data repeatability, the area and thickness of a CO$_2$ accumulation above the reservoir, and the probability of identification. As CO$_2$ migrates upward, it passes into gas phase and smaller masses can be detected. Pore volumes of about 3000 m^3 can be robustly detected at the top of the Utsira reservoir, corresponding to about 2100 tonnes, while the detection threshold will be around 950 tonnes at 590 m and 315 tonnes at 490 m below sea level (Chadwick *et al.*, 2014). These numbers are the order of 10^{-3} of the yearly injection volumes and sufficient to ensure containment has occurred (Chadwick *et al.*, 2014). Leakage along narrow vertical

223

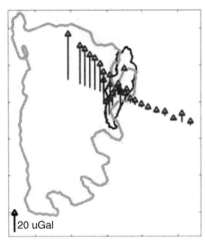 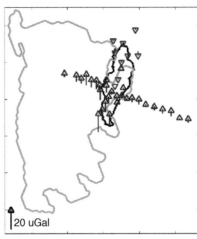

Figure 13.18 Maps of measured gravity change, outline of the Ty Fm. in green and outline of CO_2 plume in 2013 in black. (Left) Gravity changes 2009–2002. (Right) Gravity changes 2013–2009. (From Alnes, 2015.)

paths at high rates and with smaller accumulation in the overburden is possible, but unlikely.

Gravity

High-precision gravity monitoring offers direct measurements of mass changes. In 2002, 30 seafloor benchmarks were installed in a main profile over and beside the CO_2 plume, with some further aerial coverage (Figure 13.18), and a baseline was acquired using the ROVDOG technique, described in Chapter 7. Subsequent surveys were conducted in 2005, 2009, and 2013. As the CO_2 plume expanded, 13 additional benchmarks were installed in 2009 and 2013. Gravity data can give information that is complementary to the seismic data, with their limited ability to quantify CO_2 saturations.

Eroding currents causing seabed scouring has lowered some of the benchmarks more than 10 cm into the sediments since deployment. After correcting the gravity for the measured height changes, using a vertical gravity gradient derived from the data, a time-lapse gravity increase of nearly 40 µGal was observed at the westernmost station, gradually decreasing eastwards (Figure 13.18). This is explained by water inflow to the deeper gas-condensate reservoirs (Ty Fm., outlined in green in Figure 13.18). The water volumes required to explain the gravity changes agree with later acquired saturation logs in a nearby well and with 4D seismic data (Alnes *et al.*, 2008). The stations east of the CO_2 plume act as references in an area undisturbed by both CO_2 injection and gas production.

The effects of vertical benchmark movements, water influx in the deeper gas reservoirs, and CO_2 injection have been simultaneously inverted in several attempts based on different routes of data processing (Nooner *et al.*, 2007; Arts *et al.*, 2008; Alnes *et al.*, 2008, 2011; Alnes, 2015; Furre *et al.*, 2017). The estimated maximum gravity reduction caused by CO_2 injection increased from about 10 µGal in 2002–2005 to more than 20 µGal over the time interval 2002–2013 (Figure 13.19, right side). For inversions, the geometry of the CO_2 plume as imaged in 4D seismic data and the injected amount of CO_2, were further constraints.

The estimated average CO_2 density show some scatter, as discussed in Chapter 7, with the most recent values of Alnes *et al.* (2011) and Furre *et al.* (2017) based on the longest time span and most advanced analysis technique being of highest confidence. Alnes *et al.* (2011) arrived at 720 ± 80 kg/m^3 for average CO_2 density. They used the constraint on rock temperature from the Volve field water production well, discussed earlier in this chapter.

The gravimetric measurements can be used to estimate an upper bound of the dissolution rate of CO_2 in the formation water. This is important for the long-term fate of CO_2, because brine with CO_2 dissolved is heavier than pure brine and forms a more stable phase for storage. Dissolved CO_2 is seismically invisible. Alnes *et al.* (2011) estimated a dissolution rate between 0 and 1.8% per year. Hauge and Kolbjørnsen (2015) performed a Bayesian analysis of the *in situ* CO_2 density and

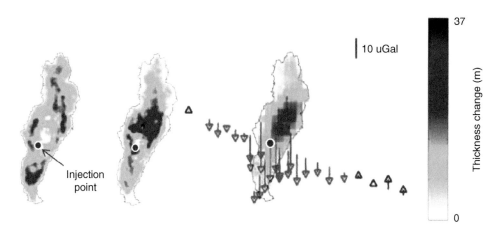

Figure 13.19 (Left) Seismic amplitude changes. (Middle) Seismic time delay. (Right) Estimated gravity changes and inverted CO$_2$ thickness, all from 2002 to 2013. (From Alnes, 2015.)

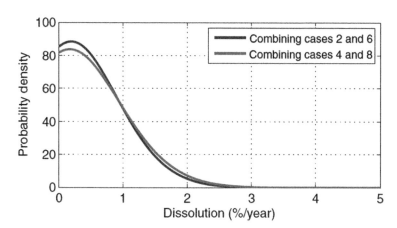

Figure 13.20 Probability distribution of dissolution rate, based on the 2002–2009 time span. (From Hauge and Kolbjørnsen, 2015.)

dissolution, using the same 2002–2005–2009 data set, with the Ty Fm. and height change effects removed. They arrived at similar conclusions, as shown in Figure 13.20. After the 2013 data were acquired, Alnes (2015) updated the dissolution estimate to less than 2.7% per year, based on the longer 2002–2013 time span. He also concluded that the accuracy is limited by uncertainty in the subtracted signal from water influx to the Ty Fm. and lack of gravity stations over the northern part of the plume in the base survey. Alnes (2015) further inverted for CO$_2$ column height, which interestingly agrees more with the one derived from seismic pushdown than from reflection amplitudes (Figure 13.19).

Controlled Source ElectroMagnetic Line

A controlled source electromagnetic (CSEM) line, centered at the long axis of the CO$_2$ plume, was acquired in 2008. Twenty-seven seabed receivers were distributed over 10 km, and the total towline length was 30 km. Detectability challenges are related to the water depth of 80 m, causing stronger airwave influence than common for most offshore CSEM surveys, and interference from the network of pipelines on the seafloor. Early studies of the data set were nonconclusive, but in the most recent analysis, Park *et al.* (2017) obtained inversion results with more detailed high-resistivity areas, which they related to

225

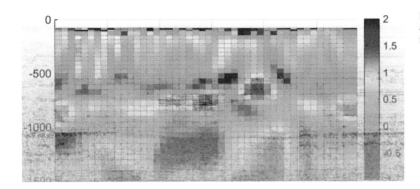

Figure 13.21 Vertical resistivity (in Ωm) after 2.5D inversion, superimposed on seismic data. (From Park *et al.*, 2017.)

the CO_2 plume (Figure 13.21). Quantitative numbers on, for example, average saturation are still imprecise and rely heavily on model assumptions. Eliasson and Romdhane (2017) initiated investigations of the value of combining CSEM inversion and FWI.

Seafloor Monitoring

The seafloor above the CO_2 plume consists of silt at 82 m water depth, planar and gently dipping about 1:1000 to the west. Bottom currents are moving the top sediments around. The seabed was surveyed with multibeam echo-sounding and side-scan sonar in 2006. In 2011 and 2012, University of Bergen used AUV-based side-scan sonar (HISAS 1030) with improved 2 × 2 cm resolution and photo mosaic to obtain improved seafloor images. None of the above techniques have indicated any leakage from the Sleipner CO_2 injection site (Pedersen, 2011). This is not surprising, as we would expect to detect a leakage on 4D seismic data long before it reached the seabed.

Bacterial mats have been found on the seafloor 40 km north of the injection site (Monastersky, 2013). These are likely fed from seepage, but highly unlikely to be connected to the injection process, because of the long distance from the injection point, the small pressure buildup in the CO_2 plume, the lack of 4D seismic signals farther than 4 km north of the injection point, and higher volumes of water production at the Volve field than CO_2 injection at Sleipner, which cause a net suction of mass into the Sleipner-Volve region. Furre *et al.* (2014) suggested the linear feature is related to gas within glacial channels that are present beneath the bacterial mats.

Flow Dynamic Interpretations

Flow modeling can help explain the 4D seismic and understand dynamic mechanisms. Lindeberg *et al.* (2000) found the first monitor survey to be in fair agreement with one premonitor model having thin shale layers as vertical barriers. While early attempts at Sleipner were few and scattered, flow modeling studies have increased in number and the matches have improved over the last decade. Bickle *et al.* (2007) calculated analytically an axisymmetric gravity flow for each layer, and obtained a match with seismically derived radii and thickness distributions, but had to reduce permeability by an order of magnitude compared to the expected 1–8 D range. A cause could be reduced relative permeability at lower CO_2 saturations. A delay in the feeding into upper layers, and high rates of accumulation prior to 1999 for the lower layers, was included in the model. Chadwick and Noy (2010) had difficulties modeling Darcy flow with a sufficiently rapid northward progression of layer 9 in the channel (Figure 13.11). Permeability needed to be increased to at least 10 D – or made anisotropic. Singh *et al.* (2010) compared invasion percolation (capillary flow) and traditional (Darcy flow) simulation, using black oil and compositional fluid descriptions. They concluded that the two approaches either propagate the CO_2 of layer 9 too far or too short and circular, compared to the seismic observations. However, the two methods provide upper and lower bounds to the expected plume behavior. Singh *et al.* (2010) also introduced the Sleipner benchmark model (www.ieaghg.org/networks/modelling-network), publicly available and based on the 4D seismic. A second benchmark model was introduced by Cavanagh and Nazarian (2014) and covers the two uppermost layers (8 and 9). Boait *et al.*

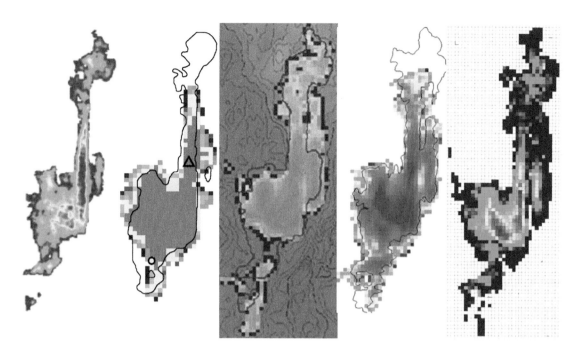

Figure 13.22 Observations (leftmost) and models of the extent of layer 9. Observed rim of plume is drawn as black or red line on top of the models. From left to right: Amplitude observations 2008, Darcy flow model with 2/10 D permeability for 2006 from Zhu *et al.* (2015), capillary flow model for 2008 from Cavanagh and Nazarian (2014), Darcy flow model for 2010 with 3/8 D permeability from Williams and Chadwick (2017). Darcy flow model for 2010 with parameter update and matching from Nilsen *et al.* (2017).

(2012) analyzed the horizon growth up to 2008, discussed models for vertical migration and dimming of the lower layers, and concluded that flows are controlled primarily by the spreading of gravity currents, moderated by topographic infilling. Cavanagh (2013) modeled layer 9 with both capillary and Darcy flow, and found that both can match the seismic observations. The latter needed to have the overpressure dissipated, however, requiring a much longer time span for the model run. The migration simulator was highly sensitive to top seal topography. Dubos-Sallée and Rasolofosaon (2011) observed the asymmetric spreading in their seismic inversion results, and suggested permeability anisotropy as the main cause.

Modeling results have been difficult to match in detail with the time-lapse seismic (e.g., Zhang *et al.*, 2014; Zhu *et al.*, 2015). Some examples are shown in Figure 13.22. This is due partly to challenges understanding the underlying physics of the CO$_2$ flow, and partly to uncertainties in geological assumptions such as the top seal topography and the vertical barriers. Cavanagh and Nazarian (2014) argued that the plume is closer to a stable distribution than previously

thought, and is likely to further stabilize after injection as a result of dissolution. Cavanagh and Haszeldine (2014) modeled all layers with capillary flow using invasion percolation physics, and were able to match the seismic only when they included fractured shale barriers. The seismic chimney was not considered part of the vertical flow, however. Bandilla *et al.* (2014) compared modeling approaches and found vertical-equilibrium models able to roughly match layer 9. However, none of their models were able to accurately match in detail, suggesting some essential physics, e.g., topography of the caprock, was incorrect. Kiær (2015) adjusted the topography of the top seal assuming gravity dominated flow, based on all seismic vintages up to 2010. The resulting topography map deviated by 5.3 m on average from the time-to-depth conversion that has been the basis for all the modeling work. This is within the mapping uncertainty (Chadwick and Noy, 2010).

Boait *et al.* (2012) found that mechanisms of CO$_2$ penetration through the mudstone layers needs to be better understood, to provide a model for plume development. If the chimney consists of broken-up shales with much increased permeability, it will be

227

a key to correct modeling of the plume growth. Williams *et al.* (2018) adjusted permeability and capillary entrance pressure to match the observed feeding into each layer. Such a "semipermeable" behavior might reflect preferential flow through pervasive small-scale fractures, as proposed by Cavanagh and Haszeldine (2014).

The modeling work has triggered a discussion of what is required of a satisfactory model-to-seismic match. Not all details may be required to match, but rather key features giving the model a stronger predictive power. Haukaas *et al.* (2013) provided a good match, but gave little detail of the flow mechanism they included and adjusted, and thus the published work has unknown predictive power. Zhu *et al.* (2015) found that local mismatches cannot be eliminated by adjusting parameters other than local topography, but resources should not be directed to such efforts. Chadwick and Noy (2015) discussed quality of matches using axisymmetric models, and concluded they had obtained a robust convergence of predictions and observations. Nilsen *et al.* (2017) automatically updated the model parameters and found a satisfying match. They also determined the parameters that had the largest influence on the mismatch. Zhang *et al.* (2017) reported a satisfactory match using a multiphase simulator, by adjusting the lateral permeability anisotropy, CH_4 in the CO_2 stream, and reservoir temperature. They extrapolated the model to 10,000 years, when both solubility and mineral trapping become important.

It is still under debate whether the upscaled flow process is dominated by viscous or gravity forces.

Cavanagh *et al.* (2015) argued that buoyant capillarity is the dominant process at Sleipner, while Williams and Chadwick (2017) revised the parameters going into Darcy flow modeling, and were able to mimic the rapid northward propagation of the CO_2 front in layer 9 with 8 Darcy permeability where the uppermost sand has thicker channels. Whereas Williams and Chadwick (2017) were able to obtain gross agreement with viscous modeling, earlier attempts were less successful. Gravity dominated models predict an affinity to travel too far north too early (Cavanagh 2013; Zhu *et al.*, 2015), and the lack of time dependence limits their power of prediction. Geological assumptions, such as the number of feeder channels up to the top layer, permeability, a correct representation of the top seal topography, and temperature distribution will all have a strong effect on the final flow pattern, and contain some uncertainty (Zhu *et al.*, 2015; Williams *et al.*, 2018).

Thermal Structure of the CO₂ Plume

In the reservoir, most of the injected CO_2 will be cooled down to the ambient reservoir temperature. However, with time a hotter core of the plume will develop. Expansion of CO_2 from the injection point up to top reservoir at constant enthalpy (i.e., no heat exchange with surroundings) would reduce the temperature from 49°C to 36.6°C. Such warm CO_2 would have much lower densities than the average 710 kg/m^3 for CO_2 cooled to reservoir temperature; below 500 kg/m^3 at the level of the injection point and possibly below 400 kg/m^3 at the reservoir top (Figure 13.23).

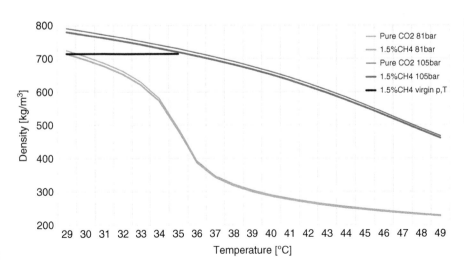

Figure 13.23 Density of pure CO_2 and mixed with 1.5% methane, at 81 bar (green line, top reservoir) and 105 bar (blue line, injection level), respectively. Black line is CO_2 with methane at virgin reservoir conditions, from top to bottom reservoir.

Legend:
— Pure CO2 81bar
— 1.5%CH4 81bar
— Pure CO2 105bar
— 1.5%CH4 105bar
— 1.5%CH4 virgin p,T

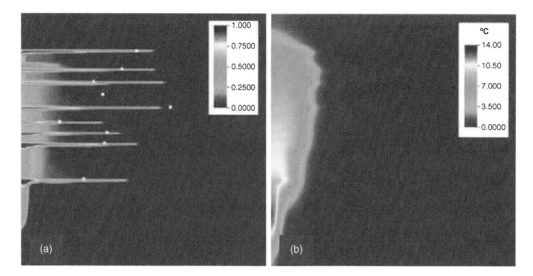

Figure 13.24 2D axisymmetric reservoir model showing the effect of injecting CO$_2$ at 48°C into a cooler reservoir. (a) CO$_2$ saturation in the reservoir. (b) Thermal anomaly. From Williams and Chadwick (2017).

A rough estimate of the temperature distribution within the CO$_2$ plume can be obtained by assuming the temperature front is sharp, i.e., that the CO$_2$ (and the Utsira reservoir) is either at initial reservoir temperature, or at the high temperature set by the injected CO$_2$. With a simple assumption of a cylindrical high-temperature region spanning the entire height of the CO$_2$ plume, a constant fraction of 7% of the CO$_2$ will be in the high-temperature state (Alnes *et al.*, 2011).

Williams *et al.* (2018) concluded that thermal effects generally need to be included in flow modeling when the injectant stream is significantly different in temperature to the reservoir, and the CO$_2$ is close to the critical point, which is the case at Sleipner. Zhu *et al.* (2015) found the spreading of layer 9 to be sensitive to temperature. However, they did not investigate the cooling effect likely to take place as the layer spreads.

Williams and Chadwick (2017) modeled the detailed temperature structure within the plume in a 2D radial axisymmetric mesh. The thermal imprint of the rising column extends in their model to a radius of around 350 m from the injection well (Figure 13.24). In the thin layers, the temperature anomaly becomes negligible at distances more than about one-third of the full layer spread. They further 3D modeled the top layer with hot CO$_2$ feeding into it. As it cooled down before reaching the main northerly channel, the propagation speed of the front in that channel did not speed up compared to isothermal modeling.

Learnings and Remaining Questions

CO$_2$ monitoring at Sleipner has been both a technical and an economic success. Important observations are that there are no indications of CO$_2$ in the overburden, the CO$_2$ plume grows and densifies steadily and, based on simple extrapolations, CO$_2$ seems unlikely to spill out of the primary structure during the injection phase, which may last another 5–15 years based on local CO$_2$ sources. Results from this pioneering CCS project have had influence beyond the technical community, on, e.g., legislators and public acceptance. 4D seismic, supported by time-lapse gravimetry, has proven that the CO$_2$ stays safely in the storage unit. Streamer seismic has proved to be a powerful and cost-efficient tool, both for containment (leakage) monitoring and for understanding the dynamic mechanisms within the reservoir. As such, the data and interpretations are crucial for building predictive flow models.

No good monitoring alternative to seismic exists. We cannot say we have the luxury of choosing from the toolbox. However, technology and parameter choices exist within 4D seismic, such as frequency of repeats, permanent seafloor receivers, or high-resolution surveys. Three-component geophones would have provided more shear wave information and permanency would improve repeatability, but likely at an order of magnitude higher cost than streamer surveys of opportunity. The survey frequency could in retrospect possibly have been reduced without much loss of information – easier to say in hindsight (Kiær *et al.*, 2015b; Nilsen *et al.*, 2017).

In spite of the excellent seismic signal/noise ratios, quantification in terms of spatial saturation distribution is difficult, still subject to research, and arguably not resolvable. Time-lapse gravimetry provides complementary data, because of their ability to detect mass changes, irrespective of small-scale homogeneous or patchy saturation distributions. Combined, geometry constraints from seismic and mass changes from gravity are more powerful. Improved joint analysis or inversion may be developed, and flow modeling may be included in the joint solutions.

The wealth of data has raised new questions, and left others yet unanswered, of which some of the more prominent are

- What are the flow mechanisms when CO_2 accumulates at low saturations instead of preferring high-saturation pathways with high relative permeability?
- What is the geological nature of the main seismic chimney, and the importance for the vertical flow and growth of the CO_2 plume? Is there more than one significant chimney?
- How well can we construct flow models that match the seismic amplitudes and pushdown? How well can we predict the next seismic survey?
- What are the uncertainty bounds on flow predictions, dissolution, and mineralization decades and centuries ahead?

Recent progress gives hope that further research will improve our understanding of the storage process in the years to come. In particular, the joint efforts of geophysicists and reservoir engineers/flow modelers on models that can satisfy all data and reasonable physical constraints seem fruitful.

References

Alnes, H. (2015). Gravity surveys over time at Sleipner. Presentation at the 10th IEAGHG Monitoring Network Meeting, June 10–12, 2015. http://ieaghg.org/docs/General_Docs/8_Mon/6_Gravity_surveys_over_time_at_SleipnerSEC.pdf

Alnes, H., Eiken, O., and Stenvold, T. (2008). Monitoring gas production and CO_2 injection at the Sleipner field using time-lapse gravimetry. *Geophysics*, 73: WA155–W161.

Alnes, H., Eiken, O., Nooner, S., Stenvold, T., and Zumberge, M. A. (2011). Results from Sleipner gravity monitoring: Updated density and temperature distribution of the CO_2 plume. *Energy Procedia*, **4**, 5504–5511 (10th International Conference on Greenhouse Gas Control Technologies).

Arts, R., Eiken, O., Chadwick, R. A., Zweigel, P., van der Meer, L., and Zinszner, B. (2004a). Monitoring of CO_2 injected at Sleipner using time-lapse seismic data. *Energy*, **29**: 1383–1393.

Arts, R., Eiken, O., Chadwick, A., Zweigel, P., Meer, B. v. d., and Kirby, G. (2004b). Seismic monitoring at the Sleipner underground CO_2 storage site (North Sea). In S. J. Baines and R. H. Worden (eds.), *Geological storage of carbon dioxide*. Special Publications 233. London: Geological Society, 181–191.

Arts, R., Chadwick, A., Eiken, O., Thibeau, S., and Nooner, S. (2008). Ten years' experience of monitoring CO_2 injection in the Utsira Sand at Sleipner, offshore Norway. *First Break*, **26**(January): 65–72.

Arts, R. J., Trani, M., Chadwick, R. A., Eiken, O., Dortland, S., and van der Meer, L. G. H. (2009). Acoustic and elastic modeling of seismic time-lapse data from the Sleipner CO_2 storage operation. In M. Grobe, J. C. Pashin, and R. L. Dodge (eds.), *Carbon dioxide sequestration in geological media: State of the science*. AAPG Studies in Geology, **59**: 391–403.

Baklid, A., Korbøl, R., and Owren, G. (1996). Sleipner Vest CO_2 disposal, CO_2 injection into a shallow underground aquifer. *SPE Annual Technical Conference and Exhibition*, Denver, CO, SPE paper 36600, 1–9.

Bandilla, K. W., Celia, M. A., and Leister, E. (2014). Impact of model complexity on CO_2 plume modeling at Sleipner. *Energy Procedia*, **63**: 3405–3415.

Bergman, P., and Chadwick, A. (2015). Volumetric bounds on subsurface fluid substitution using 4D seismic time shifts with an application at Sleipner, North Sea. *Geophysics*, **80**(5): B153–B165.

Bickle, M., Chadwick, A., Huppert, H. E., Hallworth, M., and Lyle, S. (2007). Modelling carbon dioxide accumulation at Sleipner: Implications for underground carbon storage. *Earth and Planetary Science Letters*, **255**(1–2): 164–176.

Bitrus, P. R., Iacopini, D., and Bond, C. E. (2016). Defining the 3D geometry of thin shale units in the Sleipner reservoir using seismic attributes. *Marine and Petroleum Geology*, **78**: 405–425.

Boait, F., White N., Chadwick A., Noy D., and Bickle M. (2011). Layer spreading and dimming within the CO$_2$ plume at the Sleipner Field in the North Sea. *Energy Procedia*, **4**: 3254–3261.

Boait, F. C., White, N. J., Bickle, M. J., Chadwick, R. A., Neufeld, J. A., and Huppert, H. E. (2012). Spatial and temporal evolution of injected CO$_2$ at the Sleipner Field, North Sea. *Journal of Geophysical Research*, **117**: B03309.

Borgos, H. G., Randen, T., and Sonneland, L. (2003). Super-resolution mapping of thin gas pockets. Extended Abstract, *Society of Exploration Geophysicists Annual Meeting*.

Broto, K., Ricarte, P., Jurado, F., Eitenne, G., and Le Bras, C. (2011). Improving seismic monitoring by 4D prestack traveltime tomography: Application to the Sleipner CO$_2$ storage case. Extended Abstract, *EAGE 1st Sustainable Earth Sciences Conference*.

Cavanagh, A. (2013). Benchmark calibration and prediction of the Sleipner CO$_2$ plume from 2006 to 2012. *Energy Procedia*, **37**: 3529–3545.

Cavanagh, A. J., and Haszeldine, R. S. (2014). The Sleipner storage site: Capillary flow modelling of a layered CO$_2$ plume requires fractured shale barriers within the Utsira Formation. *International Journal of Greenhouse Gas Control*, **21**: 101–112.

Cavanagh, A., and Nazarian, B. (2014). A new and extended Sleipner Benchmark model for CO$_2$ storage simulations in the Utsira Formation. *Energy Procedia*, **63**: 2831–2835.

Cavanagh, A. J., Haszeldine, R. S., and Nazarian, B. (2015). The Sleipner CO$_2$ storage site: Using a basin model to understand reservoir simulations of plume dynamics. *First Break*, **33**(June): 61–68.

Chadwick, R. A., and Eiken, O. (2013). Offshore CO$_2$ storage: Sleipner natural gas field beneath the North Sea. In J. Gluyas and S. Mathias (eds.), *Geological storage of carbon dioxide (CO$_2$): Geoscience, technologies, environmental aspects and legal frameworks.* Sawston, UK: Woodhead Publishing, 227–250.

Chadwick, R. A., and Noy, D. J. (2010). History – matching flow simulations and time-lapse seismic data from the Sleipner CO$_2$ plume. In B. A. Vining and S. C. Pickering (eds.), *Petroleum geology: From mature basins to new frontiers. Proceedings of the 7th Petroleum Geology Conference.* Petroleum Geology Conferences Ltd. London: Geological Society, 1171–1182.

Chadwick, R. A., and Noy, D. J. (2015). Underground CO$_2$ storage: Demonstrating regulatory conformance by convergence of history-matched modelled and

observed CO$_2$ plume behaviour using Sleipner time-lapse seismics. *Greenhouse Gas Science Technology*, **5**: 305–322.

Chadwick, R. A., Zweigel, P., Gregersen, U., Kirby, G. A., Johannessen, P. N., and Holloway, S. (2004a). Characterisation of a CO$_2$ storage site: The Utsira Sand, Sleipner, northern North Sea. *Energy*, **29**: 1371–1381.

Chadwick R. A., Arts, R., Eiken, O., Kirby, G. A., Lindeberg, E., and Zweigel, P. (2004b). 4D seismic imaging of an injected CO$_2$ plume at the Sleipner Field, central North Sea. In R. J. Cartwright, S. A. Stewart, M. Lappin, and J. R. Underhill (eds.), *3D seismic technology: Application to the exploration of sedimentary basins.* Geological Society, London, Memoirs. London: Geological Society, **29**: 311–320.

Chadwick, R. A., Arts, R., and Eiken, O. (2005). 4D seismic quantification of a growing CO$_2$ plume at Sleipner, North Sea. In A. G. Doré and B. A. Vining (eds.), *Petroleum Geology: North-West Europe and Global Perspectives: Proceedings of the 6th Petroleum Geology Conference*, 1385–1399.

Chadwick, A., Arts, R., Bernstone, C., May, F., Thibeau, S., and Zweigel, P. (2008). Best practice for the storage of CO$_2$ in saline aquifers: Observations and guidelines from the SACS and CO2STORE projects. *British Geological Survey Occasional Publication*, **14**: 1–267.

Chadwick, R. A., Noy, D., Arts, R., and Eiken, O. (2009). Latest time-lapse seismic data from Sleipner yield new insights into CO$_2$ plume development. *Energy Procedia*, **1**, 2103–2110.

Chadwick, R. A., Williams, G., Delepine, N., *et al.* (2010). Quantitative analysis of time-lapse seismic monitoring data at the Sleipner CO$_2$ storage operation. *Leading Edge*, **29**, February: 170–177.

Chadwick, R. A., Williams, G. A., Williams, J. D. O., and Noy, D. J. (2012). Measuring pressure performance of a large saline aquifer during industrial-scale CO$_2$ injection: The Utsira Sand, Norwegian North Sea. *International Journal of Greenhouse Gas Control*, **10**: 374–388.

Chadwick, R. A., Marchant, B. P., and Williams, G. A. (2014). CO$_2$ storage monitoring: Leakage detection and measurement in subsurface volumes from 3D seismic data at Sleipner. *Energy Procedia*, **63**: 4224–4239.

Chadwick, R. A., Williams, G. A., and White, J. C. (2016). High-resolution imaging and characterization of a CO$_2$ layer at the Sleipner CO$_2$ storage operation, North Sea using time-lapse seismics. *First Break*, **34**(February): 77–85.

Chadwick, R. A., Williams, G. A., and Noy, D. J. (2017). CO$_2$ storage: Setting a simple bound on potential leakage through the overburden in the North Sea Basin. *Energy Procedia*, **114**: 4411–4423.

Clochard, V., Delépine, N., Labat, K., and Ricarte, P. (2010). CO_2 plume imaging using pre-stack stratigraphic inversion: A case study on the Sleipner field. *First Break*, **28**(1): 91–96.

Delepine, K., Clochard, N., Labat, V., and Ricarte, P. (2011). Post-stack stratigraphic inversion workflow applied to carbon dioxide storage: Application to the saline aquifer of Sleipner field. *Geophysical Prospecting*, **59**(1): 132–144.

Dubos-Sallée, N., and Rasolofosaon, P. N. (2011). Estimation of permeability anisotropy using seismic inversion for the CO_2 geological storage site of Sleipner (North Sea). *Geophysics*, **76**(3): WA63–WA69.

Dupuy, B., Torres, V. A. C., Ghaderi, A., Querendez, E., and Mezyk, M. (2017a). Constrained AVO for CO_2 storage monitoring at Sleipner. *Energy Procedia*, **114**: 3927–3936.

Dupuy, B., Romdhane, A., Eliasson, P., Querendez, E., Yan, H., Torres, B., and Ghaderi, A. (2017b). Quantitative seismic characterization of CO_2 at the Sleipner storage site, North Sea. *Interpretation*, **5**(4): SS23–SS42.

Eiken, O., Ringrose, P., Hermanrud, C., Nazarian, B., Torp, T. A., and Høier, L. (2011). Lessons learned from 14 years of CCS operations: Sleipner, In Salah and Snøhvit. *Energy Procedia*, **4**: 5541–5548.

Eliasson, P., and Romdhane, A. (2017). Uncertainty quantification in waveform-based imaging methods: A Sleipner CO_2 monitoring study. *Energy Procedia*, **114**: 3905–3915.

Evensen, A. K., and Landrø, M. (2010). Time-lapse tomographic inversion using a Gaussian parameterization of the velocity changes. *Geophysics*, **75**(4): U29–U38.

Falcon-Suarez, I., Papageorgiou, G., Chadwick, A., North, L., Best, A. I., and Chapman, M. (2018). CO_2-brine flow-through on an Utsira Sand core sample: Experimental and modelling. Implications for the Sleipner storage field. *International Journal of Greenhouse Gas Control*, **68**: 236–246.

Furre, A.-K., and Eiken, O. (2014). Dual sensor streamer technology used in Sleipner CO_2 injection monitoring. *Geophysical Prospecting*, **62**(5): 1075–1088.

Furre, A. K., Ringrose, P., Cavanagh, A., Janbu, A. D., and Hagen, S. (2014). Characterization of a submarine glacial channel and related linear features. Extended Abstract, *EAGE Near Surface Geoscience*.

Furre, A.-K., Kiær, A., and Eiken, O. (2015). CO_2-induced seismic time shifts at Sleipner. *Interpretation*, **3**(3): SS23–SS35.

Furre, A.-K., Eiken, O., Alnes, H., Vevatne, J. N., and Kiær, A. F. (2017). 20 years of monitoring CO_2-injection at Sleipner. *Energy Procedia*, **114**: 3916–3926.

Ghaderi, A., and Landrø, M. (2009). Estimation of thickness and velocity changes of injected carbon dioxide layers from prestack time-lapse seismic data. *Geophysics*, **74**(2): O17–O28.

Ghosh, R., Sen, M. K., and Vedanti, N. (2015). Quantitative interpretation of CO_2 plume from Sleipner (North Sea), using post-stack inversion and rock physics modelling. *International Journal of Greenhouse Gas Control*, **32**: 147–158.

Gregersen, U., ed. (1998). *Saline aquifer CO_2 storage phase zero geological survey of Denmark and Greenland*. GEUS.

Gregersen, U., Michelsen, O., and Sorensen, J. C. (1997). Stratigraphy and facies distribution of the Utsira Formation and the Pliocene sequences in the northern North Sea. *Marine and Petroleum Geology*, **14**: 893–914.

Haffinger, P., Eyvazi, F. J., Doulgeris, P., Steeghs, P., Gisolf, D., and Verschuur, E. (2017). Quantitative prediction of injected CO_2 at Sleipner using wave-equation based AVO. *First Break*, **35**: 65–70.

Hansen, H., Eiken, O., and Aasum, T. O. (2005). Tracing the path of carbon dioxide from a gas-condensate reservoir, through an amine plant and back into a subsurface acquifer. Case study: The Sleipner area, Norwegian North Sea. Aberdeen, UK: SPE Offshore Europe.

Harrington, J. F., Noy. D. J., Horseman, S. T., Birchall, D. J., and Chadwick, R. A. (2010). Laboratory study of gas and water flow in the Nordland Shale, Sleipner, North Sea. In M. Grobe, J. Pashin and R. Dodge (eds.), *Carbon dioxide sequestration in geological media: State of the science. AAPG Studies in Geology*, **59**: 521–543.

Hauge, V. L., and Kobjørnsen, O. (2015). Bayesian inversion of gravimetric data and assessment of CO_2 dissolution in the Utsira Formation. *Interpretation*, **3**(2): sp1–sp10.

Haukaas, J., Nickel, M., and Sonneland, L. (2013). Successful 4D history matching of the Sleipner CO_2 plume. Extended Abstract, *EAGE Conference*.

Hermanrud, C., Andersen, T., Eiken, O., *et al.* (2009). Storage of CO_2 in saline aquifers: Lessons learned from 10 years of injection into the Utsira Formation in the Sleipner area. *Energy Procedia*, **1**: 1997–2004.

Hofmann, R. (2006). *Frequency dependent elastic and inelastic properties of clastic rocks*. PhD thesis, Colorado School of Mines.

Jullum, M., and Kolbjørnsen, O. (2016). A Gaussian-based framework for local Bayesian inversion of geophysical data to rock properties. *Geophysics*, **81**(3): R75–R87.

Kiær, A. (2011). *Trykkutvikling under CO₂-lagring*. Master's thesis, NTNU (in Norwegian).

Kiær, A. F. (2015). Fitting top seal topography and CO₂ layer thickness to time-lapse seismic amplitude maps at Sleipner. *Interpretation*, 3(2): SM47–SM55.

Kiær, A. F., Eiken, O., and Landrø, M. (2015a). Time lapse seismic amplitudes close to the rim of Sleipner CO₂ layers. In A. F. Kiær, *CO₂ fluid flow information from quantitative time-lapse seismic analysis*. PhD thesis, Norwegian University of Science and Technology, 57–72.

Kiær, A., Eiken, O., and Landrø, M. (2015b). Calendar time interpolation of amplitude maps from 4D seismic data. *Geophysical Prospecting*, 64(2): 421–430.

Labat, K., Delépine, N., Clochard V., and Ricarte, P. (2012). 4D joint stratigraphic inversion of prestack seismic data: Application to the CO₂ storage reservoir (Utsira sand formation) at Sleipner site. *Oil & Gas Science and Technology*, 67(2): 329–340.

Landrø, M., and Zumberge, M. (2017). Estimating saturation and density changes caused by CO2 injection at Sleipner: Using time-lapse seismic amplitude-variation-with-offset and time-lapse gravity. *Interpretation*, T243–T257.

Lindeberg, E. (2010). Modelling pressure and temperature profile in a CO₂ injection well. *Energy Procedia*, 4, 3935–3941.

Lindeberg, E., Zweigel, P., Bergmo, P., Ghaderi A., and Lothe A. (2000). Prediction of CO₂ distribution pattern improved by geology and reservoir simulation and verified by time lapse seismic. Expanded Abstract, *5th GHGT Conference*.

Monastersky, R. (2013). Seabed scars raise questions over carbon-storage plan. *Nature*, 504 (December 19/26): 339–340.

Neele, F. P., and Arts, R. J. (2010). Time-lapse seismic AVP analysis on the Sleipner CO₂ storage monitoring data using CFP processing. Extended Abstract, *72nd EAGE Conference*.

Nilsen, H. M., Krogstad, S., Andersen, O., Allen, R., and Lie, K.-A. (2017). Using sensitivities and vertical-equilibrium models for parameter estimation of CO₂ injection models with application to Sleipner data. *Energy Procedia*, 114: 3476–3495.

Nooner, S. L., Eiken, O., Hermanrud, C., Sasagawa, G. S., Stenvold, T., and Zumberge, M. A. (2007). Constraints on the in situ density of CO₂ within the Utsira formation from time-lapse seafloor gravity measurements. *International Journal of Greenhouse Gas Control*, 1: 198–214.

Park, J., Sauvin, G., and Vöge, M. (2017). 2.5D inversion and joint interpretation of CSEM data at Sleipner CO₂ storage. *Energy Procedia*, 114: 3989–3996.

Pedersen, R. B. (2011). Annual Report 2011. Center for Geobiology, University of Bergen.

Queißer, M., and Singh, S. C. (2013a). Full waveform inversion in the time lapse mode applied to CO₂ storage at Sleipner. *Geophysical Prospecting*, 61(3): 537–555.

Queisser, M., and Singh, S. C., (2013b). Localizing CO₂ at Sleipner:Seismic images versus P-wave velocities from waveform inversion. *Geophysics*, 78(3): B131–B146.

Rabben, T. E., and Ursin, B. (2011). AVA inversion of the top Utsira Sand reflection at the Sleipner field. *Geophysics*, 76(3): C53–C63.

Raknes, E. B., Weibull, W., and Arntsen, B. (2015). Three-dimensional elastic full waveform inversion using seismic data from the Sleipner area. *Geophysical Journal International*, 202(3): 1877–1894.

Romdhane, A., and Querendez, E. (2014). CO₂ characterization at the Sleipner field with full waveform inversion: Application to synthetic and real data. *Energy Procedia*, 63: 4358–4365.

Rossi, G., Chadwick, R. A., and Williams, G. A. (2011). Traveltime and attenuation tomography of CO₂ plume at Sleipner. Extended Abstract, *73rd EAGE Conference & Exhibition*.

Rubino, J. G., and Velis, D. R. (2011). Seismic characterization of thin beds containing patchy carbon dioxide-brine distributions: A study based on numerical simulations. *Geophysics*, 76: R57–R67.

Rubino, J. G., Velis, D. R., and Sacchi, M. D. (2011). Numerical analysis of wave-induced fluid flow effects on seismic data: Application to monitoring of CO₂ at the Sleipner field. *Journal of Geophysical Research*, 116: B03 306.

Singh, V., Cavanagh, A., Hansen, H., Nazarian, B., Iding, M., and Ringrose, P. (2010). Reservoir modeling of CO₂ plume behavior calibrated against monitoring data from Sleipner, Norway. *SPE*, 134891: 1–18.

White, J. C., Williams, G. A., and Chadwick, R. A. (2013). Thin layer detectability in a growing CO₂ plume: Testing the limits of time-lapse seismic resolution. *Energy Procedia*, 37: 4356–4365.

Williams, G., and Chadwick, A. (2012). Quantitative seismic analysis of a thin layer of CO₂ in the Sleipner injection plume. *Geophysics*, 77(6): R245–R256.

Williams, G., and Chadwick, R. A. (2017). An improved history-match for layer spreading within the Sleipner plume including thermal propagation effects. *Energy Procedia*, 114: 2856–2870.

Williams, G. A., Chadwick, R. A., and Vosper, H. (2018). Some thoughts on Darcy-type flow simulation for modelling underground CO₂ storage based on the Sleipner CO₂ storage operation. *International Journal of Greenhouse Gas Control*, 68: 164–175.

Zhang, G., Lu, P., and Zhu, C. (2014). Model predictions via history matching of CO_2 plume migration at the Sleipner Project, Norwegian North Sea. *Energy Procedia*, **63**: 3000–3011.

Zhang, G., Lu, P., Ji, X., and Zhu, C. (2017). CO_2 plume migration and fate at Sleipner, Norway: Calibration of numerical models, uncertainty analysis, and reactive transport modelling of CO_2 trapping to 10,000 years. *Energy Procedia*, **114**: 2880–2895.

Zhu, C., Zhang, G., Lu, P., Meng, L., and Ji, X. (2015). Benchmark modeling of the Sleipner CO_2 plume: Calibration to seismic data for the uppermost layer and model sensitivity analysis. *International Journal of Greenhouse Gas Control*, **43**, 233–246.

Zweigel, P., Hamborg, M., Arts, R., Lothe, A. E., Sylta, O., and Tommeras, A. (2001). Prediction of migration of CO_2 injected into an underground depository: Reservoir geology and migration modelling in the Sleipner case (North Sea). In D. Williams, I. Durie, P. McMullan, C. Paulson, and A. Smith (eds.), *Greenhouse Gas Control Technologies, Proceedings of the 5th International Conference on Greenhouse Gas Control Technologies*, 360–365.

Zweigel, P., Arts, R., Lothe, A. E., and Lindeberg, E. (2004). Reservoir geology of the Utsira Formation at the first industrial-scale underground CO_2 storage site (Sleipner area, North Sea). In S. Baines, J. Gale, and R. Worden (eds.), *Geological storage of carbon dioxide for emissions reduction*. London: Geological Society, 165–180.

Case Studies of the Value of 4D, Multicomponent Seismic Monitoring in CO₂ Enhanced Oil Recovery and Geosequestration

Thomas L. Davis, Scott Wehner, and Trevor Richards

Multicomponent seismic monitoring of carbon dioxide (CO_2) enhanced oil recovery (EOR) projects were undertaken by the Reservoir Characterization Project (RCP) at Postle Field, Oklahoma and Delhi Field, Louisiana. Both fields are being rejuvenated through CO_2 flooding. At the time of these studies Scott Wehner managed the Postle CO_2 EOR project for Whiting Oil and Gas Company, Trevor Richards managed the geophysical reservoir characterization at Delhi Field for Denbury Resources, and Thomas L. Davis was a codirector of RCP at the Colorado School of Mines.

At Postle Field miscible CO_2 is being injected into a water-alternating gas (WAG) scheme in a sandstone reservoir with an average thickness of 28 ft., at 6100 ft. depth. The flood patterns are a single vertical injection well in the middle of four producing wells; known as an inverted five-spot pattern. The RCP study area focused on 18 patterns in the northwestern area of Postle Field. Tracking the CO_2 with time-lapse seismic data is challenging owing to the thinness of the reservoir, the petrophysics of the reservoir and confining strata, the depth of investigation, and the operating practices within the flood. The results are encouraging in terms of the utilization of time-lapse data to characterize and monitor thin reservoirs with very low acoustic impedance contrast.

At Delhi Field, the flood is an immiscible process involving the continuous injection of CO_2. Reservoir depth is approx. 3300 ft., with thickness varying from 100 ft. to an updip wedge (i.e., pinch out). The reservoir was originally thought to be a blanket sand and, as a result, the flood pattern chosen was originally a line-drive with injection wells updip and downdip with producing wells in the middle. Seismic monitoring has identified distributary channels

diverting the CO_2 beyond some pattern boundaries. The result has been the potential for the bypass of oil reserves, thus requiring additional injectors and producers to be drilled to improve sweep efficiency.

To increase the efficiency of a CO_2 injection project it is essential to monitor where the CO_2 is going. Owing to the high mobility of CO_2 relative to water and oil, it moves readily into the high-permeability zones and bypasses the low. Bypassing can occur vertically as well as laterally. Optimizing contact with residual oil that has not been mobilized by primary production and secondary waterflooding is critical to achieving economic success. CO_2 flooding is a viable mechanism for EOR within most conventional reservoirs while providing an excellent mechanism for associated CO_2 geosequestration.

Postle Field

Located in the Panhandle of Oklahoma, Postle Field has approx. 300 million barrels of original oil in place (Figure 14.1). The field produces predominantly from the Morrow "A" sandstone of Pennsylvanian age. A stratigraphic column is shown in Figure 14.2 and the reservoir properties are listed in Table 14.1. The portion of the field monitored by RCP was approx. 6 square miles in the Hovey Morrow Unit (HMU) (Figure 14.3). Water injection was conducted initially to build the reservoir pressure to near-miscible pressures before the injection of CO_2.

Static Reservoir Characterization

Descriptions of cores from four wells in the HMU study area indicate the presence of four main lithofacies. These include a conglomeratic basal lag, a medium to fine-grain active channel fill, abandoned

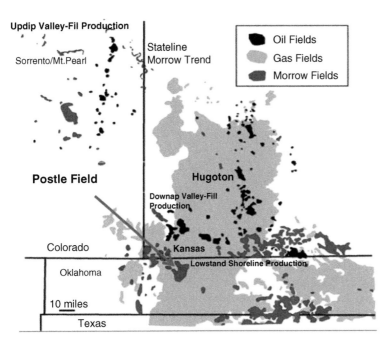

Figure 14.1 Location map, Postle Field, Oklahoma.

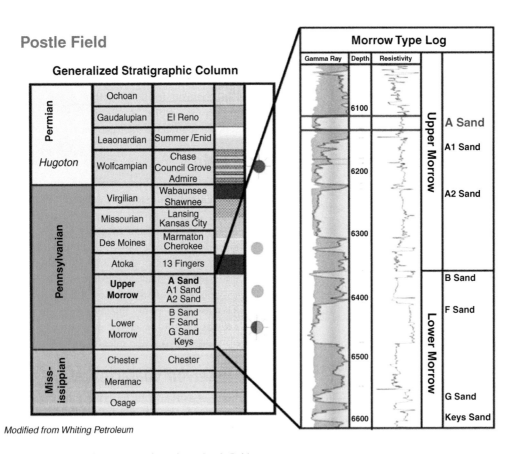

Modified from Whiting Petroleum

Figure 14.2 Generalized stratigraphic column, Postle Field.

Table 14.1 Reservoir properties, HMU, Postle Field

Postle HMU reservoir properties	
• Depth	6100 ft.
• Producing formation	Upper Morrow A – Sandstone
• Average thickness	28 ft. (10–70)
• Porosity	17% (10–23)
• Permeability	50 md (20–500)
• S_{oil}	70%
• Reservoir temperature	140°F
• Reservoir pressure initial	1630 psi
• Reservoir pressure current	2300 psi
• Minimum miscibility pressure	2100 psi
• Oil gravity	40 API
• Oil FVF orig	1.28
• Oil FVF current	1.20
• Well spacing	40 Ac (80-Ac 5-spot inj. patterns)
• OOIP	Approx. 45 MMBO

FVF, formation volume factor; OOIP, original oil-in-place.

Key Wells

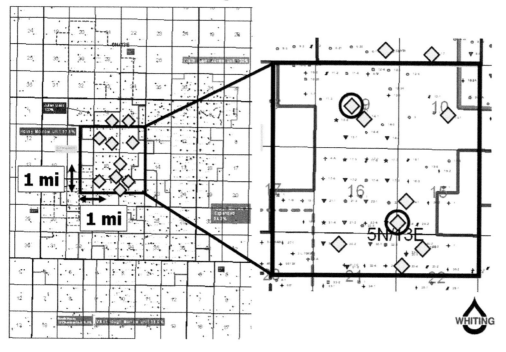

Figure 14.3 Location of the 6.25 square mile, 9-C, time-lapse multicomponent survey in the Hovey Morrow Unit (HMU), Postle Field. Key wells include VSP (circles) and location of cored wells and/or wells with sonic scanner and image logs (yellow diamonds).

channel fill, and overbank deposits. These facies occur in a depositional environment associated with braided channel deposits filling an incised valley. Provenance for the sediments is considered to be in relatively close proximity due to poor sediment sorting and the angularity and texture of rock fragments.

Secondary porosity has formed in this reservoir by dissolution of potassium feldspars. Permeability is highest in the basal lag and active channel fill facies. Within these facies permeability can vary from 20 to 500 mD. As a result, thin braided channels of high permeability can form conduits for CO_2 movement in the reservoir. Alignment of wells along the channels, although providing some early oil production, also leads to early breakthrough of the more mobile CO_2. Water alternating gas (WAG) is used to slow the displacement in the higher permeability zones and help provide better vertical conformance; pushing more CO_2 to the less permeable zones. Pattern alignment and pattern balancing are important to optimize recovery efficiency in the WAG process. The ability to proactively manage the WAG schemes has an important bearing on incremental oil recovery and conversely, CO_2 utilization, as well as its ultimate sequestration. In addition, the ability to dynamically monitor the CO_2 WAG not only provides a key to better reservoir management of a costly injectant, but it also allows monitoring for possible losses beyond the flood pattern or project complex.

How significant is this project? The HMU was the last of four CO_2 EOR expansions within Postle Field; the first starting in 1996. Prior to expanding into HMU, elevated reservoir pressures within the offset acreage (active CO_2 WAG operations) were maintained by water-curtain well management (i.e., pressure) between the project boundaries. HMU was a very mature waterflood and had only one remaining active/producing well prior to beginning CO_2 flood development. Injection began in 2008 and the flood exceeded the waterflood peak production achieved in 1971 within the first year of operation (Figure 14.4). The HMU CO_2 flood is a successful rejuvenation project due in large part to understanding the connectivity of the stacked braided channels. Seismic monitoring has played an important role in identifying the connectivity within the reservoir, which is a critical part of the EOR reservoir management operation. Establishing the field development plan in a proactive rather than reactive manner can have an

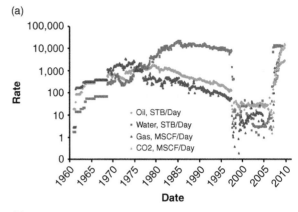

(a)

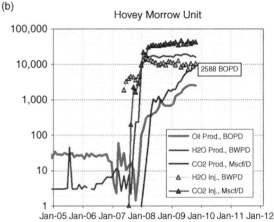

(b)

Figure 14.4 Field revitalization with miscible carbon dioxide flooding in the Hovey Morrow Unit, Postle Field.

important impact on the economic success of EOR projects. Early seismic monitoring should be an integral part of project design and reservoir management planning for both operational, as well as geosequestration purposes.

Dynamic Reservoir Characterization

The baseline seismic program was acquired in March 2008 and the first monitoring survey 9 months later in December 2008. Survey parameters are included in Table 14.2. Several wells were drilled during 2008 and little drilling went on in 2009 owing to an oil price crash carrying over from the latter part of 2008. Overall, the surface coupling conditions between the December and March surveys were comparable. Repeatability, as measured by normalized root mean square (NRMS, the square root of the

Table 14.2 Acquisition parameters used for the seismic surveys at Postle Field

4-D,9-C Seismic acquisition parameters	
Type survey	4-D, 9-C (time-lapse)
Subsurface bin size	55 ft. × 55 ft.
Number of receiver locations	1920
Number of source locations	1920
Total number of source points	5760
Type receiver spread	Stationary: 13,200 ft. × 13,200 ft.
Instrumentation	I/O System IV, 2 ms sample rate, 45 s record length
Receiver array	Single 3C digital sensor (vectorSeis)
Source (P-wave)	Vertical vibrator: 6–100 Hz linear sweep, 8 s duration, 4 sweeps
Source (S-wave)	Horizontal vibrator: 4–60 Hz linear sweep, 8 s duration, 4 sweeps, one source oriented N–S, one source oriented E–W

average of the squares of a series of measurements) after cross-equalization, is approx. 0.2, indicating excellent repeatability.

The thickness of the Morrow "A" Sandstone is well below the resolution limit of the seismic data, but detectability is dependent on the impedance of the sandstone and not just thickness. Even though the Morrow "A" is below seismic resolution, injection of CO_2 can potentially make the reservoir sandstones visible because fluid saturation change and pressure affect the impedance of the sandstones. The term "time-lapse sands" has been coined for those sandstones that become visible by introducing dynamic changes in the reservoir.

The volume of fluids injected in wells between the two surveys is shown in Figure 14.5. In total 5.10 Bscf of CO_2 was injected from the baseline to acquisition of the first seismic monitor survey. Prior to the baseline acquisition, approx. 1.65 Bscf had been injected in the southernmost wells. The pressure change in the reservoir between the two surveys was approx. 800 psi in

the vicinity of the northernmost wells, and negligibly different elsewhere. An increase in fluid pressure decreases the acoustic impedance. In addition, a change in fluid saturation occurs as CO_2 mixes with oil to form a miscible flood bank while predominantly replacing water in the pore space: a resultant of the previous waterflood operation. Laboratory rock physics measurements on cores shows the predicted seismic changes (Figure 14.6).

It is critical to understand the fluids in the reservoir at any given time when planning seismic surveys. Prior to the injection of CO_2, water is typically injected to bring pressure up to miscibility. In most cases, it is far more economical to inject water rather than CO_2 to build reservoir pressure. In the case of Postle, with a limited CO_2 supply available, it is also critical to maximize the efficiency of the injection volumes. A WAG process to improve conformance within the strata and maintain pressures once the pressure has reached miscibility, coupled with balancing of the injection and withdrawals within the flood patterns, provides a means of managing the efficiency of CO_2 EOR and associated geosequestration. Two downdip water injectors (a.k.a. water curtain) on the southwest side of the study area were used to provide pressure support and prevent the CO_2 from moving downdip outside the Unit and to maintain the pressure front eastward and northward within the reservoir.

Overall, the seismic data acquired were very good. The surface is relatively flat agricultural land and source and receiver coupling was good. There was not much variability in near surface conditions between the two surveys. A vertical seismic profile (VSP) was acquired in a well in the northern part of the survey area. It showed that the Morrow "A" sandstone is a very weak peak on P-wave seismic, but shows a much stronger peak on the S-wave data (Figure 14.7). This reflectivity difference is also shown on the surface seismic data (Figures 14.8–14.11).

Scheduling the timing of surveys is critical. A true baseline was not achieved as injection began in the southern portion of the survey area shortly before the first survey was acquired. Nevertheless, the hypothesis about CO_2 increasing the visibility of the sandstones was confirmed (Figure 14.9). The high-amplitude area in the south showed good continuity overall and matched the well information regarding presence of CO_2 indicating the potential promise that the flood

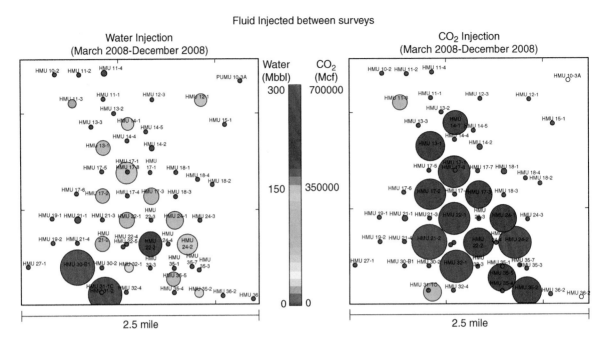

Figure 14.5 Injected fluid volumes for wells in study area from March through December, 2008.

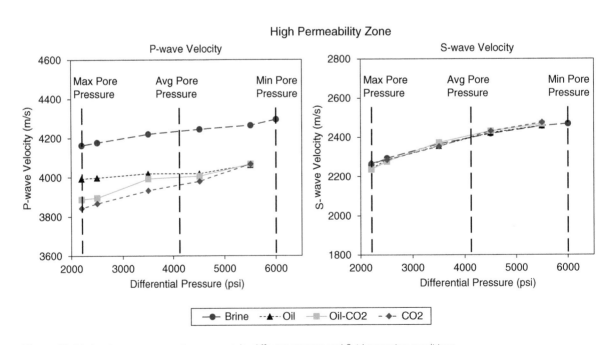

Figure 14.6 Laboratory measurements on core under different pressure and fluid saturation conditions.

Zero-Offset P-Wave VSP Data **Zero-Offset S-Wave VSP Data**

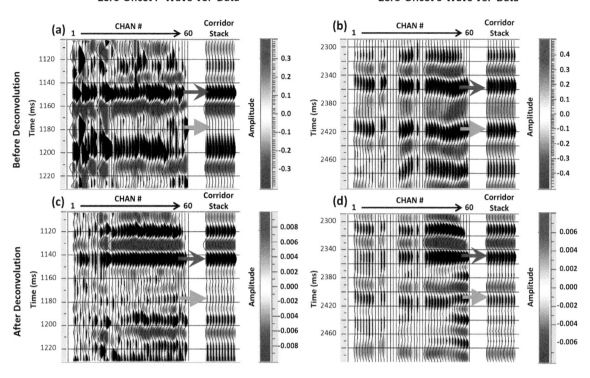

Figure 14.7 The near offset vertical seismic profile shows the reflectivity of the Morrow "A" sandstone (green arrow) on P- and S-wave data. The Atoka limestone is shown by a red arrow and is a good seismic marker on both P- and S-wave data.

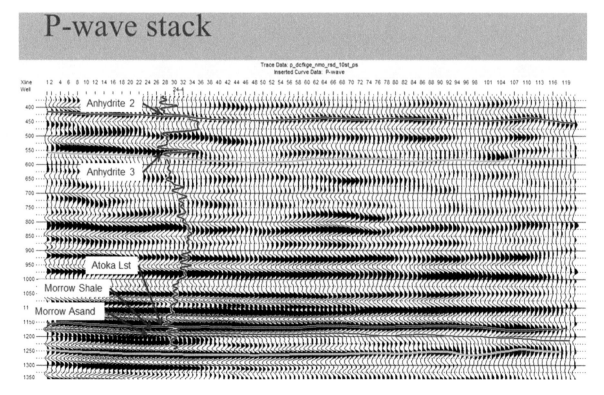

241

Figure 14.8 A P-wave line from the 3D volume shows the weak reflectivity of the Morrow "A" sandstone on P-wave data.

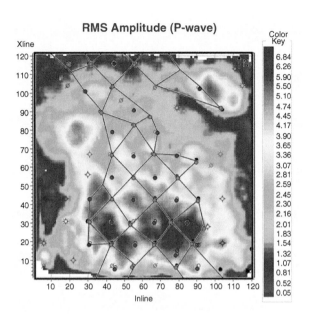

RMS Amplitude (P-wave)

Figure 14.9 RMS amplitude map from a 10-ms window centered on the Morrow "A" sandstone event on the P-wave data. The southern part of the survey area shows an increased seismic amplitude due to the presence of CO_2. By the time of this baseline survey 1.65 Bscf had already been injected in the southernmost wells. The green dots are oil wells and the red and blue are CO_2 and H_2O injection wells, respectively. Well patterns are depicted.

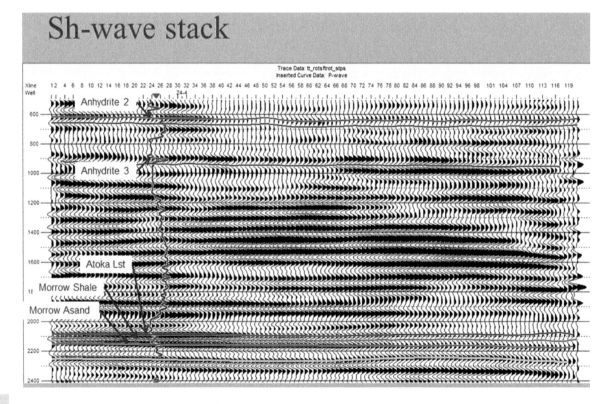

Figure 14.10 A 2D line out of the 3D volume showing the Morrow "A" sandstone as a strong seismic peak on the shear wave data. The term Sh-wave pertains to a shear wave with horizontal particle motion in a plane perpendicular to the source-receiver plane in an isotropic medium.

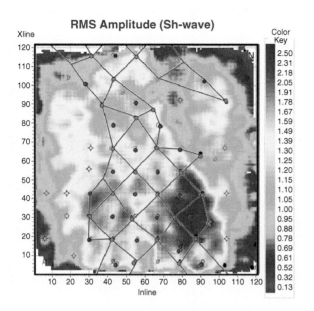

would be successful. By the time of the March survey, herein referred to as the baseline, the northern area had not had CO_2 introduced.

The weak reflectivity of the P-wave data is due to the compressibility of the sandstones. Compressibility links to porosity, texture, and to the fluids within the pore space. In the case of the Morrow "A" sandstone, the high compressibility of the sandstones lowers the seismic velocity and density such that the acoustic impedance of the sandstones nearly equals that of overlying Morrow shale rendering the sandstones as acoustically invisible to P-wave data.

Shear wave data are affected by rigidity and, because the sandstones are more rigid than the overlying shale, they are visible on S-wave data. Fluid pressures can affect the rigidity and density of the sandstones and thus fluid pressure and not fluid composition can affect shear waves.

Figure 14.12 shows a seismic-guided isopach map made from well and seismic control. Notice that as the sandstones thin to the edges of the incised valley, the seismic resolution is diminished and, therefore, weighting functions applied in the geostatistical mapping are diminished. The Unit boundaries also are a constraining factor when it comes to operational controls in the field. Constraining the CO_2 from moving out of pattern and outside the boundaries of the Unit is necessary. At Postle Field our study led to the drilling of a well in the southeast corner of our study

area that found 30 ft. of Morrow "A" sandstone very near the edge of the valley and the boundary of the Unit, thus completing the pattern in that area.

Figures 14.13 and 14.14 show the time-lapse differences for P- and S-waves respectively. The areal extent of these changes aids in mapping permeability trends in the reservoir and these trends can be then entered into the reservoir model and used to improve history matching in reservoir simulation.

P-waves are affected by fluid saturation and pressure. A small amount of CO_2 drops the P-wave velocity as shown previously in Figure 14.6. After about 10% saturation there is little change in the P-wave response other than pressure. As this is a WAG injection scheme there is very little response to the flood on the P-wave data, but more on the S-wave data, as they are only sensitive to pressure. The pressure front is more diagnostic of where the fluids are moving than where the CO_2 miscible bank is within the reservoir. By recording both P- and S-waves we can differentiate between pressure and saturation when compared.

The value of seismic monitoring at Postle Field involves proactive management of the CO_2 flood through WAG optimization, pattern orientation and pattern balancing. This is managed in spite of the absence of a true baseline survey. All of the seismic data were acquired, processed, and interpreted for less than the cost of a single well.

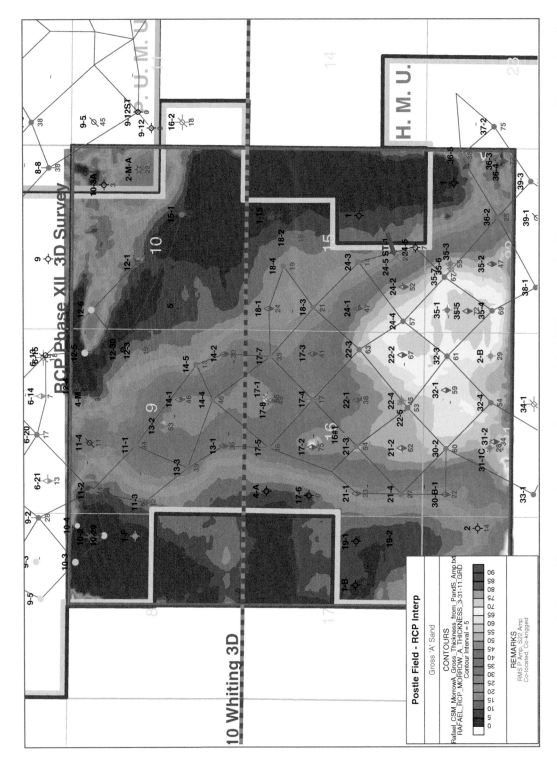

Figure 14.12 A seismic-guided Morrow "A" isopach map showing the well patterns and the HMU boundary outline. The arrow shows the location of the new-drill well.

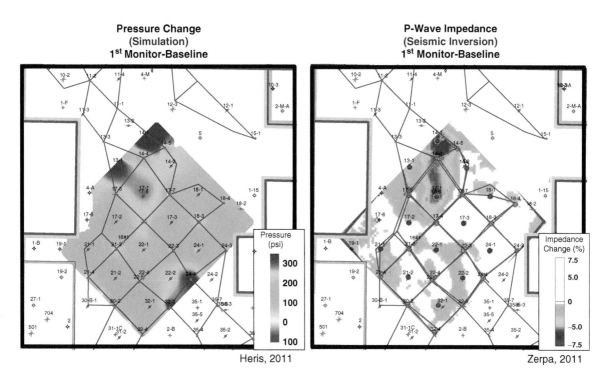

Figure 14.13 P-wave time-lapse map.

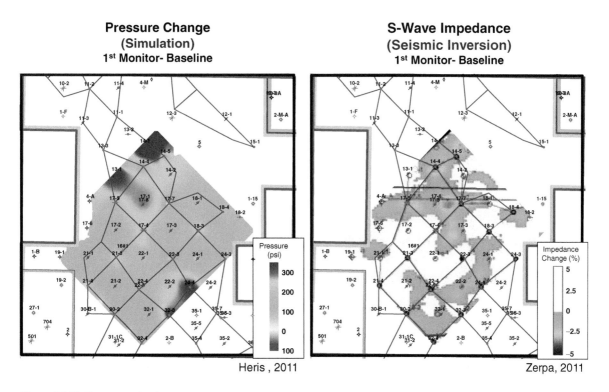

Figure 14.14 Shear wave time-lapse map.

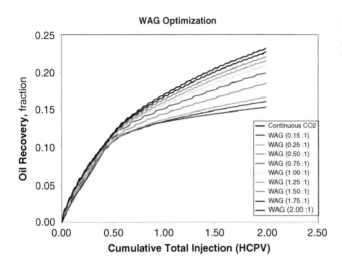

WAG Optimization

Figure 14.15 Simulation of the process suggests that an increasing WAG ratio can increase oil recovery. Conversely, decreasing the WAG ratio could result in higher associated sequestration if the total CO_2 was not held constant.

Time-lapse seismic data were used to guide the permeability distribution in the reservoir simulation model (Heris *et al.*, 2011). The reservoir simulation was used to evaluate scenarios for prediction of oil and CO_2 recovery and associated CO_2 storage under different WAG injection schemes. The results are shown in Figure 14.15. A 2:1 WAG injection scheme was recommended to the Unit operator. However, WAG optimization on oil recovery is not necessarily the only focus by an operator. WAG is more typically optimized on produced gas handling and its associated facility capital infrastructure needs, which were not part of this evaluation. Generally, the WAG is modified by "wetting it up" over years, but the cycle time is short, usually on the order of weeks, to control the CO_2 mobility.

Geosequestration

A CO_2 project is effectively a "closed loop system." From an economical sense, projects are selected with the intent of containing all the injected CO_2. Whatever is injected within the reservoir complex generally stays within the complex. The purpose of monitoring is foremost to manage the fluid flow within the system to maximize oil recovery. At the same time CO_2 stored in association with the EOR process is being geosequestered in this closed loop system. Nearly 100% of the injected volume is sequestered because what CO_2 is produced is captured and repressurized for recycle back into the reservoir complex. However, there are downhole situations that could allow movement beyond the confines of the flood patterns and multicomponent seismic offers

one monitoring technique that can provide early qualitative understanding of where CO_2 exists in the subsurface. Through proactive reservoir management techniques based on monitoring the system response, both EOR economic goals and the associated safe long-term storage of CO_2 can be optimized.

Delhi Field

Time-lapse, multicomponent seismic data were acquired during the startup of the CO_2 flood at Delhi Field, Louisiana. The purpose of the integrated study at Delhi was to identify zones of high residual oil saturation and to monitor possible flow paths in the reservoir. Contacting bypassed oil is critical for economic EOR.

Study Location

Delhi Field is located in northeastern Louisiana (Figure 14.16). The field was discovered in 1944. The reservoir consists of Upper Cretaceous (Tuscaloosa) and Lower Cretaceous (Paluxy) sandstones (Figure 14.17). As of this writing, the majority of the field is undergoing a CO_2 flood for enhanced oil recovery. Depth to the top of the producing interval averages 3250 ft., porosity averages 25%, and the average permeability is 1400 mD. Permeability in the reservoir is extremely variable and is controlled by depositional facies. Reservoir parameters are listed in Table 14.3.

The Paluxy is made up of a fluvial deltaic interval of Early Cretaceous age. The Late Cretaceous Tuscaloosa unconformably overlies the

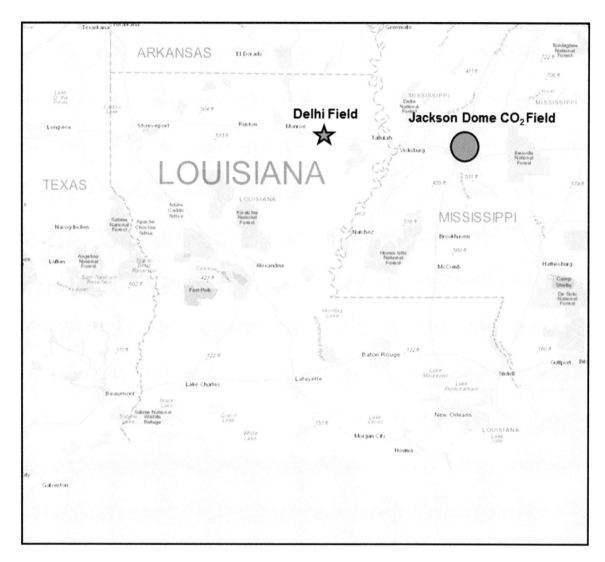

Figure 14.16 Location of Delhi Field.

Paluxy and comprises transgressive marine deposits overlain by a shallowing upward sequence of barrier bars capped by a fluvial deltaic interval at the top.

Gross thickness of the Paluxy varies from 75 to 30 ft. in the study area. The Tuscaloosa varies from 100 ft. to a wedge or pinch-out in the updip direction. Distinguishing the connectivity of reservoir units in the Paluxy is the objective of this study. It is difficult to assess the connectivity based on well control alone. Figure 14.18 shows assumed continuity of sandstone units between wells in cross-section view based entirely on well log correlation. Given the complexity of fluvial-deltaic environments we know this assumed correlation to likely be inaccurate. Also, the potential exists for connectivity between the Paluxy and the Tuscaloosa owing to overlapping hydraulic units.

Static reservoir characterization can unravel the complexity of the geology to a large degree by careful analysis of the core, well logs, and the baseline seismic survey, all of which was done at Delhi during the RCP study. The value of monitoring comes in the understanding of flow unit scale conformance and improved understanding of connectivity, i.e., dynamic reservoir characterization.

Seismic Acquisition and Processing

Prior to the monitoring seismic surveys, a pre-CO_2 baseline P-wave seismic survey was acquired in late 2007 and early 2008. Later, two multicomponent

Table 14.3 Delhi Field reservoir parameters

Reservoir parameters
Reservoir depth approx. 3300 ft. (1 km)
Initial reservoir pressure approx. 1500 psia
Reservoir pressure prior to CO_2 flood approx. 1300 psia
40° API oil, GOR approx. 400 SCF\STB, Viscosity 0.77 cp
Reservoir temperature approx. 135°F
Porosity range: 25–30%
Permeability range: 250–3000 md
Denbury to operate field approx. 1800 psi
Pressure at injectors approx. 2400 psi

seismic surveys encompassing four square miles each were centered on the CO_2 flood startup location (Figure 14.19). These surveys took place in June of 2010, six months after the startup of the CO_2 flood in November 2009, and again in August 2011. The first few months of 2010 were excessively wet, preventing acquisition of a survey. A cableless, three-component nodal receiver system was used to acquire the surveys. Dynamite was used as the source after extensive source testing showed the ineffectiveness of surface sources in acquiring data because of ground conditions in the area. Multicomponent seismic survey parameters are listed in Table 14.4 and the survey outline is shown in Figure 14.20.

The survey was designed to monitor the Paluxy interval as the initial startup involved six Paluxy injectors. Downdip from the Paluxy injectors, three Tuscaloosa injectors were also located in the seismic grid, but only two were operational during the time of our surveys. Each injection well was intended to inject

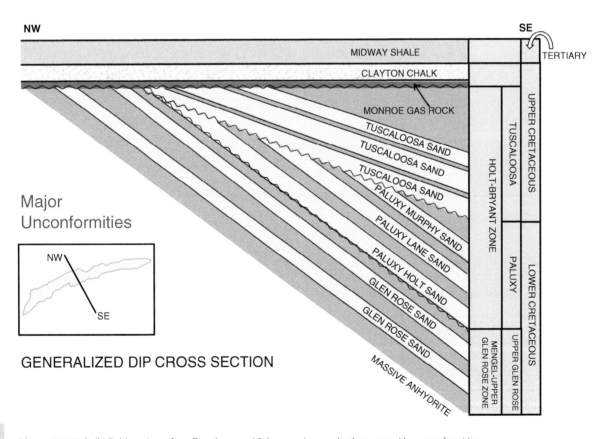

GENERALIZED DIP CROSS SECTION

Figure 14.17 Delhi Field produces from Tuscaloosa and Paluxy sandstones, both truncated by unconformities.

Cross Section

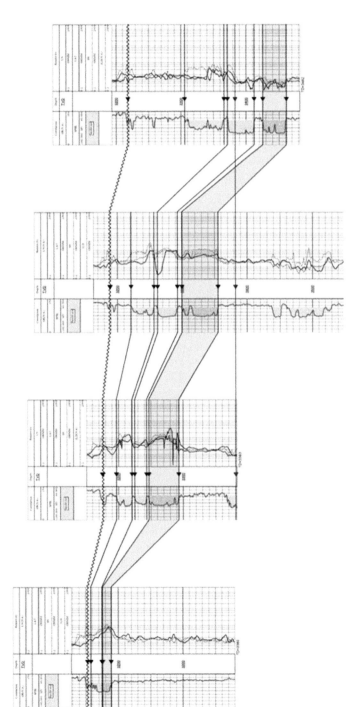

N

S

Figure 14.18 Well log cross section showing the unconformity trap at Delhi. The Paluxy is highlighted. Above the Paluxy is the Tuscaloosa.

Table 14.4 Multicomponent, time-lapse seismic survey parameters

Acquisition parameters

RCV interval	82.5 ft.
SRC interval	165 ft.
RCV line interval	495 ft.
SRC line interval	495 ft.
Receiver type	3C MEMS
Source type	Dyn (1lb @ 30 ft.)
Recording patch	18 × 108 (~4400 ft. × ~4400 ft. offsets)
Sampling rate	2 ms
Listen time	4 s
Bin size	82.5 ft. × 82.5 ft.
Trace density	663,552/sq. mi. (×3)
Acquisition by Tesla exploration	

up to 10 MMscf per day. From November 2009 until the second survey in August 2011, approx. 24 Bscf of CO_2 had been injected into the Paluxy. The minimum miscibility pressure (MMP) is 2335 psi (Chen *et al.*, 2014) and the goal was to inject up to near miscibility while keeping the reservoir below the fracture gradient, which at 3300 ft. depth is approx. 2475 psi. Monitoring the overburden became an important and integral part of the Delhi monitoring project. S-wave azimuthal anisotropy was used as a tool to monitor stress changes not only in the reservoir but also the overburden (O'Brien and Davis, 2013). Time-lapse time shifts were also studied in the overburden on both the P and PS data to ensure overburden integrity throughout the injection startup where the pressure buildup was designed to go up to, but not exceed, 1800 psi in the Tuscaloosa–Paluxy zones of interest (Bishop and Davis, 2014). Operating the immiscible flood at Delhi

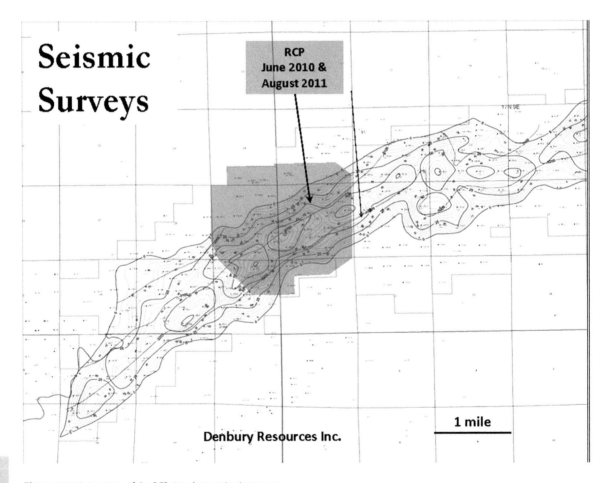

Figure 14.19 Location of the RCP time-lapse seismic surveys.

Delhi Survey Area

Figure 14.20 Survey layout.

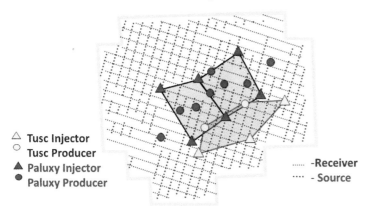

△ Tusc Injector
○ Tusc Producer
▲ Paluxy Injector
● Paluxy Producer

...... -Receiver
.... - Source

requires careful monitoring to maximize reservoir contact while at the same time keeping pressures fairly uniform and well below the fracture gradient.

(a)

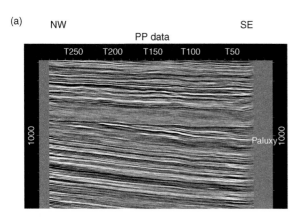

(b)

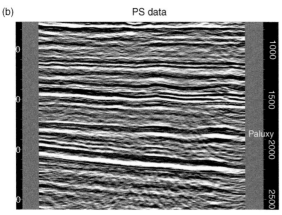

Figure 14.21 PP and PS data from the 2010 survey.

Petrophysical Data

Log data, calibrated with core data from a new-drill well in the field, were used to determine reservoir petrophysical properties. Sixty years of waterflooding had taken place and yet there is still significant residual oil in this reservoir. Seismic-assisted analysis of reservoir properties through seismic inversion was used to build a reservoir model for simulation. This model was further enhanced though time-lapse studies to map the permeability structure.

Fluid substitution modeling showed that a small amount of CO_2 would change the acoustic impedance of the reservoir. Modeling suggested a strong observable response to CO_2 injection. Both saturation change and pressure change would affect the P-wave response. The S-wave response is only influenced by pressure change.

Multicomponent Seismic Data

Multicomponent (PP and PS) time-lapse seismic data were acquired in Delhi Field in June 2010 and August 2011. Data quality was excellent even though different ground conditions existed at the time of the two surveys. The PP and PS data have been tied to the subsurface through quality well ties provided by modern dipole sonic logs (Figure 14.21).

Processing was undertaken to preserve the repeatability of the surveys. The time-lapse seismic data were processed simultaneously. The NRMS was 0.19 for P-wave data and 0.21 for the PS data. The NRMS values indicate excellent repeatability. Generally, a NRMS of 0.3 or less is considered good repeatability and below 0.2 is excellent.

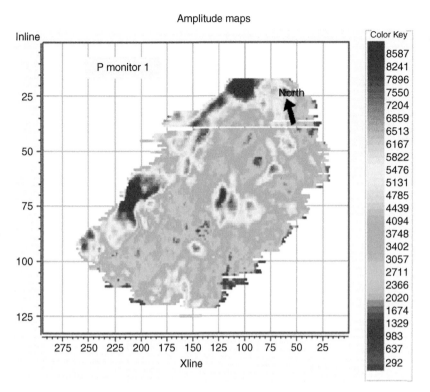

Amplitude maps

Figure 14.22 P-wave RMS amplitude map of the Paluxy over a 10-ms window.

Interpretation

The Paluxy Formation was picked on the respective seismic volumes. RMS amplitude maps of the Paluxy were generated using windows of time that coincided with the Paluxy on both volumes (Figures 14.22 and 14.23). The maps show that the P-wave data are very sensitive to fluid saturation in the sandstones of the Paluxy. The six injectors were all operational at the time of the first monitoring survey. The PS data are not sensitive to fluid saturation change, but they are sensitive to lithology and pressure. Of interest is the trend of high amplitudes that are related to the presence of distributary channels in the Paluxy. In addition, there are crevasse splays, clay plugs, and low-permeability zones affecting the flood response. The CO_2, because of its higher mobility and the fact that this is a continuous CO_2 injection flood, bypasses the lower permeability regions in the reservoir, degrading the sweep efficiency of the flood.

A P-wave, time-lapse, difference map between the two surveys, clearly shows that some CO_2 from the southern injectors in the Paluxy was moving downdip into the aquifer (Figure 14.24). The cause was theorized to be a permeability barrier or baffle preventing the CO_2 from moving updip and/or a pressure barrier from updip injection. As a result, two downdip water injectors were completed into the aquifer after our monitoring survey showed this downdip movement. The intent of the two water injectors was to provide pressure support to keep the CO_2 from moving into the aquifer. The August 2011 survey showed that the curtain was successful in creating a pressure wall in the producing area of the oil column causing CO_2 to preferentially sweep updip (Figure 14.25). A time-lapse pressure map was computed from the near-linear relationship between pressure and V_p/V_s. The map shows the effectiveness of the downdip water curtain that was implemented (Figure 14.26).

Monitoring shows lateral migration of the CO_2, but what about the vertical? Figure 14.27 shows the vertical migration from the Paluxy up into the Tuscaloosa between monitors 1 and 2. It is best seen on the V_p/V_s time-lapse illustrations. CO_2 acts as a tracer in delineating flow units in the reservoir. A sand-on-sand contact across the unconformity allows the upward migration of CO_2 into the Tuscaloosa affecting the lateral sweep in the Paluxy.

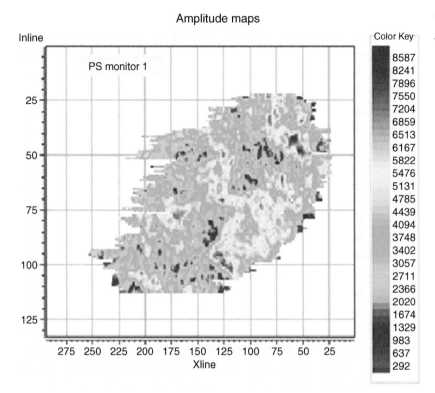

Figure 14.23 P–S RMS amplitude map of the Paluxy over a 20-ms window.

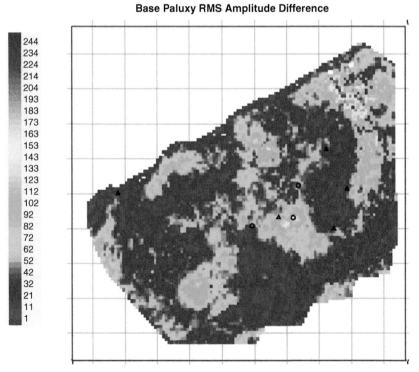

Figure 14.24 CO_2 is clearly moving downdip from the southern injectors in June, 2010.

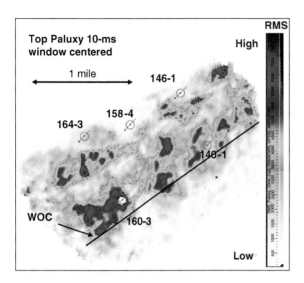

Figure 14.25 By time of the monitor 2 survey the CO_2 had moved back updip.

Production History

Delhi Field is another example of a successful CO_2 EOR project in that, by the time of the second monitoring in 2011, the field was producing 4000 barrels per day after more than a decade long hiatus of being shut-in (Figure 14.28). At the time of initiation of the CO_2 flood, 50% of the original oil-in-place (OOIP)

had been produced through primary and secondary recovery. CO_2 EOR has rejuvenated an old field and seismic monitoring could assist in optimizing and predicting the incremental recovery from the field.

Reservoir Modeling and Simulation

Downdip mobility occurs on the eastern and southern side of the study area, presumably along a distributary channel system. The time-lapse seismic data show bypass areas and clear evidence where CO_2 is moving upward from the Paluxy into the overlying Tuscaloosa. Monitoring has helped improve the reservoir characterization by defining areas of high transmissibility that can be input into a reservoir model and improve reservoir simulation and associated predictions. Forecasting based on improved reservoir simulation can lead to proactive reservoir management and greater potential for economic success. The prize is forecast to be 10–12% incremental recovery of the OOIP (Figure 14.29).

Conclusions

Time-lapse multicomponent seismic data have been used to monitor CO_2 floods at Postle Field, Oklahoma and Delhi Field, Louisiana. It is recognized that the P-wave seismic response does not significantly change after the initial fluid substitution occurs and these data show the importance of monitoring the reservoir,

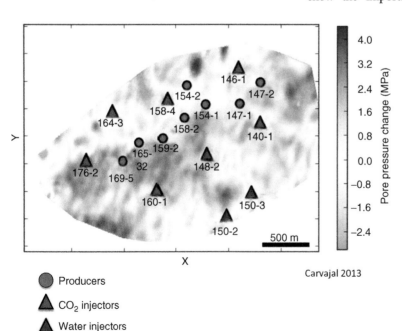

Figure 14.26 Time-lapse multicomponent pressure monitoring of the change in pore pressure between the two monitor surveys. (From Carvajal et al., 2014.)

Carvajal 2013

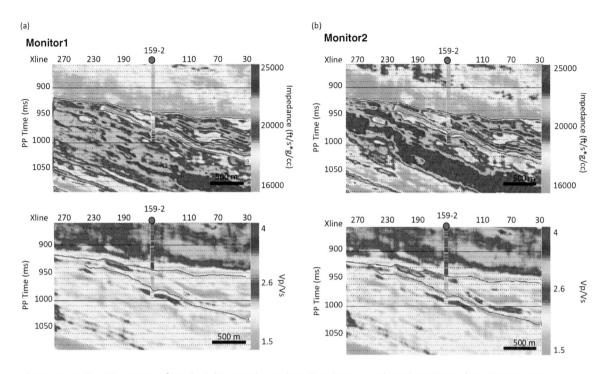

Figure 14.27 The CO_2 is moving from the Paluxy into the overlying Tuscaloosa as evidenced on the time-lapse V_p/V_s response.

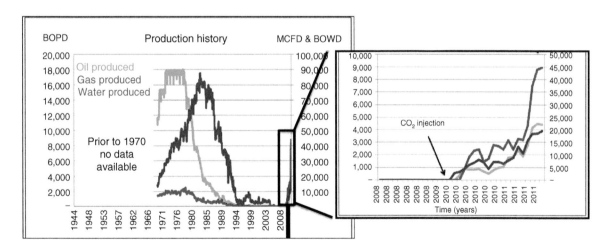

Figure 14.28 The production history of Delhi Field.

especially in the early startup of the EOR process. S-waves, on the other hand, exhibit almost linear behavior in response to pressure, making the S-wave data valuable for not only lithology prediction but also pressure monitoring. Pressure monitoring is important for the CO_2 EOR process and CO_2 geosequestration.

Case studies from Postle and Delhi Fields show that seismic can monitor movement of CO_2 in these sandstone reservoirs. CO_2 acts like a tracer to identify permeability pathways in a reservoir. CO_2 with seismic monitoring can be used to identify flow units, flow paths, bypassed oil, and potential thief zones. Based on our studies at Postle and Delhi Fields we can say with great certainty that multicomponent seismic monitoring brought value through reservoir management modifications.

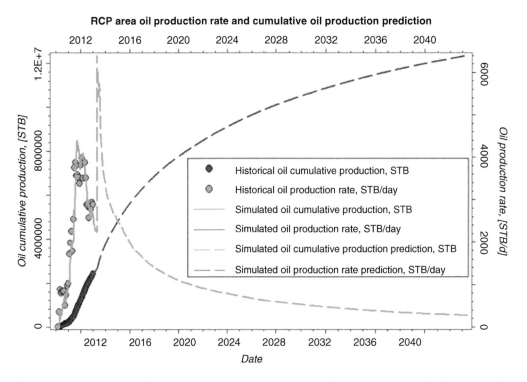

Figure 14.29 Prediction based on reservoir simulation shows the potential for 10–12% incremental oil recovery in the 4 square mile study area of the field.

Early seismic monitoring can identify less efficient reservoir processing, guiding the operator to modify reservoir management practices. A degradation in CO_2 efficiency can be found in both flood pattern management itself, and losses beyond the flood pattern. Managing the CO_2 EOR process in conjunction with seismic monitoring brings economic dividends and can assist in evaluating the associated CO_2 geosequestration potential.

References

Bishop, J. E., and Davis, T. L. (2014). Multicomponent seismic monitoring of CO_2 injection at Delhi Field, Louisiana. *First Break*, **32**(5): 43–48.

Carvajal, C., Putri, I., and Davis, T. (2014). Dynamic reservoir characterization using 4-D multicomponent seismic data and rock physics modelling in Delhi Field, Louisiana. *First Break*, **32**(2): 63–69.

Chen, T., Kazemi, H., and Davis, T. L. (2014). Integration of reservoir simulation and time-lapse seismic in Delhi Field: A continuous CO_2 injection EOR project. *SPE* 169049.

Heris, A. E., Wandler, A., Kazemi, H., and Davis, T. L. (2011). Quantitative integration of flow simulation and 4-D multicomponent seismic in a CO_2 WAG EOR project. *SPE* 146960.

O'Brien, S., and Davis, T. (2013). Time-lapse shear wave splitting analysis to monitor caprock integrity at Delhi Field, Louisiana. *First Break*, **31**(5): 75–81.

15

Integrated Geophysical Characterization and Monitoring at the Aquistore CO$_2$ Storage Site

Don White

Introduction

The Aquistore carbon dioxide (CO$_2$) storage (geosequestration) and monitoring site is located near Estevan, Saskatchewan, Canada (Figure 15.1). Aquistore is a commercial-scale storage project that utilizes a deep saline formation as the storage reservoir (Worth *et al.*, 2014). As such, it is the world's first integrated storage project associated with CO$_2$ capture from an industrial scale coal-fired power plant. As of October 31, 2016, a total of approx. 100 ktonnes of CO$_2$ had been injected at the site. The Aquistore project is conducting research into monitoring methods for CO$_2$ injection and storage within deep saline aquifers. The project aims to test and demonstrate a broad range of geophysical methods for this purpose. To date, geophysical methods have been utilized in the initial site characterization, baseline data collection, and early CO$_2$ storage monitoring. The final goal is to assess the relative effectiveness of the various monitoring methods, what role they may play in a storage program, and to integrate them where possible.

In this chapter, the results from the variety of geophysical methods that have been utilized at the Aquistore site are presented. The list comprises various seismic techniques (four-dimensional [4D] surface, 4D vertical seismic profiles [VSPs], passive), electromagnetic methods (magnetotelluric [MT], controlled-source electromagnetic [CSEM]), gravity, surface-deformation measurements (inteferometric synthetic aperture radar [InSAR], GPS, tilt), and a number of emerging methodologies (borehole-to-surface electromagnetic [BSEM], distributed acoustic sensing [DAS], seismic interferometry, downhole gravimetry) that have been tested. The chapter begins with the geological characterization of the site using three-dimensional (3D) seismic data and well logs. This is followed by the various geophysical baseline measurements, emerging technology tests, and finally by the first time-lapse measurements. A summary of the geophysical methods employed and some key results are provided in Table 15.1.

Background Geology

The Aquistore storage site lies within the northern Williston Basin, an intracratonic basin in central North America (Kent and Christopher, 1994). At the Aquistore site, sedimentary strata have a cumulative thickness of approx. 3350 m and comprise a lowermost Cambro-Ordovician clastic sequence overlain by predominantly carbonates and evaporites of Ordovician to Mississippian age. Above these strata are Mesozoic age shales, siltstones, and sandstones lying beneath thin Tertiary and Pleistocene clastic deposits.

The CO$_2$ storage reservoir is located within the lowermost Paleozoic section immediately above the Precambrian basement. The Deadwood and overlying Winnipeg formations form an approx. 200 m thick interval with approx. 50% of the interval having potential for CO$_2$ injection and storage. These formations have been the target of oil production and waste water disposal elsewhere in the basin (Brunskill, 2004). The Deadwood Formation is a sandstone unit that contains silty-to-shaley interbeds with subordinate calcareous layers. The overlying Winnipeg Formation includes a lower sandstone unit (Black Island Member) and an upper shale unit (Icebox Member) that constitutes the reservoir caprock and primary CO$_2$ storage seal. Regionally, Winnipeg–Deadwood sandstones have permeabilities of approx. 100–1000 mD and porosities of approx. 11–17% (Vigrass *et al.*, 2007). At the Aquistore site, increased shale content within this interval reduces permeabilities and porosities to 0.05–300 mD and 5–8%, respectively. The Middle Devonian Prairie Formation is an

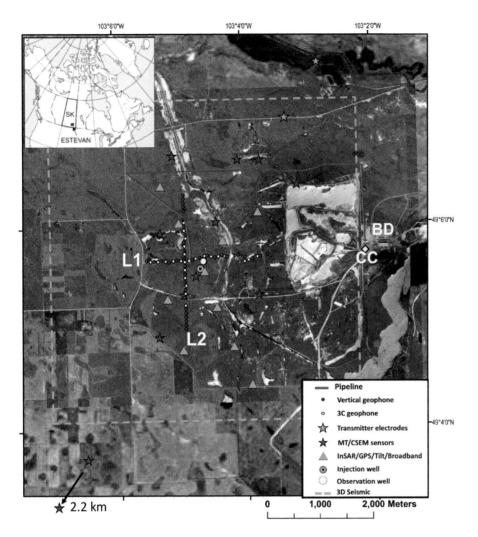

Figure 15.1 Aquistore Project site map showing the location of the Boundary Dam Power Plant (BD), the CO_2 capture facility (CC), and the CO_2 injection well and observation well. Also shown are locations of the surface monitoring stations, short-period geophone arrays (L1 and L2), area covered by the baseline 3D seismic survey, MT/CSEM sites, and bipole transmitter. Each surface monitoring station includes some combination of GPS, InSAR reflectors, tilt meters, and broadband seismographs. Inset shows site location in Canada. Note that there is an additional MT/CSEM sensor located 2.2 km to the southeast as indicated by the lower left symbol outside of the map frame. (© Her Majesty the Queen in Right of Canada, as represented by the Minister of Natural Resources, 2017.)

approx. 200 m thick evaporite sequence at 2,500 m depth that forms a competent regional aquitard (Marsh and Love, 2014) and as such represents an excellent regional secondary sealing unit for CO_2 storage at the site (Khan and Rostron, 2005; Palombi, 2008).

CO_2 Capture and Injection

CO$_2$ capture at the nearby SaskPower Boundary Dam power plant (Figure 15.1) began in October,

2014 at a rate of approx. 2400 tonnes/day, whereas CO_2 injection at the Aquistore site was initiated in April, 2015. The bulk of the captured CO_2 is delivered to the Weyburn oil field via pipeline for the purpose of enhanced oil recovery (EOR), with any surplus CO_2 being sent to the Aquistore site for storage. The Aquistore storage site anticipates receiving 250–300 ktonnes of CO_2 over the first few years of operation at variable rates of up to 800 tonnes per day. As of October 31, 2016, approx. 100 ktonnes of CO_2 had been injected.

Table 15.1 Summary of geophysical measurements made at the Aquistore site and key results

Method	Results
4.0 Characterization and baseline	
4.1 3D Surface seismic	Seals are intact. Local fault reactivation unlikely. Suitable storage site.
4.2 Surface electromagnetics	Regional 1D resistivity structure. Method suited for monitoring shallow CO_2 leakage or 100s Mtonnes injection at approx. 3000 m depth.
4.3 Crosswell seismic tomography	Detailed image of the reservoir between injection and observation wells.
4.4 Logging	Characterization of reservoir properties and CO_2 partitioning.
4.5 InSAR, GPS, tilt meters	Background surface deformation of 0.2–1.0 cm/year and 0.1–2.0 μrad tilt is comparable to expected signal levels for 0.5 CO_2 Mtonnes/year.
4.6 Surface gravity	Background variability of ±20 μGal makes method suited for monitoring shallow CO_2 leakage or 100s Mtonnes injection at 3000 m depth.
5.0 Technology/methodology tests	
5.1 Distributed acoustic sensing VSP	VSP images obtained are comparable to conventional geophones VSP.
5.2 Borehole gravity	No significant borehole gravity density difference with 6500 tonnes injected 150 m away.
5.3 Borehole to surface electromagnetics	Measured SNR compared with modeled response suggests approx. 100 ktonnes CO_2 detectable.
5.4 Seismic reflection interferometry	Promising but requires improvements for time-lapse monitoring at 3000 m depth.
6.0 Monitoring	
6.1 3D surface seismic	200 m wide anomaly in the upper Deadwood unit corresponding to approx. 16 ktonnes CO_2. No CO_2 observed in Black Island or lower Deadwood where approx. 4 and 16 ktonnes injected.
6.2 3D VSP	Similar to 3D surface results.
6.3 Passive seismic monitoring	No significant injection-related seismicity ($M_w > -1$) detected during first 17 months of CO_2 injection, and no smaller magnitude seismicity ($M_w > -3$) during the first 8 months.

CO_2 is transported to the Aquistore site by pipeline from the capture unit, which is located approx. 3 km to the east (Figure 15.1). As designated by the provincial regulating agency, CO_2 is injected at pressures of less than 90% of the estimated formation fracturing pressure (gradient of 14.9 kPa/m). The initial ambient pressure and temperature conditions in the reservoir were approx. 115°C and approx. 35 MPa. The temperature of the injected CO_2 is variable depending on the time of year. Well perforations are located over four depth intervals within the reservoir corresponding to the zones of highest permeability/porosity as determined by well logging.

3D Geological Characterization and Baseline Measurements

Design of the CO_2 storage and monitoring program for the Aquistore site required determination of how large an area would be appropriate for site characterization and subsequent active monitoring during CO_2 injection operations. The regional geological architecture near the storage site was well known from large-scale compilations (e.g., Whittaker et al., 2004 and references therein). Initial characterization of the prospective CO_2 storage site was conducted through interpretation of an existing set of five 2D seismic profiles in conjunction with geophysical/geological logs from nearby wells (RPS Boyd PetroSearch, 2011). The seismic profiles covered an area of approx. 140 km^2 and showed no evidence for features that would disqualify the site for storage. A 3D geological model was then constructed that formed the basis for fluid-flow simulations using maximum anticipated injection rates for 25 years (e.g., Whittaker and Worth, 2011). The resultant CO_2 footprint from these simulations was approx. 3 km in diameter for injection rates of up to 1600 tonnes/day. Based on this result, detailed geophysical monitoring was focused on an area of approx. 3 × 3 km.

Specific details of the geology at the storage site have been obtained by 3D seismic imaging (White *et al.*, 2016), magnetotellurics (McLeod *et al.*, 2014), crosswell tomography, analysis of core samples (EERC, 2014), well logs, and injection tests from the two wells (Rostron *et al.*, 2014). In addition to characterization, these various surveys provide baseline measurements (see next section) for the purposes of time-lapse measurements. Results from each of these studies are described in the text that follows.

3D Surface Seismic Data

3D seismic data were acquired at the Aquistore site in 2012 to provide the lithostratigraphic framework for a 30-km^2 area (Figure 15.1) around the injection well (see White *et al.*, 2016 for details). The specific objectives were (1) to map the Precambrian basement structure and overlying sedimentary section; (2) to delineate the storage reservoir and associated caprock/primary seal; (3) to image the regional continuity of the secondary seal (Prairie Formation); and (4) to look for subvertical pathways for potential upward migration of CO_2 from the reservoir.

An east–west section from the 3D seismic volume is shown in Figure 15.2. Interpretation of the 3D data was aided by correlation with the Aquistore well logs (Figure 15.3). Along with consideration of the local seismotectonics, this interpretation led White *et al.* (2016) to conclude that the site has the essential geological features required for large-scale CO_2 storage. The conclusion was based on the following: (1) The site is located in a region of very low natural seismicity with the nearest demonstrably seismogenic fault zone located approx. 200 km away. (2) The clastic reservoir is part of a regionally extensive formation and is at least 200 m thick at the storage site. (3) The primary reservoir seal – a 15 m thick shale caprock— is laterally

continuous and shows no evidence for vertical faulting, as is also true for the regional evaporitic secondary seal (Prairie Formation) that is greater than 150 m thick and has no salt dissolution features. (4) Above the Prairie Formation lie 1500 m of Middle Devonian to Lower Cretaceous strata that underlie an additional 1000 m of Upper Cretaceous and younger sedimentary rocks including thick tertiary sealing units: the Watrous Formation (approx. 120 m), Colorado Group (>185 m), and Bearpaw Formation. (5) Evidence is absent of any vertical faults that extend above the Silurian section (or approx. 2700 m depth). (6) Whereas a local subvertical basement fault has been interpreted at the site, it is oriented at an azimuth of 75°–85° relative to the regional maximum horizontal stress, making reactivation during CO_2 injection unlikely. A flexure observed in overlying Cambrian to Silurian sedimentary rocks does not appear to be ruptured or faulted.

Surface-Based Electromagnetic Methods

Surface-based magnetotelluric (MT; e.g., Vozoff, 1991) and controlled-source electromagnetic (CSEM) methods provide an alternative to seismic methods for defining the properties of the host geological structures and for monitoring underground CO_2 storage, although with generally lower spatial resolution (e.g., Gasperikova and Hoversten, 2006). Preinjection broadband MT, audio-frequency MT (AMT), and CSEM surveys were conducted at the Aquistore site in August, 2013, November, 2014 (MT and AMT only), and November, 2015 with the specific objectives of imaging the host geological structure, defining the background EM noise levels and testing the resolution and repeatability of natural-source and controlled-source EM methods (McLeod *et al.*, 2013; McLeod *et al.*, 2016). The MT and CSEM recording sites (Figure 15.1) were located along a profile passing near the CO_2 injection well. The CSEM utilized a 30 Amp, horizontal bipole transmitter that was located approx. 3.5 km northeast of the injection well. The transmitter produced a rectangular waveform with frequencies between 0.008 and 32 Hz. Additional MT sites were located off the central part of the profile to provide areal coverage.

The observed MT responses are consistent with previous studies within the Williston Basin (e.g., Gowan *et al.*, 2009). The 1D resistivity from log-based inversion of the MT data is shown in

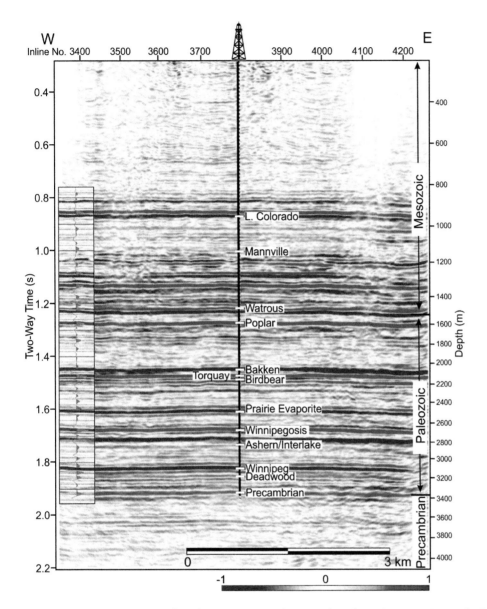

Figure 15.2 West-to-east section from the 3D time migrated seismic volume located approx. 100 m north of the injection well. The approximate location of the injection well is marked by the derrick symbol. Inset (left) shows the tie between the 3D seismic data and the well-based synthetic seismogram. Seismic horizons identified by correlation with injection well geology are labeled. (Modified from White *et al.* [2016]. © Her Majesty the Queen in Right of Canada, as represented by the Minister of Natural Resources, 2017.)

Figure 15.4. McLeod (2016) concluded that (1) MT impedance estimates obtained from the field measurements were repeatable to within 1–3% under favorable conditions; (2) nonrepeatability can be attributed mainly to the effects of ambient electromagnetic noise rather than difficulties in replicating the field work; (3) modeling showed that the MT measurements generally do not have the sensitivity or resolution for monitoring CO_2 at the reservoir level, but that the CSEM method is suitable for monitoring large quantities of injected CO_2. For example, a 240-m thick, 50 Ωm (CO_2-saturated) reservoir interval would need to have lateral dimensions of approx. 3–5 km for the scattered electromagnetic fields to

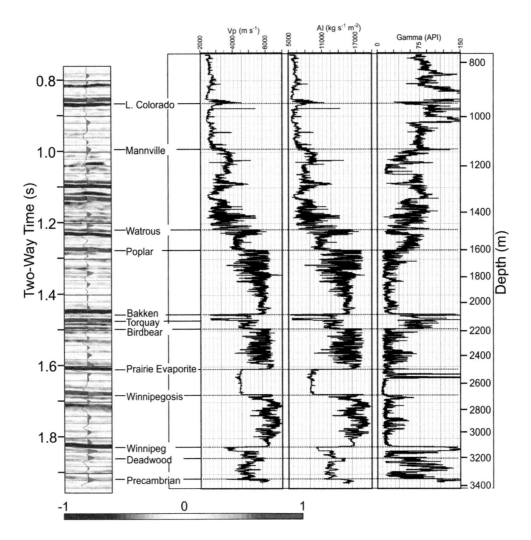

Figure 15.3 Seismic well tie. From left to right are log-based synthetic seismic trace (gray) superposed on the 3D seismic data at the injection well location, tops of a subset of geological units, and logs of compressional wave velocity (V_p), acoustic impedance (AI), and natural gamma ray. The nonlinear depth scale is determined from the two-way travel times using the velocities from the well-tie process. (Modified from White *et al.* [2016]. © Her Majesty the Queen in Right of Canada, as represented by the Minister of Natural Resources, 2017.)

produce detectable phase anomalies of 0.8°–1.9°. This suggests that the surface CSEM methodology is better suited for monitoring for CO₂ leakage into shallower, more conductive geological formations.

Crosswell Seismic Tomography

The highest resolution seismic images of the reservoir are provided by crosswell tomography (e.g., Sato *et al.*, 2011). At the Aquistore site, the tomographic velocity

image (Figure 15.5) provides meter-scale definition of the reservoir but is limited to the 150-m-wide 2D zone between the injection and monitoring wells. In a more complex reservoir setting, details on this scale might be useful in defining reservoir heterogeneity, but in the Aquistore setting the primary utility of crosswell imaging is in monitoring details of how the CO₂ plume initially spreads. Although saturation logs (e.g., Reservoir Saturation Tool [RST] or Pulsed Neutron Decay [PND] logs) in the observation well are capable

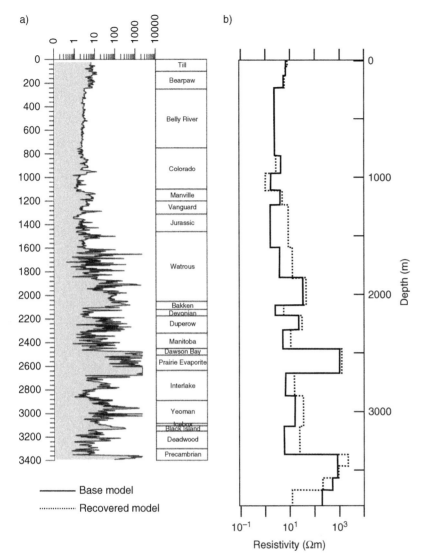

a) Base model
.... Recovered model

Figure 15.4 (a) Resistivity log from observation well with geological units (reproduced from McCleod, 2016). (b) Resistivity obtained by 1D iterative linearized inversion of MT data using the base model as the starting model. (Modified from Figure 8.7 of McLeod (2016). © Her Majesty the Queen in Right of Canada, as represented by the Minister of Natural Resources, 2017.)

of determining CO_2 arrival and time-lapse saturation profiles, crosswell seismic tomography and reflection imaging are potentially capable of providing a CO_2 saturation cross-section between the wells. Other downhole methods (gravity, EM) are also capable of providing this type of information. Thus, the main objective in acquiring the crosswell seismic tomogram was to provide a baseline survey for further time-lapse imaging.

The baseline crosswell seismic survey was acquired in March, 2013 (White *et al.*, 2014). The crosswell data were recorded using a 20-level hydrophone array

deployed in the injection well and a vibratory source in the observation well. Source and receiver positions were occupied at 1.5-m intervals over the depth range of approx. 2890–3375 m. At each source location, the source was operated eight times with a 2.6-second linear upsweep of 100–800 Hz. The resultant eight records were stacked to increase signal-to-noise ratio (SNR). Multiple passes of the source over the entire depth range (2889–3374 m) were required for each 28.5-m interval covered by the receiver array between 2953 and 3372 m.

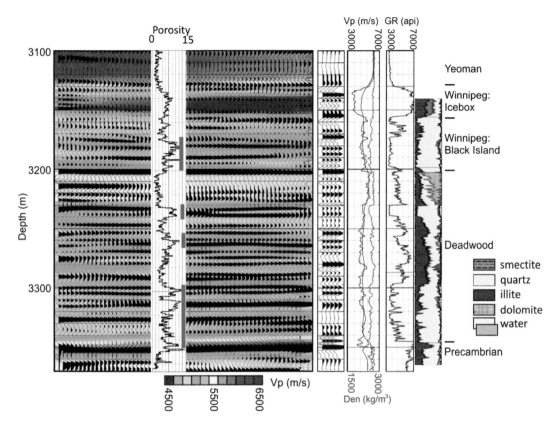

Figure 15.5 Composite 2D seismic reflection/tomographic velocity image for the zone between injection and observation wells. Horizontal exaggeration is approx. 1.5:1.0. Porosity log and perforated intervals (blue bars) are overlain. Also shown (from left to right) for the reservoir interval are a log-based synthetic seismogram, V_p (black) and density (red) logs with a $V_p(z)$ curve (blue) derived from the velocity image superposed, the gamma log, and a lithologic log. (© Her Majesty the Queen in Right of Canada, as represented by the Minister of Natural Resources, 2017.)

Tube wave energy was strong in the raw correlated data and had to be removed during processing. 3D anisotropic travel-time tomography (e.g., Washbourne et al., 2002) was applied using travel time picks for both first breaks and reflection events. The starting structural model used for tomography was derived from the formation tops in the two wells. Reflection events could be recognized and were picked on amplitude versus angle (AVA) gathers. A complementary reflection image was obtained by processing the upcoming reflection events.

Logging

An extensive suite of geophysical well logs was acquired in both the injection and observation wells before and after casing and cementation of the wells (Rostron et al., 2014). The purpose of

acquiring these logs was to characterize rocks within the storage complex and to determine baseline properties prior to the start of CO_2 injection. The log suite comprised resistivity, gamma ray, dipole sonic, density, neutron porosity, elemental and natural gamma ray spectroscopy, and nuclear magnetic resonance. Baseline CO_2 saturation (pulsed neutron) logs were acquired over the reservoir interval. In situ stresses were partially characterized by measurement of reservoir formation pressures and two mini-frac tests. Temperature–depth profiles were measured continuously using the fiber optic distributed temperature sensing (DTS) systems attached to the tubing string in the injection well and behind casing in the observation well.

Figure 15.5 shows examples of the logs over the reservoir interval. As can be seen, the injection well perforation intervals are located within the better quality sand units. Downhole CO_2 flow measurements from a spinner survey conducted shortly after the start of injection indicated that approx. 10%, 45%, and 45% of the injected CO_2 was flowing respectively into the Winnipeg, upper Deadwood, and lower Deadwood formations.

InSAR, GPS, Tilt Meters

Ground deformation monitoring has proven to be a powerful tool for identifying nonconformance in large-scale CO_2 storage projects (e.g., Rutqvist et al., 2010; Vasco et al., 2010). A total of nine surface-deformation monitoring sites (Figure 15.1) covering a total area of approx. 8 km^2 were deployed at the Aquistore site and comprise some combination of GPS receivers, InSAR reflectors, and tilt meters. These stations were deployed in 2012–2013 providing two to three years of preinjection baseline data.

From June 2012 to October 2014, space-borne differential interferometric synthetic aperture radar (DInSAR) monitoring allowed measurement of the surface deformation field associated with background natural and anthropogenic processes (Samsonov et al., 2015; Czarnogorska et al., 2016). Vertical and horizontal ground deformation was measured with local maximum rates of ±1.0 and ±0.5 cm/year and with precision of 0.3 and 0.2 cm/year (2σ), respectively. Areas of high background elevation changes are associated with processes such as erosion, groundwater withdrawal and recharge, mining operations, and land reclamation. Average vertical velocities determined from GPS monitoring are -2.0 mm/year (Craymer et al., 2015), consistent with the regional background velocity predicted from a Canada-wide velocity solution (Craymer et al., 2011). The GPS and InSAR velocities are generally in good agreement.

Simple poroelastic analytic modeling (Mathias et al., 2009; Xu et al., 2012a) was conducted for the Aquistore site by Samsonov et al. (2015), who estimated that the maximum expected vertical deformation will be centered on the injection well and will be less than 1.6 cm for CO_2 injection at a rate of 0.5 Mtonnes/year. Vertical deformation in close proximity to the injection well exceeded 0.5 cm as early as one year after the start of injection, and vertical deformation exceeding 1.0 cm is predicted to distances of approx. 6000 m after 25 years of injection.

Model-based tilt estimates exceed 10 µrad, which is significantly greater than background tilt noise (Earth tide and precipitation effects) of 0.1–2.0 µrad measured at the site (Wang, 2015). These estimated signal levels are small but detectable and are based on CO_2 and pressure communication being restricted to the reservoir zone. If this is not the case (e.g., CO_2 migrates or pressure communicates with shallower depths) then greater ground deformation can be expected.

Gravity

Time-lapse gravity measurements have been utilized for monitoring water-table levels (Liard et al., 2011), oil field water injection (Brady et al., 2006), and CO_2 injection monitoring (Sherlock et al., 2006; Alnes et al., 2008). The first objective of surface-based gravity measurements at the Aquistore site was to determine the maximum background seasonal gravity variations. Typically, these variations can be on the order of ±10 µGal and are generally attributed to changes in ground water levels (Liard et al., 2011). The second objective was to monitor for gravity changes associated with the injection of CO_2. Anticipated CO_2-injection-induced gravity changes are expected to be on the order of a few µGal for Mtonne-scale CO_2 quantities at the reservoir level, and thus are generally below the noise level. However, in the case of upward migration of CO_2 into the overburden, the associated gravity changes (10's of µGal; e.g., White, 2012: p.176) can significantly exceed background noise levels.

Gravity measurements were made at the Aquistore site in fall 2013, spring 2014, fall 2014, and fall 2015. Up to 11 sites were visited in each of the surveys. All measurements were made with an A10 absolute gravimeter with estimated uncertainties of approx. 5 µGal. The mean magnitude of measured gravity differences for 10 Aquistore sites from fall 2013 to spring 2014 was 18 µGal. This is significantly larger than gravity changes of 0.34 µGal/year due to regional subsidence (-0.2 cm/year) or maximum differences of ±3 µGal/year due to local vertical ground motions of ±1.0 cm/year (see previous section). The measured gravity variability at the Aquistore sites may result from local "transient" mass transfer signals associated with changes in total water storage, where 10 µGal would be equivalent to a 24 cm thick layer of water. The variability in water table levels from local monitoring wells for May 2013, October 2013, and June 2014 is 18 cm (Klappstein, pers. comm., 2016). For 20–40% porosity, this would result in gravity difference of only 2–4 μGal. Thus, changes in the water table alone are insufficient to explain the observed gravity differences. In regard to CO_2 monitoring, related gravity changes would exceed background noise only if large volumes of CO_2 migrate to depths of less than 1000 m. For example, 1.5 Mtonnes of CO_2 at 600 m depth would produce a gravity anomaly of 60 μGal (White, 2012).

Technology/Methodology Tests

Distributed Acoustic Sensing Data Acquisition

Early successful results of VSP recording using distributed acoustic sensing (DAS; e.g., Mesteyer et al., 2011) provided the incentive to include a casing-conveyed fiber optic cable in the designs for the injection and observation wells that were drilled at the Aquistore site in 2012. A multimode fiber cable was included in the original well design, but a single-mode fiber cable was added for the sole purpose of DAS acquisition. The fiber optic cable in the observation well (Figure 15.1) extends from the surface to a depth of approx. 2700 m. As was subsequently demonstrated at the Aquistore site, comparable DAS data can be acquired using either single or multimode fibers, as reported in Harris et al. (2015, 2016) and Miller et al. (2016).

A direct comparison of VSP data acquired using DAS and clamped geophones was conducted in 2013 as part of the baseline survey prior to any CO_2 injection. Figure 15.6a compares raw data for DAS and geophones. The same seismic arrivals (direct wave and reflections) can be seen in both data but the geophone data clearly have a higher SNR; 45 dB as compared to 15 dB for the DAS data (Harris et al., 2017). Furthermore, the bandwidth (for SNR >1) of the geophone data (approx. 5–150 Hz) is greater than that for the DAS data (approx. 0–80 Hz). Figure 15.7 compares 2D depth-migrated images obtained for both DAS and geophone data. Despite the lower SNR, the DAS data produce images that are comparable to the geophone data and, because of the greater deployment interval for the DAS, provide imaging capability at much shallower depths.

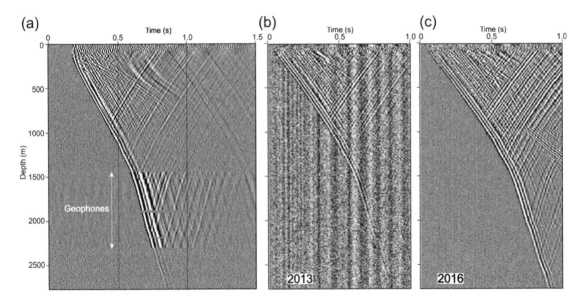

Figure 15.6 (a) Raw DAS seismic data and vertical-component geophone data for a single shot gather in 2013. The shot point is located 470 m from the observation well. Geophone data are substituted for DAS data in the depth range of 1455–2295 m. There are 963 single-mode DAS traces spaced at 2-m intervals and 57 vertical-component geophone traces spaced at 15-m intervals. (b) A comparison of RAW DAS data for shots at the same shot location in 2013 and 2016 surveys (SP4608-3800). (© Her Majesty the Queen in Right of Canada, as represented by the Minister of Natural Resources, 2017.)

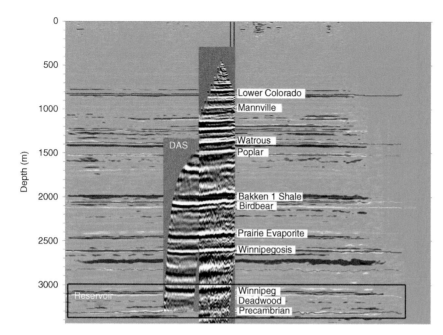

Figure 15.7 Eight-fold 2D-migrated VSP images for the 2013 DAS and vertical-component geophone data superposed on a slice from the migrated 3D surface data (green/blue/red) along an east–west line. Geological formations are labeled. The geophone image shows the same area as the DAS image, but has been shifted left to allow comparison. (Modified from Harris *et al.*, 2016. © Her Majesty the Queen in Right of Canada, as represented by the Minister of Natural Resources, 2017.)

The first monitor VSP was acquired in February 2016, again using both DAS and downhole geophones. The improvement in the DAS acquisition technology from 2013 to 2016 is clearly visible in Figure 15.6b and 15.6c, where the SNR is much improved. The results of the time-lapse imaging from this survey are discussed in the "4D VSP" section.

In addition, a 35-day short-term test of passive seismic monitoring was conducted using DAS in April 2015. The purpose of this experiment was to assess the utility of DAS as a potential alternative to downhole geophone systems for microseismic monitoring.

Downhole Gravity

Borehole gravity measurements for the purposes of CO_2 monitoring have been investigated in a number of studies to date (e.g., Sherlock *et al.*, 2006; Gasperikova and Hoversten, 2008; Dodds *et al.*, 2015). Borehole gravity measurements were made in the Aquistore observation well during 2014–2015. After several attempts, data were successfully acquired with a prototype Bluecap[TM] borehole gravity meter in August 2015 with resolution of less than 1 µGal and a repeatability of 10 µGal (Black *et al.*, 2016). Notably, this gravimeter was operating under high-temperature conditions (115°C). The survey occurred four months after the start of CO_2 injection when a total of approx. 6500 tonnes had been injected. The borehole survey was tied to absolute gravity by making a measurement at a surface station occupied by an A5X absolute gravity meter. Tool depth positions are accurate to 1–2 cm, resulting in depth-related gravity errors of approx. 1–2 µGal.

In the absence of preinjection baseline measurements, it is not possible to consider time-lapse gravity changes for the August 2015 survey. As a substitute, Black *et al.* (2016) compared borehole gravity (BHG) densities calculated for the postinjection survey with open hole $\gamma\gamma$ log densities acquired prior to any CO_2 injection. The density differences (Figure 15.8) represent a combination of mass changes away from the wellbore and near-wellbore fluid invasion-related changes. At the time of the survey, PND logs showed no CO_2 saturation at the observation well, and thus the presence of any CO_2 would contribute to mass changes only away from the wellbore. Brine and supercritical CO_2 have a density difference of 520 kg/m³ at reservoir P, T conditions and thus zones where CO_2 replaces brine will result in a local mass deficit. The BHG-$\gamma\gamma$ density differences (Figure 15.8b) range from –50 to 40 kg/m³. Zones where CO_2 replaces brine away from the wellbore should produce a negative density difference that would reach a maximum of –46 kg/m³ when the CO_2 arrived at the observation well. The positive density differences provide a measure of non-CO_2-related variability in the BHG-$\gamma\gamma$ density differences, as only negative differences would be expected due to the presence of CO_2. This indicates that the uncertainties are

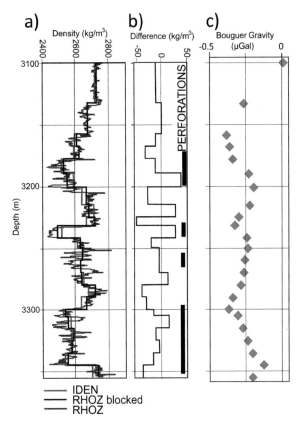

Figure 15.8 (a) Comparison of preinjection open-hole γγ density logs (RHOZ, RHOZ blocked) and the borehole gravity (BHG) density values (IDEN) for the observation well. (b) Density differences between BHG density values for the postinjection gravity survey and the open hole γγ densities for (RHOZ blocked) logs acquired prior to any CO_2 injection. Perforated intervals are indicated by the black bars. (c) Interval Bouguer gravity measurements (μGal) for a density of 2620.8 kg/m³. Both (a) and (b) are modified from Black et al. (2016).

comparable to the signal (maximum of –46 kg/m³) we are looking for. Also, a comparison of the injection intervals with zones of negative density difference shows no correlation, suggesting that the gravity effect of 6500 tonnes of CO_2 injected 150 m away from the observation well was too small to detect. Further validation of gravity (or density) differences with the approach of CO_2 will have to wait for the first monitor survey, which will use the 2015 survey as a baseline.

Controlled-Source Borehole to Surface EM Survey

A variety of controlled-source borehole-based EM methods were considered for application at the

Aquistore site (e.g., Daley et al., 2015). Early plans to deploy casing-conveyed borehole electrodes were abandoned owing to complexities of deploying casing at the reservoir depth (>3200 m). As an alternative, a borehole-to-surface electromagnetic (BSEM) method was chosen for testing that utilizes surface-based source and receivers (Hibbs, 2015). The high conductivity steel well casing is utilized to channel electric current from a surface source to the deep reservoir. The purpose of the survey was to test the suitability of this methodology for time-lapse CO_2 monitoring – specifically, whether adequate SNR could be achieved to allow detection of the very small signals that would be expected due to CO_2 injection.

A BSEM survey was conducted in August to September of 2013 (Hibbs, 2015) prior to the start of CO_2 injection. Capacitively coupled electric-field sensors were deployed along two 1000-m lines emanating radially from the location of the CO_2 injection well (Figure 15.9), each with 13–15 sensor locations. A 20 A source transmitted a square wave with fundamental frequency of 0.5 Hz for at least one hour for each receiver position. Hibbs (2015) used noise measurements made during the survey to assess whether the surface measurements would have enough sensitivity to detect a CO_2 saturated zone within the reservoir. The measured noise was compared with the DC response at the surface predicted by modeling (method of Schenkel and Morrison, 1994) for a model that included a 150-m radius, 30 m thick CO_2 saturated zone. A uniform background conductivity of approx. 0.4 S/m (2.6 Ωm) was used consistent with the conductivity at the reservoir depth (3200–3400 m) from well logs. The DC response for a 20 A source predicted a difference of 0.1–0.2% which, after adjustment based on skin depth, predicts that a change of approx. 0.01% will be observed at the surface owing to the presence of CO_2 in the reservoir. Accounting for the very high SNR achieved with the survey geometry, a SNR ≥ 10 is achieved out to 240 m offset from the injection well and a detectable signal (SNR ≥ 2) continues out to approx. 600 m. Notably, the recorded signal was reduced by a factor of 10 or less only over the distance of 80–1140 m from the injection well. This suggests that this technique might be applied over a much broader area if higher SNR can be achieved by increasing the transmitter current and/or extending the signal averaging time.

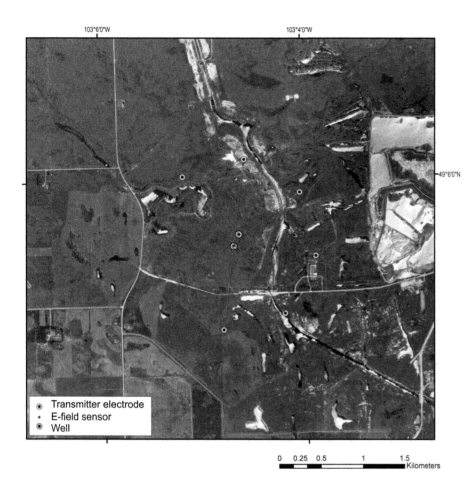

Legend:
- Transmitter electrode
- E-field sensor
- Well

0   0.25   0.5   1   1.5
Kilometers

Seismic Interferometry

Feasibility studies of seismic interferometry for CO_2 storage monitoring are limited (e.g., Boullenger *et al.*, 2015; Gassenmeier *et al.*, 2015; Xu *et al.*, 2012b). Passive seismic reflection interferometry has been conducted at the Aquistore CO_2 storage site to test its ability to image the subsurface to the depth of the storage reservoir (3200 m) and to assess its potential for time-lapse imaging (Cheraghi *et al.*, 2017). Data from two passive monitoring lines (L1 and L2 in Figure 15.1) were analyzed for 23 days of recording in June 2014 and 13 days in February 2015. Virtual shot gathers retrieved for noise panels of up to one hour in length were subjected to a common midpoint (CMP) processing sequence to obtain 2D reflection images. The best image obtained using this approach is shown in Figure 15.10 where it is compared with a colocated controlled-source image from Roach *et al.* (2015) and a log-based synthetic seismogram.

The passive 2D seismic image (Figure 15.10) bears moderate resemblance to the 2D section extracted from a controlled-source 3D seismic cube. The ambient noise image and well-based synthetic seismograms show moderate correlation values of approx. 0.5–0.65 in the two-way time range of 0.8–1.5 s. The quality of the ambient noise image is limited by the unidirectional and weak nature of the local noise source, which is most likely the local power plant and nearby associated industrial activity, all located approx. 3 km to the east. The best ambient noise images were achieved by maximizing the duration of the utilized noise panels. Significant differences observed in passive images – at times separated by eight months – indicate that substantial improvement in the repeatability of the images will be required before this methodology can be used for time-lapse imaging at the Aquistore site. Cheraghi *et al.* (2017) suggest that possible improvements

269

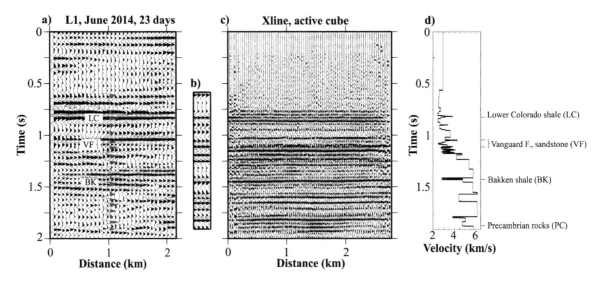

Figure 15.10 (a) 1D log-based synthetic seismogram superposed on L1 ambient noise image (from Cheraghi *et al.*, 2017). (b) Synthetic seismogram generated from a wavelet extracted from the image shown in (a) convolved with borehole log data from the Aquistore site. The superposed wiggle trace in (a) is the same trace shown in (b). (c) Cross-line slice from the 3D active-source migrated seismic cube (from Roach *et al.*, 2015) at location of L1. The passive and controlled-source images have been filtered to 30Hz and the shape of the amplitude spectra were generally matched within the passband. The active-source images shown in (c) were generally devoid of frequencies lower than 10 Hz prior to filtering. (d) Velocity log used to generate the synthetic traces in (b). (Figure is modified from Cheraghi *et al.* [2017]. © Her Majesty the Queen in Right of Canada, as represented by the Minister of Natural Resources, 2017.)

might be achieved by using the areal array of geophones of the Aquistore permanent array for 3D imaging and by strategic selection of time periods for analysis to better match the spectral characteristics of the predominant noise sources.

Monitoring Results
3D Time-Lapse Surface Seismic Data

A permanent sparse areal array of buried geophones (Figure 15.11) deployed at the Aquistore CO_2 site has been used for 4D seismic monitoring (White *et al.*, 2015). The use of permanent buried geophones was designed to improve the repeatability of the time-lapse data by reducing the effects of near-surface variability, ensuring consistent receiver coupling, eliminating intersurvey positioning errors and increasing the SNR. The sparse permanent array also economizes monitoring by minimizing mobilization and deployment costs, allowing use of the geophones for multiple purposes (e.g., controlled-source surveys and passive monitoring) and accommodating flexible on-demand surveys.

Prior to the start of CO_2 injection, the permanent array was used to acquire a total of three 3D

dynamite seismic surveys in March 2012, May 2013, and November 2013 (referred to as Baseline, Monitor 1, and Monitor 2 surveys). The objective of acquiring these data was to assess the data repeatability and overall performance of the permanent array. Comparing the raw data from the first two of these surveys (Baseline and Monitor 1) with a conventional high-resolution 3D Vibroseis survey, White *et al.* (2015) concluded that (1) a 6- to 7-dB increase in SNR is achieved for the buried geophones relative to surface-deployed geophones and an additional 20-dB increase is observed for single dynamite shots compared to a Vibroseis source; (2) the permanent array data has excellent repeatability with a mean normalized root mean square (NRMS) value of 57% for the raw baseline – monitor difference; (3) shot gathers have a higher NRMS variance (18%) than do receiver gathers (7%); (4) raw data repeatability is improved by a factor of 3 over comparable surface geophone data acquired at a nearby location. The use of a permanent sparse buried array demonstrably achieved reduced ambient noise levels and increased data repeatability, both of which are keys to successful 4D seismic monitoring.

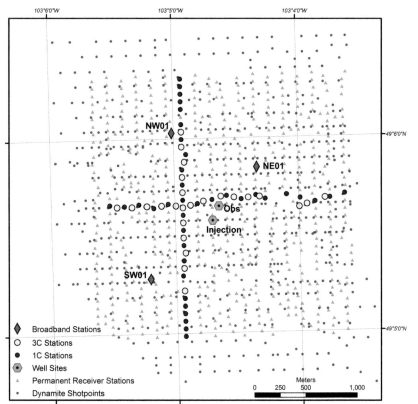

Aquistore Seismic Monitoring Components

Broadband Stations
3C Stations
1C Stations
Well Sites
Permanent Receiver Stations
Dynamite Shotpoints

Meters
0   250   500   1,000

Figure 15.11 Aquistore permanent geophone array. The positions of sensors forming the 2D permanent array are shown along with the shot points used for the 3D time-lapse surface and VSP surveys. Also shown are the continuous-recording passive monitoring lines (L1 and L2), and the locations of the surface broadband seismograph stations. (© Her Majesty the Queen in Right of Canada, as represented by the Minister of Natural Resources, 2017.)

The first post-CO_2-injection 3D time-lapse seismic survey (Monitor 3) was conducted at the Aquistore CO_2 storage site in February, 2016 (Roach *et al.*, 2017). By this date, 36 ktonnes of CO_2 had been injected within the reservoir. Monitor 3 data were processed using the same "4D-friendly simultaneous" processing flow used by Roach *et al.* (2015) for the previous surveys. Excellent repeatability was achieved among all surveys with global NRMS (GNRMS) values of 113–119% for the raw prestack data relative to the baseline data. GNRMS values decreased during processing to values of approx. 10% for all of the monitor final cross-equalized migrated data volumes.

Figure 15.12 shows Monitor 3 NRMS difference maps for three levels of the reservoir. Inspection of corresponding pre-CO_2-injection monitoring maps (included in Roach *et al.*, 2017) indicates that the NRMS maps are most reliable for the upper two reservoir units (Black Island and upper Deadwood) whereas the lower Deadwood map has some zones of

reduced repeatability (labeled as *N* in panel c). As summarized in Roach *et al.* (2017), the following observations can be made in respect to Figure 15.12: (1) No significant NRMS changes occur within the Black Island unit. (2) A 200-m-wide zone in the immediate vicinity of the injection well occurs within the upper Deadwood unit, where NRMS values range from 10% up to 25%. (3) In the lower Deadwood unit, no significant NRMS changes are observed near the injection well. Large-amplitude NRMS changes at distance from the injection well (labeled N) are zones of reduced reliability and thus are discounted as not being significant. In summary, there is clear evidence from the time-lapse seismic for CO_2 within the upper Deadwood zone, whereas evidence is absent for CO_2 residing in either the Black Island or lower Deadwood intervals. The latter could be the result of only small quantities of CO_2 residing within these zones, a distribution of CO_2 that is below the seismic detection threshold, and/or the depth-dependent sensitivity and reliability of the data (Roach *et al.*, 2017).

271

a)

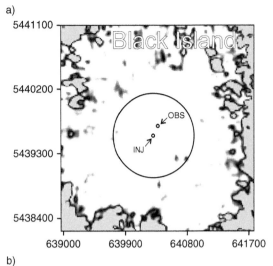

b)

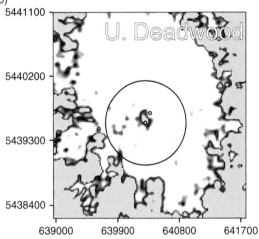

c)

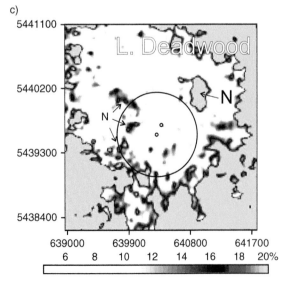

Figure 15.12 NRMS amplitude difference values determined within the (a) Black Island Formation, (b) upper Deadwood Formation, and (c) lower Deadwood Formation. Shown are the mean values within a 10-ms window of the 10-ms NRMS after 3 × 3 smoothing has been applied. The injection (INJ) and observation (OBS) wells are labeled in a). A 600 m radius circle is shown for scale. "N" labels in (c) identify anomalies that are due to time-lapse noise at this level. (Modified from Roach *et al.*, 2017.) The high NRMS values toward the edges of the images are an expected result of the survey geometry and represent an increase in time-lapse noise due primarily to lower data fold toward the periphery of the area. Northing and easting coordinates (in metres) are labelled on the figure axes. (© Her Majesty the Queen in Right of Canada, as represented by the Minister of Natural Resources, 2017.)

As the cumulative amount of injected CO_2 increases, the associated time-lapse seismic changes will more generally exceed detection thresholds and thus should provide a more definitive picture of how the CO_2 is distributed within and among the various reservoir units. Further assessment of the existing time-lapse results can be made by considering the partitioning of the 36 ktonnes of CO_2 according to the spinner log results (see the section "Logging"). The time-lapse seismic results can be interpreted in the following manner. CO_2 is undetected in the Black Island unit where approx. 4 ktonnes (10% of total) of CO_2 has been injected. This amount is clearly below the seismic detection threshold. In contrast, the pronounced time-lapse anomaly observed in the upper Deadwood unit corresponds to approx. 16 ktonnes (45% of total) of CO_2. Minimum CO_2 plume thickness estimates for this zone based on 1D seismic modeling are 4–10 m for assumed CO_2 saturations of 50–100% (Roach et al., 2017). The lower Deadwood unit has similarly received approx. 16 ktonnes of CO_2, but no significant time-lapse difference is observed. The absence of an observed time-lapse difference in the lower Deadwood interval is likely due to a combination of lower seismic repeatability in this zone and CO_2 injection occurring over a larger depth interval.

4D Vertical Seismic Profile

3D vertical seismic profile (VSP) data were acquired at the same times as the November 2013 and the February 2016 3D surface surveys (Harris et al., 2016, 2017). The number of dynamite shots was increased for the VSP surveys to 680 as compared to 260 for the surface-only 3D surveys (Figure 15.11). For the VSP data, the November 2013 survey constitutes the Baseline, and the February 2016 survey is the Monitor 1 survey. These correspond to the Monitor 2 and Monitor 3 3D surface seismic surveys of Roach et al. (2017). The VSP survey parameters were designed to reliably image a subsurface CO_2 plume residing within a radius of approx. 500 m from the injection well. The VSP data were acquired with a DAS system using a casing-conveyed fiber optic cable permanently cemented in place during well completion (Harris et al., 2016). 4D VSP monitoring has the same general objectives as the 4D surface seismic but it is designed to provide higher resolution near-borehole images of the reservoir zone during the initial stages of injection (e.g.,

O'Brien et al., 2004). VSP also allows the ability to directly tie reflections to geological horizons at the wellbore, to verify velocity–depth profiles using zero-offset shot direct arrivals, to estimate the downgoing wavelet, and reduce ghosting effects (Hardage, 2000; Kuzmiski et al., 2009).

Results for the VSP-Monitor 1 survey (Figure 15.13) are compared to the corresponding 3D surface results for a horizon within the lower part of the reservoir (upper Deadwood Formation). Near the center of the images in the vicinity of the CO_2 injection well, a clear NRMS amplitude difference is observed in both of the images where the amplitude difference is well above the background noise level. This zone likely represents the CO_2 plume where the product of CO_2 saturation and plume thickness achieve the threshold level required to produce seismic amplitude differences. 3D seismic modeling of CO_2 flow simulation results for 27 ktonnes of CO_2 injected have been conducted by Harris et al. (2016). The seismic modeling results show that a 10 m thick zone in the upper Deadwood with CO_2 saturations of 50% or more produces NRMS values exceeding 1.0 in this zone of the corresponding simulated seismic volume. Note that the amplitude of the NRMS anomaly is much higher in the VSP image than the surface-seismic image owing to the higher resolution of the VSP. However, the repeatability of the VSP is lower than the surface seismic largely due to the lower SNR of the 2013 DAS system (cf. Figure 15.6).

The seismically imaged CO_2 plume is asymmetric about the injection well and extends 200 m northward. The observed asymmetry in the CO_2 plume at this level contrasts with early flow simulations (Jiang et al., 2016) and will require adjustment of the permeability characteristics of the existing geological model. The VSP image suggests that the CO_2 plume may reach the observation well (in contrast to the surface image) within the upper Deadwood interval, but with lower saturation based on the strength of the anomaly. PND logs acquired at the time of the February monitor survey suggest that a thin zone of CO_2 had reached the observation well within the upper Deadwood interval.

Passive Seismic Monitoring

Passive seismic monitoring has been ongoing at the Aquistore site since August, 2012. This provides a record of almost three years of local background **273**

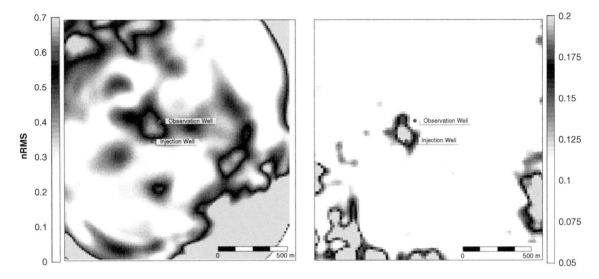

Figure 15.13 Comparison of NRMS difference maps (fractional rather than percentage) for (a) 3D VSP and (b) surface 3D for a horizon within the upper Deadwood formation. (Modified from Harris *et al.*, 2017.) The dashed circle (for scale) has a radius of 150 m. The high NRMS amplitude differences observed toward the edges of the VSP image represent an increase in time-lapse noise toward the periphery of the imaging area. (© Her Majesty the Queen in Right of Canada, as represented by the Minister of Natural Resources, 2017.)

seismicity in advance of CO_2 injection that started in April, 2015 and continuous monitoring since then (Stork *et al.*, 2016; White *et al.*, 2017). Passive monitoring has been conducted using a number of different systems each tailored for distinct purposes. First, three telemetering broadband surface seismographs (Guralp CMG-40 T 0.1–50 Hz seismometers) are deployed within approx. 1 km of the injection well (Figure 15.11). These stations are designed to provide near real-time event detection capability for either microseismic activity ($M_w > 0$–1) or potentially larger magnitude induced seismic events. Second, are two orthogonal lines (L1 and L2 in Figure 15.11, 2.5 km each) comprising 65 10-Hz geophones (1C and 3C in Figure 15.11) buried at 20 m depth. Data from these stations are retrieved once per month. This continuously monitoring array is designed to provide a complete record of local seismicity for magnitudes of $M_w > -1$. These data are amenable to multichannel data enhancement methods improving the overall sensitivity of the monitoring system. Third, a downhole geophone array was deployed during the first eight months of CO_2 injection (Nixon *et al.*, 2017). Five 3C (15-Hz) clamping geophones were deployed in the observation well (Obs in Figure 15.11) over a depth of 2850–2910 m (that is approx. 250 m above the reservoir) and operated

semicontinuously at internal temperatures of 115°C. Near-surface orientation shots (1 kg dynamite) were detonated to determine the orientation of the geophones in the well. The downhole monitoring was intended to extend the magnitude threshold for event detection to $M_w > -3$ so that a complete record of microseismic activity could be achieved for comparison with geomechanical modeling and reservoir response to injection.

Event detection methods including direct inspection, application of short-term average/long-term average (STA/LTA) detection algorithms, and multichannel stacking methods have variably been applied to the data from each of the monitoring systems itemized earlier. Although many teleseismic, regional, and local industrial seismic events have been detected during four years of recording, no significant injection-related seismicity ($M_w > -1$) was detected during the first 17 months of CO_2 injection, nor smaller magnitude seismicity ($M_w > -3$) during the first eight months when the downhole array was active. This result is in stark contrast to some other CO_2 injection projects where swarms of microseismic events are detected (e.g., Decatur; Kaven *et al.*, 2015), but is not unlike the low levels of seismicity associated with injection activities at the nearby Weyburn EOR field (Verdon *et al.*, 2016).

Conclusions

A broad spectrum of geophysical methods has been employed at the Aquistore CO_2 storage site for the purposes of site characterization, baseline definition, and monitoring. 3D seismic data interpreted in conjunction with well logs have been used to define the architecture of the CO_2 storage container and assess its suitability for long-term storage. These data show that over an area of 30 km^2 the reservoir zone and caprock are generally continuous and devoid of major structures. A basement fault is identified with a flexure in the overlying Cambrian to Silurian strata, including the reservoir and caprock. These units do not appear to be ruptured or faulted. Evidence is absent of any vertical faults that extend above the Silurian section (or approx. 2700 m depth) where the regional evaporitic secondary seal (Prairie Formation at 2500 m) is at least 150 m thick and has no salt dissolution features. The subvertical basement fault is oriented nearly orthogonal to the regional maximum horizontal stress making reactivation during CO_2 injection unlikely. Naturally occurring seismicity in the area is very low. The nearest seismogenic fault zone of any significance is located approx. 200 km away.

Baseline measurements made for many of the geophysical techniques have identified the natural background noise levels at the site. These are compared with modeled CO_2-related signals to assess whether CO_2-produced signals will exceed or be distinguished from background time-lapse noise. This is considered in the context of either outlining the CO_2 plume within the reservoir or detecting CO_2 that may have migrated to shallower levels. In all cases, to varying degrees, the ability of the various geophysical monitoring methods to detect CO_2 is dependent on the amount and depth of the CO_2.

Time-lapse seismic methods are shown to provide the most detailed maps of the CO_2 plume within the reservoir. This is due to the inherent resolving power of the technique and the good to excellent repeatability at the Aquistore site. Repeatability of 3D surface seismic was excellent (NRMS of approx. 0.10) and was good (approx. 0.30) for 3D DAS-based VSP. The 3D DAS-VSP repeatability is expected to improve over time, as the technology has improved significantly since the baseline acquisition in 2013. The first time-lapse seismic monitoring results have demonstrated

the ability to detect or image a zone containing less than approx. 16 ktonnes in the reservoir at greater than 3000 m depth and to distinguish the interval within which this CO_2 resides.

The nonseismic surface measurement techniques employed at the Aquistore site have resolving power that is generally inversely proportional to the depth of the resident CO_2. That is, the deeper the CO_2, the larger the injected quantity required to produce a discernible signal at the surface. This limits the effectiveness of these methods for plume monitoring at the Aquistore site where the reservoir (at 3150 m depth) is very deep as compared to most CO_2 storage sites. Very large quantities of CO_2 within the reservoir are required for the plume to be monitored using either surface-based electromagnetic methods or gravity based on a comparison of modeling results and background noise levels measured at the Aquistore site. For example, CSEM modeling of a CO_2-saturated, 240 m thick 50 Ωm reservoir interval estimates that the plume would have to be approx. 3–5 km in diameter (or 200–600 Mtonnes for 5% porosity) to produce a detectable response. The low sensitivity of these methods makes them better suited to leakage detection of CO_2 that has migrated to shallower depths. For example, 1.5 Mtonnes of CO_2 at 600 m depth would produce a gravity anomaly of approx. 60 µGal which significantly exceeds the ±20 µGal background noise signal at the Aquistore site.

Background surface deformation measured prior to the injection of CO_2 includes regional vertical subsidence of –0.2 cm/year, local vertical deformation of up to 1.0 ± 0.3 cm/year, and background tilt noise of up to 2.0 µrad. These values compare with modeled maximum vertical deformation of < 2 cm and tilt of approx. 10 µrad for CO_2 injection at a rate of approx. 0.5 Mtonnes/year. The maximum annual injection rate at the Aquistore site achieved in the first 20 months of operation was approx. 0.1 Mtonnes/year, and thus any associated surface uplift due to deep injection during this period is likely below the detection threshold.

New or developing geophysical methods have been tested at the Aquistore site with mixed results. DAS-based VSP surveys have proven to be suitable for time-lapse imaging of the CO_2 plume. As compared to VSP acquired using conventional wire-line geophones, DAS provided higher spatial sampling and a larger recording aperture, but generally lower SNR.

275

However, the achievable quality of DAS data should improve significantly for future surveys as a result of continued technological improvements since 2013 that significantly increase SNR and broaden the frequency bandwidth. Downhole borehole gravity measurements demonstrated resolution of less than 1 µGal and a repeatability of 10 µGal. Although these values are comparable to surface-based gravity measurements, the proximity of the measurements to the CO_2-associated density anomalies and large distance from near-surface seasonal effects increases both the repeatability and the expected CO_2-related signal. Controlled-source borehole-to-surface EM data acquired at the Aquistore site compared to the modeled surface DC response for a 30 m thick, 150 m radius CO_2 plume in the reservoir suggest that detection is possible. This provides a detection level that exceeds that of surface-based EM methods by two to three orders of magnitude. Seismic reflection interferometry applied to passive data from the Aquistore site produced images of the subsurface having moderate correlation values (approx. 0.5–0.65) at depths of 900–2200 m (0.8–1.5 s) with well-based synthetic seismograms. However, the image quality was poor at the reservoir depth (3200 m), indicating that substantial improvement in the repeatability of the images is required before this methodology can be used for time-lapse imaging.

The first post-CO_2 injection 3D time-lapse seismic monitoring data were acquired in February, 2016. At this time, a total of 36 ktonnes of CO_2 had been injected. Zones of CO_2 within the upper Deadwood Formation were imaged in both the 3D surface and VSP time-lapse data volumes. It is estimated that the imaged plume corresponds to approx. 16 ktonnes of CO_2 within this zone. The seismically imaged CO_2 plume (representing a minimum size) extends for a distance of approx. 200 m from the injection well and is consistent with a relatively thin zone (4–10 m) for assumed CO_2 saturations of 50–100%. The imaged plume is asymmetric about the injection well with apparent preferential flow up-dip to the NNW and subparallel with structural fabric in the underlying Precambrian basement. This is in contrast to initial reservoir simulation models, which are characterized by CO_2 distributions that generally show axial symmetry about the injection well. During four years of passive seismic recording at the Aquistore site, no significant local seismicity has been detected.

No injection-related seismicity ($M_w > -1$) was detected during the first 17 months of CO_2 injection using a surface based array, nor was any smaller magnitude seismicity ($M_w > -3$) during the first eight months when a downhole array was operating within 250 m of the reservoir. This result is consistent with the low levels of seismicity associated with injection activities at the nearby Weyburn EOR field.

Acknowledgments

The seismic interpretation presented here was based partly on an internal report provided by RPS Boyd Petrosearch. The Aquistore CO_2 Storage Project is managed by the Petroleum Technology Research Centre, Regina, Canada. Funding for this research was provided in part by Natural Resources Canada, Sustainable Development Technology Canada, Saskatchewan Go Green Fund, Enbridge, SaskPower, Schlumberger Carbon Services, Korea National Oil Corporation, SaskEnergy, and Research Institute of Innovation Technology for the Earth. Contributions to the work described in this chapter were made by M. Craymer, S. Samsonov, J. Craven, J. Silliker, B. Roberts, M. Czarnogorska, J. Henton, I. Ferguson, J. McCleod, K. Harris, L. Roach, C. Samson, S. Cheraghi, A. Stork, D. Schmitt, C. Nixon, K. Worth, A. Leniuk, T. Daley, M. Robertson, D. Miller, B. Freifeld, J. Cocker, I. Marsden, R. Chalaturnyk, G. Zambrano, C. Hawkes, B. Rostron, and G. Klappstein.

References

Alnes, H., Eiken, O., and Stenvold, T. (2008). Monitoring gas production and CO_2 injection at the Sleipner field using time-lapse gravimetry. *Geophysics*, **73**(6): 155–161.

Bachu, S., and Hitchon, B. (1996). Regional-scale flow of formation waters in the Williston Basin. *AAPG Bulletin*, **80**(2): 248–264.

Black, A., Hare, J., and MacQueen, J. (2016). Borehole gravity monitoring in the Aquistore Boundary Dam CO_2 sequestration well. Expanded Abstract, *Society of Exploration Geophysicists*.

Boullenger, B., Verdel, A., Paap, B., Thorbecke, J., and Draganov, D. (2015). Studying CO_2 storage with ambient-noise seismic interferometry: A combined numerical feasibility study and field-data example for Ketzin, Germany. *Geophysics*, **80**(1): Q1–Q13.

Brady, J. L., Hare, J. L., Ferguson, J. F., *et al.* (2006). Results of the world's first 4D microgravity surveillance of a waterflood – Prudhoe Bay, Alaska. *SPE Paper 101762, 2006 SPE Annual Technical Conference and Exhibition.*

Brunskill, B. (2004). CO_2 disposal potential in the deep subsurface of southeast Saskatchewan. Internal report, prepared for the University of Regina and Saskatchewan Power Corporation.

Cheraghi, S., White, D. J., Draganov, D., Bellefleur, G., Craven, J. A., and Roberts, B. (2017). Passive seismic reflection interferometry: A case study from the Aquistore CO_2 storage site, Saskatchewan, Canada. *Geophysics*, **82**(3): B79–B93.

Craymer, M., Henton, J., Piraszewski, M., and Lapelle, E. (2011). An updated velocity field for Canada. *Eos Transactions*, AGU, **92**(51), Fall Meeting Supplement, Abstract G21A-0793.

Craymer, M., White, D., Pirazewksi, M., Zhao, Y., Henton, J., Silliker, J., and Samsonov, S. (2015). First results of geodetic deformation monitoring after commencement of CO_2 injection at the Aquistore underground CO_2 storage site. *Eos Transactions*, American Geophysical Fall Meeting, Paper G33A-1132.

Czarnogorska, M., Samsonov, S., and White, D. (2016). Airborne and spaceborne remote sensing for Aquistore carbon capture and storage site characterization. *Canadian Journal of Remote Sensing*, **42**(3): 274–290. DOI:10.1080/07038992.2016.1171131.

Daley, T. M., Smithy, J. T., Beyer, J. H., and LaBrecque, D. (2015). Borehole EM monitoring at Aquistore with a downhole source. In K. F. Gerdes (ed.), *Carbon dioxide capture for storage in deep geological formations: Results from the CO_2 Capture Project*, Vol. **4**. Thatcham, Berks, UK: CPL Press, 733–758.

Dodds, K., Krahenbuhl, R., Reitz, A., Li, Y., and Hovorka, S. (2015). Evaluating time-lapse borehole gravity for CO_2 plume detection at SECARB Cranfield. In K. F. Gerdes (ed.), *Carbon dioxide capture for storage in deep geological formations: Results from the CO_2 Capture Project*, Vol. **4**. Thatcham, Berks, UK: CPL Press, 651–664.

Energy and Environmental Research Center. (2014). Geologic modeling and simulation report for the Aquistore Project: University of North Dakota & U.S. Department of Energy National Energy Technology Laboratory.

Gasperikova, E., and Hoversten, G. M. (2006). A feasibility study of nonseismic geophysical methods for monitoring geologic CO_2 sequestration. *Leading Edge*, **25**, 1282–1288. DOI:10.1190/1.2360621.

Gasperikova E., and Hoversten, G. M. (2008). Gravity monitoring of CO_2 movement during sequestration: Model studies. *Geophysics*, **73**(6): WA105–WA112.

Gassenmeier, M., Sens-Schönfelder, C., Delatre, M., and Korn, M. (2015). Monitoring of environmental influences on seismic velocity at the geological storage site for CO_2 in Ketzin (Germany) with ambient seismic noise. *Geophysical Journal International*, **200**: 524–533.

Gowan, E. J., Ferguson, I. J., Jones, A. G., and Craven, J. A. (2009). Geoelectric structure of the northeastern Williston basin and underlying Precambrian lithosphere. *Canadian Journal of Earth Sciences*, **46**: 441–464.

Hardage, B. A. (2000). *Vertical seismic profiling: Principles.* Oxford: Elsevier Science.

Harris, K., White, D., Samson, C., Daley, T., and Miller, D. (2015). Evaluation of distributed acoustic sensing for 3D time-lapse VSP monitoring of the Aquistore CO_2 storage site, GeoConvention, May 4–8, Calgary, Expanded Abstract.

Harris, K., White, D., Melanson, D., Samson, C. and Daley, T. (2016). Feasibility of time-lapse VSP monitoring at the Aquistore CO_2 storage site using a distributed acoustic sensing system. *International Journal of Greenhouse Gas Control*, **50**: 248–260. http://dx.doi.org/10.1016/j.ijggc.2016.04.016.

Harris, K., White, D., Samson, C. (2017). 4D VSP monitoring at the Aquistore CO_2 storage site. *Geophysics*, **82**(6), M81–M96.

Hibbs, A. D. (2015). Test of a new BSEM configuration at Aquistore, and its application to mapping injected CO_2. In K. F. Gerdes (ed.), *Carbon dioxide capture for storage in deep geological formations: Results from the CO_2 Capture Project*, Vol. **4**. Thatcham, Berks, UK: CPL Press, 759–776.

Jiang, T., Pekot, L. J., Jin, L., *et al.* (2016). Numerical modelling of the Aquistore CO_2 Project: GHGT-13 Proceedings. *Energy Procedia*, **114**.

Kaven, J. O., Hickman, S. H., McGarr, A. F., and Ellsworth, W. L. (2015). Surface monitoring of microseismicity at the Decatur, Illinois, CO_2 sequestration demonstration site. *Seismological Research Letters*, **86**(4): 1–6. DOI:10.1785é0220150062.

Kent, D. M., and Christopher, J. E. (1994). Geological history of the Williston Basin and Sweetgrass River Arch. In G. D. Mossop and I. Shetsen (comp.), *Geological atlas of the Western Canada sedimentary basin.* Canadian Society of Petroleum Geologists and Alberta Research Council, 421–430.

Khan, D. K., and Rostron, B. J. (2005). Regional hydrogeological investigation around the IEA Weyburn CO_2 Monitoring and Storage Project Site. In E. S. Rubin, D. W. Keith, and C. F. Gilboy (eds.), *Proceedings of the 7th International Conference on Greenhouse Gas Control Technologies*, September 5–9,

2004, Vancouver, Canada, Vol. 1: *Peer Reviewed Papers and Overviews*, 741–750.

Kuzmiski, L., Charters, B., and Galbraith, M. (2009). Processing considerations for 3D VSP: *CSEG Recorder*, **34**(4). https://csegrecorder.com/articles/view/processing-considerations-for-3d-vsp

Liard, J., Huang, J., Silliker, J., Jobin, D., Wand, S. and Doherty, A. (2011). Detecting groundwater storage change using micro-gravity survey in Waterloo Moraine, Proceeding. *Geohydro*.

Marsh, A., and Love, M. (2014). Middle Devonian prairie evaporite: Isopach map, map 132. In Regional Stratigraphic Framework of the Phanerozoic in Saskatchewan, Saskatchewan Phanerozoic Fluids and Petroleum Systems Project, Saskatchewan Ministry of the Economy, Saskatchewan Geological Survey, Open File 2014–1, set of 156 maps.

Mathias, S., Hardisty, P., Trudell, M., and Zimmerman, R. (2009). Approximate solutions for pressure buildup during CO_2 injection in brine aquifers. *Transport in Porous Media*, **79**: 265–284.

McLeod, J. (2016). *Magnetotelluric and controlled-source electromagnetic pre-injection study of Aquistore CO_2 sequestration site, near Estevan, Saskatchewan, Canada*. M.Sc. thesis, University of Manitoba.

McLeod, J., Craven, J. A., Ferguson, I. J., Roberts, B. J., Bancroft, B., and Liveda, T (2014). Overview of the 2013 baseline magnetotelluric and controlled-source electromagnetic geophysical study of CO_2 sequestration at the Aquistore site near Estevan, Saskatchewan. Geological Survey of Canada, Open File 7617. DOI:10.4095/293921.

McLeod, J., Craven, J. A., Ferguson, I. J., and Roberts, B. J. (2016). Overview of the 2013 and 2014 baseline magnetotelluric and controlled-source electromagnetic studies of CO_2 sequestration at the Aquistore site near Estevan, Saskatchewan. Geological Survey of Canada, Open File 8101.

Mestayer, J., Cox, B., Wills, P. (2011). Field trials of distributed acousticsensing for geophysical monitoring. In 81st Annual International Meeting of the Society of Exploration Geophysics, Expanded Abstract, 4253–4257.

Miller, D. E., Daley, T. M., White, D., *et al.* (2016). Simultaneous acquisition of distributed acoustic sensing VSP with multi-mode and singlemode fibre optic cables and 3 C-geophones at the Aquistore CO_2 storage site. *CSEG Recorder*, 28–33.

Nixon, C. G., Schmitt, D. R., Kofman, R. S., *et al.* (2017). Experiences in deep downhole digital micro-seismic monitoring near 3 km at the PTRC Aquistore CO_2 Sequestration Project, 2017 Geoconvention, Expanded Abstract.

O'Brien, J., Kilbride, F., and Lim, F. (2004). Time-lapse VSP reservoir monitoring. *Geophysics*, **23**(11), 1178–1184.

Palombi, D. D. (2008). *Regional hydrogeological characterization of the northeastern margin in the Williston Basin*. M.Sc. thesis, University of Alberta, Edmonton, Alberta.

Roach, L. A. N., White, D. J., and Roberts, B. (2015). Assessment of 4D seismic repeatability and CO_2 detection limits using a sparse permanent land array at the Aquistore CO_2 storage site. *Geophysics*, **80**(2): WA1–WA13.

Roach, L. A. N., White, D. J., Roberts, B., and Angus, D. (2017). Initial 4D seismic results after CO_2 injection start-up at the Aquistore storage site. *Geophysics*, **82**(3). http://dx.doi.org/10.1190/geo2016-0488.1

Rostron, B., White, D., Hawkes, C., and Chalaturnyk, R. (2014). Characterization of the Aquistore CO_2 Project storage site, Saskatchewan, Canada. In 12th International Conference on Greenhouse Gas Control Technologies, GHGT-12. *Energy Procedia*, **63**: 2977–2984, DOI:10.1016/j.egypro.2014.11.320.

RPS Boyd PetroSearch. (2011). Interpretation of 2D seismic data for the Aquistore Project near Estevan, Saskatchewan: Internal Report for Petroleum Technology Research Centre, Regina, 54.

Rutqvist, J., Vasco, D., and Myer, L. (2010). Coupled reservoir geomechanical analysis of CO_2 injection and ground deformation at In Salah, Algeria. *International Journal of Greenhouse Gas Control*, **4**(2): 225–230.

Samsonov, S., Czarnogorska, M., and White, D. (2015). Satellite interferometry for high-precision detection of ground deformation at a carbon dioxide storage site. *International Journal of Greenhouse Gas Control*, **42**: 188–199, DOI:10.1016/j.ijggc.2015.07.034.

Sato, K., Mito, S., Horie, T., et al. (2011). Monitoring and simulation studies for assessing macro- and meso-scale migration of CO_2 sequestered in an onshore aquifer: Experiences from the Nagaoka pilot site, Japan. *International Journal of Greenhouse Gas Control*, **5**: 125–137, DOI:10.1016/j.ijggc.2010.03.003.

Schenkel, C. J., and Morrison, H. F. (1994). Electrical resistivity measurement through metal casing. *Geophysics*, **59**(7): 1072–1082.

Sherlock, D. A., Toomey, A., Hoversten, M., Gasperikova, E., and Dodds, K. (2006). Gravity monitoring of CO_2 storage in a depleted gas field: A sensitivity study. *Exploration Geophysics*, **37**: 37–43.

Stork, A. L, Nixon, C., Schmitt, D. R., White, D. J., Kendall, J.-M., and Worth, K. (2016). The seismic response at the Aquistore CO_2 injection project, Saskatchewan, Canada. *Seismological Research Letters*, **87**(2B): 477. DOI:10.1785/0220160046.

Streich, R. (2016). Controlled-source electromagnetic approaches for hydrocarbon exploration and monitoring on land. *Surveys in Geophysics*, **37**: 47–80.

Vasco, D., Rucci, A., Ferretti, A., *et al.* (2010). Satellite-based measurements of surface deformation revealfluid flow associated with the geological storage of carbon dioxide. *Geophysics Research Letters*, **37**(3): L03303, 1–5.

Vasco, D. W., Ferretti, A., Rucci, A., *et al.* (2016). *Geodetic monitoring of the geological storage of greenhouse gas emissions*. Submitted to AGU books.

Verdon, J. P., Kendall, J.-M., Horleston, A. C., and Stork, A. (2016). Subsurface fluid injection and induced seismicity in southeast Saskatchewan. *International Journal of Greenhouse Gas Control* (in press).

Vigrass, L., Jessop, A., and Brunskill, B. (2007). Regina Geothermal Project. In *Summary of Investigations 2007*, Vol. 1, Saskatchewan Geological Survey, Saskatchewan Industry Resources, Misc. Rep. 2007–4.1, CD-ROM, Paper A-2.

Vozoff, K. (1991). The magnetotelluric method. In M. N. Nabighian, (ed.), *Electromagnetic methods in applied geophysics*, Vol. 2: *Applications*. Tulsa, OK: Society of Exploration Geophysicists, 641–711.

Wang, Y. (2015). *Design, deployment, performance and baseline data assessment of surface tiltmeter array technology in Aquistore geologic CO_2 storage project*. M. Sc. thesis, University of Alberta.

Washbourne, J. K., Rector, J. W., and Bube, K. P. (2002). Crosswell traveltime tomography in three dimensions. *Geophysics*, **67**(3) (May–June): 853–871, DOI 10.1190/1.1484529.

White, D. (2012). Geophysical monitoring. In B. Hitchon (ed.), *Best practices for validating CO_2 geological storage*. Sherwood Park, AB: Geoscience Publishing, 155–210.

White, D. J. (2013a). Seismic characterization and time-lapse imaging during seven years of CO_2 flood in the Weyburn Field, Saskatchewan, Canada. *International Journal of Greenhouse Gas Control*, **16S**: S78–S94.

White, D. J. (2013b). Toward quantitative CO_2 storage estimates from time-lapse 3D seismic travel times: An example from the IEA GHG Weyburn–Midale CO_2 monitoring and storage project. *International Journal of Greenhouse Gas Control*, **16S**: S95–S102.

White, D., Roach, L. A. N., Roberts, B., and Daley, T. M. (2014). Initial results from seismic monitoring at the Aquistore CO_2 storage site, Saskatchewan, Canada. *Energy Procedia*, **63**: 4418–4423. DOI:10.1016/j.egypro.2014.11.477.

White, D. J., Roach, L. A. N., and Roberts, B. (2015). Time-lapse seismic performance of a sparse permanent array: Experience from the Aquistore CO_2 storage site. *Geophysics*, **80**(2): WA35–WA48.

White, D. J., Hawkes, C. D., and Rostron, B. J. (2016). Geological characterization of the Aquistore CO_2 storage site from 3D seismic data. *International Journal of Greenhouse Gas Control*, **54**(1), 330–344.

White, D., Harris, K., Roach, L., *et al.* (2017). Monitoring results after 36 ktonnes of deep CO_2 injection at the Aquistore CO_2 storage site, Saskatchewan, Canada. In 13th International Conference on Greenhouse Gas Control Technologies, GHGT-13. *Energy Procedia*, **114**.

Whittaker, S., and Worth, K. (2011). Aquistore: A fully integrated demonstration of the capture, transportation and geologic storage of CO_2. *Energy Procedia*, **4**: 5607–5614.

Whittaker, S., Rostron, B., Khan, D., *et al.* (2004). Theme 1: Geological characterization. In M. Wilson and M. Monea (eds.), IEA GHG Weyburn CO_2 Monitoring and Storage Project summary report 2000–2004, from the *Proceedings of the 7th International Conference on Greenhouse Gas Control Technologies*. Regina: Petroleum Technology Research Centre, 1–72.

Worth, K., White, D., Chalaturnyk, R., *et al.* (2014). Aquistore Project measurement, monitoring and verification: From concept to CO_2 injection. In 12th International Conference on Greenhouse Gas Control Technologies, GHGT-12. *Energy Procedia*, **63**: 3202–3208. DOI:10.1016/j.egypro.2014.11.345.

Xu, Z., Fang, Y., Scheibe, T., and Bonneville, A. (2012a). A fluid pressure and deformation analysis for geological sequestration of carbon dioxide. *Computational Geosciences*, **46**: 31–37.

Xu, Z., Juhlin, C., Gudmundsson, O., *et al.* (2012b). Reconstruction of subsurface structure from ambient seismic noise: An example from Ketzin, Germany. *Geophysical Journal International*, **189**: 1085–1102.

Chapter 16

Development and Analysis of a Geostatic Model for Shallow CO$_2$ Injection at the Field Research Station, Southern Alberta, Canada

Donald C. Lawton, Jessica Dongas, Kirk Osadetz, Amin Saeedfar, and Marie Macquet

Introduction

Storage of carbon dioxide (CO$_2$) in geological formations has become increasingly well recognized as a key, viable long-term method to reduce emissions of CO$_2$ into the atmosphere. Carbon capture and storage (CCS) is a strategy that is being demonstrated in western Canada by the Quest CCS Project developed by Shell and partners (Rock *et al.*, 2015) and the Aquistore Project in Saskatchewan (Rostron, 2014). The fate of an injected CO$_2$ volume is ultimately site-specific, and is determined by the dominant trapping mechanisms present. The geochemical and geomechanical properties of the reservoir and seal intervals govern how the reactive CO$_2$ volume will behave.

The Containment and Monitoring Institute (CaMI) of CMC Research Institutes Inc., in conjunction with the University of Calgary, has developed a comprehensive Field Research Station (FRS) in southern Alberta, Canada (Figure 16.1), to facilitate and accelerate research and development leading to improved understandings and technologies for geological containment and secure storage of CO$_2$. Of particular importance is the ability to detect vertical migration of CO$_2$ out of the storage complex and a change in phase of the CO$_2$ from liquid to gas as it enters the shallow subsurface. Key goals for the FRS program are to develop and refine monitoring technologies to determine the detection threshold of gas-phase CO$_2$ in the subsurface and to develop geophysical and other monitoring technologies for optimizing CO$_2$ enhanced oil recovery (CO$_2$-EOR) and to de-risk CCS in general. New approaches and innovative technologies need to be developed for sampling, measurement, and monitoring methodologies in order to provide comprehensive models of the subsurface and to identify early loss of containment or conformance of the injected CO$_2$.

At the FRS, small volumes (<600 tonnes/year) will be injected into the Basal Belly River Sandstone Formation at a depth of 296 m below the ground surface, designed to simulate CO$_2$ leakage from a deeper, large-scale CO$_2$ storage site into the overburden. Currently Alberta legislation and regulations permit geological sequestration of CO$_2$ only at depths greater than 1000 m below the ground surface, where CO$_2$ is a supercritical fluid. In the unlikely event that CO$_2$ migrates above the storage complex due to a failure of containment and conformance, it will be most easily detected at shallower depths in a gas phase, prior to its contact with the groundwater protection zone. To address these challenges, the FRS has been established to undertake research into the efficacy and evaluation of monitoring technologies at shallow to intermediate depth in a realistic field setting. The facility will be used to test new measurement, monitoring, and verification (MMV) technologies as they are developed and commercialized (e.g., fiber optic devices, slim wells, new analytical instruments for air and water analyses) with an emphasis on integrated geophysical, geochemical, and geodetic surveys as well as new approaches to the integration of volume-based geophysical data sets (e.g., 3D seismic volumes) with high-resolution point measurements (wells). This chapter provides an overview of the general development of a geostatic model for CO$_2$ injection at the FRS and outcomes of reservoir simulation of CO$_2$ injection using the physical properties of the geomodel as input. A particular focus of the chapter is on numerical modeling of the time-lapse seismic response of CO$_2$ injection over five years of injection, followed by five years of postclosure seismic monitoring.

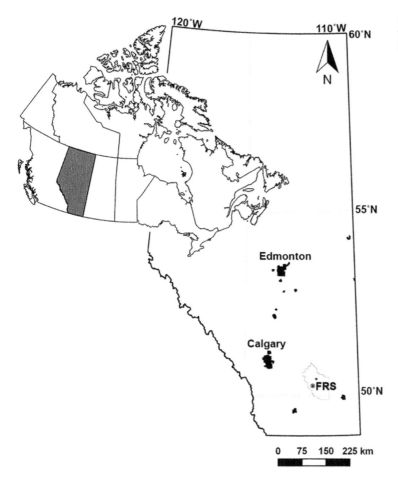

Figure 16.1 Map of Alberta, Canada showing the location of the Containment and Monitoring Institute Field Research Station (FRS).

Project Contribution

This chapter presents a case history demonstrating a methodological workflow to incorporate integrated data sets to characterize the reservoir and caprock sequence for a CO_2 injection and monitoring experiment. The work provides baseline characterization and automatic workflows that have been established to enable the static model to be updated and refined on acquisition of newer data sets during the CO_2 injection program. The simulation and modeling results will also provide insight into the anticipated plume behavior, and to guide the implementation of monitoring technologies and sampling strategy to be used for the going monitoring program at the FRS.

Site Selection and Characterization

The FRS has been constructed in Newell County, southwest of Brooks, Alberta, Canada on a 200-hectare

site with surface access provided through a 10-year lease from Cenovus Energy (now transfered to Torxen). The site is currently in its final construction stage and both the initial injection well (10-22-17-16W4) and two 350 m deep monitoring wells and three water wells all less than 90 m deep have been completed. Research activities at the FRS will be undertaken for at least 10 years.

Building a geomodel in a petroleum field that has been previously explored and produced can be highly advantageous in providing existing data sets to understand the geological and geophysical attributes of a reservoir. Such is the case for the QUEST project by Shell and partners in central Alberta, and also for enhanced oil recovery (EOR) projects such as the Teapot Dome project in the United States, and the Weyburn-Midale project in southern Saskatchewan, Canada. A desktop study known as the Wabamun Area Sequestration CO_2 Project (WASP), located west of Edmonton in Central Alberta, was conducted in a similar way to this project. In that project, a geomodel was built based on 96 wells with existing geophysical logs, seismic data sets, and limited core measurements that penetrate the reservoir in Paleozoic carbonate Nisku Formation (Eisinger and Jensen, 2009). The stratigraphic framework and 3D model grid was defined by the wireline logs and seismic data. Using petrophysical models of porosity and permeability, the model grid was then populated and upscaled using geostatistical algorithms to define the spatial distribution of the subsurface properties. Some other examples of reservoir characterization studies that have been undertaken for CO_2 injection and/or storage projects include the Sleipner project offshore Norway (Chadwick et al., 2004), potential sites in Korea described by Kim et al. (2014), and in the Gippsland Basin offshore Australia (Gibson-Poole et al., 2008).

Geological Setting

The FRS is located in the Countess Oil Field between the Bow and Red Deer Rivers, in a mixed grassland region on the Alberta Plains (Figure 16.1). This glaciated landscape is characterized by Laurentide (continental) till-covered plains of low topographic relief and internal drainage between incised river valleys where Upper Cretaceous bedrock strata crop out (Fenton et al., 1994). The FRS is located in the Interior Platform Structural Province, a nondiastrophic region of predominantly undeformed, unconformity-bounded Phanerozoic stratigraphic successions that are in places locally deformed through dissolution of Paleozoic evaporates in underlying strata, differential sedimentary compaction, and rare faults in the Precambrian basement.

At the FRS the shallow stratigraphy is composed of approx. 1000 m of interbedded sandstone and shale-dominated strata of Cretaceous age, deposited during successive regressive and transgressive sequences, associated with Laramide orogenic foreland progradation into the Cretaceous Interior seaway. The stratigraphic succession of the Upper Cretaceous strata is shown in Figure 16.2. The storage complex lies within the Belly River Group, with the injection zone being at the base of this unit, in the Brosseau Member of the Foremost Formation, also known as the Basal Belly River Sandstone (BBRS).

The Cretaceous strata unconformably overly an eroded carbonate and evaporate-dominated Paleozoic succession, approx. 1200 m thick, that itself lies unconformably over Precambrian basement rocks (Wright et al., 1994). Regional seismic data indicate minor local structure due to Paleozoic salt dissolution in the easternmost part of the study area, but the Phanerozoic sedimentary successions at the CO_2 injection site and within the anticipated extent of the CO_2 plume are locally undeformed. Petroleum is produced locally from Lower Cretaceous sandstones that occur below the horizons of interest for the CO_2 injection and monitoring program. The first injection well (Countess 10-22-17-16W4) was drilled to a depth of 550 m to characterize the overburden and the underburden within the FRS. This well has subsequently been completed as an injection well in the BBRS.

The three wells drilled at the FRS are completed in a typical Upper Cretaceous succession for southeastern Alberta (Glombick, 2010a,b, 2011a,b, 2014a,b; Glombick and Mumpy, 2014a,b). The site lies at the eastern edge of the Bow City coalfield north of Bow River and the Countess 10-22 well lies just east of the erosional edge of the Bearpaw Shale and encounters the Lethbridge Coal Zone and Dinosaur Park Fm. at the bedrock surface. The eroded top of Dinosaur Park Fm. is unconformably overlain by a granitic granule lag at the base of Holocene glacial fluvial sediments and soils. Shales of the Bearpaw Fm. outcrop on the north side of the Bow River about 5 km south of the FRS. The Dinosaur Park Fm. crops out in Dinosaur Provincial Park in the Red Deer River valley northeast of Brooks.

Belly River Group strata were deposited gradationally and conformably on top of the underlying Pakowki mudstones in shoreface marine and nonmarine depositional environments (Figure 16.2). The BBRS injection zone is at a depth of 295 m below ground surface and it is a 7-m-thick, upper shoreface, fine to medium-grained sandstone of poorly to well sorted, angular to subangular grains respectively, at the base of the Foremost Fm. (Hamblin and Abrahamson, 1996). They found significant presence of diagenetic clays and calcite cements that reduce reservoir permeability and that could provide some local barriers to vertical and lateral flow.

The overlying sealing succession of the storage complex is provided by interbedded heterolithic stratification of meter-scale interbeds of mudstone, fine-

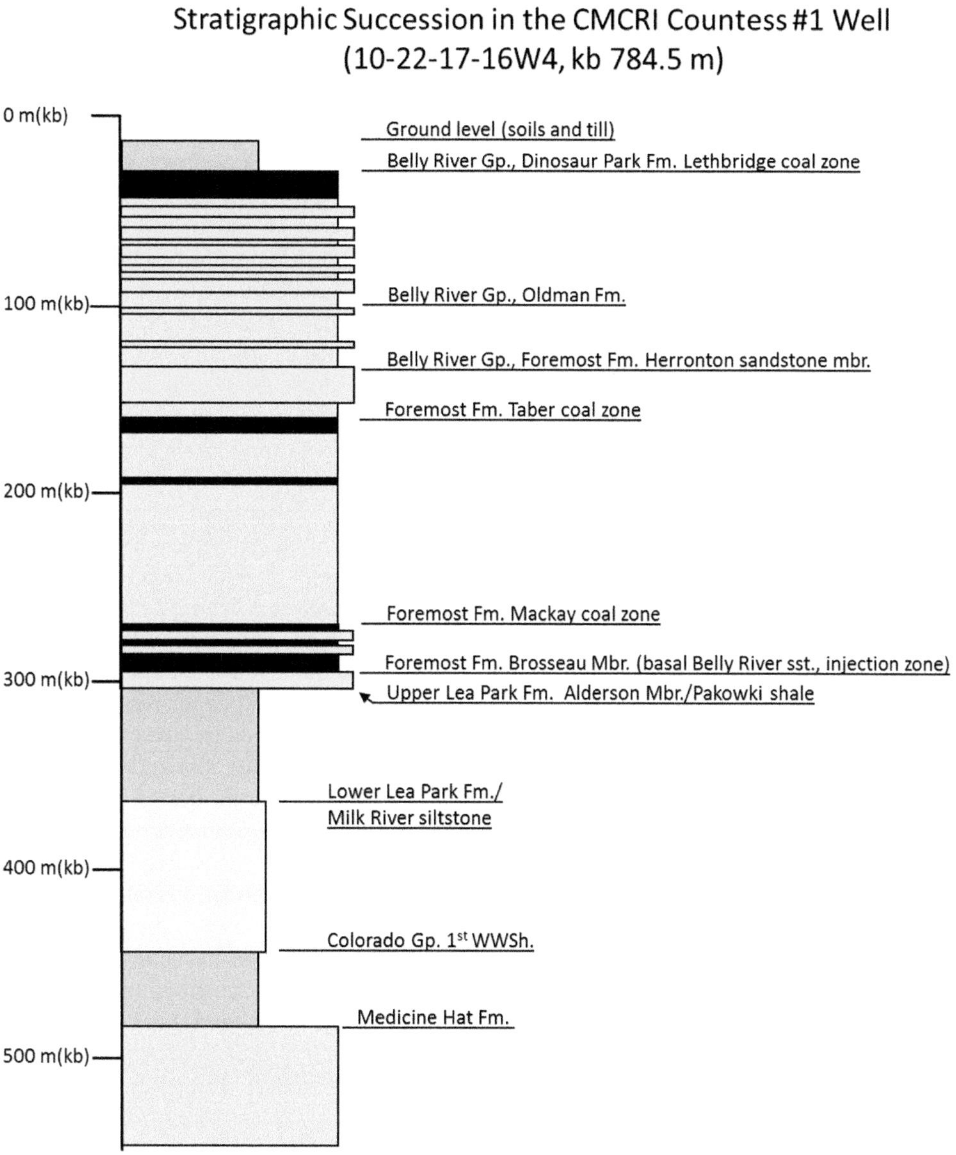

Stratigraphic Succession in the CMCRI Countess #1 Well
(10-22-17-16W4, kb 784.5 m)

Figure 16.2 Stratigraphic succession in the CMCRI Countess #1 Well (10-22-17-16W4, kb 784.5 m).

grained sandstone, and uncleated coals within the McKay coal zone that directly overlies the injection zone (Figure 16.2). Additional seals are provided by the stratigraphically higher Taber Coal zone, of similarly interbedded mudstones, fine sandstones, and uncleated coals. Together, the potential top seal has a combined average net thickness of 225 m in the study area. Groundwater and petroleum studies

(Beaton, 2003) suggest that the superior Foremost Fm. overlying the injection area has very low permeability and that it will act as an effective top seal generally, but especially at the FRS, where a 1 m thick mudstone bed directly overlies the injection zone in all the wells.

Occasionally natural gas, predominantly methane, occurs in shallow aquifers, associated with near-

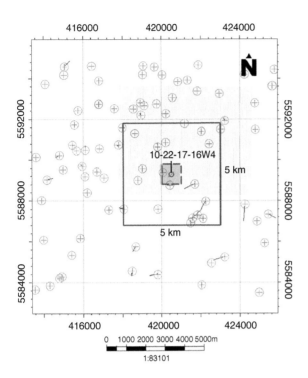

Figure 16.3 Surface map showing locations of wells and seismic surveys used to develop the geomodel. The red square depicts the 5 km × 5 km area of the initial geomodel and the dashed square depicts the detailed 1 km × 1 km geomodel that was extracted for reservoir simulation and time-lapse analysis. Colors of the wells are based on weighting distance from the 10-22 well.

surface coal beds. Some of the coal zones have much higher hydraulic conductivities compared to the clay-rich, fine-grained sandstones and mudstones within which they are interbedded. The base of groundwater protection occurs at 220 m below ground level and is above the base of the surface casing in the injection well, within the overlying seals of the storage complex.

Geomodel Development

For the development of the geomodel, legacy wells and an available seismic volume within a 13 km × 11 km area around the FRS were examined initially. This area includes 88 wells and a 3D seismic volume, as shown in Figure 16.3. Most of the legacy wells include wireline log suites with gamma ray, caliper, spontaneous potential, compressional sonic, shallow-deep resistivity/induction, bulk density, density-por-osity, and neutron-porosity well logs. These well logs were interpreted using geostatistical methods to obtain general petrophysical parameters of the storage

complex that were input into a regional geomodel and subsequently used in the reservoir simulator to model the CO_2 injection program and development of the CO_2 plume.

Two 3D seismic reflection volumes were available for the study. A 30-km^2 regional survey was provided courtesy of Cenovus Energy, and had been acquired in 1997. Figure 16.4 shows an interpreted south–north seismic section extracted from this processed seismic volume. The shallow reflectors, including the BBRS event, are reasonably flat, but deeper horizons (approx. 750 ms) show evidence of structural collapse due to dissolution of evaporites, although these structures do not appear to propagate up-section to the level of the storage complex. In order to obtain higher resolution seismic data, particularly for the shallow part of the section, a 1 km × 1 km multicomponent seismic survey was collected in 2014 by CMC Research Institutes as part of baseline studies for the FRS. A map showing the acquisition layout is shown in Figure 16.5, which also includes the layout of wells and other surface monitor-ing technologies that have been deployed. An orthogonal geometry was used for the seismic pro-gram, with shot lines oriented east–west and receiver lines oriented north–south. The shot and receiver line intervals were nominally 100 m, decreasing to 50 m within the inner 500 m × 500 m survey area, in order to ensure optimal imaging of the storage complex at 300 m depth. Shot and receiver intervals were both 10 m, yielding a 5 m × 5 m bin size. The sources used for the survey were two Envirovibes sweeping from 10 to 150 Hz over 16 s. For development of the geomodel, primarily the PP volume was used and a view of the processed data cube is displayed in Figure 16.6.

In 2015, CMC Research Institutes drilled the Countess 10-22 well and acquired a comprehensive log suite including a dipole sonic log, and also cored the storage complex. Following the core and log ana-lysis and the interpretation of the regional and newly acquired seismic data, a 5 km × 5 km geomodel was developed within the area shown in Figure 16.3, in order to place greater weighting of the petrophysical data on wells local to the FRS. Detailed reservoir and geophysical modeling was subsequently undertaken in a 1 km × 1 km area centered on the 10-22 well.

Structural and Stratigraphic Framework

Within the 5 km × 5 km surface area of the geomodel, the local structural and stratigraphic framework

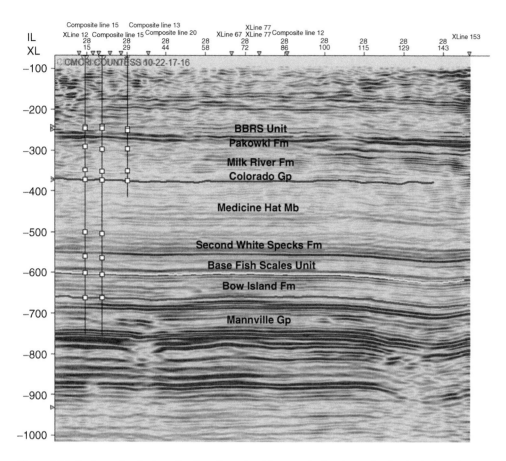

Figure 16.4 South–north seismic section from the regional P-wave seismic volume, showing key horizons interpreted and also the well-ties. The short well is the 10-22 well drilled in the center of the 1 km × 1 km detailed geomodel. The extent of the seismic survey is shown in Figure 16.3.

determined from well data was confirmed and improved through incorporating the interpretation of the seismic data sets. A miss-tie step was required to remove a 5 ms difference in seismic horizon interpretation between the regional and new 1 km × 1 km seismic volumes. Horizon tops were then verified with the depth conversion of the two 3D seismic PP volumes through log-derived velocity modeling. The two multilayered models were computed for each seismic volume, within which velocity was treated velocity as a linear function of depth over discrete intervals. The interval velocities of each subsurface zone were calculated from the time–depth relationship (TDR), a direct output from completing the seismic-to-well tie. Three seismic-to-well ties were completed, using the compressional sonic log, bulk density log, and zero-phase extracted statistical wavelets at each respective well location. No check-shot

data were available for this project; thus the TDR was developed through the completion of a sonic calibration step. The velocity model for the 2014 local 3D volume utilized the TDR produced from the 10-22 well tie, and the velocity model for the 1997 regional volume utilized the TDRs produced from the adjacent well ties. Each model was run separately to calculate the interval velocities for each subsurface horizon zone below the seismic reflection datum of 800 m above sea level.

The final framework that defines the stratigraphy in the FRS study region combines the formation tops and interpretations of the seismic horizons. The weighting regime was tailored to consider a 70% weighting on the 1997 regional seismic volume and 30% on the 2014 local seismic volume, where the formation top interpretation guided interpolation between well locations. Isopach and isochron maps

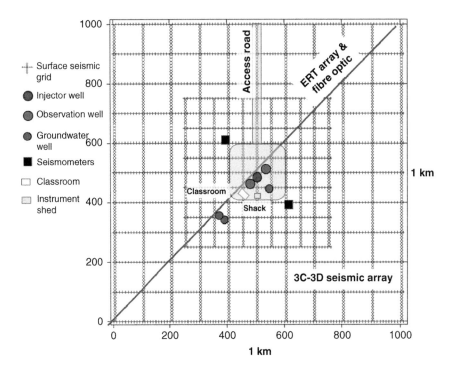

Figure 16.5 Layout of the Containment and Monitoring Institute Field Research Station, showing the seismic acquisition grid and locations of wells and monitoring equipment installations.

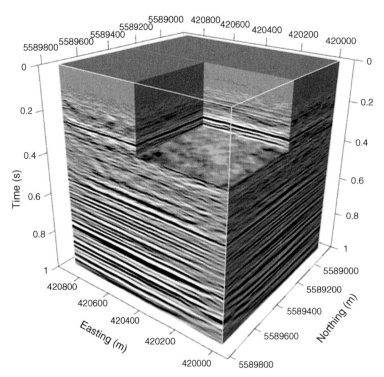

Figure 16.6 Processed PP volume from the 2014 seismic survey. Base of chair display illustrates the time slice at the top of the Basal Belly River Formation.

Table 16.1 Formation tops, two-way seismic travel times, and formation thicknesses

Formation top	TWT (ms)	Depth (m)	Thickness (m)
Foremost Fm.	—	143	152
BBRS member	245	295	7
Pakowki Fm.	250	302	61
Milk River Fm.	298	363	76
Colorado Group.	353	439	43
Medicine Hat Fm.	373	482	217
Second White Speckled Shale Fm.	500	699	88
Base Fish Scales Fm.,	560	787	40
Bow Island Fm.	600	827	112
Mannville Group	658	939	

were used to determine subsurface formation thicknesses, defined by the depth or reflection time to the top of each formation, respectively; the general relationships between seismic reflection travel times and formation tops are listed in Table 16.1.

Petrophysical Analysis

Two main reservoir properties used to characterize the BBRS and Foremost Fm. were effective porosity and intrinsic permeability. The wireline well logs of Countess 10-22 were studied using Schlumberger's elemental log analysis (ELAN) that provided calculated depth profiles of effective porosity, permeability, and rock fraction composition. The wireline logs in Countess 10-22 are displayed in Figure 16.7. The sandstone reservoir demonstrates an approximate 45 API signature on the gamma ray log, with trailing neutron– (NPSS) and density–porosity (DPSS) logs indicative of the mild clay content

characterizing the BBRS. The lack of separation in the shallow to deep induction logs (AF10-90) in both the Foremost Fm. and BBRS are generally indicative of relatively low permeability. Routine core analysis and tight rock analysis provided sample-based porosity and permeability results that permitted a well log-to-core calibration of inferred rock properties in the reservoir and seal intervals. Once calibrated and modeled, the BBRS averaged an effective porosity of 11% and intrinsic permeability of between 0.5 mD and 2 mD.

The ELAN mineralogical-derived data was unavailable at the offset wells, and thus porosity was calculated using a volume of shale (v-shale) model, where effective porosity is defined as the total porosity less the clay-bound fluid content. Intrinsic permeability is an isotropic property and was calculated using the free-fluid Timur–Coates equations (Timur, 1968; Coates and Dumanoir, 1974). Average effective porosity and intrinsic permeability values of the Foremost Fm. could not be provided for the entire 152-m-thick unit owing to heterogeneity; however, well logs and core measurements confirm low permeability values of 1.5E-4 mD for the shale caprock immediately above the BBRS.

Facies Modeling

The coal zones within the Foremost Fm. are of specific interest, as they are estimated to be one of the main laterally continuous sealing units above the BBRS. The wireline log character of the coals demonstrated artificially high neutron and low-density responses, resulting in skewed values of effective porosity and intrinsic permeability. To our knowledge of the local FRS study region, the coals are known to be water-saturated and have adsorbed volumes of CO$_2$, resulting in limited relative permeability. No visible cleating systems were observed in the core slabs or routine core analysis completed.

To constrain further the petrophysical calculations, a facies model with a simple lithology legend made to mimic the ELAN mineralogical-derived data was developed to isolate the coal zones from the sandstone, silty-sand, and shale formations. In the geomodel, where a coal interval was defined, low values of porosity and permeability were assigned (0.03 and 0.001 mD respectively, from core analysis results). The lithological interpretations are shown in Figure 16.7.

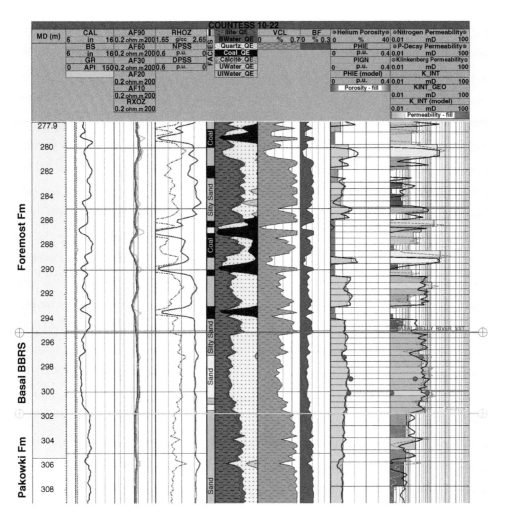

Figure 16.7 Interpreted well logs through the storage complex from the 10-22 well. Dots within the BBRS interval for the porosity and permeability columns are calibrated values from core measurements.

Geomodel Upscaling

The geomodel volume is defined by the number of cells in the Cartesian grid, and the layer thicknesses. The structural framework defined by the 3D model horizons resulted in 12 subsurface intervals. For the upscaling, the cell lateral dimensions were 25 m × 25 m and the cell thickness varied within the model, from 0.5 m in the storage complex and under-burden, to 5 m in the remaining parts of the model. The dimensions of the geomodel volume are 197 × 198 × 1557, resulting in a total of just over 60 million cells.

Variogram analysis was used in conjunction with the calculated effective porosity and intrinsic permeability well logs to populate the geomodel. The reservoir properties in each well were upscaled using an arithmetic mean, and a Gaussian Random Function Simulation (GRFS) algorithm was chosen to populate the full 3D model given the correlation statistics. It is considered to be a conditional and stationary simulation algorithm, incorporating both kriging and unconditional simulation. The conditional portion of the algorithm honors the input data and is able to model the expected variability in property distribution, whereas the unconditional simulation portion incorporates a degree of randomness with each realization, as it does not replicate the data's mean, variance, or semivariogram outputs.

In summary, the procedure for building the 3D static geomodel included the following:

- Import, management, and quality control of all data into project (wireline logs, 3D seismic volumes, well locations, deviation surveys, core measurements).

- Definition of structural framework through interpretation completed on wireline logs and 3D seismic volumes. Depth to formation tops for the upper 200 m of the section (i.e., within the zone of surface casing) was extrapolated from geological bedrock maps.

- Reservoir characterization through log analysis and petrophysical analysis to define the two main reservoir parameters, namely effective porosity and intrinsic permeability. Calculations were calibrated from core measurements with samples taken from reservoir and caprock intervals.

- Statistical analyses of the petrophysical data for effective porosity and intrinsic permeability, including normal score transformation and variograms to define the optimum distribution and range.

- Upscaling the model using the following constraints:

  . Gridding the 3D model using a vertical pillar method. The lack of any significant geological structure eliminated the need to incorporate faults into the model.

  . Defining the stratigraphy of the model from the well tops and seismic interpretation. In the final model, there were 13 model horizons that bind the top and bottom of each model zone, with a total of 12 model intervals.

  . Layering the grid cells within each interval from the bottom upwards, thus eliminating any topographic character that may propagate into the underlying zone layering. Grid cells were assigned a height of 0.5 m in the Foremost Fm., BBRS, as well as the underlying Pakowki Fm. Grid cells in the remaining model zones were assigned a height of 5 m. Each layer within a model zone is evenly distributed between the top and bottom of a model horizon.

- Locally within the model, upscaling effective porosity and intrinsic permeability well logs at well locations into the appropriate cell using an arithmetic mean method.

- Using a GFRS algorithm to populate the upscaled properties into the 3D model grid. Considered as a conditional simulation algorithm, the function honors the input data as well as inserting a degree of randomness in each model realization.

Storage Capacity of the BBRS

Storage capacity estimates were computed through a P10-50-90 framework for the 1 km^2 detailed geomodel volume, involving the Foremost Fm., BBRS, and underlying Pakowki Fm. These three cases define the conservative, typical, and optimistic estimates of the total pore volume within the BBRS. Corresponding permeability volumes were computed by kriging each 10th, 50th, 90th percentile of total pore volume. Figure 16.8 displays the P50 case for effective porosity for the 1 km^2 area extracted from the 5 km × 5 km geomodel, used as input for dynamic modeling and fluid flow simulations.

Reservoir Simulation

Reservoir simulations were undertaken to provide initial predictions of the CO$_2$ plume behavior in the reservoir and the caprock strata, based on the petrophysical properties of the geomodel. Simulations were undertaken using the GEM simulator of CMG (Computer Modelling Group Ltd.) software. GEM is an advanced equation-of-state compositional and geomechanical reservoir simulator. Perforations in the BBRS formation were placed over a vertical thickness of 7 m in the Countess 10–22 well, located at the center of the geomodel. The reservoir simulation undertaken was for CO$_2$ injection over a five-year period into the BBRS with a further five-year monitoring period to follow. The simulations involved testing multiple continuous injection scenarios on the static geomodel. Given the relatively low permeability of the reservoir, it was desired to inject at the maximum allowable bottom-hole pressure. A water injectivity test yielded a fracture closure pressure of 6.62 MPa that matches the lithostatic pressure at the BBRS depth of 300 m, based on analysis of the density log. For regulatory reasons, the maximum allowable injection pressure must be less than 90% of the fracture gradient, so in this project a maximum bottom-hole pressure of 5.8 MPa was input into the simulator. Considering this bottom-hole pressure, a minimum reservoir

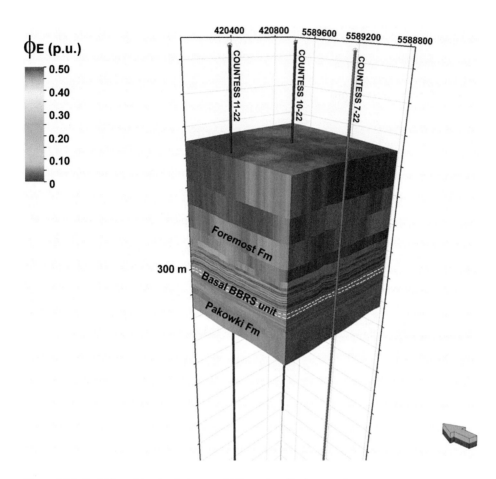

Figure 16.8 Final 1 km x 1 km detailed geomodel illustrating effective porosity.

temperature of 20°C is necessary to ensure that the CO_2 will remain in the gas phase. The injection simulation presented in this chapter used these injection parameters.

Conditioning Data

The initial 25-km^2 simulation grid consisted of more than 60 million grid cells, requiring some further upscaling to reduce computational run times. A tartan grid configuration aided in decreasing the total number of cells, as well as preserving the detailed cell dimensions along the Countess 10-22 well path. The resultant grid is defined by (124 × 124 × 73 cells), with the cell size in the X- and Y-direction set at 8 m and the vertical cell thickness remaining at 0.5 m. In the overlying and underlying intervals, the cell height increased to 51 m above and below the storage complex. Some layer discretization enabled the low

permeability coal layers to be represented in the Foremost Fm. From the static geomodel, the grid was clipped vertically to include only the BBRS and Foremost formations, as well as the underlying Pakowki Fm. seals. We did not set no-flow boundaries to the static geomodel in order to properly predict the lateral distribution and migration of the injected CO_2 plume within the geomodel, without edge effects.

Simulation Parameters

Most parameters governing the pixel-based statistical simulation were taken directly from data from the Countess 10-22 well. For the flow parameters, the estimated intrinsic permeability realizations are assumed to be isotropic laterally and the ratio of vertical to horizontal intrinsic permeability (k_v/k_h) was assumed to be 0.1, thus resulting in primarily horizontal flow of the CO_2 in the BBRS reservoir. This ratio of

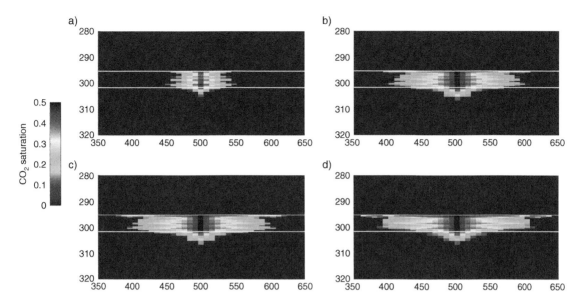

Figure 16.9 CO_2 saturation during and after the injection program. (a) After one year of injection. (b) After five years of injection. (c) After one year postclosure. (d) After five years postclosure. The top and bottom of the BBRS are shown by the white lines.

vertical to horizontal permeability is commonly used in the Alberta Basin (Schlumberger, pers. commun.). In the absence of core analysis data, the relative permeability of gaseous CO_2 and water denoting the irreducible water saturation was estimated using the Brooks–Corey approximation (Brooks and Corey, 1964) and CO_2-brine effects (Bachu and Bennion, 2008). Using capillary pressure data (Li and Horne, 2006), the minimum and critical water saturation were assumed to be 0.5. The dynamic model assumes 100% initial water saturation and does not consider geochemical or compositional phase changes, although the dynamic model does consider CO_2 gas dissolution into the formation water in the pore spaces.

Results

For the geomodel developed and the injection parameters described earlier, the reservoir simulation determined an injected CO_2 mass of 266 tonnes/year, yielding a total of 1,330 tonnes of CO_2 injected over the five-year injection program. Example sections illustrating CO_2 saturation and changes in pressure in the storage complex are displayed in Figures 16.9 and 16.10, respectively. Realizations after one year and five years of injection are shown, as well as one year and five years postinjection. The CO_2 plume shape is driven mainly by the buoyancy of

the gaseous CO_2, the k_v/k_h ratio and local variations in permeability in the model that were derived from upscaling of the well-log data.

The results illustrated in Figure 16.9 show that the injected gaseous CO_2 is contained within the BBRS. The plume has a radius of approx. 110 m after five years of injection, extending to a radius of about 140 m after five years of postinjection. The maximum saturation achieved, local to the 10-22 injector well, is close to 0.5. Interestingly, the simulations show no upward migration of the CO_2 above the BBRS due to the thin shale seal (Figure 16.7), but do demonstrate minor vertical migration downward into the Pakowki Fm. Within the first five-year injection phase, the plume migrates downward about 3 m into the Pakowki Fm. The plume shows no additional vertical migration above or below the BBRS during the five-year postclosure period.

Through the duration of the injection phase, an induced pressure plume develops in the BBRS reservoir with maximum pressure reaching the maximum bottom-hole pressure proximal to the well (Figure 16.10). The simulations show the excess pressure quickly dissipates after one-year postinjection. The pressure almost equilibrates to the baseline pressure environment after five years postinjection (Figure 16.10d).

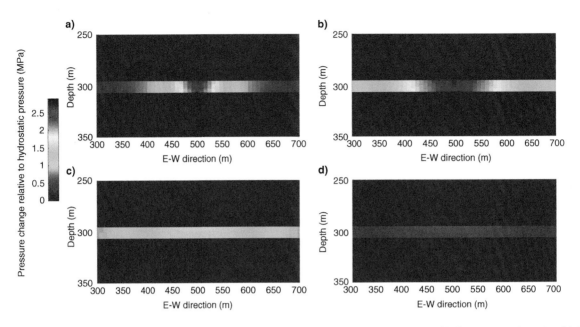

Figure 16.10 Pressure changes relative to baseline reservoir pressure due to gas-phase CO_2 injection. (a) after one year of injection. (b) After five years of injection. (c) After one year postclosure. (d) After five years postclosure.

Time-Lapse Numerical Seismic Modeling

Before starting the CO_2 injection, we modeled the feasibility of seismic monitoring of the development of the CO_2 plume using synthetic data. Initially, Gassmann fluid substitution (Gassmann, 1951) was used to generate the different elastic property models (V_p, V_s, and density). This was followed by forward seismic modeling using a finite difference approach. For this part, the geomodel needed to be rescaled to have uniform depth sampling. We used a horizontal sampling interval of 5 m and a vertical sampling interval of 1 m.

Gassmann Fluid Substitution

Gassmann fluid substitution modeling was undertaken using the approach outlined by Smith *et al.* (2003). The bulk modulus of the porous rock frame, the mineral matrix and the reservoir fluids were first computed from the dipole sonic well-log data (for V_p and V_s), density-log data, and ELAN analysis for mineral composition analysis, yielding the matrix bulk modulus. The results of the Gassmann fluid substitution are shown in Figure 16.11, as the variations of the elastic parameters between the baseline model and the model after five years of injection.

As expected, the shape of the property anomalies is highly correlated with the shape of the CO_2 saturation plume in the reservoir over the same time interval. The average decrease in P-wave velocity throughout the volume of the plume is approx. 6%, although this can reach 38% near the base of the reservoir. This large velocity change corresponds to a high-calcite/low-clay content layer at a depth of 306 m (Figure 16.7), which increases the matrix bulk modulus. The S-wave velocity increases only slightly (around 0.3%) due to an average decrease in formation density of approx. 0.6%.

Seismic Modeling

Numerical seismic data were computed using TIGER™, which is a 3D finite-difference modeling software (from SINTEF Petroleum Research). We used a simplified version of the acquisition survey shown in Figure 16.5 (36 sources and 561 receivers). The data time sampling interval was 4 ms and we used a Ricker wavelet with 40-Hz dominant frequency as the source. Data were then processed using Vista processing software through a standard processing flow including deconvolution, normal-moveout correction, common midpoint stack, and poststack migration.

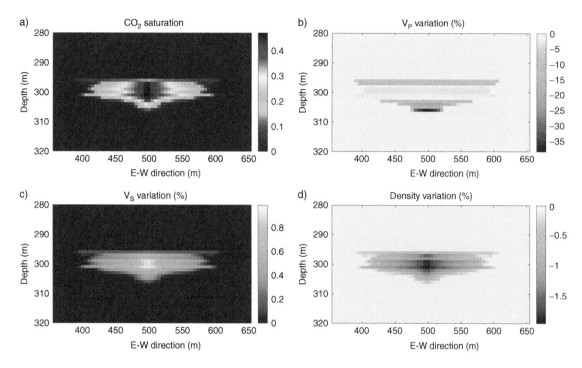

Figure 16.11 2D sections from the geomodel showing the variation of elastic parameters after five years of CO$_2$ injection, for the bottom hole pressure set to 5.8 MPa, reservoir temperature of 20°C. (a) Rescaled CO$_2$ saturation. (b) P-wave velocity change (%). (c) S-wave velocity change (%). (d) Density change (%).

Results

Figure 16.12 shows a perspective view of in-line and cross-line sections from the processed 3D numerical seismic volumes from the baseline survey and the monitor survey after five years of CO$_2$ injection. The two data sets appear to be very similar, but the time-lapse response is visible after the baseline volume is subtracted from the monitor volumes. Figure 16.13 shows amplitude difference displays of vertical sections and time slices after one year and five years of injection. The maximum lateral extent of the amplitude anomalies are shown by the black lines in Figure 16.13 and for both time periods, the lateral extent of the reflectivity anomaly in the time slices correspond well to the lateral dimensions of the CO$_2$ plume on the vertical sections. Despite the small amount of CO$_2$ injected (266 tonnes), the plume radius is 55 m after one year of injection. This plume should be detectable in the seismic data during the first year of injection. The seismic anomaly is generated by the full thickness of the CO$_2$ plume and not just by the strong P-wave velocity contrast

seen at the bottom of the reservoir (Figure 16.11a). The time slices in Figure 16.13 are were taken at a two-way reflection time of 0.21 s. Considering a mean overburden velocity of 2800 m/s, the corresponding depth is 295 m, which is the depth to the top of the BBRS reservoir.

Discussion and Conclusions

This chapter has outlined the approach and methodology used to predict the behavior of CO$_2$ injection into the Basal Belly River Sandstone at a depth of 296 m below the ground surface. This shoreface sandstone has effective porosity of 11% with intrinsic permeability of 0.5–2 mD, and is overlain by complexly interbedded shale-silt sequences with laterally extensive coal seams. The project is designed to simulate leakage of CO$_2$ from a deep sequestration site into formations at shallow to intermediate depths, and to determine the detection threshold of the CO$_2$ for various monitoring technologies. In this chapter we examined the 4D P-wave seismic response during the injection program.

293

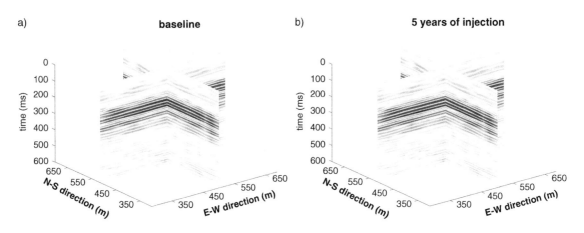

Figure 16.12 Processed 3D seismic volumes. (a) Baseline volume. (b) Monitor volume after five years of injection.

Multicomponent seismic data were recorded during the acquisition program but the converted-wave (PS) data has not yet been interpreted. Well log and core data from the Countess 10-22 well as well as petrophysical data from surrounding wells were used to characterize the BBRS reservoir and caprock strata. Reservoir simulations were then undertaken to predict the extent of the gas-phase CO_2 plume after five years of injection. With the relatively low permeability of the reservoir sandstones (0.5–2.0 mD) and the limit on the injection pressure to 5.8 MPa at the reservoir resulted in a predicted injection rate of 266 tonnes/year, this reaching a total volume of 1330 tonnes of CO_2 injected over the planned program period of five years. At this injection rate, it is predicted that the CO_2 plume will reach a diameter of approx. 220 m after five years of injection, and a diameter of 280 m five years after cessation of injection. The results demonstrate that both the pressure and CO_2 plumes should remain within the 1 km × 1 km footprint of the Field Research Station. Cumulatively, injection rates appear to gradually increase over time and reach a plateau, caused by relative permeability hysteresis. This is an effect where the relative permeability (CO_2-water) increases over time as a result of the decreasing water saturation (Bachu and Bennion, 2008). Two other factors contributing to this effect include gas compressibility and CO_2 dissolution into the BBRS formation water (salinity is approx. 8000 ppm). Note that the simulated scenarios do not take into account further complexity of reservoir compositional changes, such as chemical dissolution or precipitation of minerals.

Changes in the reservoir petrophysical properties after CO_2 injection were computed using Gassmann's equations and numerical time-lapse seismic modeling of the updated reservoir properties indicates that seismic data should be able to detect the CO_2 plume after one year of injection, despite the small amount of CO_2 (266 tonnes) and the limited radius of the plume (55 m). Actual injection of CO_2 into the BBRS is scheduled to begin in mid-2018 and monitoring results will be history matched against actual reservoir and caprock performance and monitoring surveys conducted.

Acknowledgments

We thank CMC Research Institutes, Inc. for permission to publish this chapter and for access to the well and seismic data. Funding for the construction of the FRS was provided by the Province of Alberta (Department of Environment and Parks) and from the Government of Canada (Western Economic Diversification), with strong support from the University of Calgary. Financial support from Shell Global Solutions and Klohn Crippen Berger is also gratefully acknowledged. Support for J. Dongas was provided by the industrial sponsors of the Consortium for Research in Elastic Wave Exploration Seismology (CREWES) at the University of Calgary, and the Natural Sciences and Engineering Council of Canada (NSERC)

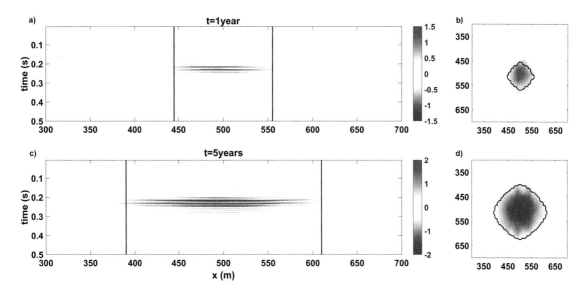

Figure 16.13 Vertical and time slice seismic amplitude difference displays from the 3D volumes. (a) Vertical section, difference between $t = 1$ year and the baseline. (c) Vertical section, difference between $t = 5$ years and the baseline. On both sections, the lateral extent of the inferred CO_2 plume is denoted by the vertical black lines. (b) Time slice at 0.21 s of the difference volume between $t = 1$ year and the baseline. (d) Time slice at 0.21 s of difference volume between $t = 5$ years and the baseline. Interpreted lateral extent of the CO_2 plume is indicated by the black outline.

through grant CRDP 379744–08. Support for M. Macquet was provided by the University of Calgary Eyes High Postdoctoral Fellowship program. Schlumberger Carbon Services is thanked for Project Management during the development and construction of the Field Research Station, including drilling and logging of wells, and also for providing access to PETREL software. The Computer Modelling Group (CMG) is thanked for providing GEM simulation software, and SINTEF is acknowledged for providing access to TIGER seismic modelling software. IHS Energy Canada provided the wireline data used in the construction and characterization of the static geomodel. Cenovus Energy has kindly provided access to the site for a period of 10 years. We also thank Dr. Malcolm Wilson and Dr. Thomas Davis for constructive reviews of this contribution.

References

Bachu, S., and Bennion, B. (2008). Effects of in-situ conditions on relative permeability characteristics of CO_2-brine systems. *Environmental Geology*, **54**(8): 1707–1722.

Beaton, A. (2003). Production potential of coalbed methane resources in Alberta. *Alberta Energy and Utilities Board*, EUB/AGS Earth Sciences Report 2003-03.

Brooks, R. H., and Corey, T. (1964). Hydraulic properties of porous media. Colorado State University, Hydrology Papers, 3, 1–24.

Chadwick, R. A., Zweigel, P., Gregerson, U., Kirby, G.A., Holloway, S., and Johannessen., P. N. (2004). Geological reservoir characterization of a CO_2 storage site: The Utsira Sand, Sleipner, northern North Sea. *Energy*, **29**: 1371–1381.

Coates, G. R., and Dumanoir, J. L. (1974). A new approach to improved log-derived permeability. *Log Analyst*, January–February: 17.

Eisinger, C. L., and Jensen, J. L. (2009). Data integration, petrophysics, and geomodelling: Wabamun Area CO_2 Sequestration Project (WASP). Energy and Environmental Systems Group, Institute for Sustainable Energy, Environment and Economy, University of Calgary.

Fenton M. M., Schreiner, B. T., Nielsen, E., and Pawlowicz, J. G. (1994). Quaternary geology of the Western Plains. In G. D. Mossop and I. Shetsen (comp.), *Geological Atlas of the Western Canada Sedimentary Basin*. Canadian Society of Petroleum Geologists and Alberta Research Council, Calgary, 413–420.

Gassmann, F. (1951). Über die Elastizität poröser Medien. Viertel. *Naturforschenden Gesellschaft Zürich*, **96**: 1–23.

Gibson-Poole, C. M., Svendsen, L., Underschultz, L., *et al.* (2008). Site characterization of a basin-scale CO_2 geological storage system: Gippsland Basin, southeast Australia. *Environmental Geology*, **54**: 1583–1606.

Glombick, P. M. (2010a). Top of the Belly River Gp. in the Alberta Plains: Subsurface stratigraphic picks and modelled surface. *Alberta Energy Regulator Alberta Geological Survey*, Open File Report 2010–10.

Glombick, P. M. (2010b). Top of the Belly River Gp. in the Alberta Plains: Subsurface stratigraphic picks and modelled surface. *Alberta Energy Regulator Alberta Geological Survey* Digital InFm. Series DIG 2010–0022.

Glombick, P. M. (2011a). Subsurface stratigraphic picks for the top of the Oldman Fm. (Base of Dinosaur Park Fm.), Alberta Plains. *Alberta Energy Regulator Alberta Geological Survey*, Open File Report 2011–13.

Glombick, P. M. (2011b). Subsurface stratigraphic picks for the top of the Oldman Fm. (Base of Dinosaur Park Fm.), Alberta Plains. *Alberta Energy Regulator Alberta Geological Survey*, Digital InFm. Series DIG 2011–0006 (tabular data, tab-delimited format, to accompany Open File Report 2011–13.

Glombick, P. M. (2014a). Subsurface stratigraphic picks for the Belly River Gp./Lea Park Fm. Transition in East-Central Alberta. *Alberta Energy Regulator Alberta Geological Survey*, Digital InFm. Series DIG 2013–0031.

Glombick, P. M. (2014b). Subsurface stratigraphic picks for the top of the Foremost Fm. (Belly River Gp.), Alberta Plains. *Alberta Energy Regulator Alberta Geological Survey*, Digital InFm. Series DIG 2013–0030.

Glombick, P. M., and Mumpy, A. J. (2014a). Subsurface stratigraphic picks for the top of the Milk River 'shoulder', Alberta Plains. *Alberta Energy Regulator Alberta Geological Survey*, Digital InFm. Series DIG 2013–0025.

Glombick, P. M., and Mumpy, A. J. (2014b). Subsurface stratigraphic picks for the Milk River 'shoulder', Alberta Plains: Including tops for the Milk River Fm. and Alderson Mbr. of the Lea Park Fm: *Alberta Energy Regulator Alberta Geological Survey, Open File Report* 2013–17.

Hamblin, A. P., and Abrahamson, B. W. (1996) Stratigraphic architecture of "Basal Belly River" cycles, Foremost Formation, Belly River Group, subsurface of southern Alberta and southwestern Saskatchewan. *Bulletin of Canadian Petroleum Geology*, **44**(4): 654–673.

Ah-Ram Kim, Gye-Chun Cho, and Tae-Hyuk Kwon. (2014). Site characterization and geotechnical aspects on geological storage of CO_2 in Korea. *Geosciences Journal*, **18**(2): 167–179.

Li, K., and Horne, R. N. (2006). Comparison of methods to calculate relative permeability from capillary pressure in consolidated water-wet porous media. *American Geophysical Union, Water Resources Research*, **42**: W06405. DOI:1029/2005WR004482.

Rock, L., Brydie, J., Jones, D., Jones, J.-P., Perkins, E., and Taylor, E. (2015). Methodology to assess groundwater quality during CO_2 injection at the Quest CCS project. *Geoconvention 2015*, Geoscience New Horizons.

Rostron, B., White, D., Hawkes, C., and Chalaturnyk, R. (2014). Characterization of the Aquistore CO_2 project storage site, Saskatchewan, Canada. *Energy Procedia*, **63**: 2977–2984.

Smith, T., Sondergeld, C., and Rai, C. (2003). Gassmann fluid substitutions: A tutorial. *Geophysics*, **68**, 430–440.

Timur, A. (1968). An investigation of permeability, porosity, and residual water saturation relationship for sandstone reservoirs. *Log Analyst*, **9**(4): 8.

Wright, G. N., McMechan, M. E., Potter, D. E. G., and Holter, M. E. (1994). Structure and architecture of the Western Canada Sedimentary Basin. In G. D. Mossop and I. Shetsen (comp), Geological atlas of the Western Canada Sedimentary Basin. Canadian Society of Petroleum Geologists and Alberta Research Council, Calgary, 25–40.

Seismic and Electrical Resistivity Tomography 3D Monitoring at the Ketzin Pilot Storage Site in Germany

17

Christopher Juhlin, Stefan Lüth, Monika Ivandic, and Peter Bergmann

Summary

Geophysical monitoring is essential for any CO_2 storage site. At the Ketzin CO_2 pilot storage site, seismic and electrical resistivity tomography (ERT) measurements were integral components of the monitoring program. The operational phase (injection) at the site lasted between June 2008 and August 2013. During this time a total of about 67 ktons of CO_2 was injected into a saline formation in the depth range of 630–650 m. For monitoring purposes, four 3D seismic surveys have been acquired to date, one baseline, two repeats during the injection period, and one postinjection survey. ERT was performed using both crosshole ERT measurements and surface-downhole ERT measurements as the main geometries. Amplitude and time-shift analysis of the seismic data allow high-resolution tracking of the gaseous CO_2 plume to be made at Ketzin. The preferred plume migration is in the northwesterly direction, nearly parallel to the anticlinal structure of the site. Heterogeneity in the reservoir formation is interpreted to control the plume migration. ERT time lapse analysis shows a significant decrease in resistivity in the vicinity of the borehole due to CO_2 injection. Even these data indicate a predominantly northwesterly migration of the CO_2 plume. Comparison of the geophysical data to fluid flow modeling of the injection process shows that the site is behaving in conformity to expectations. The amount of gaseous CO_2 imaged is in line with the modeling given that there is a high dissolution rate and that there is significant lateral spreading of the CO_2 in thin layers that have a thickness less than the seismic detection limit. Based on the geophysical data it is concluded that no measureable amounts of CO_2 have leaked out of the storage reservoir.

Introduction

Although industrial-scale CO_2 storage projects have been operational for many years (e.g., Sleipner: Torp

and Gale, 2004), pilot-scale projects, designed for the injection of small CO_2 quantities (100 ktonnes or less), are important laboratories, because they provide well-controlled conditions and allow testing various technologies at relatively low cost.

The Ketzin project, located west of Berlin (Figure 17.1), is one of these pilot sites. It was initiated in 2004 by the GFZ German Research Centre for Geosciences in collaboration with European industry and academic partners as the first European onshore storage project (Martens et al., 2014). The Ketzin project was designed to cover all life-cycle phases of a storage project, from the assessment to the postclosure and posttransfer phases. The operational phase (injection) lasted between June 2008 and August 2013. The Ketzin site was permitted according to German Mining Law. Therefore, the transfer of liability for this site will follow German Mining Law, but the research and development activities at Ketzin will continue in order to close the full life cycle of a storage site according to the EU CCS-Directive (Martens et al., 2014).

The main components of the Ketzin site infrastructure are the injection well (CO2 Ktzi 201/2007, abbreviated as Ktzi 201) and four monitoring wells. The injection well and two monitoring wells (Ktzi 200, Ktzi 202) were drilled in 2007 and reached depths of about 750–800 m. In two further drilling campaigns, a shallow monitoring well, P300, and a deep monitoring well, Ktzi 203, were drilled, reaching depths of 446 m and 701 m, respectively. The injection facility, including tanks, pumps, and heating system, was installed in 2008 and dismantled after injection closure in 2013.

CO_2 injection at Ketzin lasted from June 30, 2008 until August 29, 2013. A total mass of 67 ktonnes of CO_2 was injected into sandstone layers of the Triassic Stuttgart Formation at a depth of 630–650 m. Typical injection rates were between 1,400 and 3,250 kg CO_2/hour. Figure 17.2 shows the cumulative mass of CO_2

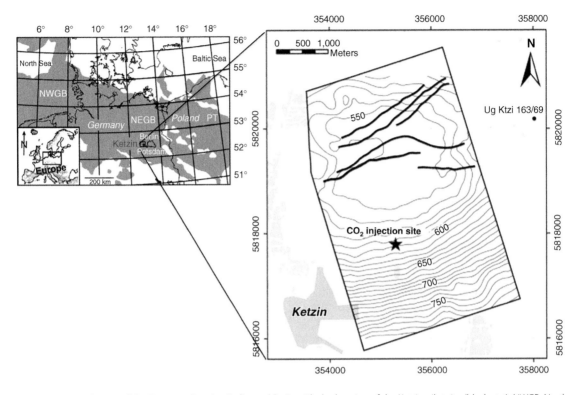

Figure 17.1 (Left) Map of the European Permian Rotliegend Basin with the location of the Ketzin pilot site (black star). NWGB, Northwest German Basin; NEGB, Northeast German Basin; PT Polish Trough. (Right) Interpreted depth of the top Stuttgart Formation (in meters below ground level) and mapped faults of the Central Graben Fault Zone (CGFZ; Juhlin *et al.*, 2007). The bold star marks the location of the Ketzin pilot site. (Figures modified from Norden and Frykman, 2013.)

injected over time. Pressure was monitored at 550 m depth in order to ensure that the injection process did not cause unsafe elevated reservoir pressure. Mainly, food grade quality CO_2 was injected (>99.9% purity), delivered by road tankers. From May to June 2011, 1.5 ktonnes of CO_2 captured from the Vattenfall oxyfuel pilot plant Schwarze Pumpe was used. This power plant CO_2 had a slightly lower degree of purity (>99.7%) than the food grade CO_2 used in the main experiment. Between March and July 2013, the injection temperature was incrementally decreased from 40°C (approximate reservoir temperature) to 10°C in order to test the feasibility of CO_2 injection without energy-intensive heating (Möller *et al.*, 2014). At the end of the injection phase, in July and August 2013, the coinjection of CO_2 and N_2 was tested in order to demonstrate the technical feasibility of continuous injection of impure CO_2 (Martens *et al.*, 2014).

After the end of the injection phase, two experiments were performed in order to investigate the

storage efficiency and the stabilization of the storage complex:

- Back-production of 240 tonnes of CO_2 and 55 m³ of brine
- Injection of 2900 tonnes of brine

The CO_2 and brine back-production experiment was performed in October 2014 (Martens *et al.*, 2015). The purpose of this experiment was to investigate reservoir and well behavior during back-production and to test the performance of monitoring methods for detecting the propagation of atmospheric CO_2 in the case of leakage through a well. The back-production test was accompanied by an extensive sequence of ERT monitoring surveys.

The brine injection experiment took place between October 2015 and January 2016 (Möller *et al.*, 2016). The purpose of the brine injection was to trigger accelerated imbibition of the reservoir rocks with brine close to the injection well. By ERT monitoring of this process key data for the quantification of

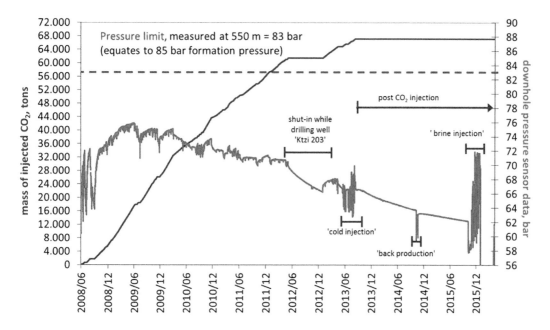

Figure 17.2 Cumulative mass of CO_2 injected over time (green curve) and pressure monitored at 550 m depth in the injection well (blue curve). Injection shut-in phases are characterized by fast reservoir pressure reduction and field tests (cold injection, back production, and brine injection) triggered strong pressure perturbations. (Figure modified from Wipki *et al.*, *Energy Procedia* 97, 2016.)

residual gas saturation was collected. Furthermore, this experiment was performed as a test of brine injection for leakage remediation in the case of a damaged leaking well.

Further site operations after the termination of the field experiments comprise the abandonment of the remaining wells (well Ktzi 202 was abandoned in 2015) and dismantling the remaining surface infrastructure at the site, executed in 2017.

Several papers have been published concerning the geophysical monitoring of the Ketzin site. Ivanova *et al.* (2012) reported on results from the first seismic repeat survey, Ivandic *et al.* (2015) from the second repeat survey, and Huang *et al.* (2017) from the third repeat survey (the first postclosure survey). Huang *et al.* (2015) presented results integrating dynamic flow simulations with seismic monitoring observations. Schmidt-Hattenberger *et al.* (2014) reviewed the extensive ERT monitoring program that has been ongoing at Ketzin. Bergmann and Chadwick (2015) estimated volumetric bounds on the amount of CO_2 that is observable in the time lapse seismic surveys. A recent comprehensive overview of the Ketzin project was presented by Bergmann *et al.* (2016). In this contribution we complement the

Bergmann *et al.* (2016) paper with more recent monitoring results and provide an overview of the most important monitoring results.

Geological Setting

Förster *et al.* (2006) give an extensive overview of the geological setting of the Ketzin site and we briefly summarize the main points here. Flow of the Upper Permian Zechstein Salt in the northeast German Basin began in the Triassic and initiated the development of the Roskow–Ketzin anticline (Kossow *et al.*, 2000) that the Ketzin site is located on. From the later part of the Triassic into the Cretaceous, salt movement continued in phases, with the two main ones being at 140 Ma and 106 Ma. The former resulted in uplift and erosion of the Jurassic successions and the latter one in erosion of Lower Cretaceous deposits (Förster *et al.*, 2006). During the Upper Cretaceous there was no deposition at the site, as it was part of a structural high (Stackebrandt and Lippstreu, 2002). Therefore, Oligocene sediments unconformably overlie the Jurassic ones (Förster *et al.*, 2006). The present day strike of the anticline axis is in the NNE–SSW direction, with flanks dipping at about 15 degrees.

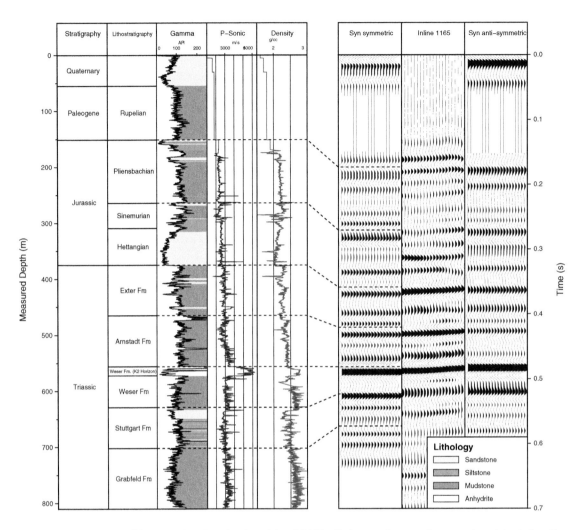

Figure 17.3 Stratigraphy of the Ktzi200/2007 based on Prevedel *et al.* (2008) with the natural gamma, P-wave sonic, and density logs. The natural gamma log is shaded by lithology as shown in Sopher *et al.* (2014). Zero-offset synthetic seismograms were generated using the velocity and density logs after averaging over 1.5-m intervals with a 40-Hz symmetric Ricker wavelet as input (left seismic panel) and an antisymmetric first-derivative Gaussian wavelet (right seismic panel). The migrated seismic data (middle seismic panel) are from Inline 1165 just to the west of the well.

The actual injection site is located on the southern flank of the eastern dome of the anticline, where sandstones, shales, and siltstones are the predominant lithologies of the Triassic–Jurassic succession (Figure 17.3). CO_2 was injected into the saline water bearing reservoir rocks of the Upper Triassic Stuttgart Formation at 630–650 m depth. This heterogeneous c. 80 m thick formation consists of both high-quality sandy channel reservoir facies and poor-quality clayey flood plain deposits (Förster *et al.*, 2006). The Weser and Arnstadt Formations, containing mainly fine grain marls and mudstones (Förster *et al.*, 2006), overlie the Stuttgart Formation and constitute the caprock. An approx.

10–20 m thick anhydrite layer (the K2 reflector) is present within the Weser formation and generates a strong seismic reflection (Juhlin *et al.*, 2007). Above the K2 reflector is the Exter Formation that consists of shallow marine deposits, forming the top of the Triassic units. Natural gas was stored in the overlying Jurassic sandstones just to the north of the Ketzin site until the year 2000 (Lange, 1966). The Tertiary Rupelton clay comprises the caprock for the former gas storage, above which lie Quaternary sediments that constitute present-day freshwater aquifers (Figure 17.3).

The K2 reflector generates the most prominent seismic response based on logging data from the

Ktzi 200 well (Figure 17.3). This is due to the large increase in the velocity and density of the anhydrite layer compared to the surrounding clastic rocks. Excellent correlation is observed between the synthetic seismic response and an inline from the 3D seismic volume (Juhlin *et al.*, 2007), regardless of whether a symmetric or antisymmetric wavelet is used as input. A good correlation between the synthetic and real data is also seen at the Top Arnstadt Formation and Top Exter Formation at this location. The Top Stuttgart Formation generates a reflection about 40 ms below the K2 reflection and represents the top of the CO_2 injection reservoir. The impedance contrast between the top of the reservoir and the overlying caprock is negative. Injected CO_2 that remains in a gaseous state will reduce the P-wave velocity and density of the reservoir further and, therefore, the amplitude of reflected waves from the reservoir portions containing CO_2 should be enhanced compared to those with no CO_2.

Data Acquisition

Seismic Data

Various seismic monitoring methods, such as 2D and 3D surface seismic surveys (Bergmann *et al.*, 2012; Ivanova *et al.*, 2012, Ivandic *et al.*, 2012), crosswell surveys (Zhang *et al.*, 2012), and moving source profiling (Yang *et al.*, 2010), have been applied at the Ketzin CO_2 pilot site to monitor the CO_2 plume behavior within the target reservoir at different scales. Among them, the time-lapse 3D surface seismic method has been an essential tool for mapping the lateral extent of the spreading gaseous plume and tracking its movement. Between autumn 2005 and autumn 2015 four full 3D seismic surveys were performed at the Ketzin site (Figure 17.4a). The first 3D seismic survey covering an area of about 12 km^2 around the injection site was acquired in summer/autumn 2005 prior to the injection. Main objectives of the survey were (1) to verify earlier geological interpretations of the structure based on vintage 2D seismic and borehole data; (2) to provide, if possible, an understanding of the structural geometry for flow pathways within the reservoir; (3) a baseline for later evaluation of the time evolution of rock properties as CO_2 is injected into the reservoir; and (4) detailed subsurface images near the injection borehole for planning of the drilling operations (Juhlin *et al.*, 2007). The 3D seismic data were acquired using

an overlapping template scheme with 5 receiver lines containing 48 active channels in each template. Five receiver lines made up one swath. In each template, 200 nominal source points were activated using an accelerated weight drop. After all sources had been activated within a template, the receiver locations were shifted by half a template when moving within a swath. A similar overlapping scheme was used between the source points of adjacent swaths. This overlap of source points and recording stations from template to template yielded an even nominal fold of 25 for the survey area. However, owing to logistical restrictions such as roads, villages, and infrastructure, the number of actual source points in each template varied.

The first two repeat 3D seismic surveys were acquired in the autumns of 2009 (Ivanova *et al.*, 2012) and 2012 (Ivandic *et al.*, 2015) during the injection phase (Figure 17.4a). The cumulative amount of injected CO_2 at the times of the two surveys was approximately 23 ktonnes and 61 ktonnes, respectively. The third repeat survey was conducted in the postinjection phase, two years after a total amount of 67 ktonnes of CO_2 had been injected.

The recording equipment and acquisition parameters used in the repeat surveys were identical to those used in the baseline (Juhlin *et al.*, 2007) to ensure high repeatability of the data sets. Table 17.1 summarizes the acquisition parameters used for the 3D surface seismic surveys.

ERT Data

Aside from seismic measurements, ERT was performed in order to monitor the injected CO_2 (Kiessling *et al.*, 2010). The use of ERT is motivated by the significant change in rock resistivity that occurs when electrically well conductive brine in the pore space is substituted by poorly conductive CO_2 (e.g., Natatsuka *et al.*, 2010). Crosshole ERT measurements and surface-downhole ERT measurements are the two main geometries that have been applied at Ketzin (Kiessling *et al.*, 2010). The technical layout primarily focused on the crosshole measurements and comprised a vertical electrical resistivity array (VERA), consisting of 45 downhole electrodes deployed in the three wells Ktzi200, Ktzi201, and Ktzi202 (Schmidt-Hattenberger *et al.*, 2011, 2012). Each well was equipped with 15 electrodes, arranged with a vertical spacing of 10 m. To suppress bypassing electrical

301

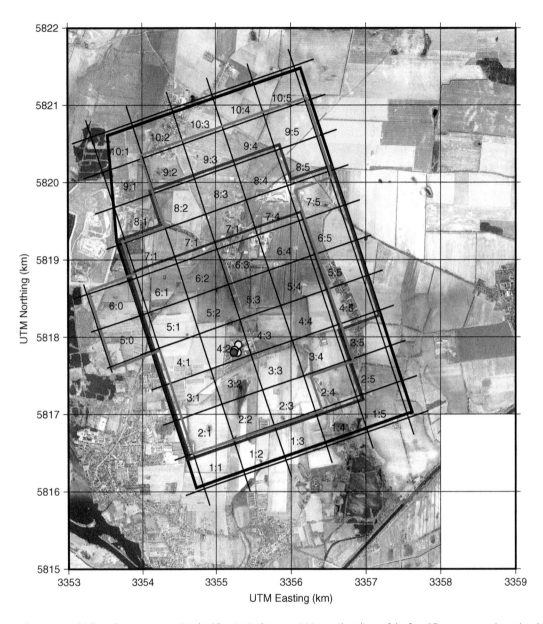

Figure 17.4 (a) Template system used in the 3D seismic data acquisition and outlines of the four 3D surveys conducted at the Ketzin site. Black polygon outlines the baseline survey area, whereas red, blue, and green polygons enclose areas of the first, second, and third repeat survey, respectively. Red and yellow dots show the locations of the injection well and three observation wells, respectively; (b) actual common depth point fold for the baseline 3D survey with inlines and crosslines marked by blue dashed lines. White dot marks the location of the injection well.

currents through the metallic well casings, the casings were covered with an insulation (Kiessling *et al.*, 2010).

In ERT data acquisition, each reading involves four electrodes (two electrodes for current injection,

and two electrodes for voltage registration). This setup is varied over the range of available electrodes. At Ketzin, bipole–bipole, dipole–dipole, and dipole–dipole across three boreholes were the setups applied for crosshole measurements (Schmidt-Hattenberger

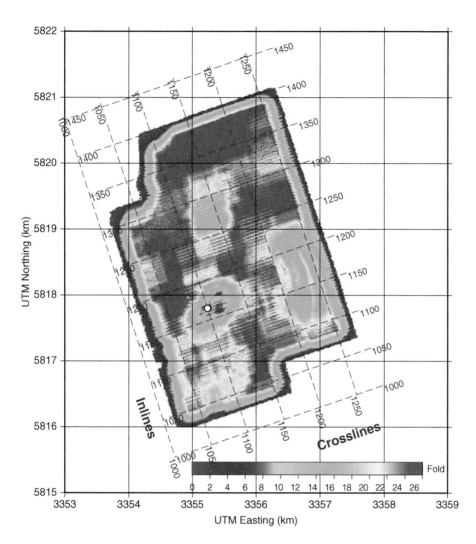

Figure 17.4 (cont.)

et al., 2016). The limit for current injection was 2.5 Ampere, which was limited by the current transmission capability of the cables that connect the electrode to the acquisition unit at the surface. A ZT-30 (Zonge Engineering) was used for current transmission and a GPD32 unit (Zonge Engineering) was used for voltage registration.

The surface-downhole ERT measurements included current injections at the surface and voltage registrations by means of the VERA electrodes (Bergmann *et al.*, 2012). For this purpose, current injections were performed at 16 surface locations, which were concentrically arranged around the injection site (Figure 17.5). In order to optimize the signal-to-noise (S/N) ratio, each current injection was performed over a period of 45–60 minutes with a maximum power of up to 10 kilowatts using a Scintrex TSQ-4 power source.

While the surface-downhole ERT measurements comprise seven surveys in total (two baseline measurements before CO_2 injection and five repeat measurements during ongoing injection), the crosshole ERT measurements were repeated with much higher frequency. That is, initial crosshole ERT measurements were performed on a daily basis. Once the CO_2 arrived at the first observation well (i.e., approx. 15 days after

Table 17.1 Template acquisition parameters for the 3D surveys at Ketzin

Parameter	Value
Receiver line spacing /number	96 m/5
Receiver station spacing /channels	24 m/48
Source line spacing /number	48 m/12
Source point spacing	24 m or 72 m
Common depth point bin size	12 m × 12 m
Nominal fold	25
Geophones	28 Hz single
Sampling rate	1 ms
Record length	3 s

Seismic source: 240 kg accel. weight drop, 8 hits per source point.

beginning injection), the repetition rate was reduced to two measurements per week. After the arrival of the CO_2 at the second observation well (March 21, 2009; Martens *et al.*, 2011; Zimmer *et al.*, 2011), the repetition rate was further reduced to one measurement per week (Schmidt-Hattenberger *et al.*, 2016).

Time-Lapse Results

Seismic Data

Almost identical processing steps as used for the baseline survey were applied to the three repeat data sets in order to optimize repeatability. Owing to the different weather and ground conditions during the surveys, only static corrections had to be recalculated separately for each repeat data set (Kashubin *et al.*, 2011; Bergmann *et al.*, 2014b). Poststack cross-calibration further minimized the time-lapse noise and enhanced the time-lapse reservoir signature (Ivanova *et al.*,

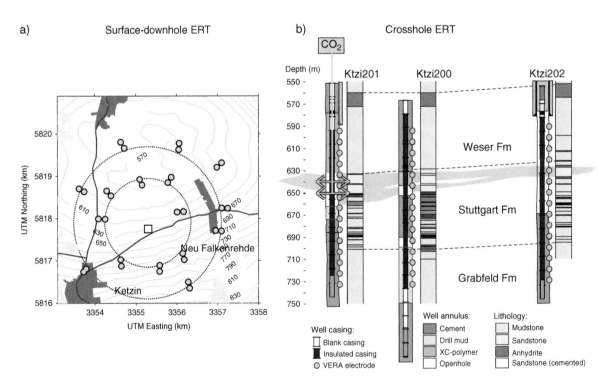

Figure 17.5 (a) Acquisition of the surface-downhole ERT surveys. Gray points indicate electrode pairs used for current injection in the surface-downhole ERT experiments. Isolines indicate depth of the storage formation (Top Stuttgart Formation). (b) Schematic of the VERA installation showing the downhole electrodes, well completions, and surrounding lithology (after Förster *et al.* 2010) at the Ketzin site. (Figure modified after Bergmann *et al.*, 2017)

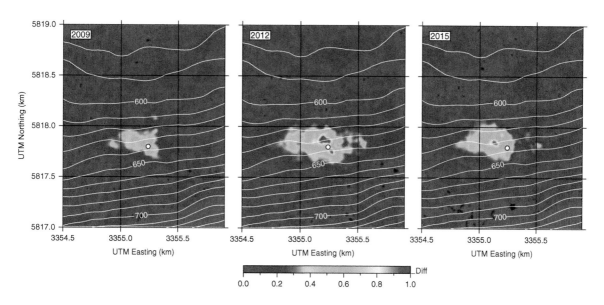

Figure 17.6 Normalized amplitude differences (baseline minus repeat) at the reservoir level for the first (left), second (middle), and third (right) repeat survey. White dot marks the locations of the injection well. Isolines indicate the elevation of top of the Stuttgart Formation.

2012; Ivandic *et al.*, 2015; Huang *et al.*, 2017). The data processing workflow is described in detail in Juhlin *et al.* (2007).

The 3D surveys imaged a sequence of clear reflections from approx. 150 ms down to 900 ms two-way-time. Reduced image quality of near-surface structures is observed only where the fold is lower (Figure 17.4b). The stronger reflections in the target reservoir area observed around the injection site in the three repeat data sets were interpreted to be due to the increased impedance contrast between the caprock and the reservoir sandstone caused by CO_2 injection. However, to image the lateral distribution of amplitude variations related to the injected CO_2, the seismic amplitudes of the baseline and repeats were normalized to the peak amplitude of the K2 reflection and subtracted. Figure 17.6 shows maps of the normalized amplitude differences at the top of the Stuttgart Formation from the repeat data sets.

At the time of the first repeat survey, a pronounced amplitude anomaly with a lateral extent of 300–400 m and a thickness of 5–20 m, was situated at the top of the reservoir near the injection well. The gaseous plume had a westerly trending tendency, revealing the heterogeneous nature of the reservoir caused by the complex sedimentary history in a fluvial environment (Ivanova *et al.*, 2012). Results from the second 3D repeat survey showed further growth and

migration of the plume. The observed amplitude anomaly was similar in shape to the one observed at the time of the first repeat survey, but larger by approx. 150 m in the N–S direction and 200 m in the W–E direction. Furthermore, it was much stronger, with the highest amplitude nearly centered at the injection well. The thickness of the plume had also increased, varying from 10 m to 30 m (Ivandic *et al.*, 2015). The latest survey, conducted in 2015 during the postinjection phase, revealed a decrease in the intensity and extent of the amplitude anomaly in both the horizontal and vertical directions. The most significant decrease in the amplitude anomaly occurs east and south of the injection site. The amplitude decrease in this area may be due to lower S/N ratio in the 2015 survey in this area, preferential movement of the brine, greater lateral flow into thin layers, or a combination of factors (Huang *et al.*, 2016).

Figure 17.7 displays vertical cross sections extracted from the three repeat volumes along inline 1166 and crossline 1100, both crossing close to the injection site. They show a prominent time-lapse signature at approx. 42 ms below the K2 reflector, which corresponds to the approximate depth of the top Stuttgart Formation. The growth of the amplitude anomaly in the reservoir during the injection phase can be observed in both the inline and crossline direction. However, as mentioned earlier, at the time of the

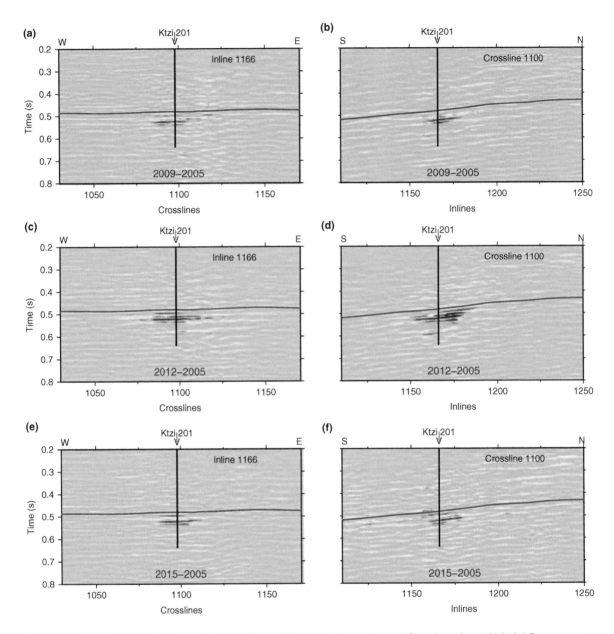

Figure 17.7 Vertical cross sections of time-lapse amplitude differences along inline 1166 (left) and crossline 1100 (right). From top to bottom: first repeat, second repeat, and first postinjection results. The red line represents the K2 horizon. The location of the injection well is marked by the black vertical line (Huang et al., 2017).

first postinjection survey, the range and intensity of the amplitude anomaly within the reservoir has decreased and the push-down effect at deeper levels is weaker compared to the second repeat survey. This has been interpreted to be mainly due to CO_2 dissolution in the formation brine and lateral spreading of the gaseous CO_2 (Huang et al., 2017), both leading to

thinning of the gaseous CO_2 component into layers thinner than the seismic detection limit. Nevertheless, as dissolution is a crucial trapping mechanism in the early postclosure phase, and spreading of the CO_2 through rather thin layers indicates a pressure relaxation in the reservoir, both processes demonstrate ongoing stabilization of the CO_2 plume at the Ketzin site (Huang *et al.*, 2017). No systematic changes in the seismic amplitudes within the overburden were detected by the three monitoring surveys, indicating that CO_2 is being contained within the storage reservoir.

Electrical Resistivity Tomography

Processing of the ERT data consists of two steps. First, preprocessing of the acquired voltage time series into resistivity values is required. For both measurement setups the voltage time series cover various cycles of the input DC signal (the number of signal cycles is chosen according to the S/N ratio conditions as well as to the number of four-electrode combinations that can be realized). This sequence is then stacked into time series of a single cycle, which undergoes despiking, filtering, and phase corrections (Labitzke *et al.*, 2012). On the basis of the peak and trough voltage amplitudes in the stacked time series, and the geometric arrangement of the involved electrodes, apparent resistivity values are determined, which represent weighted averages of the true resistivities in the volume between the electrodes. In addition, resistivity error values are estimated in order to characterize the quality of the individual readings (Schmidt-Hattenberger *et al.*, 2016).

In the second step of the ERT processing, the apparent resistivities and estimated errors are used to image the resistivity distribution in the particular subsurface region under investigation. This imaging is based on an iterative update loop (e.g., Ellis and Oldenburg, 1994), in which a numerical resistivity model, as a representation of the true subsurface, is used to generate synthetic resistivity data. In each iteration, the synthetic resistivity data set is compared against the measured resistivity data set, and the synthetic resistivity model is updated until a sufficient match between the two data sets is achieved. In the case of Ketzin, this imaging was based on finite-element forward modeling (Rücker *et al.*, 2006) and a Gauss–Newton inverse procedure for model updating (Günther *et al.*, 2006). The data errors mentioned earlier are herein used in order to weight the input resistivity data and give preference to those with higher quality.

Figure 17.8a shows sections of the resistivity models obtained from crosshole ERT. For the preinjection stage, the resistivity models show rather weak internal contrasts, because the reservoir mudstones and brine-bearing reservoir sandstones have similar ranges of electrical resistivity. However, after CO_2 injection starts the models display an increase in resistivity at the depth of the injection. For the later stages of CO_2 injection, weakening of the resistivity signature is observed that coincides with a reduction in the peak CO_2 injection rates from 3.25 tonnes/hour to 1.5 tonnes/hour (Figure 17.8c). In general, the crosshole ERT models provide a better resolution than the surface-downhole ERT models (not shown in Figure 17.8). In this context it is important to note that the imaged resistivity model generally suffers from a rapid loss in resolution with increasing distance from the wells. Although this implies that the time-lapse electrical resistivity signature cannot be compared in detail to the corresponding time-lapse seismic signature in terms of lateral details, it is in agreement about a preferential migration trend in the northwesterly direction (Figure 17.9).

Discussion

Resolution

Seismic Resolution

As indicated by the time-lapse analysis of the three repeat surveys, the amount of gaseous CO_2 observable in the Stuttgart Formation has decreased significantly from the second repeat time to the third repeat time even though more CO_2 had been injected. This decrease has been attributed to two main processes, dissolution of the CO_2 into the saline water and lateral migration (spreading of the CO_2 as a thin layer at the top of the reservoir). The former process alone is estimated to account for about 30% of the CO_2 being dissolved into the formation water. The influence of the latter process on the observations is highly dependent on the seismic acquisition parameters used (source frequency and strength, geometry repeatability, near surface changes, and ambient noise) and then how thin the CO_2 layer is. Under perfect conditions any changes in the subsurface can be detected by time-lapse surveys.

In the Ketzin case, source frequency and strength were very similar for all surveys, so variations in the source characteristics are not expected to have a great influence on the time-lapse results. Likewise, there was good repeatability on the source and receiver locations, also implying that this factor probably is

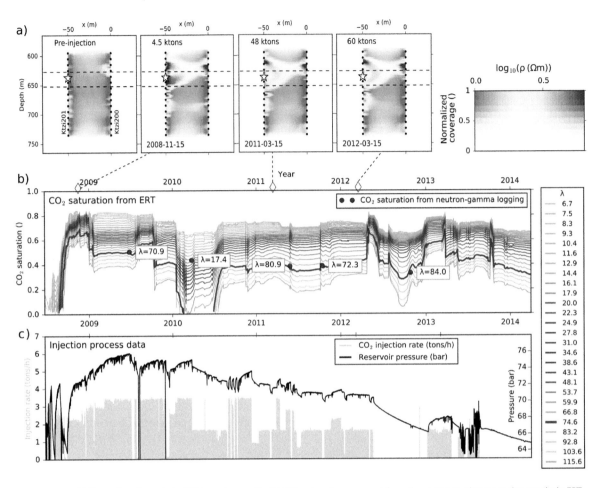

Figure 17.8 (a) Examples of crosshole ERT inversion results. (b) CO_2 saturation assessed from the resistivity change in the crosshole ERT results for variable regularization parameter. The regularization parameter is a smoothness constraint applied in the processing. Red line shows the regularization parameter that fits CO_2 saturations inferred from neutron-gamma logging (after Baumann et al., 2014, indicated by red markers) best. Text boxes show the regularization parameters that fit the logging results best. (c) CO_2 injection rate and reservoir pressure for the period of CO_2 injection (after Möller et al., 2012; Martens et al., 2014.) (Figure modified after Bergmann et al., 2017.)

not influencing the results significantly. It has been shown that the near surface conditions have varied during the different surveys due to weather (Kashubin et al., 2011). These variations influenced the velocity in the uppermost loose sediments causing different static corrections having to be applied for each survey. However, after the static corrections, the same processing sequence could be used and the final normalized root mean square values away from the injection area are quite low or can be explained by low fold. This leaves ambient noise and CO_2 layer thickness as two factors that can affect the seismic signature of the gaseous CO_2 plume. The two are related since, as the

layer thickness decreases below the tuning thickness, the amplitude on the difference section will decrease. When ambient noise is present this difference in amplitude will be masked by the noise.

In order to test how ambient noise may affect the capacity to map thin CO_2 layers in time-lapse seismic surveys, synthetic data were generated based on the logs from the Ktzi 200 borehole (Figure 17.3). The K2 reflector and reservoir were modeled using the values listed in Table 17.2 and the same convolutional seismic modeling routine was applied as in Figure 17.3 using the antisymmetric wavelet. A comparison of the seismic response using the model in Table 17.2

Table 17.2 Parameters used for the seismic modeling in Figure 17.10

Rock type	Top (m)	Bottom (m)	Velocity (m/s)	Density (g/cc)
K2 anhydrite	555	575	5500	2.75
Reservoir (saline)	630–650	650	2750	2.25
Reservoir (CO_2)	630	630.5–650	2200	2.20
Background			3300	2.40

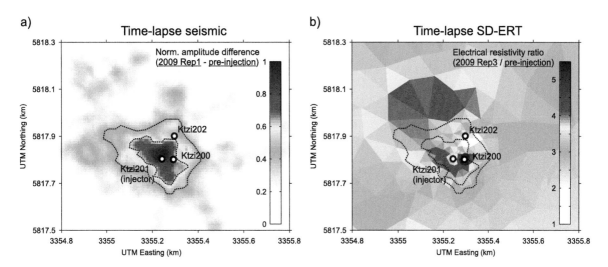

Figure 17.9 (a) Time-lapse electrical resistivity anomaly and (b) time-lapse seismic anomaly along the Near Top Stuttgart horizon. Subfigure (a) shows the resistivity ratio, obtained from the constrained inversion, of the third repeat surface-downhole ERT survey and the second baseline surface-downhole ERT survey. Subfigure (b) shows the amplitude difference of the first repeat 3D seismic survey and the baseline 3D seismic survey after Ivanova *et al.* (2012). Both methods image a northwesterly trend for the CO_2 migration. (Figure modified after Bergmann *et al.*, 2014a.)

with no CO_2 in the reservoir and the response using the log data as input shows that the simplified model gives a reasonably similar response as the logs (Figures. 17.10a and 17.10b). Given that the simplified model seems adequate, a thin CO_2 layer was gradually introduced into the model and the seismic difference response calculated, starting with a thickness of 0.5 m at the top of the reservoir and increasing to 20 m (filling the entire reservoir). On the difference section (Figure 17.10c) it is observed how the increased thickness increases the amplitude and delays the arrival of the main peak. Note also that at 20 m thickness the amplitude difference is about the same as the amplitude of the K2 reflection at a value of about 0.4. (the source wavelet had a maximum amplitude of 1). This is greater than the actual reflection coefficient of any interface and is

due to thin bed tuning generating constructive interference at the larger thicknesses.

Kashubin *et al.* (2011) analyzed the S/N ratios on source gathers at Ketzin on the baseline and the second repeat surveys by comparing the amplitudes at the reservoir level with those before the first arrivals in the offset range 300–600 m. They found that the S/N ratio was generally around 1 throughout most of the survey area. If the noise is Gaussian then stacking would increase the S/N ratio to 5 for the processed data with a fold of 25. However, the noise is probably not Gaussian and the fold is lower in the vicinity of the injection site, so a lower S/N ratio may be expected on many parts of the processed data. To test the influence of noise on the difference section (Figure 17.10c) low-pass (40 Hz) Gaussian noise was generated using Seismic Unix (www.cwp.mines.edu)

309

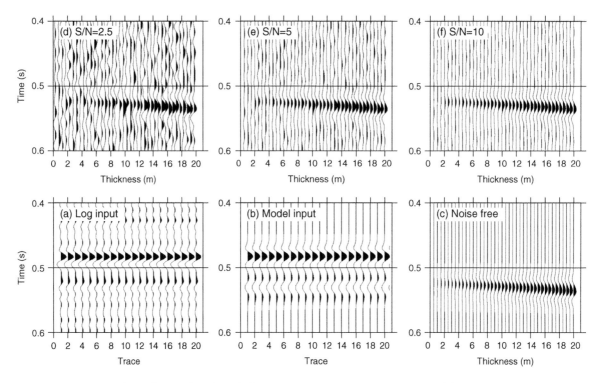

Figure 17.10 Analysis of the influence of CO_2 layer thickness and S/N ratio on the seismic response of the reservoir. Modeling was done using a first derivative Gaussian wavelet with a dominant frequency of about 40 Hz. (a) Log data from Figure 17.3 used as input to the convolutional model. (b) Values in Table 17.2 used as input. (c) Seismic response on the difference section for CO_2 layer thickness increasing from 0.5 to 20 m. (d) Same as (c), but with noise added at 2.5 S/N. (e) Same as (c), but with noise added at 5 S/N. (f) Same as (c), but with noise added at 10 S/N.

at S/N ratio levels of 2.5, 5, and 10 and added to the difference section (Figures 17.10d to 17.10f). The S/N ratio is defined based on the maximum amplitude in the section before adding it. For a high S/N ratio value of 10, CO_2 layers as thin as 2–3 m can potentially be identified. For S/N ratio values in the range of 2.5–5 the modeling indicates that CO_2 layer thickness needs to be at least in the the range of 5–8 m to be detected, with the higher value corresponding to an S/N ratio of 2.5. These values are roughly consistent with the seismic observations and will be discussed later in the mass estimate section.

Electrical Resistivity Tomography

The following quantitative interpretation is based on the assumption that the change in resistivity, which is displayed in Figure 17.8, is dominantly caused by the substitution of brine with CO_2. In other words, dissolution and mineral alteration are assumed to play minor roles, at least throughout the early years of CO_2 injection. Note that this assumption may not be valid based on the fluid flow modeling and the seismic observations (Huang *et al.*, 2015). However, the assumption allows us to convert CO_2 saturation levels from relative resistivity changes by means of an Archie equation, as presented by Kummerow and Spangenberg (2011) for the Ketzin reservoir sandstone. A second assumption, which is made in this approach, is that the electrical conductivity contributed by the clay is small compared to the conductivity of the brine. Based on the revision of the petrophysical model, Bergmann *et al.* (2017) suggest that this assumption is valid for CO_2 saturation levels of up to 70%. Converting resistivity changes into CO_2 saturation levels on the basis of this petrophysical model leads to the CO_2 saturation levels in Figure 17.8b. These levels have been further calibrated by thickness-weighted averages of the reservoir CO_2 saturations reported by Baumann *et al.* (2014) on the basis of repeated neutron-gamma loggings. It is obvious that, apart from the spreading CO_2 distribution, other processes are also of relevance for the estimated CO_2 saturation levels. Rapid saturation increases can, for instance, in several cases be correlated with short-term pressure changes and injection changes. Generally, the estimated CO_2 saturations can be seen to be in the range of 40–60% for most of the CO_2 injection period.

ERT methods generally have more limited resolution compared to that of seismic reflection methods. Without the use of the VERA electrodes, i.e., solely surface-based measurements, CO_2 monitoring in the reservoir would in fact not be feasible by means of ERT. Bergmann *et al.* (2012, 2016) estimate that the vertical resolution of the surface-downhole ERT results corresponds to approx. 2–3 electrode spacings (i.e., approx. 20–30 m). Because of the denser data coverage and diversity of electrode configurations that can be realized in the crosshole ERT measurements, the corresponding results are interpreted to have a resolution of approx. 5 m in the near-wellbore region.

Amplitude versus Offset

Amplitude versus offset (AVO), together with results of petrophysical experiments on core samples from the Ketzin reservoir, are used to compare the repeat seismic data sets with the baseline prior to CO_2 injection in order to investigate the AVO response from the sandstone reservoir during the injection and postinjection phases and obtain better insight into the time-lapse signature observed in the seismic data sets.

First, AVO modeling was done to determine whether an AVO anomaly is to be expected and, if so, which type. The input model for the baseline survey was based on well logging data and lithological information (Kazemeini *et al.*, 2010). The thickness of the reservoir was assumed to be 20 m. After applying Gassmann fluid substitution in the reservoir by replacing brine with a range of CO_2 saturation levels, synthetic seismic traces were generated for both the baseline and repeat cases. The wavelet used for the synthetic trace was extracted from the seismic data.

For the analysis of the real data sets, supergathers were generated by averaging over 4 common depth points (CDPs) to improve the S/N ratio. In spite of the varying weather conditions and variations in the acquisition geometries, a comparison of the baseline and repeat AVO volumes reveals good repeatability for the three repeat seismic data sets (Figure 17.11). A prominent decrease in the intercept values is observed only at the top of the reservoir around the injection well in all three repeat data sets (Figure 17.11). This correlates well with the spatial variations of the amplitude response at the top of the reservoir displayed in Figure 17.6. Based on the AVO/AVA modeling by Ivanova *et al.* (2013) and the modeling conducted in this study, the observed anomaly is interpreted to be caused by increased CO_2 saturation. The same modeling studies also show that both elevated reservoir pressure and increased CO_2 saturation would result in somewhat higher gradient values. However, petrophysical experiments on core samples from the Ketzin reservoir indicate that the pressure increase measured during the injection phase could only have had a rather minor impact on the seismic amplitudes. Moreover, because the injection was temporarily stopped during the acquisition of the second repeat survey and the third survey was conducted during the postinjection phase, the slight increase in gradient values observed in those data are most likely due to CO_2 saturation.

Patchy versus Homogeneous Saturation

The 4D seismic response considered earlier is an expression of changes in seismic wave velocity and bulk density that is due to the displacement of brine by CO_2 in the pore space. While the density, and thereby the density change, can be simply determined from the densities and volume fractions of the constituents (i.e., rock minerals, brine, and CO_2), the same approach is not valid for seismic wave velocity. Seismic wave velocity is depending on the density and the rock's bulk modulus (K), the latter of which specifies the rock's volume change when being exposed to uniaxial force. We refer to the dependency of seismic velocity and CO_2 saturation here as a velocity–saturation relationship, which is typically described by means of the Gassmann model (Gassmann, 1951).

This model assumes that the bulk modulus of the two-phase pore fluid can be averaged from the bulk moduli of brine and CO_2. This means that the mixture of CO_2 and brine is treated as a single homogeneous fluid, which is referred to as uniform saturation (Mavko *et al.*, 2003). For the Ketzin case, the solid line in Figure 17.12 shows the

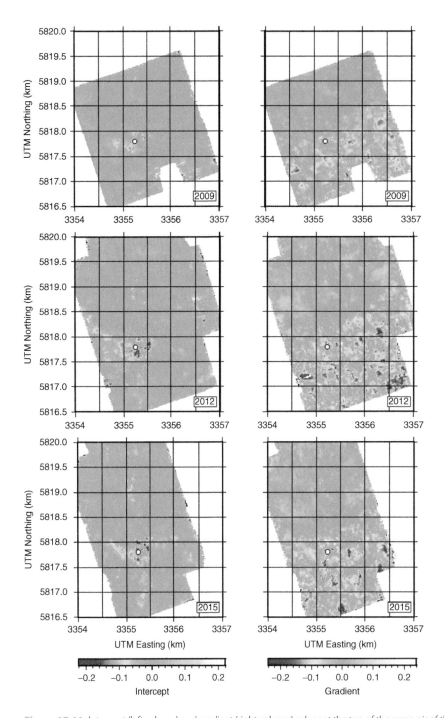

Figure 17.11 Intercept (left column) and gradient (right column) values at the top of the reservoir of the three repeat volumes. The injection well is marked by the white dot.

velocity–saturation relationship by Kazemeini *et al.* (2010) computed on the basis of the Gassmann model. This velocity–saturation relationship shows a significant decrease of the compressional wave velocity (V_p) for CO_2 saturations of approx. up to 5%, which is explained by the fact that the bulk

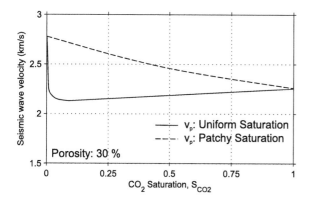

Figure 17.12 Compressional wave velocity (Vp) as functions of CO_2 saturation for uniform and patchy saturation for the Ketzin reservoir model. (Modified after Kazemeini et al., 2010).

modulus of the gaseous CO_2 at Ketzin is very small compared to that of brine (K_{CO2} = 0.00832 GPa, K_{brine} = 3.68 GPa for Ketzin reservoir conditions). For larger CO_2 saturations, a gradual rebound in the compressional wave velocity is expected, which is due to the progressive decrease in density (Whitman and Towle, 1992).

Aside from the assumption of uniform saturation, the Gassmann model is also based on the assumption that the rock is homogeneous and isotropic, and the rock pores are hydraulically well connected. These assumptions are, however, in the context of CO_2 storage prevalently not met. This is conditioned by the fact that a region, once it has been swept, has an increased effective permeability to flow, which leads to spatial variations in fluid mobility and preferential migration paths. In most cases, reservoir rocks are further characterized by spatial variations in wettability, permeability, or shaliness, which can cause finger-type CO_2 migration (Asveth, 2009). The occurrence of nonuniform CO_2 migration has been observed in a wide range of experimental studies (e.g., Shi et al., 2011; Alemu et al., 2013; Nakagawa et al., 2013), which indicate that (1) nonuniform CO_2 migration is most likely the relevant case and (2) that the degree of saturation heterogeneity is scale dependent. This scale dependency has direct implications on the relationship of seismic velocity and CO_2 saturation, as a passing seismic wave causes local pore-pressure differences. Given the circumstance that heterogeneities in the fluid mixing and porosity variations hamper the equilibration of wave-induced pore-pressure differences, the assumption of uniform saturation is

not obeyed. The Gassmann model can then be considered to be valid only locally, i.e., in regions that comprise sufficiently homogeneous fluid mixing and permeability conditions. This case is referred to as the patchy saturation model (Dvorkin and Nur, 1998; Eid et al., 2015), and the corresponding bulk modulus has to be computed from a weighted average that is also taking the rock's shear modulus into consideration (Hill, 1963; Berryman and Milton, 1991). The dashed line in Figure 17.12 shows the corresponding velocity–saturation relationship by Kazemeini et al. (2010) computed on the basis of the patchy saturation model. In comparison to the uniform saturation case, it shows a linear relationship for the velocity–saturation relationship. In fact, both models constitute the upper and lower velocity bounds for the range of possible velocity–saturation relationships (Mavko et al., 2003). On the basis of laboratory measurements of Ketzin core samples, Kummerow and Spangenberg (2011) observed that ultrasonic velocities show trends from both models for different CO_2 saturation ranges (Bergmann et al., 2016). Based on these laboratory measurements, Kummerow and Spangenberg (2011) report velocity decreases from approx. 3 km/s to approx. 2.77 km/s.

Mass Estimates and Conformity of Observed and Simulated Reservoir Behavior

Estimating the quantity of stored CO_2 that has been detected by geophysical observations is an important task for assuring containment and/or quantifying any potential leakage (i.e., unexpected migration of CO_2 out of the storage complex). For a large scale storage site, Chadwick et al. (2005) demonstrated that it is possible to quantify the mass of CO_2 stored with some uncertainty remaining. Pilot scale storage sites typically contain small CO_2 quantities that are in the same order of magnitude as would be potential accumulations of CO_2 leaking/migrating out of the storage complex. A reliable quantification of leakage is a prerequisite for determining appropriate remediation measures. Comprehensive geophysical monitoring and petrophysical investigations at the Ketzin pilot site provide the data for an integrated mass estimation of the CO_2 in the storage formation that has been imaged by time-lapse seismic monitoring.

The integrated mass estimation aims at attributing the mass of CO_2 to the normalized amplitude differences detected by the three seismic monitoring

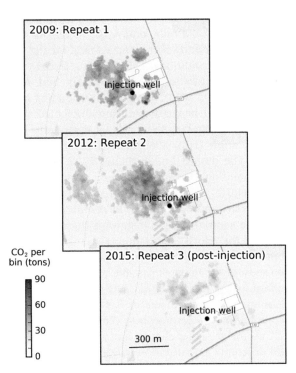

2009: Repeat 1

Injection well

2012: Repeat 2

Injection well

CO_2 per bin (tons)

90

60

30

0

2015: Repeat 3 (post-injection)

Injection well

300 m

(after Huang *et al.*, 2017).

Figure 17.13 Estimated CO_2 mass for each common depth point bin for the first repeat survey (after Ivanova *et al.*, 2012), second repeat survey (after Ivandic *et al.*, 2015), and third repeat survey. (after Huang *et al.*, 2017)

surveys acquired in the years 2009, 2012, and 2015 (Ivanova *et al.*, 2012; Ivandic *et al.*, 2015; Huang *et al.*, 2017). The integrated mass estimation used the following results from geophysical and petrophysical investigations: Normalized difference amplitudes from 3D repeat seismic surveys, time shifts of seismic reflections from below the reservoir, saturation profiles at well locations from Pulsed Neutron Gamma (PNG) logging measurements, and seismic velocity changes due to CO_2 replacing brine observed on core samples from the storage formation. Difference amplitudes and saturation profiles were used to estimate the lateral distribution of average CO_2 saturation within the CO_2 plume. Furthermore, time shifts and CO_2-related velocity changes were used to determine the thickness of the CO_2 plume. Using average values for reservoir porosity and CO_2 density, the total mass of gaseous CO_2 was then computed by summing up inferred CO_2 masses from each image point in the seismic data. Based on this approach, 23.83 kilotonnes and 52.51 kilotonnes of

CO_2 were estimated to be present for the first and second repeat surveys, respectively (Figure 17.13; Ivanova *et al.*, 2012; Ivandic *et al.*, 2015). This is to be compared to 22–25 kilotonnes of actually injected CO_2 by the first repeat survey and 61 kilotonnes of actually injected CO_2 by the second repeat survey. For the second and third repeat surveys, i.e., the surveys before and after CO_2 injection ended, Huang *et al.* (2017) estimated the amount of CO_2 to lie between 55.7 and 10.8 ktonnes (Fig. 17.13), which is a significant reduction that cannot be explained completely by reservoir stabilization processes such as dissolution and CO_2 propagation in very thin layers (fingering). Another effect adding uncertainty to the mass estimation may be the potential variability of CO_2 density in the reservoir.

The integrative approach suffers from a number of uncertainties that are discussed in the aforementioned publications. Saturation and porosity distributions are assumed to be represented by average values that may be incorrect for large areas of the reservoir, and the petrophysical model, derived from rock core experiments, may be incorrect for reservoir scale surface seismic observations (see also discussion in the preceding section on homogeneous vs. patch saturation). Also, an incorrect assumption of initial seismic V_p velocity in the reservoir before CO_2 injection (full brine saturation) may result in a significant uncertainty of the mass estimation.

Huang *et al.* (2017) further addressed the uncertainties in the velocity–saturation relationship by revisiting the previous quantification studies by means of an alternative approach. This is of particular relevance because the petrophysical measurements, from which the velocity–saturation relationship is interpreted, were performed only on a limited number of core samples (Kummerow and Spangenberg, 2011). In contrast to the approach presented in Ivanova *et al.* (2012) and Ivandic *et al.* (2015), which is based on time-lapse amplitudes and 4D time shifts, the alternative quantification approach provides bounds on the amount of CO_2 present based on 4D time shifts only (Bergmann and Chadwick, 2015). Huang *et al.* (2017) compared the CO_2 mass bounds to the true amount of CO_2 injected in order to validate alternative velocity–saturation relationships that have not been considered in the previous interpretations. Thereby, they showed that the set of possible velocity–saturation relationships is not confined to patchy fluid mixing models, but contains also moderate Gassmann-type models.

Both quantification approaches have demonstrated that uncertainties are significant and, therefore, only a rough estimate of the amount of gaseous CO_2 imaged by time-lapse seismic measurements can be provided. Mass estimations alone are therefore of rather limited use for demonstrating full containment. They can be used for quantifying leakage by providing reasonable minimum to maximum estimates of the detected CO_2 mass.

An important requirement for a safe long-term storage site operation, in particular in the postclosure and pretransfer phase of the site, is demonstrating conformity between observed (from monitoring) and predicted (from reservoir simulations) reservoir behavior. This requirement is defined in carbon capture and storage regulations, but to date no specific instructions have been made available indicating under which conditions conformity between observed and predicted behavior is achieved. An important monitoring parameter assuring long-term stability of the storage complex is pressure, and matching simulated pressure history to observational data can be regarded as a valid approach to demonstrate conformity. However, at various storage sites, pressure measurements may not be available, or may be available at only a few locations. The availability of seismic monitoring data and full reservoir scale simulations have allowed an evaluation of the performance conformity for the Ketzin storage site involving the full CO_2 plume extent to have been made. For the quantitative assessment of conformity between seismic observations and reservoir simulations, different performance parameters can be used describing geometrical properties of the CO_2 plume such as plume footprint area, plume volume, and lateral migration distance (Chadwick and Noy, 2015). A direct comparison of the actual plume shapes can be quantified using the "similarity index" (Lüth *et al.*, 2015). Applying these performance parameters to the seismic data and modeling results for Ketzin revealed high quantitative mismatches between observations and simulation. For example, in terms of the plume footprint area, seismic monitoring was able to detect only between 30% and 85% of the footprint area predicted by history-matched simulations (Lüth *et al.*, 2015). Comparing simulation results with a range of minimum thickness thresholds showed that the seismic

observations were in good conformity with the simulated plume footprint for a minimum thickness of 5–7 m. Equivalent observations were made for the similarity index comparing plume shapes from seismic data and simulations where best results were achieved for the central part of the simulated CO_2 plume with a minimum thickness of 5 m. This case study showed that several performance parameters are available providing stable quantitative measures for the conformity (or nonconformity) of observed and simulated CO_2 plume behavior. However, for an implementation of this approach of quantitative assessment, reference data from real large scale sites as well as from generic modeling studies are needed as a basis for translating quantitative assessment results into a catalog of criteria for the decision making whether observed site behavior is within the expectations or not.

Conclusions

Both seismic and electrical resistivity tomography (ERT) measurements have been employed at the Ketzin site for monitoring the injected CO_2. In this paper we have focused on the 3D time-lapse seismic surveys and the crosshole ERT measurements. The seismic measurements allow the lateral extent of the gaseous CO_2 plume to be mapped in high spatial resolution, while the crosshole ERT measurements allow the CO_2 distribution in the near wellbore volume to be mapped in high temporal resolution. Both data sets show a predominantly northwesterly migration of the CO_2 plume, interpreted as due to the heterogeneity of the reservoir. Comparison of the geophysical data to fluid flow modeling of the injection process shows that the site is behaving in conformity to expectations. The amount of gaseous CO_2 imaged is in line with the modeling given that there is a high dissolution rate (30%) and that there is significant lateral spreading of the CO_2 in thin layers that have a thickness less than the seismic detection limit. Based on the geophysical data, it is concluded that no significant amounts of CO_2 have leaked out of the storage reservoir.

Acknowledgments

The authors gratefully acknowledge the funding for the Ketzin project received from the European Commission (6th and 7th Framework Program), two German ministries (the Federal Ministry of

Economics and Technology and the Federal Ministry of Education and Research), and industry since 2004. The ongoing R&D activities are funded within the project COMPLETE by the Federal Ministry of Education and Research. Further funding is received by VGS, RWE, Vattenfall, Statoil, OMV, and the Norwegian CLIMIT program.

References

Alemu, B. L., Aker, E., Soldal, M., Johnsen, Ø., and Aagaard, P. (2013). Effect of sub-core scale heterogeneities on acoustic and electrical properties of a reservoir rock: A CO_2 flooding experiment of brine saturated sandstone in a computed tomography scanner. *Geophysical Prospecting*, **61**(1): 235–250.

Asveth, P. (2009). Exploration rock physics. In K. Bjørlykke (ed.), *Petroleum geoscience: From sedimentary environments to rock physics*, Berlin/Heidelberg: Springer-Verlag.

Baumann, G., Henninges, J., and Lucia, M. D. (2014). Monitoring of saturation changes and salt precipitation during CO_2 injection using pulsed neutron-gamma logging at the Ketzin pilot site. *International Journal of Greenhouse Gas Control*, **28**: 134–146.

Bergmann, P., and Chadwick, A. (2015). Volumetric bounds on subsurface fluid substitution using 4D seismic time shifts with an application at Sleipner, North Sea. *Geophysics*, **80**: B153–B165.

Bergmann, P., Schmidt-Hattenberger, C., Kiessling, D., *et al.* (2012). Surface-downhole electrical resistivity tomography applied to monitoring of CO_2 storage at Ketzin, Germany. *Geophysics*, **77**: B253–B267.

Bergmann, P., Ivandic, M., Norden, B., *et al.* (2014a). Combination of seismic reflection and constrained resistivity inversion with an application to 4D imaging of the CO_2 storage site, Ketzin, Germany. *Geophysics*, **79**(2): B37–B50.

Bergmann, P., Kashubin, A., Ivandic, M., Lüth, S., and Juhlin, C. (2014b). Time-lapse difference static correction using prestack crosscorrelations: 4D seismic image enhancement case from Ketzin. *Geophysics*, **79**: B243–B252.

Bergmann, P., Diersch, M., Götz, J., *et al.* (2016). Review on geophysical monitoring of CO_2 injection at Ketzin, Germany. *Journal of Petroleum Science and Engineering*, **139**: 112–136.

Bergmann, P., Schmidt-Hattenberger, C., Labitzke, T., *et al.* (2017). Fluid injection monitoring using electrical resistivity tomography: Five years of CO_2 injection at Ketzin, Germany. *Geophysical Prospecting*, **65**(3): 859–875.

Berryman, J., and Milton, G. (1991). Exact results for generalized Gassmann's equations in composite porous media with two constituents. *Geophysics*, **56**: 1950–1960.

Chadwick, R. A., and Noy, D. J. (2015). Underground CO2 storage: demonstrating regulatory conformance by convergence of history-matched modeled and observed CO2 plume behavior using Sleipner time-lapse seismics. *Greenhouse Gas Science and Technology*, **5**: 305–322. DOI:10.1002/ghg.1488.

Chadwick, R. A., Arts, R., and Eiken, O. (2005). 4D seismic quantification of a growing CO_2 plume at Sleipner, North Sea. Geological Society, London, Petroleum Geology Conference series, 6, 1385-1399. https://doi.org/10.1144/0061385

Dvorkin, J., and Nur, A. (1998). Acoustic signatures of patchy saturation. *International Journal of Solids and Structures*, **35**(34): 4803–4810.

Eid, R., Ziolkowski, A., Naylor, M., and Pickup, G. (2015). Seismic monitoring of CO_2 plume growth, evolution and migration in a heterogeneous reservoir: Role, impact and importance of patchy saturation. *International Journal of Greenhouse Gas Control*, **43**: 70–81.

Ellis, R., and Oldenburg, D. (1994). Applied geophysical inversion. *Geophysical Journal International*, **116**: 5–11.

Förster, A., Norden, B., Zinck-Jørgensen, K., *et al.* (2006). Baseline characterization of the CO_2 SINK geological storage site at Ketzin, Germany. *Environmental Geosciences*, **13**: 145–161.

Förster A., Schöner R., Förster H., *et al.* (2010). Reservoir characterization of a CO_2 storage aquifer: The upper triassic Stuttgart Formation in the Northeast German Basin. *Marine and Petroleum Geology*, **27**(10): 2156–2172.

Gassmann, F. (1951). Über die Elastizitgät poröser Medien. *Vierteljahresschrift der Naturforschenden Gesellschaft Zürich*, **96**: 1–24.

Günther, T., Rücker, C., and Spitzer, K. (2006). Three-dimensional modelling and inversion of DC resistivity data incorporating topography II: Inversion. *Geophysical Journal International*, **166**: 506–517.

Hill, R. (1963). Elastic properties of reinforced solids: Some theoretical principles. *Journal of the Mechanics and Physics of Solids*, **11**: 357–372.

Huang F., Juhlin, C., Kempka, T., Norden, B., and Zhang, F. (2015). Modeling 3D time-lapse seismic response induced by CO_2 by integrating borehole and 3D seismic data: A case study at the Ketzin pilot site, Germany. *International Journal of Greenhouse Gas Control*, **36**: 66–77.

Huang, F., Bergmann, P., Juhlin, C., *et al.* (2017). The first post-injection seismic monitor Survey at the Ketzin pilot CO_2 storage site: Results from time-lapse analysis. *Geophysical Prospecting*, **66**(1): 62–84.

Ivandic, M., Yang, C., Lüth, S., Cosma, C., and Juhlin, C. (2012). Time-lapse analysis of sparse 3D seismic data from the CO_2 storage pilot site at Ketzin, Germany. *Journal of Applied Geophysics*, **84**: 14–28.

Ivandic, M., Juhlin, C., Lüth, S., *et al.* (2015). Geophysical monitoring at the Ketzin pilot site for CO_2 storage: New insights into the plume evolution. *International Journal of Greenhouse Gas Control*, **32**: 90–105.

Ivanova, A., Kashubin, A., Juhojuntti, N., *et al.* (2012). Monitoring and volumetric estimation of injected CO_2 using 4D seismic, petrophysical data, core measurements and well logging: a case study at Ketzin, Germany. *Geophysical Prospecting*, **60**(5): 957–973, http://dx.doi.org/10.1111/j.1365–2478.2012.01045.x

Juhlin, C., Giese, R., Zinck-Jørgensen, K., *et al.* (2007). 3D baseline seismics at Ketzin, Germany: The CO_2 SINK project. *Geophysics*, **72**: B121–B132.

Kashubin, A., Juhlin, C., Malehmir, A., Lüth, S., Ivanova, A., and Juhojuntti, N. (2011). A footprint of rainfall on land seismic data repeatability at the CO_2 storage pilot site, Ketzin, Germany. In 81st Annual International Meeting, Expanded Abstracts, *Society of Exploration Geophysicists*.

Kazemeini, S. H., Juhlin, C., and Fomel, S. (2010). Monitoring CO_2 response on surface seismic data: A rock physics and seismic modeling feasibility study at the CO_2 sequestration site, Ketzin, Germany. *Journal of Applied Geophysics*, **71**(4): 109–124.

Kiessling, D., Schmidt-Hattenberger, C., Schütt, H., *et al.* and the CO2SINK Group. (2010). Geoelectrical methods for monitoring geological CO2 storage: First results from cross-hole and surface-downhole measurements from the CO2SINK test site at Ketzin (Germany). *International Journal of Greenhouse Gas Control*, **4**: 816–826. DOI:10.1016/j.ijggc.2012.05.001.

Kossow, D., Krawczyk, C., McCann, T., Strecker, M., and Negendank, J. F. W. (2000). Style and evolution of salt pillows and related structures in the northern part of the Northeast German Basin. *International Journal of Earth Science*, **89**: 652–664.

Kummerow, J., and Spangenberg, E. (2011). Experimental evaluation of the impact of the interactions of CO_2–SO_2, brine, and reservoir rock on petrophysical properties: A case study from the Ketzin test site, Germany. *Geochemistry Geophysics Geosystems*, **12**: Q05–Q010.

Labitzke, T., Bergmann, P., Kiessling, D., and Schmidt-Hattenberger, C. (2012). 3D Surface-downhole electrical resistivity tomography data sets of the Ketzin CO_2 storage pilot from the CO_2 SINK project phase. *GFZ Scientific Technical Report*, **10**(5) (available online).

Lange, W. (1966). Geologisch-lagerstättenphysikalische und förder-technische Fragen bei der Erkundung des Untergrundspeichers Ketzin. *Zeitschrift für Angewandte Geologie*, **12**(1): 27–34.

Martens, S., Liebscher, A., Möller, F., Würdemann, H., Schilling, F., and Kühn, M. (2011). Progress report on the first European on-shore CO_2 storage site at Ketzin (Germany): Second year of injection. *Energy Procedia*, **4**: 3246–3253.

Martens, S., Möller, F., Streibel, M., Liebscher, A., and the Ketzin Group. (2014). Completion of five years of safe CO_2 injection and transition to the post-closure phase at the Ketzin pilot site. *Energy Procedia*, **59**: 190–197. DOI:10.1016/j.egypro.2014.10.366.

Martens, S., Kempka, T., Liebscher, A., *et al.* (2015). Field experiment on CO_2 back-production at the Ketzin pilot site. *Energy Procedia*, **76**: 519–527. DOI:10.1016/j.egypro.2015.07.902.

Mavko, G., Mukerji, T., and Dvorkin, J. (2003). *The rock physics handbook: Tools for seismic analysis of porous media*. Cambridge: Cambridge University Press.

Möller, F., Liebscher, A., Martens, S., Schmidt-Hattenberger, C., and Streibel, M. (2014). Injection of CO_2 at ambient temperature conditions: Pressure and temperature results of the "cold injection" experiment at the Ketzin pilot site. *Energy Procedia*, **63**: 6289–6297.

Möller, F., Liebscher, A., and Schmidt-Hattenberger, C. (2016). Report on the dataset of the Brine Injection at the CO_2 Storage Pilot Site Ketzin, Germany: Scientific Technical Report STR; 16/05, Potsdam: GFZ German Research Centre for Geosciences. DOI:10.2312/GFZ. b103-16059.

Nakagawa, S., Kneafsey, T. J., Daley, T. M., Freifeld, B. M., and Rees, E. V. (2013). Laboratory seismic monitoring of supercritical CO_2 flooding in sandstone cores using the Split Hopkinson Resonant Bar technique with concurrent x-ray computed tomography imaging. *Geophysical Prospecting*, **61**(2): 254–269.

Natatsuka, Y., Xue, Z., Garcia, H., and Matsuoka, T. (2010). Experimental study on CO_2 monitoring and quantification of stored CO_2 in saline formations using resistivity measurements. *International Journal of Greenhouse Gas Control*, **4**(2): 209–216.

Norden, B., and Frykman, P. (2013). Geological modelling of the Triassic Stuttgart Formation at the Ketzin CO_2 storage site, Germany. *International Journal of Greenhouse Gas Control*, **19**. DOI:10.1016/j.ijggc.2013.04.019.

Prevedel, B., Wohlgemuth, L., Henninges, J., Krüger, K., Norden, B., and Förster, A. (2008). The CO2SINK boreholes for geological storage testing. *Scientific Drilling*, **6**: 32–37.

Rücker, C., Günther, T., and Spitzer, K. (2006). Three-dimensional modelling and inversion of DC resistivity data incorporating topography I: Modelling. *Geophysical Journal International*, **166**(2): 495–505.

Schmidt-Hattenberger, C., Bergmann, P., Kießling, D., Krüger, K., Rücker, C., and Schütt, H. (2011). Application of a vertical electrical resistivity array (VERA) for monitoring CO_2 migration at the Ketzin site: First performance evaluation. *Energy Procedia*, **4**: 3363–3370.

Schmidt-Hattenberger, C., Bergmann, P., Labitzke, T., *et al.* (2012). A modular geoelectrical monitoring system as part of the surveillance concept in CO_2 storage projects. *Energy Procedia*, **23**: 400–407.

Schmidt-Hattenberger, C., Bergmann, P., Labitzke, T., and Wagner, F. (2014). CO_2 migration monitoring by means of electrical resistivity tomography (ERT): Review on five years of operation of a permanent ERT system at the Ketzin pilot site. *Energy Procedia*, **63**: 4366–4373.

Schmidt-Hattenberger, C., Bergmann, P., Labitzke, T., Wagner, F., and Rippe, D. (2016). Permanent crosshole electrical resistivity tomography (ERT) as an established method for the long-term CO_2 monitoring at the Ketzin pilot site. *International Journal of Greenhouse Gas Control*, **52**: 432–448.

Shi, J., Xue, Z., and Durucan, S. (2011). Supercritical CO_2 core flooding and imbibition in Tako sandstone: Influence of sub-core scale heterogeneity. *International Journal of Greenhouse Gas Control*, **5**(1): 75–87.

Sopher, D., Juhlin, C., Huang, F., Ivandic, M., and Lüth, S. (2014). Quantitative assessment of seismic source performance: Feasibility of small and affordable seismic sources for long term monitoring at the Ketzin CO_2 storage site, Germany. *Journal of Applied Geophysics*, **107**: 171–186.

Stackebrandt, W., and Lippstreu, L. (2002). Zur geologischen Entwicklung Brandenburgs. In W. Stackebrandt and V. Manhenke (eds.), *Atlas zur Geologie von Brandenburg im Maßstab 1:1,000 000*. Kleinmachnow, Germany: Landesamt für Geowissenschaften und Rohstoffe Brandenburg, 13–18.

Torp, T., and Gale, J. (2004). Demonstrating storage of CO_2 in geological reservoirs: The Sleipner and SACS projects. *Energy*, **29**(9–10): 1361–1369.

Whitman, W., and Towle, G. (1992). The influence of elastic and density properties on the behavior of the Gassmann relation. *Log Analyst*, **33**(6): 500–506.

Wipki, M., Ivanova, A., Liebscher, A., *et al.* (2016). Monitoring Concept for CO2 storage at the Ketzin pilot site, Germany: Post-injection continuation towards transfer of liability. *Energy Procedia*, **97**: 348–355.

Yang, C., Juhlin, C., Enescu, N., Cosma, C., and Lüth, S. (2010). Moving source pro-file data processing, modelling and comparison with 3D surface seismic data at the CO2SINK project site, Ketzin, Germany. *Near Surface Geophysics*, **8**: 601–610.

Zhang, F., Juhlin, C., Cosma, C., Tryggvason, A., and Pratt, R. G. (2012). Cross-well seismic waveform tomography for monitoring CO_2 injection: A case study from the Ketzin Site, Germany. *Geophysical Journal International*, **189**: 629–646, http://dx.doi.org/10.1111/j.1365-246X.2012.05375.x

Zimmer, M., Erzinger, J., and, Kujawa, C. (2011). The gas membrane sensor (GMS): A new method for gas measurements in deep boreholes applied at the CO2 SINK site. *International Journal of Greenhouse Gas Control*, **5**(4): 995–1001.

Chapter 18

Time-Lapse Seismic Analysis of the CO$_2$ Injection into the Tubåen Formation at Snøhvit

Sissel Grude and Martin Landrø

Introduction

At the Snøhvit field (Figure 18.1) in the Norwegian sector of the Barents Sea, the produced gas contains between 5% and 8% carbon dioxide (CO$_2$). The CO$_2$ is separated from the sales gas at the Melkøya LNG plant, piped back and, in the period between 2008 and 2011, the CO$_2$ was injected into the saline Tubåen Formation (Fm.). This injection into the Tubåen Fm. was monitored by repeated 3D seismic with data sets acquired in 2003 and 2009 in addition to downhole pressure measurements. The faults and depositional environment limit the volume of the Tubåen Fm., which can be regarded as semiclosed. The limited formation volume and lack of near well injectivity led to a pore pressure buildup and low injection rates. The injection into the Tubåen Fm. was stopped in 2011. Today, CO$_2$ is injected into the Stø Fm. at the Snøhvit field. In total, 1.6 Mtonnes of CO$_2$ was injected into the Tubåen Fm. The main purpose of monitoring the injected CO$_2$ was to demonstrate safe storage and detect possible leakage. Monitoring will also give valuable information regarding CO$_2$ migration and reservoir pressure changes.

CO$_2$ migration in a saline aquifer is governed by viscous, capillary, and gravitational fluid forces at an early stage of injection (Lindeberg and Wessel-Berg, 1997). The dominant flow regime is site specific and depends, among other factors, on injection rate, aquifer volume, reservoir injectivity, reservoir depth, formation pressure, and temperature. CO$_2$ must overcome the capillary entry pressure to migrate into brine-saturated pores (Bryant *et al.*, 2008). Viscous force is pressure driven and is dependent on the injection overpressure (Class *et al.*, 2009). Supercritical CO$_2$ is less dense and less viscous than the resident formation brine. The mobility contrast between supercritical CO$_2$ and brine, combined with the effect of gravity, will lead to vertical segregation of the fluids, and a thin layer of CO$_2$ will form at the top of the formation (Nordbotten *et al.*, 2005).

A modified version of the time-lapse pressure–saturation inversion introduced by Landrø (2001) is applied to separate the pressure and saturation effects caused by CO$_2$ injection at the Snøhvit field. A first-order approximation is used here to describe the pressure effect instead of using Landrø's second-order approximation. The inverted pressure and saturation are then used to predict expected time shifts that are compared to conventional time shifts estimated directly from the time-lapse seismic data. These can be seen as two independent measurements of the time shifts and used as a quality check of the inverted pressure and saturation results.

CO$_2$ injection laboratory experiments on two core plugs from the Tubåen Fm. were performed to evaluate the effect of salt precipitation on injectivity. Analysis of bottom-hole-pressure (BHP) was performed to gain information about flow regimes acting in the reservoir. The CO$_2$ saturation inverted from time-lapse seismic was compared to an analytical expression of CO$_2$ plume saturation and thickness for viscous dominated flow (Nordbotten and Celia, 2006).

The Snøhvit Field and the Tubåen Formation

The Tubåen Fm. is approx. 110 m thick and 2500 m wide close to the injection well, compartmentalized by a series of E–W trending faults (Figure 18.2). The formation can be vertically subdivided into four sandstone units, Tubåen 1–4, separated by interbedded shale units that potentially act as barriers to vertical flow. Overall net to gross of the formation is 0.7. Gamma ray, porosity, density, V_p, and V_s well logs

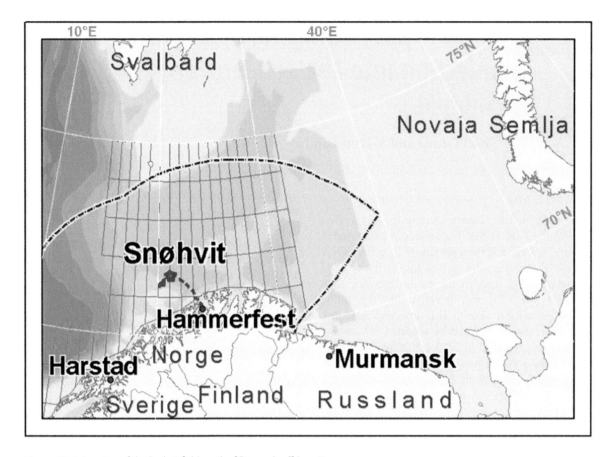

Figure 18.1 Location of the Snøhvit field, north of Finnmark, offshore Norway.

from the injection well are displayed in Figure 18.3. The base of the reservoir is located at 2670 m true vertical depth (TVD), corresponding to 2800 m measured depth, as the well is slightly deviated. The deepest 10–15 m thick Tubåen 1 sandstone unit is interpreted to be distributary channels oriented in a north–south direction (Hansen *et al.*, 2013) with porosity up to 20% and permeability up to 12 000 mD; the lateral extent of the good sandstones is uncertain (Eiken *et al.*, 2011). Quartz cement reaches 20% in parts of the formation. Initial reservoir temperature was 95°C, brine salinity 14%, overburden pressure 56 MPa, and initial pore pressure 29 MPa, which increased to 35 MPa by the time the monitor seismic survey was acquired.

CO$_2$ was originally injected through three perforation intervals covering the Tubåen 1–3 sandstone units (marked by the three red bars in Figure 18.3), with an additional perforation interval covering the

uppermost Tubåen 4 unit at a later stage. The fourth perforation interval did not improve the injectivity (Hansen *et al.*, 2011). A production logging tool was run in 2011. This showed that 80% of the injected CO$_2$ migrated into the Tubåen 1 unit and the remaining 20% into the Tubåen 2 and 3 units (Hansen *et al.*, 2013).

Fluid-Pressure Discrimination

Landrø (2001) exploits the time-lapse near and far offset seismic stacks as independent measurements, and uses these to invert for fluid and pressure changes in the seismic amplitude. The method is based on the Smith and Gidlow (1987) linearized amplitude versus offset (AVO) equation for P-wave reflectivity as a function of incidence angle, which estimates the reflectivity at a boundary surface in a two-layer model. The change in P-wave reflectivity, $\Delta R\,(\theta)$,

Time (ms)

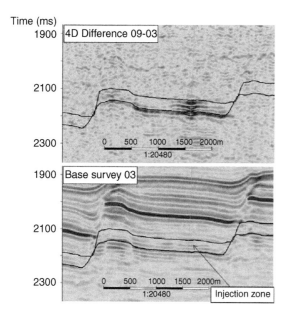

Figure 18.2 Cross section through the reservoir zone. The uppermost picture shows time-lapse difference between the seismic surveys. The lowermost picture is from the 2003 survey. The top and base of the reservoir zone are shown by the black lines and the injection well is the blue vertical line. The injection zone is marked by the red arrow.

caused by pressure and fluid changes in the reservoir, can be found by subtracting the reflectivity before any injection (base) from the reflectivity after injection (monitor), given the assumption that the changes in P-wave velocity (α), S-wave velocity (β), and density (ρ) occur only in the reservoir layer. These measurements, for pressure (ΔR^P) and saturation (ΔR^F), change respectively:

$$\Delta R^F(\theta) \approx -\frac{1}{2}\left(\frac{\Delta\rho^F}{\rho}+\frac{\Delta\alpha^F}{\alpha}\right)-\frac{\Delta\alpha^F}{2\alpha}\tan^2\theta \quad (18.1)$$

$$\Delta R^P(\theta) \approx -\frac{1}{2}\frac{\Delta\alpha^P}{\alpha}+\frac{4\beta^2}{\alpha^2}\frac{\Delta\beta^P}{\beta}\sin^2\theta-\frac{\Delta\alpha^P}{2\alpha}\tan^2\theta \quad (18.2)$$

This combines to

$$\Delta R(\theta) \approx -\frac{1}{2}\left(\frac{\Delta\rho^F}{\rho}+\frac{\Delta\alpha^F}{\alpha}+\frac{\Delta\alpha^P}{\alpha}\right)+\frac{4\beta^2}{\alpha^2}\frac{\Delta\beta^P}{\beta}$$
$$\sin^2\theta-\frac{1}{2}\left(\frac{\Delta\alpha^F}{\alpha}+\frac{\Delta\alpha^P}{\alpha}\right)\tan^2\theta \quad (18.3)$$

$\Delta\rho^F$, $\Delta\alpha^F$, $\Delta\alpha^P$, and $\Delta\beta^P$ are parameter contrasts from fluid and pressure changes in layer 1. $\Delta\alpha^P = \alpha^P_{Post\ inj} - \alpha^P_{Pre\ inj}$. $\alpha = (\alpha_1 + \alpha_2)/2$ etc. Subscripts 1

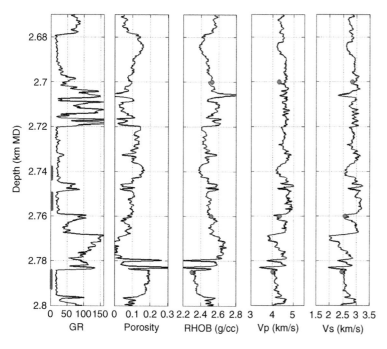

Figure 18.3 Gamma log (GR), porosity, bulk density (RHOB), P-wave- and S-wave velocity logs from the injection well. Vertical axis shows measured depth in kilometers; the well is slightly deviated. The reservoir zone is located between 2.68 and 2.8 km. Depth, V_p, V_s, and density of brine substituted core measurements are plotted as orange circles. Perforated intervals are marked by red vertical lines.

321

and 2 denote the two layers in the model. It is assumed that the shear modulus is constant for fluid substitution and that the density change from pressure changes is negligible (see Landrø [2001] for further details). The reflectivity between the base of the reservoir and the underlying strata is estimated. This gives an opposite sign on the reflectivity compared to the Landrø original method where the reflectivity between the top of the reservoir and the caprock was evaluated. Relative changes in P-wave velocity, S-wave velocity, and density, due to pressure and saturation changes, are determined from rock physics. The relationship between saturation changes and P-wave velocity is estimated from Gassmann (1951) fluid substitution. Landrø (2001) made a linear fit to the velocity–saturation curve that is appropriate when the fluids have similar compressibilities. Meadows (2001) expanded the velocity–saturation curve to a parabolic form, which more accurately represents a range of fluid compressibilities such as for CO_2 and water, but introduces nonlinearity into the inversion. In our method, Landrø's original straight-line approximation is used. Dry core measurements are used to find a relationship between the pressure changes and velocities. Landrø (2001) used a second-order approximation to describe the observed pressure–velocity dependency. Effective pressure in the Tubåen Fm. is 27 MPa, meaning that the relevant section of the pressure–velocity curve is almost linear, Figure 18.6. CO_2 injection increases the pore pressure but reduces the effective pressure to 21 MPa according to well measurements. We find that a straight-line pressure approximation is still valid in this case, which removes the second-order pressure term in Landrø's method. Expressions for the relative changes in velocities and density from saturation and pressure changes can be written as

$$\frac{\Delta \alpha}{\alpha} \approx k_\alpha \Delta S + l_\alpha \Delta P \qquad (18.4)$$

$$\frac{\Delta \beta}{\beta} \approx k_\beta \Delta S + l_\beta \Delta P \qquad (18.5)$$

$$\frac{\Delta \rho}{\rho} \approx k_\rho \Delta S \qquad (18.6)$$

By inserting Eqs. (18.4)–(18.6) into Eq. (18.3) and assuming $\tan^2 (\theta) \approx \sin^2 (\theta)$ (small-angle approximation) an expression for the change in reflectivity

expressed in terms of pressure and saturation changes is obtained:

$$\Delta R_{\text{inj}}(\theta) \approx -\frac{1}{2}(k_\rho \Delta S + k_\alpha \Delta S + l_\alpha \Delta P)$$
$$-\left(\frac{1}{2}(k_\alpha \Delta S + l_\alpha \Delta P) - \frac{4\beta^2 l_\beta \Delta P}{\alpha^2}\right)\sin^2\theta \quad (18.7)$$

With AVO intercept and gradient terms:

$$\Delta R_0 \approx -\frac{1}{2}(k_\rho \Delta S + k_\alpha \Delta S + l_\alpha \Delta P) \qquad (18.8)$$

$$\Delta G \approx -\left(\frac{1}{2}(k_\alpha \Delta S + l_\alpha \Delta P) - \frac{4\beta^2 l_\beta \Delta P}{\alpha^2}\right) \qquad (18.9)$$

Relative change in two-way travel time induced by variation in seismic velocity can be expressed as

$$\frac{\Delta T}{T} \approx -\frac{\Delta \alpha}{\alpha} \approx -(k_\alpha \Delta S + l_\alpha \Delta P) \qquad (18.10)$$

(Landrø, 2002). T denotes the layer thickness in milliseconds and ΔT the change in layer thickness caused by injection, also in milliseconds. Reservoir thickness is assumed constant, i.e., no expansion due to the injection. Inversion-predicted time shift is used to validate the inverted pressure and saturation 3D cubes, by comparison with the time shift measured from conventional cross-correlation.

The Top Fuglen Fm., a strong reflector situated above the Tubåen Fm., is used to calibrate the reflectivity due to the absence of coherent reflectors in the reservoir zone. A maximum error of 13% is found between calibrated reflectivity on the seismic data and estimated reflectivity from the Smith and Gidlow (1987) approximation after thin layer tuning correction (Lin and Phair, 1993) (Figure 18.4). A thin layer tuning effect (Lin and Phair, 1993) is applied to the 10–15 m thick layer. Frequency matched near (1–17°) and far (34–51°) offset seismic stacks are used as input to invert for pressure and saturation effects. There will always be a tradeoff using far offset data; it makes the measurements more independent, but violates the assumptions behind the reflectivity measured from the Smith and Gidlow (1987) approximation of small angles (up to 30°). Also, for angles above 45°, the $\tan^2(\theta)$-term in the Smith and Gidlow (1987) approximation, (Eq. (18.3)) will deviate from the $\sin^2(\theta)$ term and cause erroneous time-lapse AVO

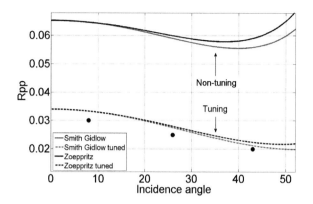

Figure 18.4 P-wave reflectivity between the Tubåen Fm. and underlying unit from the Zoeppritz equation (black) and the Smith and Gidlow approximation (red). Dashed lines is estimated reflectivity after thin-layer tuning correction. The tuning thickness is 13 ms. Calibrated reflectivity from near, mid and far seismic stacks is marked by black circles.

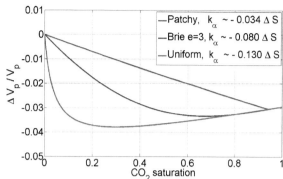

Figure 18.5 Relative changes in P-wave velocity from CO_2 saturation change estimated from Gassmann fluid substitution. The red curve is CO_2 distributed in large patches in the pores. Blue curve is partial patchy CO_2 distribution according to Brie's equation with exponent $e = 3$. The green curve is uniform CO_2 distribution in the pores.

trends. The error introduced by applying the Smith and Gidlow (1987) approximation to the Zoeppritz equation increases with offset and is at a maximum of 8% for the far offset seismic stack (Figure 18.4).

Rock Physics Input

Relative change in velocity caused by saturation changes is estimated from Gassmann fluid substitution (Gassmann, 1951). This model assumes that the bulk modulus is sufficient to characterize the pore fluid. Assessing this number is not straightforward in systems in which one has a mixture of fluids, which is the case when CO_2 is injected into a saline aquifer (Sen and Dvorkin, 2011). Effective bulk modulus of fluids distributed in large patches (patchy saturation) taking residual water saturation into account can be described from Mavko *et al.* (2009) and Sengupta (2000). Uniform fluid distribution in the pores is estimated from the harmonic average of the fluid bulk moduli. The difference between the bounds can be large. Studies undertaken by Konishi *et al.* (2008) and Caspari *et al.* (2011) indicate a partially patchy fluid distribution for the Nagaoka data set where CO_2 was injected into a saline aquifer, measured at log frequencies (10–100 kHz). In our study the patchy and uniform fluid distributions are used as upper and lower bounds for the fluid bulk moduli, and compared to a partial patchy fluid saturation described from Brie's equation (Brie *et al.*, 1995) using an exponent $e = 3$ (Figure 18.5). There are no

available measurements from Snøhvit that might indicate whether the saturation is patchy or uniform. The choice of Brie, $e = 3$, is based on geological knowledge (heterogeneities at several scales), and the results from the Nagaoka data set. It should be stressed, however, that the patchiness in the Nagaoka data set is observed for frequencies well above seismic frequencies. The second order approximation to the saturation change, as introduced by Meadows (2001), would better describe the observed variation in the relative velocity for uniform fluid distribution. The straight line approximation introduces an error of 0.5% to the data assuming a CO_2 saturation of 0.3 for patchy fluid distribution in the pore space. This error increases to 5% for partially patchy (Brie equation) fluid distribution and 50% for uniform fluid distribution, where the velocity–saturation curve becomes highly nonlinear and a straight line approximation is no longer sufficient to fit the data.

The pressure impact on the velocity is estimated from an average of three dry core measurements taken from the Tubåen Fm. Depths, V_p, V_s, and density of these, after brine substitution, is plotted in orange in the well logs in Figure 18.3. Effective pressure before injection (27 MPa) is reduced to 21 MPa by the increase in pore pressure (Figure 18.6, upper figure). The reduction in pressure is comparable to a pressure effect with a Hertz–Mindlin exponent of (1/12) for both the P and S velocities (Figure 18.6, lower diagram). The Hertz–Mindlin model (Mindlin, 1949) describes the effective elastic properties of a precompacted granular rock of 36% porosity and

323

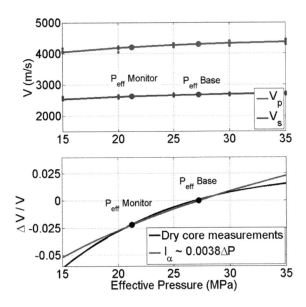

Figure 18.6 (Upper) P-wave- (red) and S-wave (blue) velocity as a function of effective pressure estimated from dry core measurements in the Tubåen Fm. (Lower) relative change in velocity from the dry core measurements compared to the relative change in velocity estimated with a Mindlin exponent of 1/12. The P- wave and S-wave show approximately the same pressure sensitivity, and is assumed similar. Effective pressure for base and monitor survey is indicated in the figures.

six grain-to-grain contacts. The Hertz–Mindlin model gives a P- and S-wave velocity proportional to the different stress raised to a power of (1/6). Hertz–Mindlin is known not to fit real pressure data well (Saul and Lumley, 2013), and is here used only to describe the grain arrangement, which is an indication of the stiffness of the rock. Duffaut and Landrø (2007) showed that by increasing the number of grain-to-grain contacts in the Hertz–Mindlin model they were able to represent the rock consolidation. Field examples from the Gullfaks and the Statfjord fields showed that Hertz–Mindlin exponents of (1/10) and (1/15) correlated well with dry core measurements. The higher amount of grain-to-grain contacts at the Statfjord field was due to more poorly sorted sandstone.

Results of Fluid-Pressure Discrimination Process

Time-lapse difference between the base and monitor surveys observed in Figure 18.2 is a combined effect of both pressure and CO_2 saturation increase. The pressure effect dominates the time-lapse seismic except in the near well area where a fluid effect is also observed (Figure 18.7). The calibration of the reflectivity was performed on the bottom reservoir

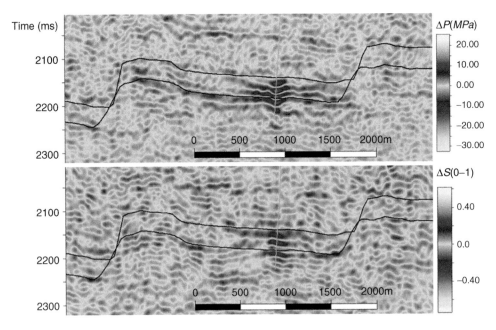

Figure 18.7 Cross section through the reservoir zone for the inverted pressure measured in MPa (upper) and the inverted saturation measured in fraction of 1 (lowermost). Negative ΔS is caused by sidelobe effects. Top and base reservoir zone are shown by the black lines and the injection well is the blue vertical line. A maximum pressure of 15 MPa is measured and CO_2 saturation of 0.22 for partially patchy fluid distribution is measured for the lowermost horizon close to the well.

and is hence the horizon showing representative numerical values. The pressure increase is strong in the near well area (up to 15 MPa) and decreases with distance from the well. A continuous pressure buildup (4–5 MPa) terminates against the faults for the lowermost sandstone unit. A CO_2 saturation of 0.22 is observed in the near well area for an intermediate fluid distribution according to Brie's equation with an exponent $e = 3$. The anomalies for the uppermost

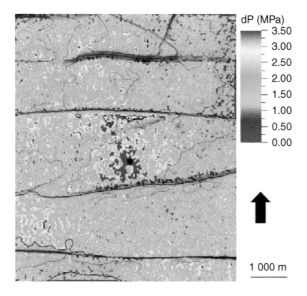

Figure 18.8 RMS amplitude measured in the Tubåen Fm. on the inverted pressure 3D cube. The CO₂ injection well is shown by the black circle, where a maximum pressure of 9 MPa is measured. The north direction is indicated by the black arrow.

parts of the Tubåen Fm. are limited to the near well area as compared to the lowermost unit, where the effect reaches the faults. The variation between the sandstone units is an indication of minor vertical flow between the units. The formation was perforated in three different intervals (red vertical lines in Figure 18.3), and the observed anomalies show different behavior for the layers.

Scaling of the input data would have an impact on the magnitude of the saturation and pressure. Here it is reasonable to assume some uncertainty in the scaling, as the predicted pressure increase from the BHP data is maximum 10 MPa and the estimated value of 15 MPa exceeds that.

Root mean square (RMS) seismic amplitudes measured within the Tubåen Fm. (Figure 18.8) indicate that the fluvial channels may restrict horizontal pressure distribution. The sandstone channel(s) from southeast to northwest act as high-permeability pathways for pressure, which terminate against the faults. An amplitude response is observed in the western corner. This response may be caused by pressure propagation westward inside the reservoir block and into the neighboring block, where the fault fades into a ramp and is, therefore, not limiting the migration of pressure any longer, or it may be an artifact. Amplitude maps from the Fuglen Fm. above the reservoir exclude acquisition nonrepeatability as an effect causing this anomaly. Changing the fluid distribution in the pores, from uniform to patchy mixing, leads to scaling of the saturation effect, with no change in the lateral extent (Figure 18.9). RMS amplitude maps (for

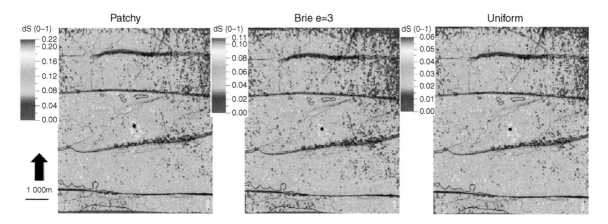

Figure 18.9 RMS amplitude measured in the Tubåen Fm. on the inverted saturation 3D cube, for fluids distributed in large patches (left), according to the Brie theory with Brie $e = 3$ (center) and uniform fluid distribution (right) as described above the figures. The CO₂ injection well is shown by the black circle, where a maximum saturation of 0.37 (Patchy), 0.17 (Brie $e = 3$), and 0.10 (Uniform) measured in the well. The north direction is indicated by the black arrow.

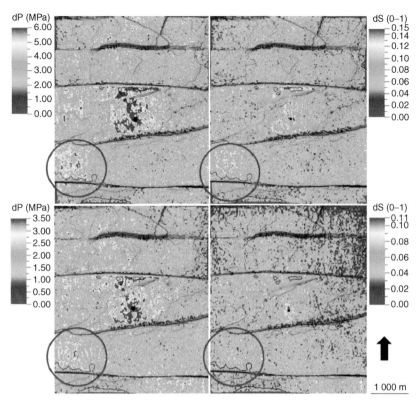

Figure 18.10 RMS amplitude measured in the Tubåen Fm. on the inverted pressure (left) and saturation (right) 3D cubes based on near and mid seismic stack (uppermost cubes) and near and far seismic stack (lowermost cubes). The CO_2 injection well is shown by the black circle. The north direction is indicated by the black arrow. The red circle focuses on the amplitude effect in the neighbor fault block to the injection block, where the noise is far less for the near-far seismic stacks (lowermost figures). The leakage between the inverted pressure and saturation can be seen for the near-mid seismic stacks (uppermost figures).

the entire reservoir thickness) of the estimated pressure and saturation change are expected to show the extent of the fluid and pressure distribution. However, the actual numerical values are then reduced, since the effects are spread over the entire reservoir thickness. The estimated pressure increase of 15 MPa in the well is reduced to 9 MPa when RMS amplitudes are estimated for the entire reservoir thickness. Estimated CO_2 saturation of 0.22 is reduced to 0.17 for RMS amplitudes with a partially patchy fluid distribution in the pore space. Background RMS seismic amplitudes measured above the Tubåen Fm. show a mean noise level of $S_{CO_2} = 0.04$ and a corresponding noise level of 1.5 MPa in pressure change.

Figure 18.10 shows pressure and saturation changes using near and mid seismic stacks (18–34°) as input to the inversion instead of near and far seismic stacks; the remaining workflow is identical. Clear leakage between the pressure and saturation 3D cubes can be seen. The leakage shows the importance of having a large separation between the seismic stacks to treat them as separate measurements and

make offset inversion more stable. The reader should especially note the amplitude effect observed in the southwest corner on both the pressure and saturation amplitudes for the near and mid seismic stack. This anomaly is absent on the saturation inverted from the near and far offset seismic stacks (lower Figure 18.10).

A comparison of time shift estimated from cross-correlation and inverted from the pressure and saturation can be seen in Figure 18.11. A partially patchy fluid distribution according to Brie $e = 3$ is used as an input. The time shifts show a good correlation, both in lateral extent and in magnitude. Based on this, one can assume that the saturation and/or the pressure changes occurred in the lowermost layer. Using the RMS values and not assuming that peak values are valid for the interval gives a reasonable quantitative fit. This fit also indicates a fluid distribution somewhere between pure uniform and large patches. An anomaly is observed in the neighboring reservoir segment (lower western corner) on the inverted pressure that is not visible on the time-shift map. This indicates that the anomaly observed on the

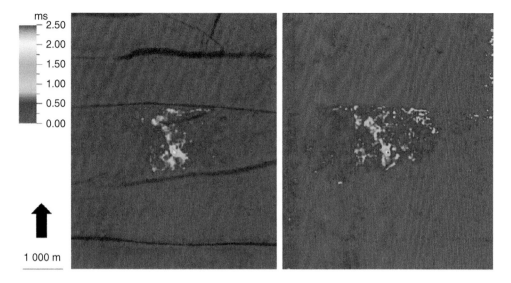

Figure 18.11 Estimated time shift from the inverted saturation and pressure (left) compared to time shift estimated from cross-correlation (right). A cross-correlation window of 200 ms covering the reservoir zone is applied. The CO₂ injection well is located in the middle of the reservoir block where the maximum time-shift value is present. The north direction is indicated by the black arrows.

inverted pressure (marked by the red ring in Figure 18.10) is either caused by a very thin pressurized layer or that it is an artifact.

Uncertainties

Uncertainties in the inverted parameters will depend on the uncertainty in the input parameters (Landrø, 2002). A detailed quantitative study of the uncertainties is not performed here, but the following qualitative uncertainties of some of the input parameters are listed:

- Reservoir thickness, T: The thickness of the reservoir experiencing the time shift caused by injection may not be proportional to the reservoir thickness. Intervening shale units act as barriers to vertical flow so the CO₂ may not have reached the top of the reservoir. The pressure front terminates against impermeable shale units. The top reservoir is impermeable, and considered as the maximum thickness.
- Time shift, ΔT: Different cross-correlation windows result in different time shifts. The average and standard deviation from 11 different cross-correlation windows is used in this study.
- Seismic 3D cubes for ΔR_0 and ΔG: A small-angle approximation is used in the calibration of the input zero-offset reflectivity and gradient 3D cubes. The small angle approximation gives an

error of a maximum of 8% at far offset seismic stack as compared to the Zoeppritz equation. A mean value is chosen to represent the entire span of the near (1–17°) and far offset (34–51°) seismic stacks. Other weighting would lead to different zero-offset reflectivity and gradient reflectivity 3D cubes.
- Linear fit to the velocity–saturation curve: The linear fit to the velocity–saturation curve as shown in Figure 18.5 is valid for a fluid distribution between the uniform and patchy bounds up to S_{CO_2} = 0.3. No unique CO₂ saturation can be found from the relative change in velocity for a uniform fluid distribution in the pore space.
- Uncertainties in zero-offset reflectivity and gradient reflectivity are not evaluated here. In general, the zero-offset reflectivity is expected to be fairly stable (approx. 2 × NRMS), while the gradient reflectivity will be more uncertain (approx. 5 × NRMS) (David Lumley, pers. commun.).

Comparison between the fluid effect in Figure 18.9 and Figure 18.10 shows the importance of a wide separation between the near and far offset seismic stacks to have two independent measurements, and reduce the leakage between the inverted saturation and the pressure. The far offset seismic stack violates the assumption of θ below 30° in the Smith and

Gidlow (1987) approximation. The deviation between the Smith and Gidlow approximation and the Zoeppritz equation is 3–8% for the far offset seismic stack (Figure 18.4).

Pressure and fluid changes are expected to dominate the time-lapse signature and are the only effects that are evaluated in this study. Other changes caused by injection, such as thermal fracturing, may also influence the time-lapse signature. The injected CO_2 has a temperature of approx. 26°C, while the temperature in the formation is 95°C. The injected CO_2 will cause a significant cooling of the rock that may lead to thermal fracturing, followed by a thermal expansion of the CO_2 plume as it heats up.

Reservoir Injectivity

Reservoir injectivity characterizes the ease with which fluid can be injected into a geological formation without pressure building up (Benson et al., 2005). The injectivity is primarily controlled by the permeability and available pore volume. Permeabilities above 100 mD are recommended to ensure sufficient access to the pore space (Cooper, 2009). The permeability around the wellbore may be different from the formation permeability, due to "skin effects" caused by drilling, pore clogging with drilling mud and chemical reactions. A positive skin represents a zone of reduced permeability around the wellbore, while a negative skin represents increased permeability (Hurst et al., 1969). Salt precipitation may clog pore throats and reduce the porosity and permeability. This effect is probably limited to the dry-out front, a small radius around the injection well (Pruess and Müller, 2009; Bacci et al., 2011). Weekly injection of a 90:10 mixture of methyl ethyl glycol and water (MEG) was implemented in the Tubåen Fm. to mitigate salt precipitation (Hansen et al., 2013).

Laboratory Measurements

Supercritical CO_2 was injected into two brine filled core plugs from the Tubåen Fm. to understand how precipitated salt could reduce the porosity and permeability, and impact the P-wave velocity. Experimental parameters were chosen to mimic the reservoir conditions in the Tubåen Fm., given the capacity of the laboratory equipment. The dry samples were preheated to 60°C, before the loading pressure curve was acquired, up to 28 MPa confining pressure.

Synthetic brine of similar salinity as the formation brine (14%), and with the main salt components, was then injected into the sample. The confining pressure (up to 56 MPa) and pore pressure (up to 28 MPa) were increased simultaneously to keep a fixed differential pressure (ΔP) of 28 MPa. Supercritical CO_2 was then injected at a rate of approx. 1 ml/min, until no more brine was produced at the outlet. The chosen injection rate was based on the capacity of the laboratory equipment, and does not represent the reservoir condition. Finally, an unloading sequence was applied, and measurements of porosity and permeability on CO_2 dry samples at the residual water saturation were performed. Reservoir temperature in the Tubåen Fm. at Snøhvit is 95°C. The laboratory experiment was first described by Grude et al. (2013a) and is rephrased here for completeness.

Precipitated salt reduced the porosity and permeability for both the horizontal (Q-274h) and vertical (Q-274v) core plugs. The impact is greatest in the vertical core plug (Figure 18.12). Preinjection pressure sensitivity of the P-wave velocity is more significant than the post injection, especially for the horizontal core plug with pressure below 10 MPa (Figure 18.12). A cycle of loading–unloading of the confining pressure was performed prior to injection to reduce the impact of nonelastic deformation during loading and unloading cycles on the vertical core plug. The reduction in porosity due to precipitated salt and grain rearrangement will lead to increased P-wave velocity. A porosity–velocity relationship from well log measurements was used to correct the laboratory measured velocity (Figure 18.12, magenta triangles). Porosity corrections make the pre- and postvelocity measurements comparable in pressure sensitivity at ΔP above 20 MPa (Grude et al., 2013b).

Relative change in P-wave velocity ($\delta V_p / V_p$) due to pressure, saturation, and salt precipitation in these two experiments is shown in Figure 18.13 and Table 18.1. Differential pressure was 27 MPa before CO_2 injection. This was reduced to 17–21 MPa in the time between the base and monitor seismic surveys. Observations from this laboratory data show that the P-wave velocity is reduced by the reduced differential pressure (marked $\Delta V_{Pressure}$ in Figure 18.13). Fluid substitution from brine to CO_2 also reduces the P-wave velocity (marked ΔV_{Fluid} in Figure 18.13). Salt precipitation is not affecting the P-wave velocity

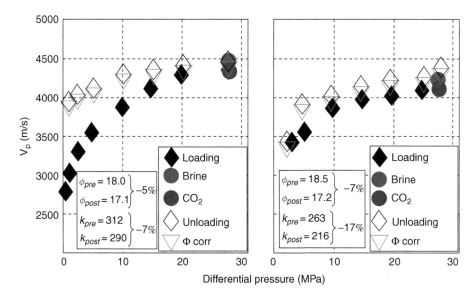

Figure 18.12 P-wave velocities as a function of differential pressure (ΔP) estimated on horizontal (left) and vertical (right) plugs from Q-274. The filled diamonds are loading of the dry plug; the open diamonds are unloading of the CO_2-dried plug. Pink triangles are unloading measurements corrected for porosity reduction. Red circles are brine-filled and blue are CO_2-filled. Porosity and permeability measured pre- and postinjection are given in the figure insets (Grude *et al.*, 2013b).

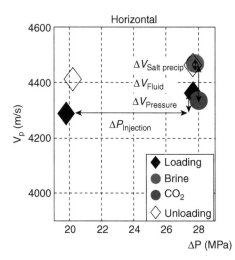
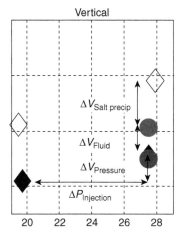

Figure 18.13 P-wave velocities as a function of differential pressure (ΔP) estimated on horizontal (left) and vertical (right) plugs from Q-274. Symbols have the same meaning as in Figure 18.12. P-wave velocity variation caused by pressure, saturation, and salt precipitation is shown in the figure. (Modified from Grude *et al.*, 2013b.)

in the horizontal core plug. It increases the P-wave velocity in the vertical core plug (marked $\Delta V_{Salt\ precip}$ in Figure 18.13). The directional variations in P-wave velocity caused by salt precipitation may be due to uncertainties in the laboratory experiment or heterogeneities at core scale.

Fall-Off Pressure Measurements

A downhole pressure gauge installed 800 m above the Tubåen Fm. monitored pressure change caused by CO_2 injection (Hansen *et al.*, 2013). The pressure measurements can be seen in Figure 18.14.

A continuous rise in pressure is observed, with minor reduction in periods of shut-in. A rapid pressure increase occurs after 5000 hours of injection. This pressure increase is interpreted as reduced injectivity caused by salt precipitation. The injectivity was partly improved by weekly injection of MEG (Hansen *et al.*, 2013).

Analysis of fall-off data can give information about flow regimes acting in the reservoir that can be associated with reservoir geometry and flow barriers. Several points should be considered for the analysis of fall-off tests. As the shut-in time is much less than the injection time, the

329

Table 18.1 Relative change in P-wave velocities ($\delta V_p/V_p$) measured on core plugs in laboratory

	Range (m)	($\delta V_p/V_p$)
Pressure, ΔP	2000 (to faults)	-0.022^a
Saturation, ΔS	200–300	-0.028^a
Salt precipitation, $\Delta\varphi$	Maximum 10	0.033 (vertical plug)

[a] Average of horizontal and vertical core plug.

Miller–Dyes–Hutchinson (MDH) method of analyzing fall-off data can be used (Bourdarot, 1998; Johnson and Lopez, 2003). That is, the time function for further analysis is considered to be the shut-in time. The pressure analysis technique should suffice because of the relatively small pressure changes seen during the pressure fall-off test. Furthermore, CO_2 is injected in the supercritical state, which makes liquid well testing analysis applicable to CO_2 fall-off testing. A stable injection rate and long shut-in period are desirable to obtain high-quality pressure data for analysis. Four shut-in periods lasting for more than 250 hours were chosen based on these criteria (Figure 18.14). The first shut-in period occurred during the high-pressure period after approx. 5000 hours of injection. Hence, the data are valuable in order to analyze the impact of precipitated salt on the injectivity and pressure behavior.

Different flow regimes can be identified by characteristic patterns on a log-log plot of logarithmic pressure derivative versus shut-in time, i.e., ($dP/d\ln(\Delta t)$). Pressure derivatives are calculated from the fall-off data using a differentiation algorithm. In general, the following flow regimes will occur:

- Early time unit slope line (Wellbore storage)
- Zero slope line (Radial flow regime)
- Half slope line (Linear flow regime)
- Late time unit slope line (pseudo-steady-state flow regime)

Wellbore storage affects the pressure immediately after shut-in and is mainly governed by wellbore hydraulics. The flow regime dominating the pressure response after wellbore storage depends on the size of the reservoir and potential flow barriers. Radial flow toward the well is most likely to occur for open homogeneous reservoirs without flow barriers. Linear flow indicates a semiclosed reservoir where faults, sandstone channels, or other geological heterogeneities channelize the flow.

Pseudo-steady state flow indicates a closed system, where the response from the well has encountered all the boundaries around the well (Johnson and Lopez, 2003). Boundary dominated flow regime is a combination of linear and pseudo-steady-state flow regimes.

Coinciding pressure and pressure derivative curves with close to unity slope were observed on logarithmic plots initially for all the shut-in periods. This is interpreted as pressure measurements

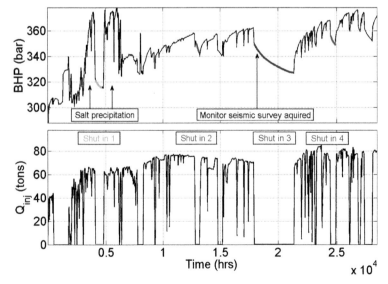

Figure 18.14 (Upper) Pressure measured 800 m above the injection interval and extrapolated to bottom-hole-pressure (BHP). (Lower) Injection rate (Q_{inj}) measured in the well. The rapid increase in BHP observed after approx. 5000 hours of injection is likely caused by reduced injectivity from precipitated salt. Four shut-in periods are used in fall-off analysis. Base survey was acquired prior to injection start. Time of monitor seismic survey is marked.

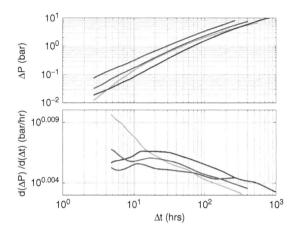

Figure 18.15 Log-log plot of fall-off pressure (upper) and derivative of the pressure (lower, $d\Delta P/d\Delta t$) versus time for the four shut-in periods. ΔP and Δt corresponds to pressure and time since shut-in. The colors correspond to the shut-in periods in Figure 18.14.

dominated by wellbore effects. The wellbore storage period is interpreted to gradually transform into boundary-dominated flow. A deviation from this general trend can be observed for the first shut-in period. This deviation is evident in a derivative plot ($d\Delta P/d\Delta t$) (Figure 18.15). A unit slope line is observed on pressure and pressure derivative plots for shut-in periods 2–4 (Figure 18.15). This can be interpreted as a pseudo-steady-state flow regime, where the pressure boundaries (channel boundaries and faults) are felt immediately. The slope is steeper for the first shut-in during the initial 30 hours. This indicates a closed reservoir with even lower injectivity initially that slows down the pressure propagation, and may be caused by flow from the formation behind the dry-out front where the salt has reduced the injectivity. The slope declines approx. 30 hours after shut-in. The steeper slope initially is not present for the shut-in periods after MEG injection, which indicates that the low injectivity was resolved by injection of MEG.

The extent of the CO₂ front between base and monitor seismic surveys (approx. 18 000 hours of injection) is interpreted to be approx. 420 m in the channel direction (Grude et al., 2013c). This length of the anomaly corresponds to horizontal CO₂ migration of approx. 0.023 m/hour. The low-permeability zone surrounding the injection well was interpreted to

affect the pressure measurements up to 30 hours after shut-in. Considering the CO₂ velocity in the reservoir this corresponds to approx. 0.7 m and can be an indication of the dry-out radius where salt precipitation reduced the injectivity.

Hansen et al. (2013) published an analysis of the pressure fall-off data from the 3 months long shut down in 2009 (corresponding to shut-in 3 in Figure 18.14) from a production logging tool (PLT) and the down-hole pressure gauge. Based on the PLT falloff data, Hansen et al. (2013) found that a flow barrier at about 3000 m from the well must be added, in order to match the permanent gauge pressure history and falloffs. The active and connected reservoir to the well then becomes rectangular in shape with no flow boundaries at 110 m, 3000 m, and 110 m distance from the well. The shallow location of the downhole pressure gauge generated challenges for the fall-off analysis, and reservoir behavior was recognized after 100 hour (Hansen et al., 2013). The analysis performed here is based on the early injection time, and the results should be treated with care owing to the location of the pressure gauge. Still, the shallow location of the pressure gauge is expected to influence all the shut-in periods equally, and valuable information can be gained by comparing these.

CO₂ Saturation and Thickness Prediction from Analytical Expression

This study focuses on the 0.4 Mtonnes of CO₂ injected into the Tubåen 1 sandstone unit prior to the monitoring survey in September 2009. Table 18.2 summarizes the reservoir parameters of the Tubåen 1 unit at the Snøhvit field, including the estimated mobility ratio and gravity factor. CO₂ and brine fluid properties are given in Table 18.3.

Capillary entry pressure for two core plugs from the Tubåen 3 sandstone unit is 0.01 MPa and the capillary pressure is 1 MPa at residual brine saturation measured in the laboratory for these samples. The capillary pressure is low compared to the injection overpressure of a maximum of 6–10 MPa. No measurement of capillary pressure was performed on core plugs from the Tubåen 1 sandstone unit, but the capillary pressure is most likely negligible based on the measurements in the Tubåen 3 unit and the average permeability of 750 mD in the Tubåen 1 unit. Laboratory measurements of the vertical and

Table 18.2 Average reservoir properties, mobility ratio, and gravity factor for the Tubåen 1 sandstone unit

Thickness	h	14	m
Porosity	φ	0.19	fraction
Permeability	k	750	mD
Residual brine saturation	S_{br}	0.13^a	fraction
Relative permeability of CO_2 at S_{br}	$k_{r,CO2}$	0.7^b	fraction
Injected volume	Q(Mtonnes/s)× t(s)	0.4	Mtonnes
Mobility ratio	λ (λ $_{CO2}$/λ $_{Brine}$)	5.65	
Gravity factor	Γ	0.6	

^{a} Average of horizontal and vertical core, estimated from Sengupta (2000), taking the porosity and permeability into consideration.

^{b} No laboratory measurement available from the Tubåen 1 unit. The value is assumed from comparison with laboratory measurements of the relative permeability on three composite cores from the Tubåen 3 sandstone unit at 2735 m measured depth (MD) and published literature (Bennion and Bachu, 2008).

Table 18.3 Fluid properties estimated at reservoir pressure of 27 MPa, temperature of 90°C, and brine salinity of 14%.

CO_2 density,	ρ_{CO2}	663	kg/m³
Brine density	ρ_{Brine}	1090	kg/m³
CO_2 viscosity	μ_{CO2}	0.055	mPa s
Brine viscosity	μ_{Brine}	0.444	mPa s

The CO_2 density is estimated from the Span and Wagner (1996) equation of state. CO_2 viscosity is estimated from Scalabrin et al. (2006). Brine density is estimated from Batzle and Wang (1992). Brine viscosity is estimated from Kestin et al. (1978).

horizontal permeability were performed on core plugs from the sandstone units. These showed that the permeability was approximately similar in both directions. Gravitational forces are expected to be small in this 10–15 m thick sandstone unit with vertical permeability of 750 mD. It is reasonable to assume that viscous forces govern the CO_2 flow in the Tubåen 1 sandstone unit.

Nordbotten et al. (2005) and Nordbotten and Celia (2006) developed an analytical expression that describes the shape of a CO_2 plume when the capillary pressure is neglected. In this model, the CO_2 is injected into a porous formation that is confined above and below by impermeable rock formations. The expression is viscous dominated, with influence from gravity. Their dimensionless gravity factor, Γ, can be used to identify when the gravity force

contributes to the fluid flow (Eq. 14 in Nordbotten et al. [2005]):

$$\Gamma = \frac{2\pi\Delta\rho g\lambda_b kB^2}{Q_{well}} \quad (18.11)$$

where $\Delta\rho = \rho_{brine} - \rho_{CO_2}$ is the difference in density between brine and CO_2; g is the gravitational constant, $\lambda_b = k_{r, b}/\mu_b$ is the mobility of brine in the formation, which depends on the relative permeability ($k_{r,b}$) and viscosity of the brine (μ_b). The saturation in the brine phase is 1 due to the assumptions of immiscible fluids; this makes $k_{r,b} = 1$ (Okwen et al., 2010). B is the thickness of the formation, k is the vertical permeability, and Q_{well} is the injection rate (Nordbotten et al., 2005). Nordbotten et al. (2005) and Nordbotten and Celia (2006) found that a lower cutoff value of 1 for the gravity factor is suitable to illustrate when gravity forces exert an influence on the CO_2 flow. Okwen et al. (2010) revised the cutoff value to 0.5, to within a 10% error. Based on this equation, one observes that the influence of gravity increases with the thickness of the formation and vertical permeability, and decreases with increasing brine viscosity and injection rate. The gravity factor is estimated to be approx. 0.6, using Eq. (18.11), in the Tubåen 1 sandstone unit. This value is at the lower cutoff for gravity influenced flow and it is reasonable to neglect this effect as CO_2 migration is expected to be viscous dominated.

Supercritical CO_2 is less dense and less viscous than the formation brine and therefore injected CO_2

will migrate upward and along the overlying shale owing to the density difference (Nordbotten and Celia, 2006). The mobility ratio between CO_2 and brine is defined as

$$\frac{\lambda_{CO_2}}{\lambda_{Brine}} = \frac{k_{r,CO_2}}{k_{r,Brine}} \frac{\mu_{Brine}}{\mu_{CO_2}} \qquad (18.12)$$

where λ_{CO_2} and λ_{Brine} are the mobilities of the CO_2 and brine respectively. Using estimates of the relative permeability and viscosities of the two fluids (Table 18.3), a mobility ratio of 5.65 is derived in the Tubåen Fm. The mobility ratio is strongly dependent on the viscosity for the Tubåen 1 sandstone unit, as the CO_2 has a much lower viscosity than the formation brine at the reservoir pressure of 27 MPa, temperature of 95°C, and brine salinity of 14%. A mobility ratio of 5.65 is relatively low compared to typical mobility ratios of 5–20 for CO_2 displacing brine (Okwen et al., 2010).

The thickness and effective saturation of the CO_2 layer (Figure 18.16) are calculated from the similarity solutions described in Nordbotten and Celia (2006), with Eq. (18.11) as input. The reader is referred to Nordbotten and Celia (2006) for further details. The negligible capillary pressure will lead to a constant saturation within the plume. The CO_2 migrates along the top of Tubåen 1 sandstone unit and leads to a reduction in effective CO_2 saturation within the Tubåen 1 sandstone unit. According to this model, the Tubåen 1 sandstone unit is filled with CO_2 to approx. 40 m from the injection well, before the CO_2 layer thins due to vertical segregation. Maximum radius of the CO_2 layer is approx. 750 m. The CO_2 saturation inside the plume is assumed to be uniformly distributed, and equal to 0.87. The residual brine saturation (S_{br}) is estimated to $S_{br} = 0.13$ for the Tubåen 1 sandstone unit. The reduction in effective CO_2 saturation is due to the thinning of the layer.

Comparing Analytical Solution and Time-Lapse Seismic Data

A CO_2 saturation (S_{CO_2}) of 0.87 is estimated within the plume from the analytical expression. This is much higher than the CO_2 saturation of 0.22 estimated close to the well from the time-lapse fluid-pressure inversion assuming an intermediate fluid distribution in the pore space. This discrepancy between the analytical solution and time-lapse seismic data is most likely caused by the linear approximation between the CO_2 saturation

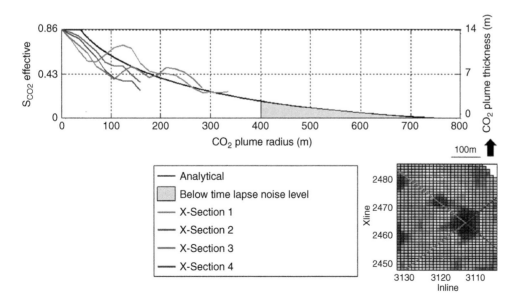

Figure 18.16 Effective CO_2 saturation and CO_2 plume thickness from the analytical solution (black) and four time-lapse seismic lines (blue, green, gray, and red). The input parameters for the analytical solution are given in Tables 18.2 and 18.3. The maximum CO_2 saturation corresponds to 0.87 (1 − residual brine saturation). The time-lapse noise level is marked in the figure. The location of the seismic lines is shown in the figure below. The north direction is indicated by the arrow.

and relative change in velocity caused by saturation (Figure 18.5). The linear approximation is valid up to approx. $S_{CO_2} = 0.1$ for uniform fluid distribution in the pore space and $S_{CO_2} = 0.3$ for an intermediate fluid distribution in the pore space. The spatial extent of the inverted CO_2 (Figures 18.7 and 18.9) is still valid for higher saturations, but no unique CO_2 saturations can be found between approx. $S_{CO_2} = 0.3$–0.87 for an intermediate fluid distribution in the pore space.

The lack of a unique solution gives the possibility of much higher saturations than originally found in Grude et al. (2013c). Comparison of the CO_2 layer thickness and effective CO_2 saturation obtained from the analytical expression and time-lapse seismic data can be made by assuming that the saturation obtained from the analytical expression is true, and scale the S_{CO_2} from the time-lapse seismic according to this. Comparison of the shape of the layer with distance from the injection well can then indicate if the assumption of viscous dominated flow is valid, and if the CO_2 saturation within the layer is close to the residual water saturation, as assumed in the analytical expression.

Comparison of the CO_2 layer thickness and effective saturation from the analytical expression and the time-lapse seismic is presented in Figure 18.16. Four cross sections are chosen for comparison, two in the channel direction (gray and red) and two perpendicular to the channel direction (blue and green). Location of the seismic line is indicated in the figure. Background RMS seismic amplitudes measured above the Tubåen Fm. correspond to approx. 18% of the maximum saturation. This relates to S_{CO_2} effective = 0.16 for the analytical expression. A CO_2 layer radius above 400 m is not detectable in the seismic data, this area is shaded blue in Figure 18.16. The decrease in CO_2 layer thickness and effective saturation inverted from time-lapse seismic follow the same trend as the analytical solution with distance from the injection well, but decreases more rapidly. There are many uncertainties in the input parameters that can explain the deviation. For instance, the mobility ratio is strongly dependent on relative permeability. A lower relative permeability would lead to a lower mobility ratio, and a more piston-like displacement of the fluid front. The CO_2 saturation in NW and SW directions can better describe the radius of the CO_2 plume than in the SE and NE directions. It is likely to assume that the CO_2 saturation is close to the residual water saturation, and constant within the plume from comparison between the analytical expression and the AVO inverted seismic data.

Discussion

Static input parameters for V_p/V_s, porosity, and pressure sensitivity are used in the estimate of pressure and fluid variations caused by CO_2 injection in the Tubåen Fm. Given the strongly heterogeneous nature of the reservoir, additional information on the variation in reservoir microstructure will give valuable input to the inversion. Averaging based on the properties in the near well area of the lowermost injection interval will introduce an error if the reservoir properties change away from the well. Studies have shown that small uncertainties in the reservoir parameters, such as porosity, can lead to large uncertainties in the inverted pressure and saturation (Trani et al., 2011; Shahraeeni, 2012). Still, the inverted time shift and time shift estimated from cross-correlation reveal a striking correlation that is an indication that the major changes occur in the lowermost reservoir section.

The pressure sensitivity on velocity is expected to be low from dry core measurements, the depth of the formation (2670 m), and the high degree of cementation. Still, a major pressure buildup is observed that extends to the faults; this is most likely due to pressurized water (Hansen et al., 2013). No pressure changes in the neighboring segments indicate that the faults are sealing, with pressure propagation westward into the neighboring reservoir block where the fault fades into a ramp.

Trani et al. (2011) showed that introducing the second-order term for the saturation effect does not improve the uncertainty in the gradient term. The validity of combining time shift measured over the reservoir thickness and the localized amplitude at the base of the reservoir is questionable for reservoirs where the saturation and/or the pressure change occur in the entire reservoir zone. This error is expected to be small in the Tubåen Fm., where most of the saturation and pressure changes occur in the deeper parts of the reservoir.

The permeability reduction caused by salt precipitation was estimated to be –7% (horizontal core plug) and −17% (vertical core plug) from laboratory experiments. This reduction in permeability is not sufficient to explain the near well plugging on the pressure data at an early stage of injection (Figure 18.15). There may be several causes for this discrepancy. Comparing laboratory experiments with fluid flow at reservoir scale is not straightforward, and the results from the laboratory measurements can at best indicate the effects observed at a larger scale. The permeability in the Tubåen Fm. varies significantly. Heterogeneities in the reservoir and clogging of important pore throats may have a major impact on the permeability that is not observable at core scale. Also, the interplay among viscous, capillary, and gravity forces in the reservoir fluid flow cannot be reproduced at laboratory scale.

The analysis is limited to early injection times, when the primary trapping mechanisms for CO_2 are stratigraphic and structural trapping. CO_2 dissolution into the formation brine, evaporation of brine into the dry CO_2, capillary trapping, and mineral trapping mechanisms are expected to dominate at a later stage (Bachu *et al.*, 2007; IPCC, 2005). The fraction of CO_2 dissolution into formation water and brine evaporation into the dry CO_2 is expected to be small at an early stage of injection (Nordbotten and Celia, 2006).

Laboratory measurements of capillary pressure on core samples from the Tubåen 3 sandstone unit indicate that capillary pressure can be ignored. These are samples taken at one location in a heterogeneous reservoir. Local heterogeneity may introduce capillary barriers that cannot be ignored and will influence the reservoir fluid flow. The derivation of the analytical solution makes numerous assumptions and simplifications, but concedes that they are not particularly restrictive and are approximately valid for most plausible injection scenarios. In this study, a radial CO_2 distribution around the injection well has been assumed. The assumption is inaccurate in this fluvial environment, where an elliptical shape better describes the CO_2 layer distribution. In addition, CO_2 at a temperature of 26°C is injected at Snøhvit, and heated to 95°C by the formation. This violates the assumption of constant temperature in the analytical expression. Furthermore, CO_2 injection in the Tubåen Fm. occasionally halted due to operational challenges at the Melkøya LNG plant (Hansen *et al.*, 2013). Hence, the injection rate was not constant. However, a constant injection rate is assumed in the model, based on the total amount of CO_2 injected into the Tubåen 1 sandstone unit between acquisition of the seismic surveys. Despite all the deviation from the assumptions made in the analytical expression, there is a good agreement in the shape of the CO_2 layer radius found from the analytical solution and the time-lapse seismic measurements.

The viscosity ratio between brine and CO_2 is approx. 8 under the pressure and temperature regimes of the Tubåen Fm. The CO_2 is less viscous than the brine, and hence has the highest mobility. Based on this significant viscous fingering effects are expected to occur and the effective CO_2 saturation of the formation is low. The viscous fingers are most likely below the resolution of the seismic data, and are difficult to detect.

Conclusions

Pressure dominates the time-lapse signature with the exception of areas near the well, where both a pressure and fluid effect is estimated. Varying the fluid distribution in the pore space leads to scaling of the fluid effect, not changes in the spatial distribution. Estimated pressure close to the well is 15 MPa for the lowermost horizon; this is reduced to approx. 4–5 MPa close to the faults. Estimated fluid saturation varies from 0.06 to 0.22 close to the well for the lowermost horizon, depending on the fluid distribution in the pores (uniform to patchy). There is a good correlation between observed time shifts from cross correlation and time shifts estimated from inverted pressure and saturation. This fit indicates a fluid distribution somewhere between pure uniform and large patches.

Results from the laboratory measurements showed that precipitated salt reduced the porosity and permeability for both of the core plugs. The P-wave velocity increased in the vertical core plug (Q-274v), while it was not affected by salt precipitation for the horizontal core plug (Q-274h). Analysis of bottom-hole-pressure measurements (BHP) from periods of shut-in indicates a low-permeability zone surrounding the well at early stage of injection. The low-permeability zone may be caused by precipitated salt reducing the injectivity. MEG was injected weekly to mitigate salt precipitation.

The distribution and saturation of CO_2 in the Tubåen 1 sandstone unit have been estimated from an analytical expression. The analytical expression assumes that capillary pressure can be ignored, and that the fluid flow is viscous dominated. Furthermore, it supposes the CO_2 distribution inside the plume is uniform and equal to $1 - S_{br} = 0.87$, while the reduction in effective CO_2 saturation is due to the thinning of the layer. The influence of gravity has been shown to be negligible on the reservoir CO_2 flow. Analytical results suggest CO_2 fills the entire sandstone unit up to approx. 50 m away from the injection well before the plume is reduced to a thin wedge that propagates along the top of the formation. The maximum radius of the CO_2 layer from the analytic expression is 750 m, with 400 m likely to be observable above the time-lapse noise level.

CO_2 layer thickness and effective CO_2 saturation obtained from the analytical expression and time-lapse seismic data are in good agreement. This is an indication of viscous dominated CO_2 flow and that S_{CO_2} is close to the residual water saturation. The influence of gravity is negligible on the reservoir CO_2 flow. The maximum observable CO_2-layer thickness inverted from time-lapse seismic is on average 405 m in the channel direction, and 273 m perpendicular to the channel direction.

Acknowledgments

We want to acknowledge Statoil and their Snøhvit license partners Petoro, Total E&P Norge, GDF SUEZ E&P Norge, and RWE Dea Norge for permission to use their data. The Norwegian Research Council is acknowledged for support to the BIGCCS center for environmentally friendly Energy Research (FME) and to the ROSE consortium at NTNU. Jack Dvorkin, Gary Mavko, Anthony Clark, Tiziana Vanorio, and the rest of the SRB group at Stanford University are acknowledged for their input and hosting Sissel Grude during the research exchange to Stanford. The laboratory measurements were done at Stanford. Bård Osdal, Olav Hansen, Philip Ringrose, Ola Eiken, Ashkan Jahanbani, Hossein Mehdi Zadeh, Jan Martin Nordbotten, Erik Lindeberg, Jon Kleppe, James C. White, Rune Martin Holt, Per Avseth, Ole Torsæter, Alexey Stovas, and Børge Arntsen are all acknowledged for direct input and discussions that made this chapter possible.

References

Bacci, G., Korre, A., and Durucan, S. (2011). An experimental and numerical investigation into the impact of dissolution/precipitation mechanisms on CO_2 injectivity in the wellbore and far field regions. *International Journal of Greenhouse Gas Control*, **5**: 579–588.

Bachu, S., Bonijoly, D., Bradshaw, J., *et al.* (2007). CO_2 storage capacity estimation: Methodology and gaps. *International Journal of Greenhouse Gas Control*, **1**: 430–443.

Batzle, M., and Wang, Z. (1992). Seismic properties of pore fluids. *Geophyics*, **57**: 1396–1408.

Bennion, D. B., and Bachu, S. (2008). Drainage and imbibition relative permeability relationships for supercritical CO_2/brine and H_2S/brine systems in intergranular sandstone, carbonate, shale, and anhydrite rocks. *SPE Reservoir Evaluation & Engineering*, **11**: 487–496.

Benson, S. M., Cook, P., Anderson, J., *et al.* (2005). Underground geological storage. IPCC Special Report on Carbon Dioxide Capture and Storage, Chapter 5: Intergovernmental Panel on Climate Change.

Bourdarot, G. (1998). *Well resting: Interpretation methods.* Paris: Technip Publications.

Brie, A., Pampuri, F., Marsala, A. F., and Meazza, O. (1995). Shear sonic interpretation in gas-bearing sands. *SPE*, **30595**: 701–710.

Bryant, S. L., Lakshminarasimhan, S., and Pope, G. A. (2008). Buoyancy-dominated multiphase flow and its effect on geological sequestration of CO_2. *SPE Journal*, **13**: 447–454.

Caspari, E., Müller, T. M., and Gurevich, B. (2011). Time-lapse sonic logs reveal patchy CO_2 saturation in-situ. *Geophysical Research Letters*, **38**: L13301.

Class, H., Ebigbo, A., Helmig, R., *et al.* (2009). A benchmark study on problems related to CO_2 storage in geologic formations. *Computational Geosciences*, **13**: 409–434.

Cooper, C. (2009). A technical basis for carbon dioxide storage. *Energy Procedia*, **1**: 1727–1733.

Duffaut, K., and Landrø, M. (2007). Vp/Vs ratio versus differential stress and rock consolidation: A comparison between rock models and time-lapse AVO data. *Geophysics*, **72**: C81–C94.

Eiken, O., Ringrose, P., Hermanrud, C., Nazarian, B., Torp, T. A., and Høier, L. (2011). Lessons learned from 14 years of CCS operations: Sleipner, In Salah and Snøhvit. *Energy Procedia*, **4**: 5541–5548.

Gassmann, F. (1951). Uber die Elastizitat poroser Medien. *Vierteljahrsschrift der Naturforschenden Gesselschaft*, **96**: 1–23.

Grude, S., Clark, A., Vanorio, T., and Landrø, M. (2013a). Changes in the rock properties and injectivity due to salt precipitation on the Snøhvit CO_2 injection site. Trondheim CCS Conference, June 4-6, Trondheim.

Grude, S., Dvorkin, J., Clark, A., Vanorio, T., and Landrø, M. (2013b). Pressure effects caused by CO_2 injection in the Snøhvit Field. *First Break*, **31**: 3.

Grude, S., Landrø, M., and Osdal, B. (2013c). Time-lapse pressure–saturation discrimination for CO_2 storage at the Snøhvit field. *International Journal of Greenhouse Gas Control*, **19**: 369–378.

Hansen, O., Eiken, O., Østmo, S., Johansen, R. I., and Smith, A. (2011). Monitoring CO_2 injection into a fluvial brine-filled sandstone formation at the Snøhvit field, Barents Sea. In 81th Annual International Meeting. Expanded Abstracts, *Society of Exploration Geophysicists,* 4092–4096.

Hansen, O., Gilding, D., Nazarian, B., *et al.* (2013). Snøhvit: The history of injecting and storing 1 Mt CO_2 in the fluvial Tubåen Fm. *Energy Procedia*, **37**: 3565–3573.

Hurst, W., Clark, J. D., and Brauer, B. (1969). The skin effect in producing wells. *Journal of Petroleum Technology*, **21**: 7.

IPCC. (2005). *Special report on CO_2 capture and storage.* Cambridge: Cambridge University Press.

Johnson, K., and Lopez, S. (2003). *The nuts and bolts of falloff testing.* Washington, DC: United States Environmental Protection Agency.

Kestin, J., Khalifa, H. E., Abe, Y., Grimes, C. E., Sookiazian, H., and Wakeham, W. A. (1978). Effect of pressure on the viscosity of aqueous sodium chloride solutions in the temperature range 20–150 degree C. *Journal of Chemical & Engineering Data*, **23**: 328–336.

Konishi, C., Azuma, H., Nobuoska, D., Xue, Z., and Watanabe, J. (2008). Estimation of CO_2 saturation considering patchy saturation at Nagaoka. In 70th Conference & Exhibition, *EAGE*, Extended Abstract, I018.

Landrø, M. (2001). Discrimination between pressure and fluid saturation changes from time-lapse seismic data. *Geophysics*, **66**: 836–844.

Landrø, M. (2002). Uncertainties in quantitative time-lapse seismic analysis. *Geophysical Prospecting*, **50**: 1–12.

Lin, T. L., and Phair, R. (1993). AVO tuning. In 63th Annual International Meeting, *Society of Exploration Geophysicists,* Expanded Abstracts, 727–730.

Lindeberg, E., and Wessel-Berg, D. (1997). Vertical convection in an aquifer column under a gas cap of CO_2. *Energy Conversion and Management*, **38** (Supplement), S229–S234.

Mavko, G., Mukerji, T., and Dvorkin, J. (2009). *The rock physics handbook: Tools for seismic analysis of porous media.* Cambridge: Cambridge University Press.

Meadows, M. A. (2001). Enhancements to Landro's method for separating time-lapse pressure and saturation changes. In 71th Annual International Meeting, *Society of Exploration Geophysicists,* Expanded Abstracts, 1652–1655.

Mindlin, R. D. (1949). Compliance of elastic bodies in contact. *ASME Journal of Applied Mechanics*, **16**: 259–268.

Nordbotten, J. M., and Celia, M. A. (2006). Similarity solutions for fluid injection into confined aquifers. *Journal of Fluid Mechanics*, **561**: 20.

Nordbotten, J., Celia, M., and Bachu, S. (2005). Injection and storage of CO_2 in deep saline aquifers: Analytical solution for CO_2 plume evolution during injection. *Transport in Porous Media*, **58**: 339–360.

Okwen, R. T., Stewart, M. T., and Cunningham, J. A. (2010). Analytical solution for estimating storage efficiency of geologic sequestration of CO_2. *International Journal of Greenhouse Gas Control*, **4**: 102–107.

Pruess, K., and Müller, N. (2009). Formation dry-out from CO_2 injection into saline aquifers. 1.Effects of solids precipitation and their mitigation. *Water Resources Research*, **45**: W03402.

Saul, M. J., and Lumley, D. E. (2013). A new velocity–pressure–compaction model for uncemented sediments. *Geophysical Journal International*. DOI:10.1093.

Scalabrin, G., Marchi, P., Finezzo, F., and Span, R. (2006). A reference multiparameter thermal conductivity equation for carbon dioxide with an optimized functional form. *Journal of Physical and Chemical Reference Data*, **35**: 1549–1575.

Sen, A., and Dvorkin, J. (2011). Fluid substitution in gas/water systems: Revisiting patchy saturation. In 81th Annual International Meeting, *Society of Exploration Geophysicists,* Expanded Abstracts, 2161–2165.

Sengupta, M. (2000). *Integrating rock physics and flow simulation to reduce uncertainties in seismic reservoir monitoring.* PhD thesis, Stanford University.

Shahraeeni, M. S. (2012). Effect of lithological uncertainty on the timelapse pressure-saturation inversion. In 74th Conference & Exhibition Incorporating *SPE EUROPEC EAGE,* Extended Abstract, Y045.

Smith, G., and Gidlow, P. M. (1987). Weighted stacking for rock property estimation and detection of gas. *Geophysical Prospecting*, **35**: 993–1014.

Span, R., and Wagner, W. (1996). A new equation of state for carbon dioxide covering the fluid region from the triple-point temperature to 1100 K at pressures up to 800 MPa. *Journal of Physical and Chemical Reference Data*, **25**: 1509–1596.

Trani, M., Arts, R., Leeuwenburgh, O., and Brouwer, J. (2011). Estimation of changes in saturation and pressure from 4D seismic AVO and time-shift analysis. *Geophysics*, **76**: C1–C17.

Chapter 19

Illinois Basin–Decatur Project

Robert A. Bauer, Robert Will, Sallie E. Greenberg, and Steven G. Whittaker

Introduction

The Illinois Basin–Decatur Project (IBDP) is an integrated bioenergy carbon capture and geological storage (BECCS) project conducted at Archer Daniels Midland Company's (ADM) corn processing plant in Decatur, Illinois, USA (Figure 19.1). Over three years approx. 1000 tonnes/day of carbon dioxide (CO_2), obtained from ethanol production at the ADM plant, was compressed, dehydrated, sent along a 1.9-km pipeline, and injected into the Mt. Simon Sandstone 2.14 km deep in the Illinois Basin. Injection took place from November 17, 2011 to November 26, 2014, with 999 215 tonnes of supercritical CO_2 injected and geologically stored (Greenberg et al., 2017).

The IBDP is led by the Illinois State Geological Survey (ISGS) through the Midwest Geological Sequestration Consortium (MGSC) that includes partners Schlumberger Carbon Services, Trimeric Corporation, ADM, and other research organizations. The MGSC is one of seven regional partnerships funded by the U.S. Department of Energy–National Energy Technology Laboratory (DOE–NETL) to build capacity and gain experience in CCS through regional characterization, pilot studies, and demonstration projects. In addition to the injection operations, the MGSC has deployed a full range of surface and subsurface monitoring verification and accounting (MVA) technologies throughout each of the project phases (preinjection, injection, and postinjection) to evaluate their effectiveness for use at long-term, large-scale CCS projects, including geophysical methods.

The IBDP site includes one injection well (CCS1), one deep monitoring well (VW1), one dedicated geophysical well (GM1), and a variety of near surface monitoring wells and equipment. For MVA, a combination of continuous pressure measurements from the injection and multilevel monitoring well, continuous passive monitoring of microseismic

activity, repeat Reservoir Saturation Tool (RST)[1] logging measurements, and repeat surface and borehole seismic surveys have been used to monitor the development of pressure and CO_2 saturation throughout the project. In addition to subsurface monitoring and characterization, the IBDP MVA program included extensive soil flux, atmospheric monitoring, and shallow groundwater monitoring activities. A unique challenge related to this site is that it is at a very active industrial site that impacts significantly on monitoring activities and strategy. This chapter will focus on the geophysical aspects of site characterization and monitoring conducted at the IBDP from 2007 to 2017 (Figure 19.2).

Geological Setting and Site Characterization

Geological site characterization at the IBDP is based on examination of more than 250 m of whole core, geophysical well logs, two-dimensional (2D) and three-dimensional (3D) seismic, and a range of tests and analyses (Freiburg *et al.*, 2014). The primary injection target of the IBDP is the Cambrian Mt. Simon Sandstone, an extensive formation that underlies much of the Midwestern United States, which is also used regionally for geological storage of natural gas. The Mt. Simon reaches its maximum thickness of approx. 790 m near the IBDP site, and is divided lithostratigraphically into Lower, Middle, and Upper sections. These major sections are further divided into units based on depositional facies interpreted using core examination and downhole geophysical logging results (Figure 19.3), and include fluvial braided river, floodplain, and alluvial plain deposits; eolian (windblown deposit) sandsheet, dune, and interdune deposits; and shallow marine deposits. The injection of CO_2 took place within a single lithological unit of

---

[1] Mark of Schlumberger.

Figure 19.1 Map of the IBDP site with the location of wells. The CCS1, VW1, and GM1 wells are infrastructure for IBDP. CCS2, VW2, and GM2 are similar well configurations installed for the Illinois Industrial Carbon Capture and Storage (ICCS) Project.

the Lower Mt. Simon Sandstone between 2129 m and 2138 m, where porosity and permeability range 18–25% and 40–380 mD, respectively; the reservoir interval overall has an average porosity of about 20% and an average permeability near 200 mD.

The Mt. Simon is overlain by the Eau Claire Formation, a 150 m thick impermeable package that hydraulically isolates the Mt. Simon from overlying strata (Palkovic, 2015). A 70-m shale unit in the lower part of the Eau Claire forms a highly effective seal to vertical fluid movement, and the upper portion of the Eau Claire is dense limestone with thin stringers of

siltstone. Together the Mt. Simon and Eau Claire form the storage complex at the IBDP.

Pre-Mt. Simon strata (informally known as the Argenta Formation) are in disconformable contact with the lower Mt. Simon and, although they are lithologically similar to the Mt. Simon, they have distinct depositional character and are generally well cemented and compacted throughout, resulting in significantly lower porosity and permeability (Figure 19.3). The Argenta deposits comprise the basal beds in the sedimentary succession at IBDP, and are in sharp, nonconformable contact with the

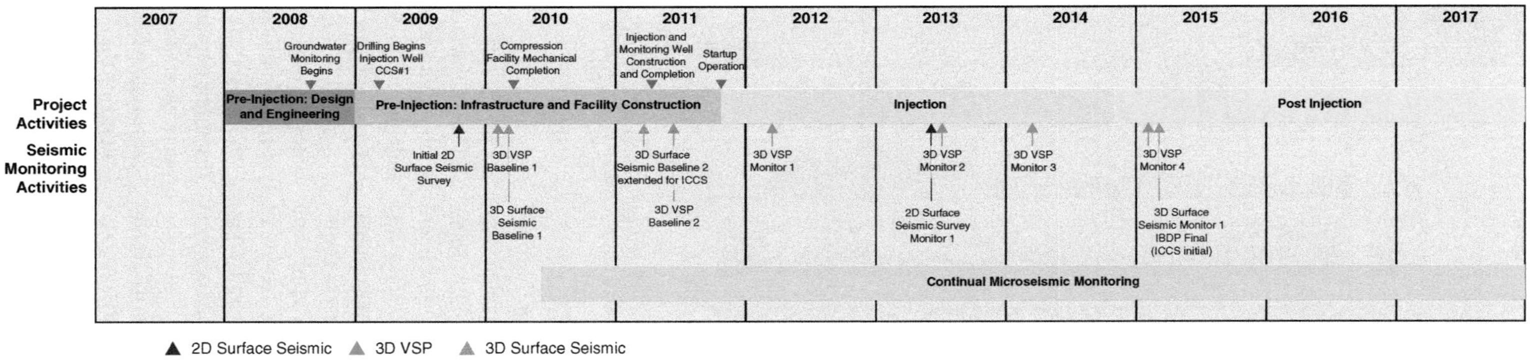

Figure 19.2 Geophysical program timeline.

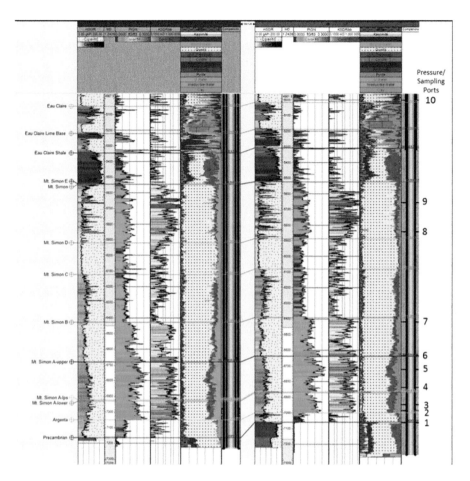

Figure 19.3 Cross section of the IBDP study site. CCS1–VW1 cross section showing the Precambrian crystalline basement, Argenta Formation (Pre-Mt. Simon), Mt. Simon Sandstone, and part of the Eau Claire Formation.

Precambrian basement. The Argenta can contain angular clasts of underlying basement rhyolite, and regional seismic data indicates more than 500 m topographic relief may exist on the Precambrian basement surface prior to Phanerozoic deposition (Leetaru and McBride, 2009). At the IBDP site, the Precambrian crystalline basement is layered rhyolite over granodiorite over granite, as shown by sidewall cores and geophysical logs (Bauer *et al.*, 2016). The basement rock can be highly fractured showing mineralization, slight alterations along the cracks, or a combination of both (Haimson and Doe, 1983).

The bedrock strata in the region of the IBDP site exhibit a slight dip of approx. 1° to the southeast, and the closest structures to the site (Figure 19.4) are minor anticlines with small structural closure about 40 km south and 30 km north.

Subsurface Configuration

The subsurface components of the IBDP geophysical site characterization and monitoring network are deployed in three multipurpose wells; the injector CCS1 (Figure 19.5), the deep verification and monitoring well VW1 (Figure 19.5), and the geophysical monitoring well GM1 (Figure 19.6). In addition to its primary function of injecting supercritical CO_2 into the Lower Mt. Simon, CCS1 hosts downhole pressure and temperature gauges, primary elements of the IBDP passive seismic monitoring array, and was an access for repeat geophysical logging, all of which are key components of the MVA program. The VW1 well, although not hosting seismic measurements, provided 11 pressure measurement and fluid sampling ports throughout the geological column extending from the Precambrian

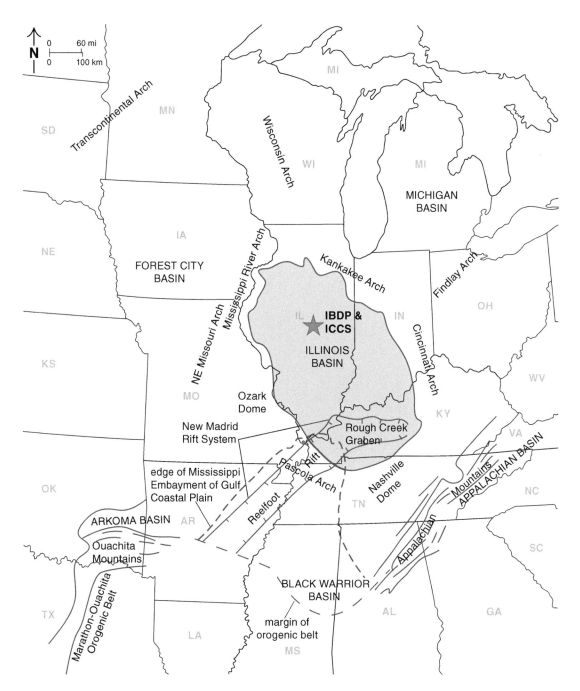

Figure 19.4 Regional fault map. The map shows regional faults with the location of the Illinois Basin (pink shaded area) and the IBDP site (blue star).

to the Underground Source of Drinking Water (USDW).[2] In addition, VW1 also was accessed for repeat geophysical logging. The GM1 well, originally intended to facilitate repeat 3D vertical

[2] The VW1 well was originally completed with a Westbay pressure/temperature and fluid sampling system. At the time of this writing, the VW1 well has been recompleted with a different monitoring system not described herein (section "Pressure Monitoring").

343

(a)

Injection Well

CCS1

(b)

In-Zone Monitor Well

VW1

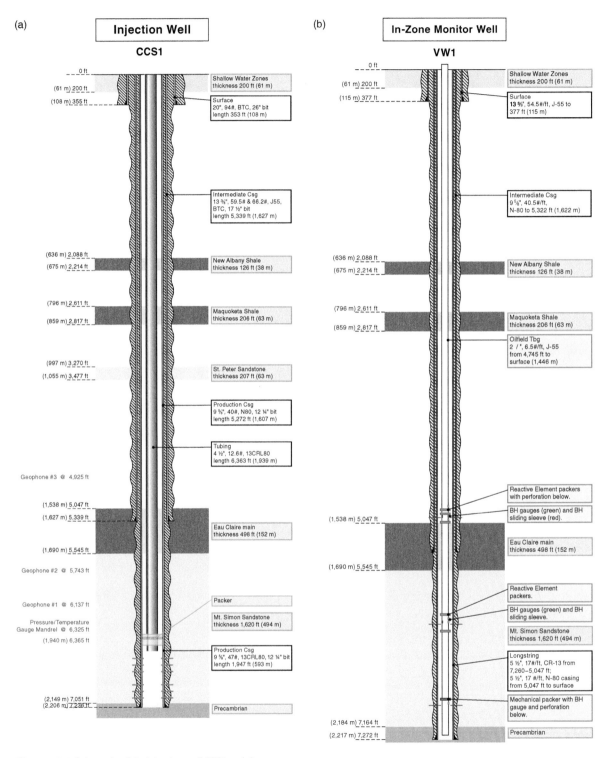

Figure 19.5 Schematic of the injection well CCS1 and deep monitor well VW1.

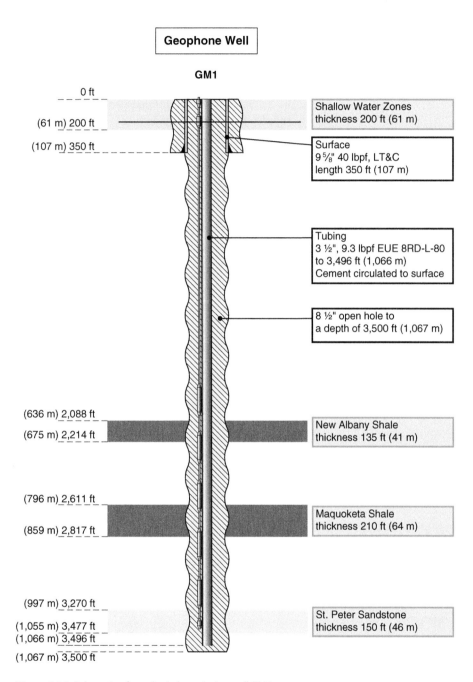

Figure 19.6 Schematic of geophysical monitoring well GM1.

seismic profile (VSP) data acquisition, is also a vital component of the passive monitoring network. The specific equipment descriptions and roles of each well will be elaborated on in the following sections on respective geophysical methods.

345

Geophysical Methods and Data Acquisition Programs

Overview

The IBDP geophysical program has provided valuable direct and indirect measurements of key elastic and petrophysical properties for characterization of the injection zone, seal, and underlying formations. A systematic geophysical site characterization effort started with acquisition of 2D seismic data to support pre-drill feasibility assessment. The site characterization was progressively refined through drilling and geophysical logging of the injection and monitoring wells, and acquisition of surface and borehole three-dimensional (3D) seismic surveys. The program entailed systematic improvement of spatial coverage and measurement resolution as the project progressed. Site characterization also included establishment of the background microseismic activity through 18 months preinjection passive monitoring.

The IBDP site characterization and monitoring program includes a combination of permanent and retrievable borehole and surface measurement techniques to acquire data both continuously and periodically. Various geophysical methods are required to estimate intrinsic properties of the storage complex for site characterization and to monitor transient reservoir state (pressure, saturation, stress) for MVA purposes. The field configuration and methodology have evolved throughout the project to achieve the best solutions to multiple technical challenges under ever changing operational and commercial conditions and constraints. The following sections describe each element of the site characterization and monitoring system. Other chapters in this book (particularly Chapter 2) provide more detail on the fundamentals of the methods used at the IBDP.

Active Seismic Methods

2D Surface Seismic Data

Before commencement of the IBDP geophysical data acquisition campaign, preexisting regional 2D seismic data were the primary source of structural information for pre-drill site characterization. Two 2D seismic profiles (ADM1 and ADM2) were acquired along the east–west and north–south roads that pass by the ADM plant in the fall of 2007. These data, along with

nearby wells, were used to reduce project risk associated with reservoir thickness, potential faulting, and storage capacity at the site. As a result of positive preliminary assessment of the area, an additional four 2D seismic data lines were acquired at the site in 2009 to obtain additional geological detail and supplement pre-drill planning. The seismic data acquisition program for site characterization is described in Table 19.1. New data further verified the absence of significant basement paleotopographic highs at the site and a very thick Mt. Simon section. These data also supported estimates of local regional trends during site characterization and development of the 3D reservoir model (Leetaru and Freiburg, 2014).

3D Surface Seismic Data

The IBDP site characterization and monitoring program involved three separate surface 3D seismic data acquisition efforts (Table 19.1). The first 3D survey conducted in 2010 was designed for detailed site characterization over the anticipated CCS1 plume area. In 2011, this survey was extended to accommodate potential additional plume monitoring needs. This extended survey (2011) also served as the baseline survey for subsequent IBDP time-lapse monitoring work. The third survey, conducted in 2015, was after completion of injection of CO_2 and this survey serves as the first time-lapse monitor survey for the IBDP plume.

Acquisition of the 3D surveys at the IBDP presented many evolving challenges. The proximity of heavy industrial activity and transportation infrastructure resulted in high levels of anthropogenic noise. New construction at the industrial site meant that each successive survey encountered additional obstructions to access and related increase in noise. The requirement to acquire surveys during the winter season owing to both agricultural activity and desirable ground coupling conditions further compounded the difficulties. The resulting progressive diminishment of achievable surface access, the accompanying progressive increase in levels of noise contamination due to surface industrial activity, and an electrical power infrastructure onsite created challenges to seismic data acquisition. These acquisition challenges were managed through a combination of acquisition design, equipment, and parameter selection. For example, the 2D seismic data was used to determine the potential frequency bandwidth and resolution of

Table 19.1 Seismic data acquisition program components

Date	Data	Purpose and comments
2007	Review of existing regional geophysical data	Increased understanding of regional geology Relatively poor quality Not site specific for project
October 2007	Two 2D seismic profiles (IA17 and IA 27)	Refined understanding of project site and regional geology
December 2009	Four high-resolution 2D seismic profiles	Validated regional dip from previous interpretations Provided confirmation that no resolvable faults were present Contributed to the understanding of local geology
January 2010	High-resolution 3D surface seismic survey	Provided detailed structural and stratigraphic characterization Used to derive rock properties for the Eau Claire and Mt. Simon
January 2011	High-resolution 3D surface seismic survey	Merged with 2010 survey to increase the seismic data coverage for structural and stratigraphic characterization; includes the ICCS Project site
January 2015	High-resolution 3D surface seismic survey	Served as postinjection time-lapse monitor survey for the IBDP and baseline survey for the ICCS Project

the new 3D seismic data set, which showed that it was feasible to acquire a high fold 3D surface seismic survey over the site despite the access limitations. However, some of the pre-survey source and receiver positions were lost once the acquisition crew started work in the field because of ground conditions and changes in infrastructure related to the industrial site. The acquisition crew worked with the survey design team to adapt the survey where surface access was an issue to ensure high fold coverage was maintained. As a result, the final post-survey fold coverage was very similar to the ideal pre-survey fold coverage. Acquisition design at the IBDP is discussed in detail by Couëslan et al. (2009).

The Schlumberger Q-Land point-receiver land seismic system was used at the IBDP, which incorporates a pattern of high-density, high-resolution 18-Hz geophone accelerometers. This acquisition configuration was intended to give the maximum amount of flexibility in attenuating the high levels of noise contamination at this industrial site, while maintaining high-frequency bandwidth during processing. The 27 216 kg vibrator trucks used a maximum displacement sweep design to generate energetic low frequencies from 4 to 100 Hz to maximize the low-frequency content in the dataset. Seismic data with frequencies between 2 and 10 Hz are particularly useful for inversion analysis, as they

can build a more robust low-frequency model less biased by sparse well-log data.

Although under ideal circumstances exact repeatability is desired in all aspects of time-lapse seismic surveying, unavoidable events result in variations in recording equipment between surveys. By maintaining consistent high-density single-receiver locations, together with specially designed vibrator sweeps, both signal and noise modes provided the fidelity required for noise reduction in data processing. The data acquisition parameters, which were maintained throughout all 3D surveys, are shown in Table 19.2. The source and receiver locations used for the 2015 time-lapse monitor survey are shown in Figure 19.7a. All of these source and receiver locations are redundant with those occupied in the 2011 survey. The fold of coverage on the processing bin size achieved in the 2015 survey within the offset limited range of 2135 m corresponding to target depth is illustrated in Figure 19.7b. The uniformity of common depth point (CDP) coverage in light of the numerous surface obstructions demonstrates the power of high-density source and receiver effort.

Borehole Seismic: Vertical Seismic Profiles

Time-lapse 3D vertical seismic profiles (VSP) were an important component of the MVA plan for the IBDP.

Table 19.2 3D surface seismic data acquisition parameters

Recording parameters

Recording system	Variable
Recording format	SEG-D, IEEE
Record length	5 s
Sample rate	2 ms
Recording filter (Hi-Cut)	100 Hz
Recording filter (Low-Cut)	4 Hz
Nominal fold	60
Receiver type	GAC-C
Line geometry	Detector interval = 10 ft.

Source Parameters

Source	Vibroseis
Source type	AHV IH 67 000 lbs.
Shotpoint interval	80 ft.
Sweep number	4
Sweep start frequency	4 Hz

Time-lapse VSPs were intended to provide information on CO_2 plume development, demonstrate containment of the CO_2 in the storage formation, and provide data to verify and update models and simulations over the life of the project (Couëslan et al., 2013). The VSP array was permanently cemented into GM1 – the dedicated geophysics monitoring well located approx. 60 m northwest of the injection well. The array consisted of 31 levels of 3C phones deployed at depths ranging from approx. 50 m to 1050 m below ground surface.

Six 3D VSP surveys were acquired at the site: two baseline surveys and four monitoring surveys (Table 19.3). Although borehole deployment of the receiver array, combined with the much-reduced energy source footprint, resulted in fewer access problems than encountered during surface seismic acquisition, the relative contribution of each VSP source point to the final image is much greater in VSP surveying. As such, surface access was a major challenge to achieving repeatability between successive VSP surveys. The VSP acquisition conditions, source parameters, and quantity of CO_2 injected for each survey are summarized in Table 19.3.

Given the use of a permanent receiver array, the greatest remaining challenge to successful time-lapse imaging is in achieving repeatability of energy

source point locations. This was particularly problematic at the IBDP site where permitting issues resulted in restricted access to major sections of the surface for the first three monitor surveys. The resulting reduction in source points resulted in reduction of image quality for these surveys and also in low-resolution time-lapse signals from these monitor surveys. Monitor 4 overcame some of the restrictions to surface access and Figures 19.8a and 8b show source and receiver location analysis for the Baseline 2-Monitor 4 (B2-M4) survey set. Green polygons in Figure 19.8a delineate the shotpoints not accessible during M1, M2, and M3 surveys. This, together with greater quantity of injected CO_2, is believed to be a major factor in the enhanced stability of time-lapse analysis for the B2–M4 survey pair (Section 5.2.1).

Passive (Microseismic) Monitoring

System Overview

The IBDP microseismic monitoring network is comprised of multiple subsurface arrays deployed in three separate wells. The subsurface data acquisition network components are described in Table 19.4. The arrays in CCS1 and GM1 were calibrated using drilling noise and perforation shots in the borehole of VW1, which Smith and Jaques (2016) describe in detail.

Data from all subsurface arrays are integrated and real-time data acquisition functions are performed using an integrated data acquisition system located at the field office. In addition to the subsurface equipment, ISGS installed five surface seismometer stations (Figure 19.9). The raw microseismic data stream from the subsurface array is recorded in 10-s SEG2 files. The rate of data sampling during the preinjection period was every 2 ms (500 samples per second). The data set composition is described in terms of date ranges, number of files, number of triggers, and events identified. The number of triggers represents the total number of possible events detected by the system, and so includes many false triggers. Given the sensitivity of the system, false triggers are caused by any combination of transient electrical glitches and well-related activity such as pipeline maintenance. A false trigger will not have a P- and S-wave signature common to microseismicity or place an event in the subsurface.

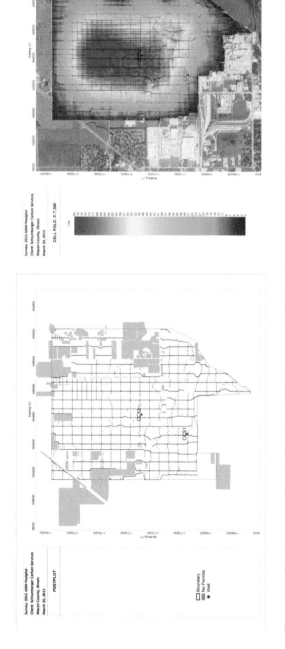

Figure 19.7 Seismic source and receiver locations with fold coverage. (a) 2015 monitor survey source and receiver postplot with surface obstructions (shaded red) and (b) resulting fold of coverage on processing bins and limited to 2135 m.

Table 19.3 Schedule of time-lapse VSP surveys and injected CO_2 quantity.

Survey	Date	Ground conditions	Vibrator sweep (Hz)	Repeated shots (relative to B2)	Amount of injected CO_2 (tonnes)
Baseline 1 (B1)	Jan. 27–30, 2010	Wet	2–100	—	Preinjection
Baseline 2 (B2)	Apr. 12–14, 2011	Dry	8–120	—	Preinjection
Monitor 1 (M1)	Feb. 11–12, 2012	Frozen/Dry	8–120	467	74 000
Monitor 2 (M2)	Apr. 4–5, 2013	Damp	8–120	385	433 000
Monitor 3 (M3)	Feb. 3–5, 2014	Frozen	8–120	378	730 000
Monitor 4 (M4)	Jan. 15–17, 2015	Frozen	8–120	458	approx. 1 000 000

Geophone Arrays

Injection Well (CCS1)

In the original monitoring plan, the passive seismic monitoring equipment in CCS1 was anticipated to provide the majority of the observations for microseismic location and characterization (Figure 19.5). A specialized deployment mechanism was used to decouple the sensors from the injection tubing string, thereby reducing flow-induced noise. As originally designed, the four sensor levels in CCS1 were set at subsurface depths spanning the interval from the Eau Claire to Unit A of the Mt. Simon (deepest part of the formation), which was intended to provide reasonable signal fidelity and location accuracy throughout the primary seal and injection intervals. Damage to two sensors during deployment resulted in only the two sensors located nearest the injection zone being usable. The eight channels of data from these two sensors are recorded via the 96-channel DAU located at the nearby field office. Preinjection data acquisition began in May 2010.

Geophysical Monitoring Well (GM1)

The dedicated geophysical monitoring well, GM1, is 1067 m deep and contains a 31-level array of 3-component tools permanently cemented into place (outside the casing) during construction (Figure 19.6). The geophones are approx. 500 m shallower than those in CCS1, and 29 of the geophones are located between 623 and 1049 m depth with the remaining two at 41 and 108 m (Will *et al.*, 2016a).

The primary objective of the GM1 array was to obtain high repeatability time-lapse 3D VSP surveys for plume monitoring. The main well construction and array design considerations focused on VSP imaging objectives, with secondary objectives to support

microseismic monitoring. However, given the reduction of observations in CCS1, as mentioned earlier, the significance of observations from the GM1 array for microseismic characterization greatly increased. Data recording in GM1 also commenced in May 2010. Data from a selected 28-level (86-channel) subset of the 31-level array are recorded to the 96-channel DAUs, along with data from CCS1.

Geophones Orientation

Locating microseismic events requires precise knowledge of the position of the geophones in *x*, *y*, and *z* directions, and the orientation of the geophones within that position.

Tool orientation calculations were performed on the various IBDP subsurface arrays at different times during the monitoring project as new subsurface components were added and energy sources became available because of field activities. Because the array in CCS1 provides the primary observations constraining the azimuth used in calculating microseismic event locations, verifying accurate orientation of these geophones was attempted during all orientation efforts.

Surface Passive Monitoring Stations

In 2013, a network of five 3-component seismometers were installed in shallow, near-surface vaults on ADM property and away from buildings (Figure 19.9). Four of the surface seismic stations were deployed in approximately the four cardinal directions from the CCS1 well at radial distances of 1066.8 to 1828.8 m. The fifth station is located outside the National Sequestration Education Center building at Richland Community College (labeled as Seis3 RCC in Figure 19.9) where real-time data monitoring is

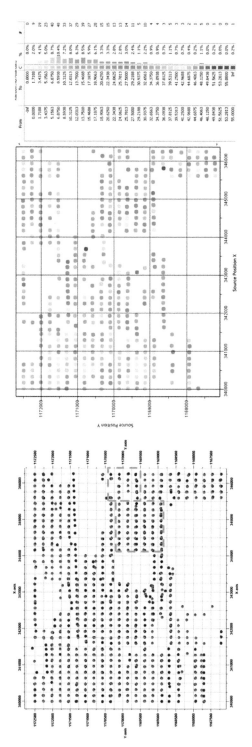

Figure 19.8 VSP source and receiver locations. (a) Source locations from B2 and M4 surveys. Orange dots are B2 and Blue are M4 shot points. Green boxes show the shots that were not available in M1, M2, and M3 surveys. (b) Extracted collocated sources, survey B2coM4. The color of circles indicated the relative distance from the corresponding source location from M4 survey.

Table 19.4 Microseismic system components

Well	Configuration	Recording
CCS1	2-level array of a 4-component WellWatcher PS3* passive seismic sensing system with 4-Geospace 2400 geophones in tetrahedral configuration (two levels functioning)	96-channel data acquisition unit (DAU) located at the field office near CCS1
GM1	A 31-level array of 3 Geospace 2400 geophones in orthogonal configuration	96-channel DAU located at the field office near CCS1
VW2: Operated by the ICCS Project	5-level array of a 3-component multilevel downhole seismic array, Sercel array installed temporarily (Sept. 26, 2013 and removed Feb. 25, 2015)	24-channel DAU located at the wellhead

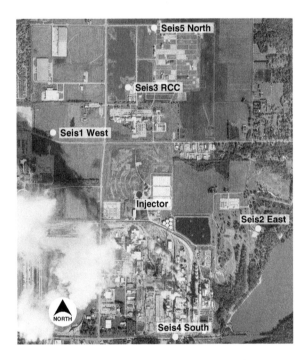

Figure 19.9 Location of five surface seismometers at the IBDP site.

performed for this station. This near-surface network was intended to be a backup for the downhole microseismic systems for large magnitude events. The instruments in the near-surface system will continue to record on-scale if a seismic event, large enough to saturate the downhole microseismic systems, were to occur near the IBDP site. If a large event were to occur, the sensors in the downhole systems would be able to detect the event and determine its 3D location, although they may not accurately measure the event magnitude. Though the near-surface data are not integrated with downhole data in event

processing, the data have been used for independent verification of larger events.

Nonseismic Measurements

Geophysical Logging

A comprehensive suite of geophysical logs were acquired in boreholes CCS1, VW1, and GM1, along with physical borehole and core tests in CCS1 and VW1. In addition to measurements required for geological characterization, the logging program was designed to support advanced geophysical and geomechanical integration workflows. Elastic properties, mineral fractions, and fluid saturations were measured or computed for seismic inversion and time-lapse integration. For geomechanical characterization and microseismic analysis, *in situ* stress and direction were documented by observation of tensile-induced drilling fractures and breakouts as shown by the Formation MicroImager (FMI)[3] fullbore formation microimager and a minifrac, injection and step rate tests. Some of the specialized well logs, tests, and resulting measured and calculated properties used for geophysical integration at the IBDP are shown in Table 19.5.

Repeat Pulse Neutron Logs (RST)

RST logging at IBDP was performed for operational and verification monitoring; CCS1, VW1, and GM1 are logged at least annually to verify the mechanical integrity of each well. The logs are also used to monitor the development of the CO_2 plume adjacent to CCS1 and VW1 and to constrain the dynamic reservoir modeling. Two baseline surveys were acquired in CCS1 and VW1; one shortly after each well was drilled

---

[3] Mark of Schlumberger.

Table 19.5 Petrophysical logs run in boreholes with measured and calculated properties.

Petrophysical logging tests	Measured and calculated properties
Spontaneous potential	Density
Neutron porosity	Clay content
Resistivity	Static Young's modulus
Microresistivity imaging	Static Poisson's ratio
Sonic velocities	Static shear modulus
Elemental capture spectroscopy	Biot's coefficient
Natural gamma ray	Frictional angle
Magnetic resonance	Tensile strength
Thermal neutron decay	Unconfined compressive strength
Fullbore formation microimager*	Overburden stress Minimum horizontal stress Maximum horizontal stress Azimuth minimum horizontal stress Tensile-induced drilling fractures and breakouts Mineral fractions Fluid saturations

* Mark of Schlumberger.

and a second some months after to give the drilling fluids a chance to migrate away from the wellbores.

Pressure Monitoring

VW1 was originally completed with a multilevel Westbay monitoring and sampling completion system optimized specifically for CCS applications and depth of operation. The Westbay system completion continuously monitored pressure and temperature using the Modular Subsurface Data Acquisition System (MOSDAX™, Westbay Instruments, North Vancouver, British Columbia) modular subsurface data acquisition system probes at 11 different depths in the well. Each sampling zone was isolated from the other ports through redundant packers (Figure 19.5). Fluid samples were also taken from each port at discrete time intervals for geochemical analysis when the pressure and temperature probes are pulled from the well.

Results

Surface Seismic Data

Imaging for Site Characterization

The main processing objectives for the 2010 and 2011 3D seismic surveys focused on site characterization. Processing flows were designed for

- Merging the 2011 extension with the 2010 survey
- Producing a data set suitable for acoustic and elastic impedance inversion
- Producing a volume suitable for structural interpretation
- Imaging any faults or fractures in the zone of interest (0.8–1.1 s)

The main data processing challenge for the surface 3D data set was the significant amount of random noise due to coupling issues and from anthropogenic activity, mainly roads throughout the acquisition area. Unfiltered prestack migrated gathers in Figure 19.10a illustrate the extent of noise contamination. The same gathers after a series of premigration filtering processes designed to maintain frequency and phase integrity of coherent signals are shown in Figure 19.10b. The corresponding stacked sections in which imaging is greatly improved throughout the vertical section are shown in Figures 19.11a and 11b.

Time-Lapse Processing

Data Processing

The 2011 and 2015 surveys were reprocessed for four-dimensional (4D) time-lapse detection purposes. As discussed, although it is desired to maintain identical acquisition parameters for time-lapse surveys, many factors affect acquisition such as changes of surface accessibility, random noise patterns, and changes in ground conditions. To reveal the true 4D difference caused by CO_2 injection in seismic datasets, it is important to remove other inferences as much as possible. Both the baseline and monitor datasets at IBDP were processed (coprocessed) using the same processing sequence (Figure 19.12). For several key steps, such as refraction tomography statics solution, residual reflection statics solution, and deconvolution, tests were run for both a joint solution and a standalone solution. Also, throughout processing, 4D quality control (QC) procedures were conducted to ensure non 4D

353

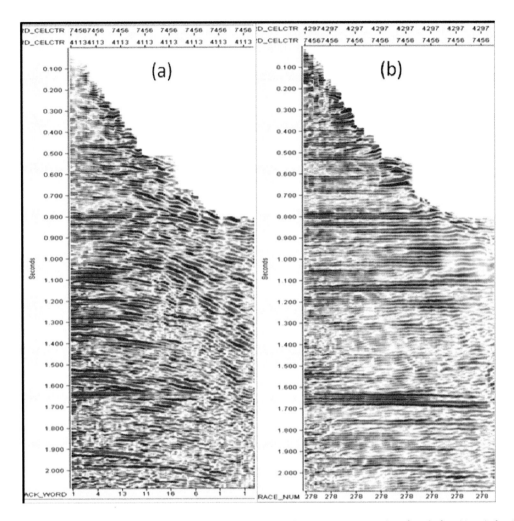

Figure 19.10 Prestack migrated 2011 3D seismic gather. Prestack migrated seismic gathers from before (a) and after (b) multichannel filtering.

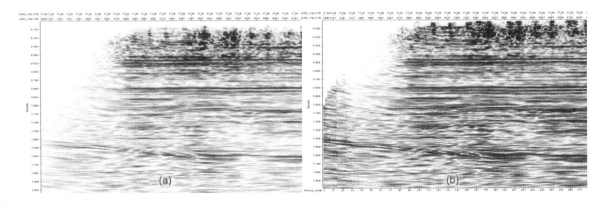

Figure 19.11 Stack 2011 3D seismic gather. Stack before (a) and after (b) multichannel filtering from gathers corresponding to Figure 19.10.

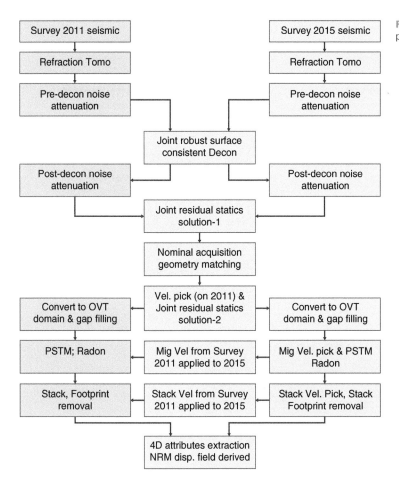

Figure 19.12 4D surface seismic data processing flow.

inferences were reduced as processing progressed. Because of the processing workflow, very subtle changes in migrated data (Figure 19.13) can be identified that are the result of time-lapse attribute analysis.

Time-Lapse Attribute Extraction

Two time windows were defined for 4D attribute extraction at IBDP (Figure 19.14). It is assumed that strata within the time Window-1 (400–700 ms) above the caprock should not be affected by CO_2 injection. Strata within the time window-2 (900–1300 ms) that includes the CO_2 injection zone could be impacted in several ways.

Replacement of one fluid by another with different density/properties changes the velocity and density of the reservoir interval, resulting in modification of reservoir acoustic impedance and/or causes small changes in reflector timing beneath the reservoir. Pore pressure changes also modify the seismic velocities within the reservoir, resulting in modification of

reservoir acoustic impedance and/or causes changes in reflector timing beneath the reservoir. Pore pressure changes may also cause gas dissolution or exsolution, producing a large change in acoustic impedance, or cause stress changes above, and possibly below, the reservoir; producing a change in velocity causes time shifts between time-lapse volumes below the reservoir.

These changes to the storage system could be detected through examination of characteristics of seismic data, and 4D attributes extracted from seismic stack volumes. Time-lapse attributes such as reliability (Figure 19.15) and normalized root mean square (NRMS) (Figure 19.16) were extracted from different time windows and were compared to detect any 4D changes caused by CO_2 injection. In both cases, the attributes from window 1 (above the injection zone) indicate good time-lapse repeatability while those from window 2 (including the injection zone) show very similar time-lapse differences, which are attributed to fluid replacement.

355

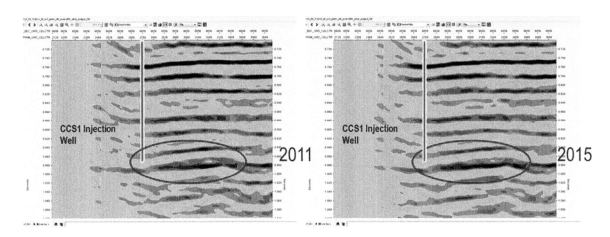

Figure 19.13 Subtle changes in migrated data. Zoom-in proximity of CO_2 injection spot (yellow bar indicates the injection well path). Notice the very subtle changes in amplitude and waveform in seismic.

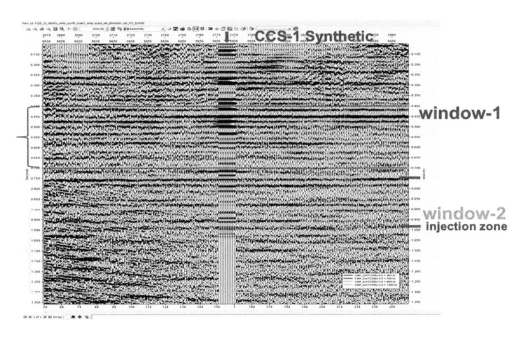

Figure 19.14 Survey 2011, 4D attribute extraction window definition. Window-1, above the CO_2 injection zone; Window-2, including the injection zone.

An additional time-lapse interpretation tool is the nonrigid matching (NRM) attribute. An NRM uses a nonrigid matching method time- or depth- to align two cubes of seismic data by generating a 3D displacement field that compares the input data to a reference volume. Owing to the fluid effect on velocity we expect the NRM attribute to show anomalies below the plume where arrival times have been impacted by the overlying fluids. The NRM computed using the 2011 baseline survey as the reference volume is shown in Figure 19.17. An NRM anomaly in the vicinity of the injection well is clearly evident.

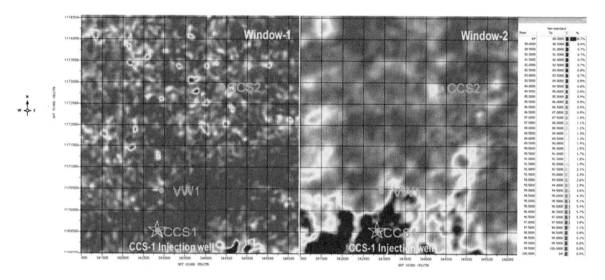

Figure 19.15 Attribute predictability comparison. Red indicates high similarity. Blue indicates low similarity.

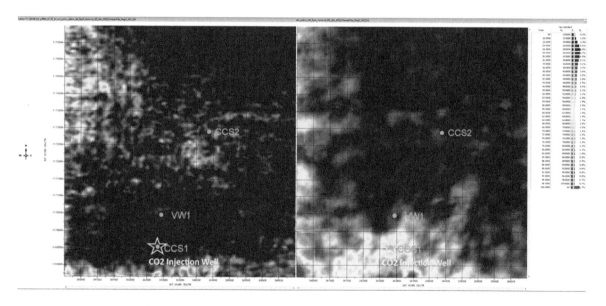

Figure 19.16 Attribute normalized root mean square comparison. Red indicates large difference in amplitude. Blue indicates small difference in amplitude.

3D VSP

Vertical Component VSP Analysis

Baseline Surveys

The first baseline 3D VSP was acquired in February 2009, after the 3D surface seismic survey, but while ground conditions were damp and not ideal for seismic signal propagation. In addition, there was significant 60-Hz noise from nearby electrical sources, such as power lines and an electrical substation, recorded in the data although this noise was attenuated through processing. The acquisition footprint of the baseline 3D VSP also suffers from the same challenges posed by surface infrastructure obstructions to

357

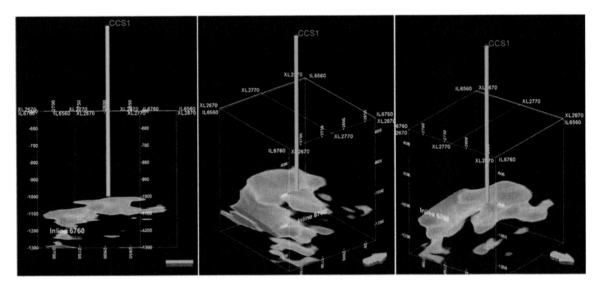

Figure 19.17 3D NRM displacement field zoomed in near the CCS1 injection zone.

energy source points. Holes in the acquisition footprint surface seismic data can be compensated by adding in-fill source and receiver locations; however, 3D VSP data is not as flexible given the geometry of the ray paths. As a result, there are identifiable artifacts in the 3D VSP images related to holes in their acquisition footprint. In 2011, a second baseline 3D VSP survey was acquired under dry ground conditions, and an isolation transformer with a Faraday shield was used to reduce the 60-Hz noise.

Monitor Surveys

Because of the favorable acquisition conditions of Baseline 2, all monitor surveys were collocated and coprocessed relative to Baseline 2 through a vertical component processing workflow (Figure 19.18). Preprocessing included receiver selection, 60-Hz notch filter, cross-equalization based trace-by-trace amplitude scaling, linear radon transform for wavefield separation, deterministic shot-by-shot reference trace wave-shape deconvolution. Collocated energy source records (15 m tolerance) were cross-equalized to allow for residual matching of the amplitude and phase spectra, and then processed for wavefield separation and VSP deconvolution. The same 1D anisotropic model that was built using zero offset VSP and 3D VSP travel times from the baseline survey was used for imaging both the datasets (Figure 19.19). After migration, nonrigid matching was applied to

both data sets to further reduce the differences between the data sets related to noise and acquisition artifacts while maintaining the differences related to CO_2 injection.

Time-Lapse Attributes

Analysis of time-lapse VSP data utilizes many of the same tools as are used for analysis of surface time-lapse data. Results from the Monitor 1 and Monitor 2 surveys were ambiguous, likely from a combination of acquisition conditions and limited quantities of CO_2 injected. However, the Monitor 3 (Figure 19.20) and Monitor 4 (Figure 19.21) surveys yield coherent time-lapse anomalies when coprocessed with Baseline 2. NRMS maps of the overburden interval show no definite trend indicating good repeatability, whereas in the injection interval the highest NRMS values are in the southeast near where injection well CCS1 is located (Figures 19.20 and 19.21).

Microseismic Monitoring

Microseismicity Pre-CO₂ Injection

Preinjection microseismic activity was monitored by the arrays in CCS1 and GM1 from May 1, 2010 until start of injection on November 15, 2011, during which they recorded more than 68,000 triggered events. Most events were related to surface noise, drilling of the VW1 borehole, well activities, VW1 perforation

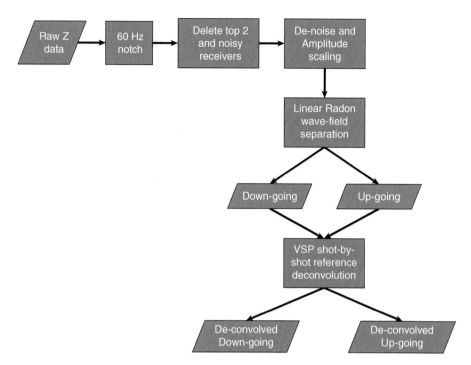

Figure 19.18 Adopted processing flow of single-axis 3D VSP processing.

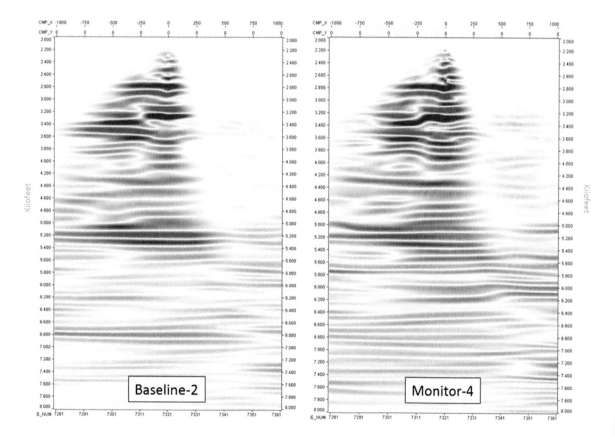

Figure 19.19 Migrated images B2 and M4.

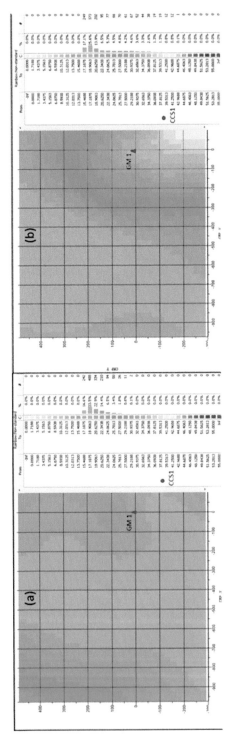

Figure 19.20 Time-lapse NRMS attribute comparison of B2-M3. (a) B2-M3 NRMS attribute in the overburden zone. (b) B2-M3 NRMS in the reservoir.

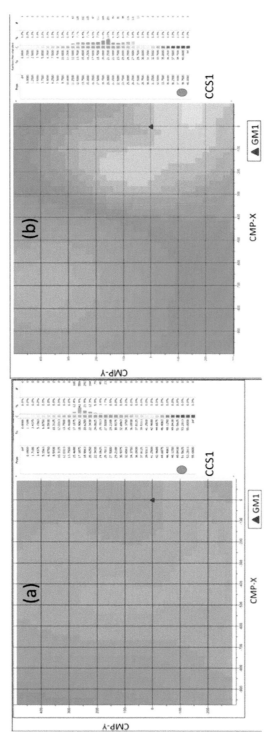

Figure 19.21 Time-lapse NRMS attribute comparison of B2-M4. (a) B2-M4 NRMS attribute in the overburden zone. (b) B2-M4 NRMS in the reservoir zone showing a distinct time-lapse anomaly in the vicinity of the injection well.

shots, and many distant events assumed to be mine blasts because of distance and time of day (8:00 to 17:00). Twenty-one distant natural earthquakes (12 in the United States Geological Survey catalog) and eight local microseismic events were recorded, which appear to be unrelated to well activity and presumed to be background events. These eight background events were located in the Mt. Simon and Argenta; six were located within 457 m of CCS1 and two were located 1646–2408 m from CCS1. The magnitude range for the eight background events was −2.16 to −1.52 (average of −1.83), with a similar range for events related to drilling and well activity of −2.6 to −1.6 (Smith and Jaques, 2016).

Microseismicity during CO$_2$ Injection

Locatable microseismic events started one month after the start of injection, with two events in December 2011 and 13 in January 2012. During injection, 4848 events were located with an average of four per day having an average magnitude of −0.8. Ninety-four percent of events were less than magnitude 0 but five events were above 1 ranging from 1.01 to 1.17. (The amount of energy released represented by a −1 magnitude is about that of the typical inch long, pencil-thin firecracker.) Using Zoback and Gorelick's (2012) magnitude versus slip surface size and displacement diagram, a magnitude −1 represents a slip of less than a millimeter on a plane of a couple of meters in size to

a slip of a fraction of a tenth of a millimeter on a larger plane of tens of meters.

The range of magnitudes of events is nearly identical in each major geological unit in which events were located: the Precambrian had the highest magnitudes of 1.16 and 1.147, whereas magnitude highs in the mid- to lower 0.8s, were recorded in the Argenta and Mt. Simon, respectively. The lowest moment magnitudes detected in the Precambrian were −2.02 and in the Cambrian sandstones −2.13. The very low magnitudes of microseismic events, their locations and estimated slip plane size, indicate there is little to no risk that induced seismicity at IBDP could cause fault slippage through the caprock and compromise the seal.

About eighty percent of the microseismic events are in the Precambrian crystalline basement, and the remainder in the Mt. Simon and Argenta Formations. The first locatable events were about 600 m from CCS1 and as events continued at this distance, new ones started about 300 m away. Around each of these distances, separate elongated clusters of events began to emerge. Eventually 18 clusters of microseismic events were identified at IBDP, with no correlation existing between time of cluster development and distance from the injection well (Figure 19.22). For example, clusters are numbered in order of appearance, and Cluster 1 averages 548 m from CCS1, Cluster 2 is nearly half that distance (335 m), Cluster 3 averages 457 m, and Cluster 4 averages 823 m from CCS1. Later clusters also

Figure 19.22 Locations of microseismic clusters.

developed out of sequence with increasing distance from the injection well.

Clusters at IBDP continued to add microseismic events during injection and transient shut-in periods, and most developed as an elongated pattern oriented SW–NE. Likely factors other than pore pressure changes influenced the sequence of cluster development, such as the angle of preexisting planes of weakness or defects in relation to maximum horizontal stress direction. Preexisting weak planes or defects close to 30° from the maximum horizontal stress (S_{Hmax}) direction are optimally oriented in the direction expected for strike-slip movements.

Evidence related to microseismic events induced by hydraulic fracturing suggest they are triggered along preexisting natural fractures that are favorably oriented for slip (Pearson 1981; Rutledge and Phillips, 2003; Shapiro et al., 2006). Baig et al. (2012) observed that with hydraulic fracturing of some horizontal wells events occurred in two main trends: early events were roughly 30° from S_{Hmax} and late events were approximately parallel to S_{Hmax}. Yang et al. (2013) found similar trends for hydraulic fracturing in the Bakken formation where detected microseismic events from many stages trend approx. 30° from the direction of S_{Hmax}. This relation appears to be present at IBDP through the sequence of cluster development, with Cluster 1 at 28° from S_{Hmax}, Cluster 2 at 11°, Cluster 3 at 9°, and Cluster 4 at 3° (Figure 19.23). The strike of each plane was developed by a best-fit plane through the 3D cloud of

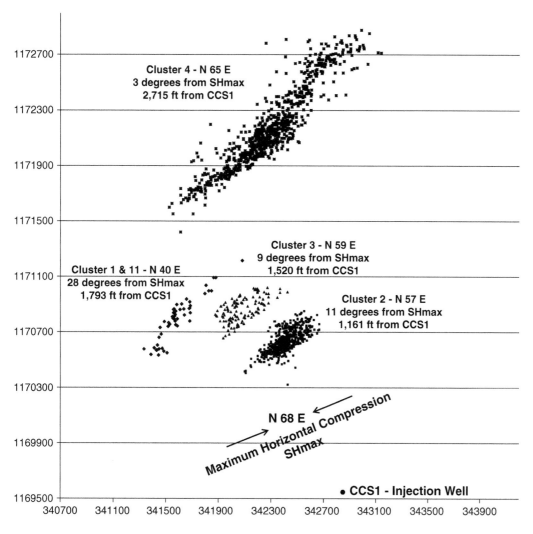

Figure 19.23 Clusters 1–4 directions in relation to maximum horizontal *in situ* stress and average distance from injection well CCS1. Axes are northing and eastings in feet. (From Bauer *et al.*, 2016.)

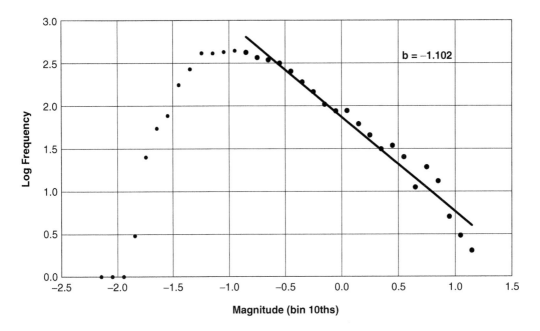

Figure 19.24 Gutenberg–Richter plot of all the microseismicity during the three years of injection. (From Bauer *et al.*, 2016.)

events. The strike of the planes in relation to the *in situ* S_{Hmax} fits the first motion analysis indicating right lateral strike-slip motions on many of the clusters (Couëslan *et al.*, 2014). These planes are interpreted to be the reactivation of preexisting features as shown by the Gutenberg–Richter (Gutenberg and Richter, 1956) plots for all the microseismicity, which show *b*-values close to 1 (Figure 19.24). Will *et al.* (2016b) present an in-depth analysis of the data integration and reservoir response resulting in microseismicity.

The total event population was separated into subsets on the basis of spatial clustering and the fault plane solution (FPS) analysis was performed on individual event subsets. Figure 19.25 shows the FPS result for one of the most developed and coherent event clusters (#4), which exhibits a very good fit to the strike-slip model for an azimuth of N45E. Figure 19.26 shows the FPS superimposed on event clusters.

Monitoring Strategy and Discussion

Integrated Modeling

Each component of the IBDP geophysical program has made a unique contribution to the understanding of key aspects of site characterization such as geological structure, stratigraphy, and reservoir properties dictating well injectivity, storage capacity, and

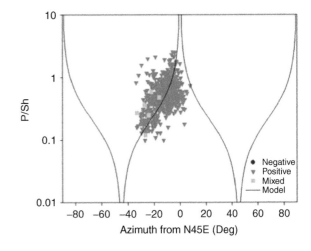

Figure 19.25 FPS for a selected event cluster (#4) using a strike-slip assumption and azimuth.

containment. However, because of intrinsic characteristics such as spatial coverage, resolution, and phenomenological ambiguity, in order to realize its full value, seismic data must be integrated with other direct measurement types such as borehole logs, core, and well tests. Therefore, the IBDP site characterization and monitoring program has included an extensive integrated geologic modeling effort. The various contributions of active seismic methods

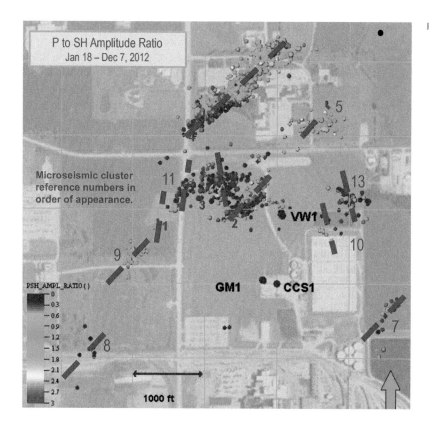

Figure 19.26 FPS results for all clusters.

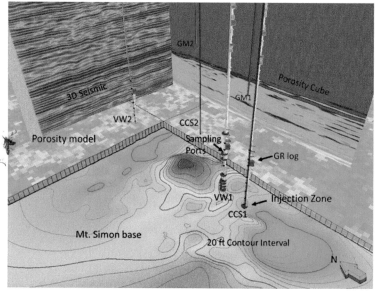

Figure 19.27 Contribution of seismic data to the static geologic model.

to the IBDP geologic model through interpretation of stratigraphic horizons, acoustic impedance and seismic porosity inversion, time-depth domain conversion through integration with acoustic and density well logs, and colocated cokriging of reservoir properties through integration with petrophysical well logs are shown in Figure 19.27.

Geophysical methods have also been instrumental in monitoring transient (both natural and injection induced) changes within the storage unit as part of the

MVA program. Active and passive seismic measurements, such as time-lapse surveys and passive microseismic monitoring, provide valuable *indirect* measurements of hydrodynamic and mechanical state within the reservoir. However, while these measurements provide valuable information about the timing and location of changes in reservoir state, each on its own is insufficient to yield an unambiguous physical characterization of the responsible phenomena. Characterizing these phenomena requires an understanding of local transient hydrodynamic and stress state as provided by fluid flow and geomechanical models. Fluid flow and geomechanical models were part of early stage IBDP characterization and have been updated as the project evolves.

The following sections describe the pivotal role of the IBDP integrated model in interpretation of time-lapse seismic and passive monitoring data for understanding of plume development and microseismic mechanisms. The geologic and transient process models have been periodically updated to incorporate new characterization and calibration data (Table 19.6).

Table 19.6 IBDP integrated model development history

	Structure/stratigraphy/ discrete features	Hydrodynamic properties	Use
2008 Preliminary	• Layer cake stratigraphy defined by well tops from analog well 60 miles away • No discrete features	• Uniform zonal porosity and permeability • Assigned using logs from analog well 60 miles away	• Site characterization • Basis for initial reservoir simulation model
2010 Update	• Layer cake stratigraphy: well tops from CCS1 • No discrete features	• Stochastic zonal porosity and permeability • Conditioned to CCS1 well logs	• Updated site characterization • Basis for initial reservoir simulation plume predictions
2011 Update	• Stratigraphy: 2010 3D seismic survey and well top control from CCS1 and VW1 • No discrete features	• Stochastic zonal porosity and permeability • Conditioned to CCS1 and VW1 well logs and 2010 seismic inversion products	• Update site characterization • Basis for final Class VI permit reservoir simulation area of review • Basis for preliminary finite element model (FEM)
2013 Update	• Stratigraphy: 2011 extended 3D seismic survey and well top control from CCS1, VW1, and VW2 • Provisional fault interpretation • Mechanical features inferred from microseismic data	• Stochastic zonal porosity and permeability • Conditioned to CCS1, VW1, and VW2 well logs and 2011 seismic inversion products	• Update site characterization • Basis for updated FEM and preliminary microseismic prediction research
2016 Update	• Stratigraphy: 2015 seismic survey was completed to be a time-lapse survey for CCS1 injection and baseline survey for CCS2 injection with a 400-m extension to the north	• Updated stochastic zonal porosity and permeability based on the additional updated structure • Conditioned to CCS2, CCS1, VW1, and VW2 well logs and 2011 seismic inversion products	• Update site characterization for reservoir simulations

Contributions include stratigraphic horizons, seismic porosity inversion, and colocated–cokriged reservoir properties.

Mapping the CO_2 Plume

Injection of CO_2 produces variations in elastic properties in the reservoir that change the time-lapse seismic response; these changes are often subtle and are nonunique with respect to saturation and pressure effects. At the IBDP, a "simulation-to-seismic" workflow has been used for model based integration of time-lapse seismic data and direct measurements of reservoir state using the geological and reservoir simulation models developed for site characterization and plume predictions.

The reservoir model was calibrated to injector and multilevel monitoring well pressures to compute the isotropic effective elastic properties for the reservoir at the times of seismic surveys. Thermodynamic properties for brine were computed using Batzle–Wang equations (Batzle and Wang, 1992), and CO_2 properties were computed using new analytical expressions in the National Institute of Standards and Technology Reference Fluid Thermodynamic and Transport Properties Database (NIST–REFPROP) software. An average of Hashin–Shtrikman (H–S) bounds were used to calculate the bulk and shear moduli of solid rock using rock compositions derived from elemental analysis on available logs. Effective elastic properties of a fluid-filled formation were computed using

Gassmann fluid substitution model and used to forward model attribute differences corresponding to field survey times.

Before time-lapse integration, the calibrated reservoir simulation model predicted an approximately radial plume centered on the injection well (Senel *et al.* 2014). This simulation was calibrated to 15 months of injection and monitoring well pressure and verified with pulse neutron logs. Upon extraction of time-lapse attributes (Figures 19.15–19.17) from the 2011–2015 surface seismic survey pair, a significant discrepancy was noted in the shape and extent of the current plume prediction and time-lapse observation. While differences are expected owing to vertical plume geometry and seismic detection limits, the significant deviation in aerial geometry (in particular the truncation of northward plume migration) indicated fundamental deficiencies in the simulation model. Examining the 2011 survey through seismic attribute and inversion analysis revealed that features exist in the baseline seismic data suggestive of enhanced reservoir quality and a possible flow barrier having geometrical characteristics consistent with the time-lapse anomaly. These features were not incorporated in the original model due to use of a global interpolation scheme. In an updated model, the reservoir quality was modified and a hypothetical flow barrier was inserted corresponding to the features identified in Figure 19.28. Figure 19.29 shows the resulting simulated plume and forward modeled time-lapse acoustic impedance change along with

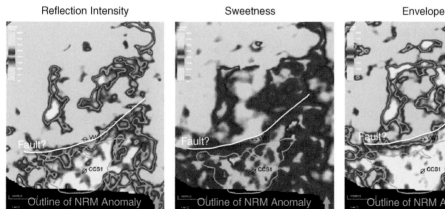
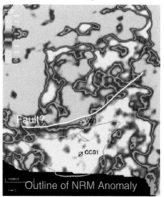

Figure 19.28 Seismic attributes used as the basis for modification of the reservoir model for time-lapse seismic history matching. (From Will *et al.*, 2017.)

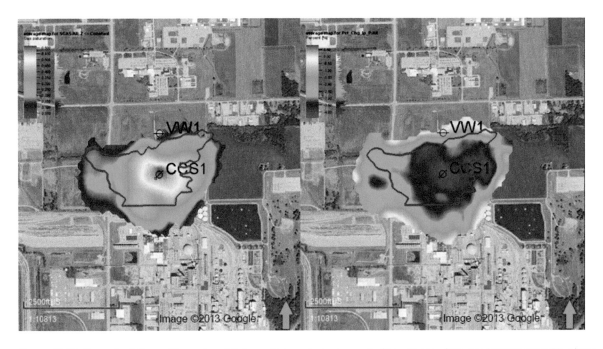

Figure 19.29 Plume simulation and forward-modeled acoustic impedance change. (Left) Aerial view of simulated 2015 monitor CO_2 plume and outline of 2001–2015 surface seismic time-lapse NRM attribute (from Figure 19.17). (Right) Aerial view of forward modeled time-lapse acoustic impedance change and outline of 2001–2015 surface seismic time-lapse NRM attribute. (From Will *et al.*, 2017.)

the outline of the NRM anomaly (from Figure 19.17) in which discrepancies have been greatly reduced. Qualitative analysis of this new result suggests a detectability limit of approx. 3–4% change in acoustic impedance. Similar efforts are being deployed toward addressing saturation inversion.

Summary

Integrating geophysical and geological data is necessary to develop robust geological, fluid flow, and geomechanical models, and to provide essential support for feasibility assessment, operational planning, and regulatory compliance. The IBDP demonstrates practical application of existing acquisition and analytical geophysical methods to achieve the goals of a large-scale CO_2 storage demonstration project. Seismic geophysical methods comprise an extensive set of tools that may be used to fulfill many site characterization and monitoring objectives for regulatory compliance or scientific investigations. Data acquisition techniques may be tailored for the operational environment and the specific imaging objective throughout the life of a CO_2 storage project from pre-drill site selection to plume monitoring during and after injection. Sophisticated noise attenuation and imaging data processing techniques help to overcome many of the challenges encountered in typical operational scenarios, providing input to structural interpretation, reservoir property estimation, and monitoring. Geophysical monitoring and analysis of the IBDP will continue throughout the postinjection and postinjection site care compliance monitoring phase (expected end date 2020).

References

Baig, A. M., Urbancic, T., and Viegas, G. (2012). Do hydraulic fractures induce events large enough to be felt on surface? *CSEG Recorder*, **10**(October 2012): 40–46.

Batzle, M., and Wang, Z. (1992). Seismic properties of pore fluids. *Geophysics*, **57**: 1396–1408.

Bauer, R. A., Carney, M., and Finley, R. J. (2016). Overview of microseismic response to CO_2 injection into the Mt. Simon saline reservoir at the Illinois Basin-Decatur Project. *International Journal of Greenhouse Gas Control*, **54**(1): 378–388.

Couëslan, M. L., Leetaru, H. E., Brice, T., Leaney, W. S., and McBride, J. H. (2009). Designing a seismic program for an industrial CCS site: Trials and tribulations. *Energy Procedia*, **1**(1): 2193–2200.

Couëslan, M. L., Ali, S., Campbell, A., *et al.* (2013). Monitoring CO_2 injection for carbon capture and storage using time-lapse 3D VSPs. *Leading Edge*, **32**(10): 1268–1276.

Couëslan, M. L., Butsch, R., Will, R., and Locke, R. A. II (2014). Integrated reservoir monitoring at the Illinois Basin-Decatur Project. *Energy Procedia*, **63**: 2836–2847.

Freiburg, J. T., Morse, D. G., Leetaru, H. E., Hoss, R. P., and Yan, Q. (2014). A depositional and diagenetic characterization of the Mt. Simon Sandstone at the Illinois Basin-Decatur Project carbon capture and storage site, Decatur, Illinois, USA, Illinois State Geological Survey Circular 583.

Greenberg, S. G., Bauer, R., Will, R., *et al.* (2017). Geologic carbon storage at a one million tonne demonstration project: Lessons learned from the Illinois Basin–Decatur Project. *Energy Procedia*, **114**: 5529–5539.

Gutenberg, B., and Richter, C. F. (1956). Magnitude and energy of earthquakes. *Annals of Geophysics*, **9**(1): 1–15.

Haimson, B. Z., and Doe, T. W. (1983). State of stress, permeability, and fractures in the Precambrian granite of Northern Illinois. *Journal of Geophysical Research*, **88**(B9): 7355–7371.

Leetaru, H. E., and Freiburg, J. T. (2014). Litho-facies and reservoir characterization of the Mt Simon Sandstone at the Illinois Basin–Decatur Project. *Greenhouse Gases: Science and Technology*, **4**(5): 580–595.

Leetaru, H. E., and McBride, J. H. (2009). Reservoir uncertainty, Precambrian topography, and carbon sequestration in the Mt. Simon Sandstone, Illinois Basin. *Environmental Geosciences*, **16**(4): 235–243.

Palkovic, M. (2015). *Depositional characterization of the Eau Claire Formation at the Illinois Basin–Decatur Project: Facies, mineralogy and geochemistry*. M.S. thesis, University of Illinois at Urbana-Champaign.

Pearson, C. (1981). The relationship between microseismicity and high pore pressures during hydraulic stimulation experiments in low permeability granitic rocks. *Journal of Geophysical Research*, **86**: 7855–7864.

Rutledge, J. T., and Phillips, W. S. (2003). Hydraulic stimulation of natural fractures as revealed by induced mircroearthquakes, Carthage Cotton Valley gas field, east Texas. *Geophysics*, **86**(2): 441–452.

Senel, O., Will, R. and Butsch, R. J. (2014). Integrated reservoir modeling at the Illinois Basin–Decatur Project. *Greenhouse Gases: Science and Technology*, **4**(5): 662–684.

Shapiro, S. A., Dinske, C., and Rothert, E. (2006). Hydraulic-fracturing controlled dynamics of microseismic clouds. *Geophysical Research Letters*, **33**: L14312.

Smith, V., and Jaques, P. (2016). Illinois Basin–Decatur Project pre-injection microseismic analysis. *International Journal of Greenhouse Gas Control*, **54**(1): 362–377.

Will, R., El-Kaseeh, G., Jaques, P., Carney, M., Greenberg, S., and Finley, R. (2016a). Microseismic data acquisition, processing, and event characterization at the Illinois Basin–Decatur Project. *International Journal of Greenhouse Gas Control*, **54**(1): 404–420.

Will, R., Smith, V., Lee, D., and Senel, O. (2016b). Data integration, reservoir response, and application. *International Journal of Greenhouse Gas Control*, **54**(1): 389–403.

Will, R., El-Kaseeh, G., Leetaru, H., Greenberg, S., Zaluski, W., and Lee S.-Y. (2017). Quantitative integration of time lapse seismic data for reservoir simulator calibration: Illinois Basin–Decatur Project. Carbon Capture, Utilization & Storage Conference, Chicago, IL, April 10–13.

Yang, Y., Zoback, M., Simon, M., and Dohmen, T. (2013). An integrated geomechanical and microseismic study of multi-well hydraulic fracture stimulation in the Bakken Formation. SPE paper 168778 presented at the Unconventional Resources Technology Conference, Denver, CO, August 12–14.

Zoback, M. D., and Gorelick, S. M. (2012). Earthquake triggering and large-scale geologic storage of carbon dioxide. *Proceedings of the National Academy of Sciences of the USA*, **109**(26): 10,164–10,168.

What Next?

Thomas L. Davis, Martin Landrø, and Malcolm Wilson

Perhaps the most significant international event recently that affects the capture and use of carbon dioxide (CO_2), and hence the topic of this book, was the United Nations Framework Convention on Climate Change (UNFCCC) Paris Agreement. This Agreement was adopted in December 2015 and came into force less than a year later in November 2016. The basis of this agreement was to work cooperatively to restrict global temperature rise to between 1.5°C and 2°C above pre-1850 levels. It is clear from groups like the Intergovernmental Panel on Climate Change (IPCC) and the International Energy Agency (IEA) that meeting such aggressive targets will require the large-scale deployment of CO_2 Capture, Transport and Geological Storage or Geosequestration (both as storage in saline aquifers and in mature oil fields as enhanced oil recovery [EOR]). Such technology will be applicable to large point sources of emissions such as electrical generation facilities, refineries, cement plants, and gas processing facilities.

The idea of putting CO_2 back into the subsurface has now matured for decades. However, the progress is slow, mainly because of cost issues related to the capture process. One hundred and twenty years after the famous publication of Svante Arrhenius in *Philosophical Magazine*, where he suggested that the increase in atmospheric CO_2 could alter the Earth's surface temperature, there has probably been more spent on research to determine the veracity of his finding than on remediation. It is always a good idea to look back in order to predict the future. From the technical and scientific point of view, there is no doubt that underground storage of CO_2 is feasible and that we, with a high degree of confidence, can monitor and ensure that the CO_2 is stored safely. The Sleipner project clearly shows that a tax level of 30–50 USD per tonne CO_2 is sufficient to make such a project economically feasible, given that the CO_2 is captured directly from the methane gas produced at

the field. Indeed, pipeline specifications preclude high concentrations of CO_2 in natural gas streams, so the produced gas must be processed first to remove most of the CO_2 entrained in the gas stream. Instead of venting, this CO_2 is compressed and injected into the expansive Utsira Formation.

The Norwegian government has now embarked on a more challenging project compared to the Sleipner case: to capture CO_2 from onshore high emission CO_2-sources such as a waste recycling plant and a cement plant. The idea is to transport the captured CO_2 via ship to a subsurface storage site offshore. This is a costly project, which will not be possible without a willingness to spend government money for this purpose. Presently, Equinor (formerly Statoil) has been given the operator responsibility for this project and has teamed up with Shell and Total as partners. The final investment decision is scheduled for 2019, with the Front-End Engineering Design (FEED) study in progress, and the future will show if this project can be realized or not.

Norway is not the only country to take leadership with projects that improve our understanding and provide the technological platforms for the continued development of the processes mentioned in the text that follows. Here we see developments in Canada (Weyburn and Quest) and in the United States (several Regional Sequestration Partnerships). Both of the Canadian projects are fully integrated with capture plants; Boundary Dam (a coal-fired electrical generation facility) is linked to Weyburn, an EOR project, and Quest (a capture unit on a refinery/upgrader) is linked to a saline aquifer storage facility. Excess CO_2 produced from Boundary Dam can be routed to a saline aquifer geosequestration site to prevent any venting to the atmosphere.

Geophysical methods related to safe storage of CO_2 underground are crucial, both in ensuring that the CO_2 is actually stored where we want and also in detecting potential leakage or pressure buildups as early as possible. The rapid development of cheaper geophysical listening equipment is critical in this respect. Of particular interest here is Distributed Acquisition Systems (DAS) technology, enabling the use of cheap fiber optic cables to be used as receivers. Currently, most of the testing for this technology has been in wells (see Chapters 15 by White and 16 by Lawton *et al.*, this volume). So far, the best response has been achieved when the fiber measures seismic waves traveling in the same direction as the fiber-optic cable. There are solutions being tested where the cable has a helical shape to mitigate the weak signal in the perpendicular direction. We think that in the future it might be possible to monitor CO_2 storage by deploying two-dimensional (2D) fiber optic cables at the seabed or the land surface for monitoring purposes. Such systems will be significantly cheaper than the Permanent Reservoir Monitoring (PRM; for example, at Aquistore in Chapter 15 by White, this volume) solutions that are presently used for geosequestation sites and hydrocarbon reservoirs. Horizontal fiber optic cables at the surface will probably record refracted waves and diving waves sufficiently accurately to be exploited for monitoring purposes. The time-lapse refraction method has been tested for several examples (North Sea and onshore fields in North America) so far, demonstrating that this method has a potential to detect and map shallow gas migration or CO_2 migration through the subsurface. This method is particularly sensitive to thin accumulations of CO_2 with large horizontal extension. It is therefore very likely that we will see a rapid development related to cheaper and more cost-effective methods for monitoring the shallow subsurface.

In the last two decades there has been a tremendous increase within the field of near surface geophysics (see Chapters 16 by Lawton *et al.* and 17 by Juhlin *et al.*, this volume). Conferences and workshops dedicated to this topic have attracted a large number of participants in recent years. This interest is encouraging and it is very likely that this will attract vendors and new companies developing geophysical instruments and methods dedicated to near surface mapping and monitoring. The near-surface work will be an essential component for the demonstration of safe storage in the deeper subsurface and early warning of any leaks that may occur in either geosequestration sites or EOR projects. Indeed, this will be essential if we are to assess the use of CO_2 in shallow horizons for increasing oil recovery.

There are interesting developments related to other geophysical methods, such as gravimetric and electromagnetic methods (Chapters 7 by Eiken and 9 by Gasperikova and Commer, this volume). Gravity is sensitive for density variations in the subsurface and time-lapse gravity has now been tested and proved for several fields. It is attractive both as a standalone monitoring tool and as a complement to seismic methods. In contrast to the seismic method, this method is sensitive to only one geophysical parameter: density. Time-lapse electromagnetic methods are sensitive to subsurface saturation changes, and are still under development. Again, there is a significant future potential for this method, both as a standalone tool and used in combination with other geophysical methods.

The In Salah CO_2 storage project demonstrated that satellite altimetry offers a nice opportunity to monitor onshore injection projects. Surface movements were measured with a precision of millimeters. Combined with geomechanical modeling such measurements give us reliable and very useful monitoring data. This technique is being tested in a more challenging environment at Aquistore (Chapter 15 by White, this volume). Such data are also used for monitoring of landslides and movements of large rock volumes and faults in mountains.

EOR has generally taken the lead in the development of geophysical techniques because of the economic return to the producer. This has led to most of the advances over the last number of decades. Many CO_2 EOR projects involve water alternating gas (WAG) injection which is very difficult to monitor unless one looks specifically at stress change and not saturation change alone. Detection of stress change versus fluid saturation change has been largely ignored except in some cases in the Norwegian North Sea, where time-lapse amplitude versus offset (AVO) and time-shift studies have been conducted to separate the effects of stress change from saturation change or in the case of multicomponent seismic surveying conducted by the Colorado School of Mines. The opportunity exists to enhance the value of seismic information by designing surveys specifically to include wide-azimuth multicomponent permanent reservoir monitoring (PRM) to monitor stress and saturation change. In doing so, we foresee the opportunity to record data that will enable more accurate quantitative interpretation. Already we are seeing the value of improved data interpretation in determining stress directions and stress changes in the subsurface as a means of guiding re-fracking operations. In the future, this will guide the application of CO_2 as an enhanced recovery agent and for geosequestration in fractured formations.

Although the number of storage projects worldwide is not increasing rapidly at present, we think that the number of projects will, however, increase in the years to come. After all, carbon capture and storage (CCS) offers a possibility to store relatively large volumes of CO_2. The political willingness to do so

has been strengthened by the Paris agreement. However, political actions related to implementation of tax systems for CO_2 emissions are not really in place yet (although many countries and subnational jurisdictions have taken some leadership in this regard, following Norway's example). We think this will come, and that this will be an enabler for CCS technology in the future, and hence also for geophysical monitoring of the CCS process.

It is hoped that the reader will find the chapters in this book to be a guide to the current state-of-the-art and that it will provide a sense of where mainstream geophysics will be headed in the future as it relates to geosequestration.

Index

375